Handbook of Glass in Construction

Other McGraw-Hill Reference Books of Interest

Handbook of Glass in Construction

Joseph S. Amstock

President, Professional Adhesive and Sealant Systems
Huntingdon Valley, Pennsylvania

McGraw-Hill

New York San Francisco Washington, D.C. Auckland Bogotá Caracas Lisbon London Madrid Mexico City Milan Montreal New Delhi San Juan Singapore Sydney Tokyo Toronto

Library of Congress Cataloging-in-Publication Data

Amstock, Joseph S.
Handbook of glass in construction / Joseph S. Amstock.
p. cm.
Includes bibliographical references.
ISBN 0-07-001619-4
1. Glass. 2. Glass construction. I. Title.
TA450.A55 1997
624.1′838—dc21 97-1273
CIP

McGraw-Hill
A Division of The ***McGraw-Hill*** *Companies*

3 4 5 6 7 8 9 0 FGR/FGR 9 0 2 1 0 9 8

ISBN 0-07-001619-4

The sponsoring editor for this book was Larry Hager, the editing supervisor was Bernard Onken, and the production supervisor was Suzanne W. B. Rapcavage. It was set in Century Schoolbook by Victoria Khavkina of McGraw-Hill's Professional Book Group composition unit.

Printed and bound by Quebecor/Fairfield.

This book is printed on recycled, acid-free paper containing a minimum of 50% recycled, de-inked fiber.

To Margaret, Gary, and Mark

Contents

Preface

This handbook deals with the materials and the methods of manufacturing glass and insulating glass in the construction industry. It is intended as an educational tool for designers, architects, engineers, and contractors, as well as the many students studying both architecture and glass.

It is the author's intention to treat the complicated chemistry, manufacturing process, and installation of both flat glass and insulating glass, so that the uninitiated will have an improved understanding of this subject.

In most cases, architects, builders, and building owners are interested in the performance aspects of glass, not necessarily the chemistry or the physical properties. This handbook attempts to assemble a wealth of information on the performance of a variety of engineering materials and to set it down in an organized form so that it can be used systematically. It is organized to lead the reader from source, to need, to design, to specific properties, and, finally, to installation.

It would be impossible to thank everyone who has assisted in the production of this handbook. Virtually every manufacturer of glass, sealants, other accessories, and equipment has contributed in the forms of specific technical data, photographs, published papers, and other relevant materials.

A special thanks to AFG Industries, Edgetech, FRD Designs, Guardian, Hygrade Metals Moulding Mfg., LOF-Pilkington, PPG Industries, Schott, Tremco, and UOP Molecular Sieves for their assistance with technical information. The author is especially grateful to the eight contributors listed for their talent, advice, and time.

I would like to express my sincerest thanks to Gary J. Amstock, my son, for his unselfish contribution of the hundreds of hours he spent in preparing the charts, graphics, and other illustrations required to make this handbook a useful tool.

Joseph S. Amstock

Contributors

Thomas J. Dangieri, *Development Specialist,* UOP Molecular Sieves, Des Plaines, IL.

Randi Ernst, *President,* FRD Design Inc., Buffalo, MN.

Donald E. Holte, P. Eng., *Senior Vice President—Marketing,* Visionwall Technologies Inc., Edmonton, Alberta, Canada.

G. Bernard Lowe, Ph.D., *Manager, Technical Marketing,* Morton International, Ltd., Coventry, England.

Steven J. Nadel, *Chief Technologist,* BOC Coating Technology, Fairfield, CA.

I. Douglas Sherman, *Vice Chairman,* Hygrade Metal Moulding Mfg. Corp., Farmingdale, NY.

Ramesh Srinivasan, *Senior Engineer,* Tremco, an RMP-owned Company, Beachwood, OH.

Kevin Zuege, *Technical Service Manager,* Tremco, an RPM-owned Company, Beachwood, OH.

Chapter

1

Introduction

Joseph S. Amstock

President, Professional Adhesive and Sealant Systems

The glass revolution has surely been in full swing in this modern building era. However, a cursory study of construction history is enough to show that, if this is not the earliest form of glass-making, it is one of the oldest technologies. Figure 1.1 covers the stages in the making of crown glass in these eighteenth-century engravings.[1]

1. The glassblower blows a piece of glass into shape.
2. The end of the glass "balloon" is cut open, and the lips are peeled back before heating.
3. After reheating, the glass is spun outward on the blowing iron to form a circular sheet of glass. A "bull's eye" will have been left in the center, where the blowing iron has been attached.

The primary readers of this handbook will be architects, designers, contractors, and specification writers who are interested in the performance of glass and other aspects of window components and how they are utilized in structural applications. See Figure 1.2 for a cross-sectional view of a four-sided, structurally glazed, insulating glass unit.

With the advent of the float glass process, the availability of glass has been enhanced along with definite economic advantages. When surveying the many uses of glass, the question always presents itself as to what factors influence a choice of a particular use of a glass component. A few of these advantages worthy of mention are accompanied by some examples.

This handbook is designed to aid our understanding of the nature of glass and, specifically, its performance. Included in this study are

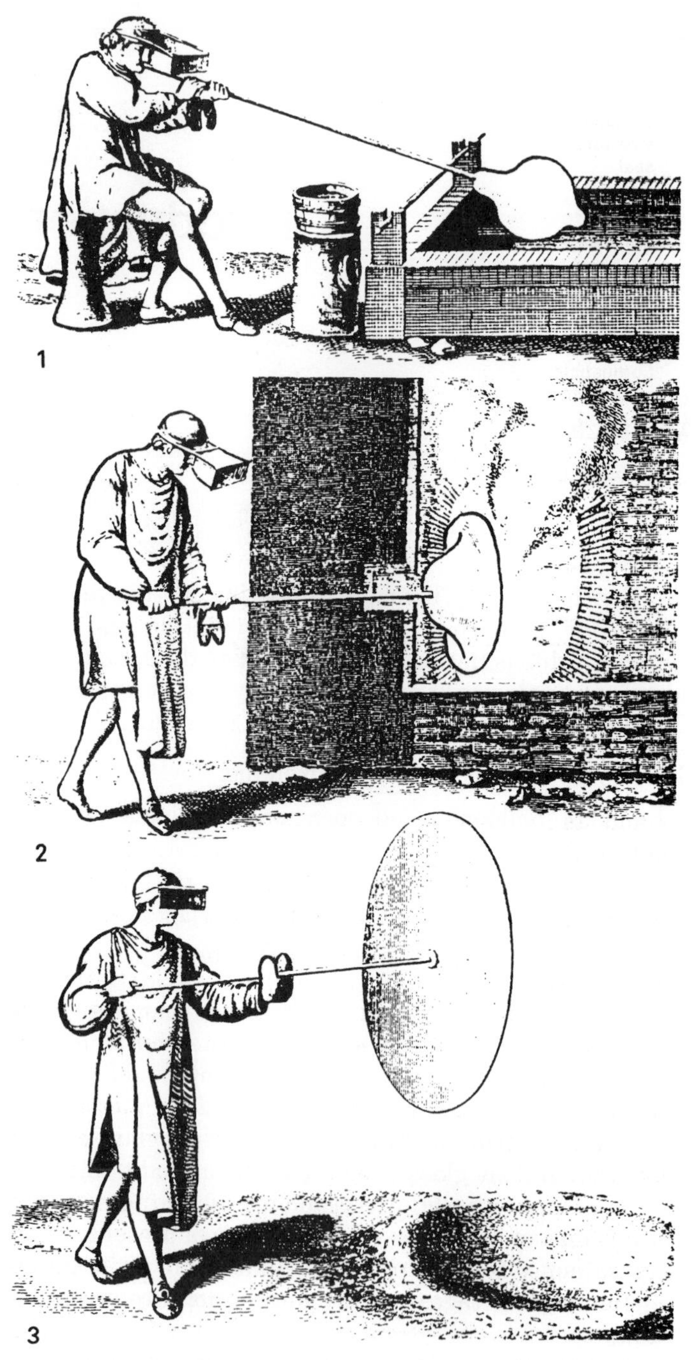

Figure 1.1 Manufacture of crown glass. (*Courtesy of PPG Industries.*)

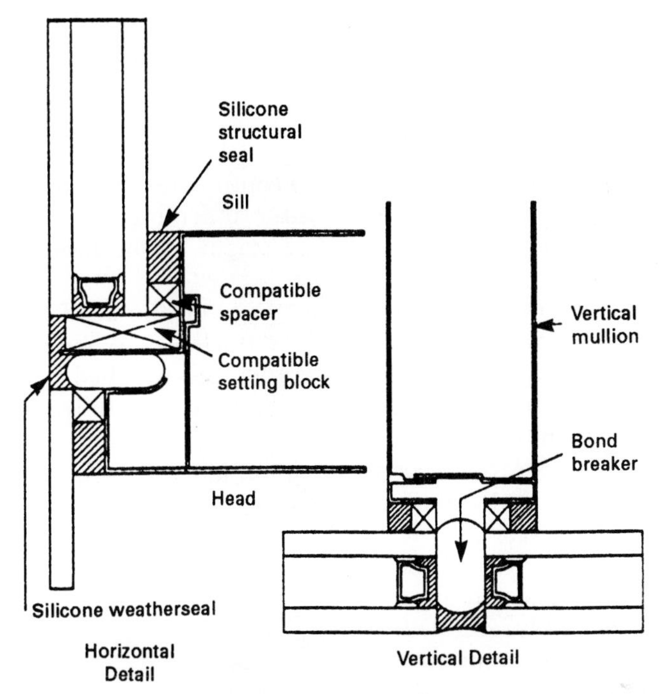

Figure 1.2 Four-sided structurally glazed unit.

insulating glass and its numerous components and the evolution we have gone through in the last 50 or so years. A brief overview is provided to give the reader a basic knowledge of the various processes used for manufacturing glass. Plate, float, bent, and laminated glasses are just a few that will be discussed in greater detail.

Paul Schaebart envisioned a dramatic change in architecture by introducing glass architecture, which allows in the light of the sun, moon, and stars, through walls entirely made of glass. Today this idea is almost possible.

We have outlined the evolution of architectural glass and its uses principally in sheet glass and insulating glass. The future outlook is even brighter for architectural glass usage with the new inventions, such as "smart glass or windows," and other new technologies that have created savings for the consumer and the manufacturer.

Modern float glass factories operate furnaces which generate more than 500 tons of molten glass per day and assembly lines which move the heated substance across the "bath of molten tin," through the cooling (annealing) process to a ribbon of hardening glass. With the large backlog in sales, there have been major breakthroughs in engineering and worker skills.

These plants are expensive; a single float line costs upward of $150 million to construct. Once the furnace is fired up, it must be operated without interruption 24 hours per day, 7 days a week, 365 days per year. Worldwide, there are more than 75 float glass plants. A brief description of some of the glass-making processes follows.

Polished plate glass is produced by rolling a continuous sheet or by casting and rolling large sheets separately. After annealing and cooling, the surfaces are made flat, parallel, and bright by mechanical grinding and polishing.

In the float glass process, the molten glass floats on the dead flat surface of molten tin where it flows to a uniform thickness. When the sheet is drawn off, both surfaces are flat, parallel, fire-polished, and sufficiently cooled to remain undamaged by the rollers used to transport it through subsequent operations.

Rolled glass includes patterned, corrugated, and plate glass blanks and wired glass. Figure 1.3 details a picture of patterned glass coming off a manufacturing line.

To construct and operate a manufacturing plant, dollars and energy are of great importance and require more than engineering and technical skills. Some of the largest U.S. firms dealing with glass are: AFG Industries, Guardian Industries, Libbey–Owens–Ford Company, PPG Industries, and Ford Motor Company.

Figure 1.3 Manufacture of patterned glass.

By 1955, the gamble to go "float" was taken by Pilkington of England. Their researchers found that the key to the float glass process was in controlling the atmosphere inside the float bath chambers. Molten tin reacts with any oxygen in the float chamber, causing damage to the ribbon surface through physical and chemical interaction at the surface between the glass and tin.

Insulating glass came into its own in the United States with LOF's Thermopane® unit and a welded glass edge unit. The glass edge had its limitation as to air-space thickness and overall size.

PPG was marketing a unit called Twindow® that had a superior mastic metal edge seal and also a glass edge unit. The author is aware of a Twindow® unit that has been in a Sigma weatherometer since 1963, still performing to the ASTM-E-773/774 specification. Figures 1.4, 1.5, and 1.6 give a cross section of a couple of earlier designed insulating glass units.[2]

With the organizing of SIGMA (Sealed Insulating Glass Manufacturers Association) the glass, sealant, and window industry

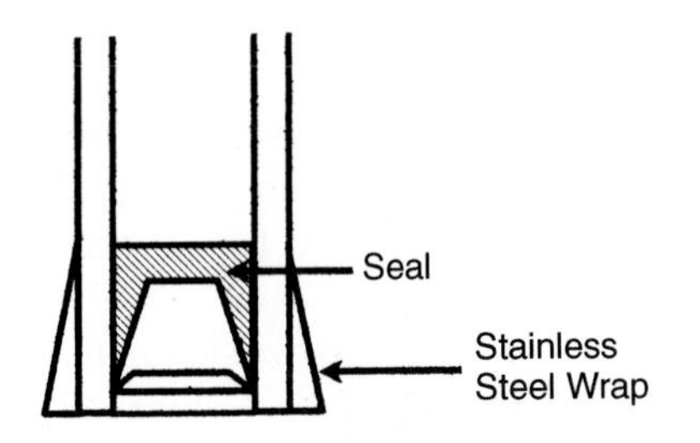

Figure 1.4 Metal edge seal.

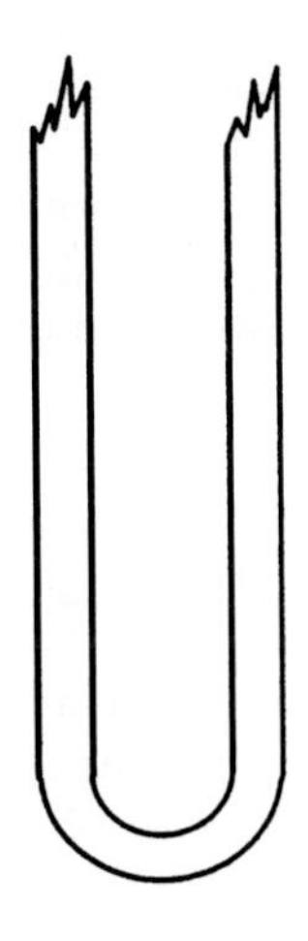

Figure 1.5 Glass edge seal.

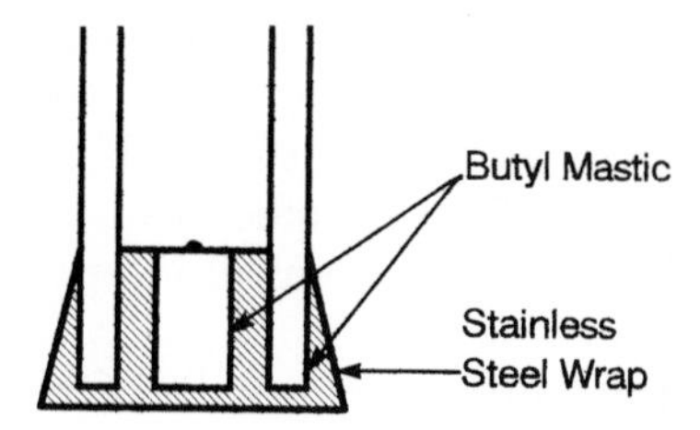

Figure 1.6 Mastic edge seal.

was able to develop standards of performance to test the new edge configurations. These will be discussed in detail in Chapter 23. Polysulfides were the first type of organic sealant used as an edge seal in early 1956. This sealant was one specified by the U.S. government as a fuel tank sealant for aircraft. Today, improved techniques have given the industry polyurethanes, silicones, hot melts, etc. Even more recently, we have seen the elimination of spacer bars with such materials as desiccated silicone foam and desiccated butyl mastic and an open U-shaped spacer with a desiccant binder on the interior of the U.

Other new inventions are low-E coated glass, drop-in coated films, and/or variations of both. To improve the R-values, we have seen insulating glass units filled with inert gases, such as argon and krypton, again, to improve overall insulating values. The new "smart window" is the newest contribution to improved products for giving the consumer an added advantage in the home. See Figure 1.7 illustrating "Smart Glass."[3]

Both private and institutional groups continue to work on new challenges. The University of Texas, The University of Missouri-Rolla, Laurence Berkeley Laboratories of the University of California, The U.S. Department of Energy, just to name a few, are instrumental in continuing research for improved qualities. In Canada, Norway, Japan, and other cold countries around the world, companies and governments are striving for added improvements.

The other significant data deals with the basic formulations, chemistry, and properties of glass compositions, staring with a pure soda-lime formula through the more sophisticated, such as borosilicate glass and those used in x-ray protection.

Sealants play a most important role in the manufacture of a sealed insulating glass unit and in the final installation of the sheet glass and/or insulating glass in a building. Sealing materials are categorized into organic, such as polysulfides or polyurethanes, and inorganic, such as a silicone backbone material. New edge-seal materials have evolved in recent years with the products, such as Super Spacer™, Swiggle Seal™, and Intercept™, all having backbones of a

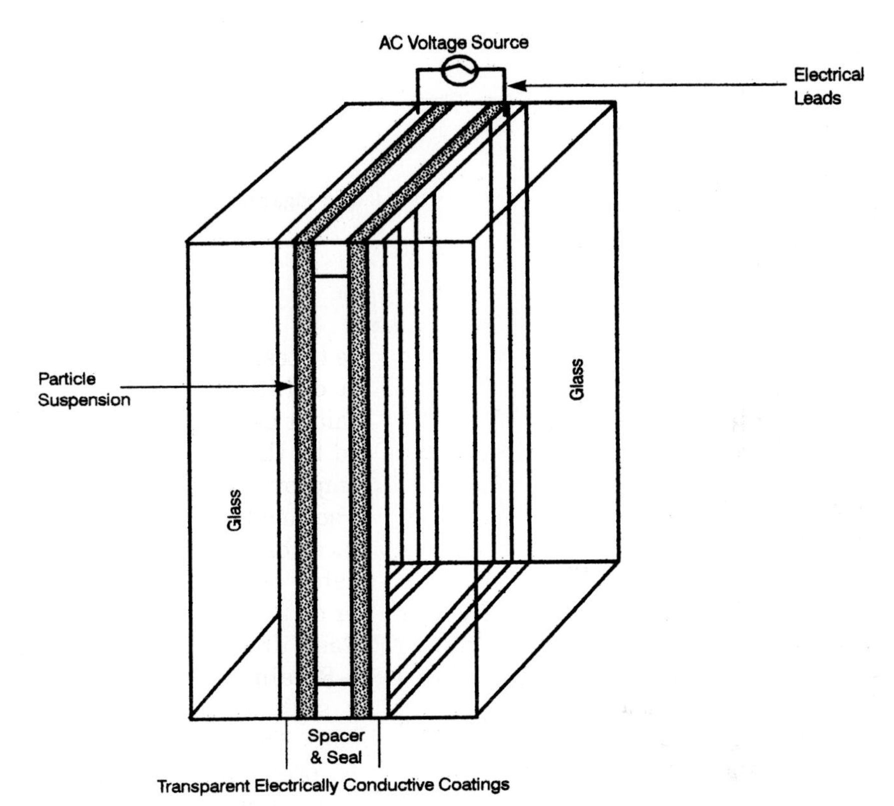

Figure 1.7 "Smart Glass." (*Courtesy of Research Frontiers, Inc.*)

proprietary nature. The edge configuration is known in the industry as *warm-edge technology*. These systems have made a marked improvement in the R-value of sealed insulating glass units.

Other components and techniques have also improved overall performance making it easier for the architect to design buildings of high value. These new developments, which also aid the consumer and manufacturers of windows, include low-emissivity glass filled with exotic gases mentioned earlier. Gas filling has improved the insulation value remarkably and Figure 1.8 shows the type used by many insulated glass producers.[4]

Other chapters of interest cover glass preparation and the manufacture of mirrors. Mirrors are one of the finest enhancements in our homes and offices.

To complete this handbook, we have reviewed a variety of subjects, such as new techniques, drop-in films, and multifilm insertions, that

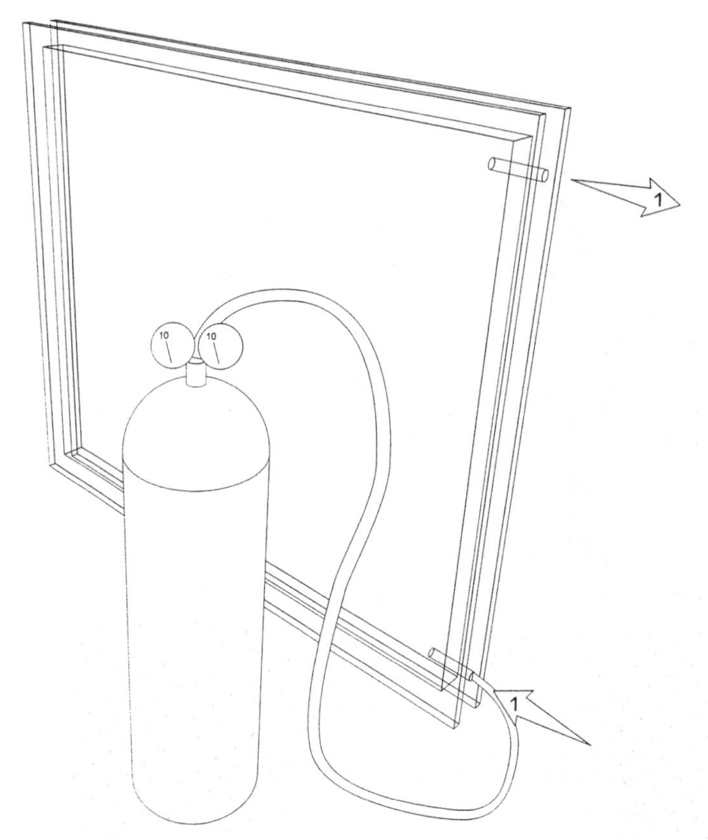

Figure 1.8 Gas filling process. (*Courtesy of FDR Designs.*)

greatly improve the R-value. For those who need assistance in developing specification and installation details, these subjects are also covered extensively.

The efforts of all the suppliers of glass, sealants, desiccants, and the variety of components for glass and insulating glass will bring satisfaction and rewards when you see the architect's design, constructed and inhabited by consumers. Figure 1.9 is one of the new all-glass structures.[5]

We will continue to look for new challenges, new achievements, and future successes of the flat glass, insulating glass, and window industries.

Figure 1.9 All-glass facade. (*Courtesy of Schott Glassworks—Germany.*)

References

1. Courtesy of PPG Industries, Inc., Pittsburgh, PA.
2. Amstock, J. S. and Markham, D. W., "Product Improvement of Insulating Glass," *Glass Digest,* Dec. 1967.
3. Courtesy of Research Frontiers, Inc., Woodbury, NY.
4. Courtesy of FDR Designs, Inc., Buffalo, MN.
5. Courtesy of Schott Glassworks, Germany.

Note: Thermopane is a registered trademark of LOF, Toledo, OH. Twindow is a registered trademark of PPG Industries, Pittsburgh, PA.

Chapter

2

History of Glass

Joseph S. Amstock

President, Professional Adhesive and Sealant Systems

Introduction

Glass is probably the oldest man-made material, used without interruption since the beginning of recorded history. It is not known exactly when glass was first discovered. The oldest finds date to 7000 B.C., in the Neolithic period. It is an amorphous substance produced by heating a mixture of such materials as sand, soda (sodium carbonate), and limestone to a temperature of about 2400°F (1300°C).[1] Temperatures only slightly lower than this can be achieved with a wood fire, which was the traditional fuel for melting. Today, glass is made continuously in large tanks. Raw materials are fed in at one end, and a molten ribbon of glass emerges at the other end. The glass is heated by oil or gas flames over the melt, but, since these fuels were becoming more expensive, the electrical glass-melting process, which employs metal electrodes in the melt to generate heat, is becoming more widely used. Substances called fluxes are added to the mix to make it melt at lower temperatures. Windows, containers, light bulbs, optical instruments, and many specialty items of glass are an essential part of modern life.

Glass was first used in Egypt for decorative objects before 3000 B.C., mainly as a colored glaze on stone, pottery, and beads. The art of making glass was perfected about 1500 B.C. in Egypt and the Near East. Figure 2.1 shows an early pre-Egyptian vase.[2] To make vessels for salves or oils, a ceramic negative core was used.

A craftsman dipped the form attached to a rod into molten glass, and the first usable hollowware was created. The goblet of Thutmose

Figure 2.1 Early pre-Egyptian vase. (*Courtesy of the Metropolitan Museum of Art, New York, NY.*)

III was made by this process, about 1490 B.C., and now is housed at New York's Metropolitan Museum of Art.

Constant rotation of the core in the molten glass caused the glass to adhere to the form. Rolling it on a flat stone slab smoothed the surface. Handles or rings were added for convenience in carrying. Copper and cobalt added to the molten glass composition yielded bluish tints. Dating from 669–626 B.C., the clay tablet library of Assyrian King Ashurbanipal contains cuneiform texts with glass formulas, the oldest of which reads, approximately: "Take 60 parts of sand, 180 parts of sea plants, 50 parts chalk—and you will get glass." This glass composition contains, essentially, all the materials used in today's formulations, although not in the same proportions. The low amount of sand, however, leads to the conclusion that melting temperatures achievable during the last millennium B.C. were not very high, and it was possible only to make soft glass suitable for fashioning simple vessels.[3] Glassblowing, as illustrated in Figure 2.2, in 50 B.C. in Phoenicia, greatly extended the type of objects that could be made of

Figure 2.2 Glassblowing.

glass.[4] It also made them easier to fabricate and more transparent. The art of glass-making spread rapidly throughout the Roman Empire, and special centers of glass-making were established in Phoenicia, Rome, Egypt, the Rhineland, and the Rhone valley. Glassware became common and relatively cheap.[5]

For many centuries after the fall of the Roman Empire, glass-making decreased in importance in Western Europe, as did many other technologies and arts, and artistic glass almost disappeared. In Byzantium, however, Greek and Syrian glass centers continued to prosper.

This new invention began to spread more rapidly and the craft experienced its first period of growth. Pliny the Elder (A.D. 23–79) described the composition and manufacturing process of glass in his encyclopedia, *Naturalis Historia.*[5] The art of coloring glass flourished. He also noted, in 79 B.C., that fine glass cups were replacing precious metals as a status symbol among the rich Romans. Glass, however, did not replace shutters on the windows of Roman homes. The Romans tried but failed to cast transparent flat glass to enclose or ornament their homes. Slabs ½ inch thick have been excavated, including a 32- by 44-inch piece at Pompeii. But the Romans did not discover the art of polishing or grinding cast glass to make it transparent. Instead of glass, the rich used thin, translucent sheets of alabaster to enclose wall openings.

Both labor and raw materials played an important role in the location of the Roman glassworks. Logos of the respective firms' products were identified as early as the first century A.D. and the products were sold throughout the Empire. The Phoenicians were the first to create small glass mirrors with a tin underlay. However, since the glass did not have a level surface, glass did not develop for this purpose. It was not until the thirteenth century in Germany, when the back side of a flat piece of glass was coated with a lead-antimony layer, that a quality mirror was successfully made of glass. Later, Venetian artisans improved upon the process as we know it today. Only after the discovery of the plate pouring process in 1688 in France, under Louis XIV, could large-surface mirrors be created.

Beginning in the eleventh century, several new centers of glass-making arose in Western Europe. In Bohemia, ash from plants (potash, which is high in potassium) was used as a raw material to make glass with a lower melting point. The most important European center of glass-making developed near Venice, where new compositions, colors, forming techniques, and artistic skills developed. The Venetians added manganese, in the form of the mineral pyrolucite, to oxidize iron impurities in glass and remove the green and brown tint caused by the reduced state of iron. By adding lead, borate, and more soda to glass, they increased its working temperature range and were able to make more intricate shapes, thinner blown glass, and finer enamels. Special additives were also used to color glass. At one point, more than 8000 artisans were employed in the Venetian glass industry. Glass-making in the city reached its peak between the fifteenth and seventeen centuries. Venetian glassware designers were strongly influenced by many aspects of Islamic art. Venice jealously guarded its glass recipes, especially that for crystal glass. At one point, the glass makers, who were housed on the island of Murano, faced the threat of death if they disclosed a formula.

About 1675, so-called crystal glass was developed to compete with Venetian glass. Purer materials, oxidation of iron, and the addition of lead gave a more transparent glass; this transparency, together with the higher index of refraction resulting from the addition of lead, gave a sparkle to faceted cut glass, resulting in a heavier, more desirable glass. An important center of glass-making at that time was London. In England, where deforestation was a problem as early as the fifteenth century, glass makers were required to use coal instead of wood in the glass-making process.

Flat glass for windows was still rare during much of the seventeenth and eighteenth centuries, as Figure 2.3 illustrates.[6]

Figure 2.3 Crown glass.

Small panes were made by blowing a large glob of glass, removing it from the blowing iron, and then rotating the glass quickly so it would spread and flatten. Such glass had a dimple in its center, many air bubbles, and a pattern of concentric circles, but it was transparent and effective in keeping out the weather. At the end of the seventeenth century, the French learned how to grind and polish cast glass to produce plate glass. Even the rich still shuttered their windows, and the Middle English word for windows—"wind eyes"—underlined the fact that wall openings enclosed in glass were, for all practical purposes, nonexistent.

Chronology of Glass Technology

The entire history of glass is characterized by the efforts of the individuals who perfected and further developed production processes and products.[8]

In 1679, Johann Kunkel (1630–1703), the director of the glassworks established near Potsdam by Friedrich Wilhelm of Prussia, the Great Elector, includes texts from his own experience and those of others in his handbook, *Ars vitrateria experimentalis.* This publication was recognized as the scientific basis of the German art of glass-making until the nineteenth century.

In Munich, Joseph Fraunhofer (1787–1826), the son of a glass master and trained mirror maker, intensively studied glass-making technology. He succeeded in producing glasses with qualities suitable for optical instruments. His telescopes and microscopes were renowned.

Otto Schott (1851–1935), a chemist and technologist from a family of glass makers, investigated the dependence of the physical qualities of glass and its composition by scientific methods. In a manner of speaking, glass was rediscovered.

In 1876, Schott contacted Ernst Abbe (1840–1905), who was a professor at the University of Jean and was the principal scientific figure behind the Carl Zeiss Optics Company. Abbe needed suitable glasses for his lenses to be used in high-quality instruments. The glasses had to be free from defects and of the highest purity. After a number of disappointing attempts, Otto Schott succeeded in producing a glass of the required quality on his ninety-third trial melt. In 1884, Schott moved to Jean and, along with Abbe, Carl Zeiss, and Zeiss' son Roderick, established the *Glastechnisches Laboratorium Schott und Genossen.*

Shortly before 1900, an American, Michael Owens (1859–1923), invented the automatic bottle blowing machine which was introduced in Europe after the turn of the century. Somewhat later, the processes for mechanical production of flat glass became available, without which the quickly rising demand for architectural glass could not have been met. Three hundred thousand standardized panes of glass were used for the wall panels of the Crystal Palace built by Paxton in London in 1851 for the World Exhibition. This was one of the earliest examples of the use of glass as a structural material.

Shortly after the end of World War II in 1945, the American Army transferred key personnel from the Zeiss and Schott firms to West Germany. The present-day corporate headquarters of the Schott Group, the *Schott Glaswerke,* was set up in Mainz in 1952.

Glass-Making in the United States

A. Mixed traditions of English and German glass-making.
B. Earliest efforts were colonial enterprises.
   1. Jamestown—failed enterprise in the 1600s.
   2. Annealing—1740s.
   3. Wistar—South Jersey Waldglas industry in the 1780s.

C. Glass-making in the new nation—three major centers.
   Nineteenth century
   1. South Jersey—Waldglas artistic ware.
   2. Pittsburgh and Ohio—frontier glass makers.
      Pittsburgh and Ohio—Emphasis on energy and potash—forest products.
   3. Northern New York/New England—Glasshouses (Boston and Salwich—Ressel Glass)—Technical innovations for artistic and consumer glass—forerunners to Corning.

D. Glass-making in the twentieth century.
   1. Growth of the American optical glass industry—Corning, Bausch and Lomb, American Optical, Pittsburgh Plate Glass.[7]
   2. Industrial and consumer glasses—Owens-Illinois, PPG, Libbey, others.

Glass-making came to America with the first colonists and persisted through the formative years of the United States with little change. In 1608, eight Polish and German glassblowers, members of Capt. John Smith's first colonizing group, landed on the shores of the James River at what was later the site of Jamestown, Virginia. Crude glass furnaces were built, thus establishing glass manufacturing as the first industrial enterprise on American soil.

The venture failed, as did early attempts by pioneers such as Stiegel, the first American glass maker, whose work today is highly valued by museums and collectors. Despite these failures, the glass plants of early America continued to rebuild, each new endeavor moving the industry ahead.

Glass was the among the last of all industries to become modernized, simply because it has always been in the hands of a comparatively small group of master craftsmen. Hand labor still characterized glass-making even at a time when mechanization had taken a firm hold in other industries. In 1900, for instance, glass-making was not essentially different from what it had been 300 years previously. At the turn of the century, when American business was about to launch one the world's most notable achievements in transportation and machinery, the automobile, glass was still standing still.

In 1916, Michael J. Owens was to make an indelible mark in the glass makers' struggle to bring mechanization to the industry. He mechanized the production of glass containers, and then perfected the world's first machine for flat drawn window glass, an act which did more in a few years to revolutionize the art of flat glass-making than anything had done for thousands of years previously. It set the stage for the great achievements we know today.

The Great Names of Glass in America

In 1881 or 1882 John Pitcairn[9] had a conversation with Capt. John B. Ford about his plans to build a plate glass factory in Creighton, Pennsylvania. It was natural for him to suggest that J. B. Ford try piping gas from a nearby well to provide steady and dependable heat. This led to the establishment of Pittsburgh Plate Glass Company.[7] The first factory at Creighton soon became insufficient to keep up with the demand for superior glass, and two more plants were built by Ford and Pitcairn. Then, in 1895, a large-scale consolidation brought ten factories under the PPG umbrella. In 1897, Ford and his two sons, Edward, President of PPG, and Emory L., its treasurer, differing with Pitcairn chiefly on financial matters, picked up stakes and moved to Wyandotte, Michigan, where there were large deposits of salt for making soda ash, one of the main ingredients for making glass. There they established the Edward Ford Plate Glass Company, which merged in 1932 to form the Libbey–Owens–Ford Company,[8] a strong competitor today to PPG.

John Baptiste Ford, an unschooled man taught to read by his wife, became intrigued with the idea of glass-making. Learning that no plate glass was being made in America, he decided to pioneer in plate glass in the United States. He failed twice in his endeavor, but, with boundless persistence, at age 70, he tried again in Camden, New Jersey. With three financiers, in 1871, the New York City Plate Glass Company was formed. At a later date, this company transferred its assets to become the Pittsburgh Plate Glass Company.

The name Libbey–Owens–Ford Company perpetuates the names of men who pioneered change in industrial processes. Their lives and fortunes were devoted to producing better glass products in America. All three died between 1920 and 1925. Their companies merged in 1930 to form Libbey–Owens–Ford, which today is one of the largest flat glass companies in the world. Edward Drummond Libbey began his career in the glass industry in 1874, when, at the age of 20, he took a minor position with the New England Glass Company, a family concern manufacturing tableware. Through hard work and perseverance, he rose to head up the company in 1883.

In search of better markets, better raw material sources and fuel, he moved to Toledo in 1888 and founded the Libbey Glass Company. After several years of financial difficulty, Mr. Libbey daringly invested all his capital to gain public notice of his products in an exhibit at the 1893 Chicago World's Fair. Being successful in this venture, the growing sale of his fine cut glassware enabled him, in later years, to finance the Owens Bottle Company and the Libbey–Owens Sheet Glass Company, moving him into the field of flat glass manufacture. In contrast to Mr. Libbey, Michael J. Owens belonged to the rough-

and-ready school of American industrialists, who rose from shirt sleeves to leadership with extraordinary vigor in organization and invention. He entered the world of glass-making at the age of 10 in a Wheeling, West Virginia, glass plant and joined the fledgling Libbey Glass company in Toledo, Ohio, in 1888. In short order he became plant manager. With the support of Mr. Libbey, who became his life long friend, Mr. Owens employed his inventive genius to develop machines for making glass tumblers and an automatic bottle machine which revolutionized the glass container industry.

In 1912, Libbey and Owens secured the patents and services of Irving W. Colburn, who invested his funds in an attempt to draw window glass in flat form. Finally, after four trials spearheaded by Mr. Owens, the process was successfully completed. Window glass in commercial quantities could be continuously drawn in a flat sheet from a huge tank of molten glass. This was the beginning of the Libbey–Owens Sheet Glass Company.

The other of Toledo's famed glass makers, whose work comprises a part of the Libbey–Owens–Ford heritage, was Edward Ford, son of the famous Captain John B. Ford, America's pioneer plate glass maker. Born in 1843, Mr. Ford cast his future in the world of glass-making by assisting his father to establish the first plate glass factory in Albany, Kentucky, in 1867. Later they built factories in Kentucky and Indiana. However, they failed due to European competition. In 1880, Captain Ford and his sons Edward and Emory built a plant, known today as the Pittsburgh Plate Glass Company. Edward served as president and general manager. In time, they withdrew from the firm and moved west to found the Edward Ford Plate Glass Company. In 1898, Mr. Ford began building the model town of Rossford and constructing the largest plate glass plant under one roof in the United States. By 1920, the company was producing over 9 million square feet of plate glass per year and increased output to 14 million square feet in 1929. That same year, an industrial "wedding" took place, merging the two large flat plate glass manufacturing companies, the Edward Ford Plate Glass Company and the Libbey–Owens Sheet Glass Company. The two firms complemented each other, and, in 1930, the three men who literally created the American glass industry were joined in the formation of the Libbey–Owens–Ford Glass Company. Today, with annual sales over $1 billion, Libbey–Owens–Ford Glass Company is an important force in the glass industry.

References

1. *Grolier Interactive, Inc.,* Grolier Electronic Publishing, Version 7.0, New York, 1995.
2. Early pre-Egyptian vase, courtesy of The Metropolitan Museum of Art, New York.
3. Paterson, A. J., *How Glass is Made,* Threshold Books Limited, New York, 1985.

4. *Glassblowing,* Courtesy Schott Corp., Yonkers, NY, 1995.
5. Plaender, H. G., *Schott Guide to Glass,* Chapman & Hall, London, 1996.
6. Courtesy of The Construction Specifier, The Construction Specifiers Institute, Alexandria, VA, 1994.
7. *Romance of Glass,* Pittsburgh Plate Glass Industries.
8. *A Member of the Pilkington Group,* Libbey–Owens–Ford, Toledo, OH.
9. Gladish, R. R., *John Pitcairn—Uncommon Entrepreneur,* The Academy of the New Church, Bryn Athyn, PA, 1989.

Chapter

3

Raw Materials and Composition

Joseph S. Amstock

President, Professional Adhesive and Sealant Systems

Commercial Glasses

A batch of commercial glass consists of a mixture of seven to twelve individual ingredients. The bulk of the batch is made up of four to six ingredients, chosen from such materials as sand, limestone, dolomite, soda ash, boric acid, borax, feldspathic materials, and lead and barium compounds. In addition to these, cullet in the amount of 15–30% is commonly included in most commercial batches. Cullet, or scrap glass, is either purchased or results from normal production practice.[1]

The list below gives the approximate compositions and properties of the main types of glass. The percentages by weight of the main ingredients are given. Detailed properties depend on the exact composition and may be sensitive to small changes in minor ingredients. The nature of this dependence is rather complicated, and therefore, exact compositions of particular glasses and the types of minor ingredients are not given here. As one instance, glass with sodium ions in it usually conducts electricity, yet, a high-resistance lead glass is made that also contains about 6% Na_2O_2. This is possible because the glass contains potash, and the potassium ions reduce the mobility of the sodium ions. If these two are present in the right proportions, the total alkali mobility is actually less than with either present by itself.

Chemical and Physical Nature of Raw Materials

The major raw materials used in glass-making may be divided into three categories:

1. Mined or quarried materials that are milled and beneficiated in some manner after extraction to render them suitable for the market. Examples are sand, limestone, dolomite, etc.
2. Manufactured chemicals, such as soda ash, borax, boric acid, aluminum hydrate, etc.
3. By-product materials. Cullet, or scrap glass, and beneficiated blast furnace slags are the most important current materials in this class.

Raw materials used for the manufacture of glass may vary considerably in their degree of purity, and the type of a particular raw material is commonly indicative of the degree of analytical control it will provide.

Description of Glass-Making Raw Materials

Alumina (Al_20_3). Although this oxide does not represent a large part of the composition of most glasses, it is an important one.

Feldspar. For a number of reasons, it is one of the most important sources of alumina. The mineral can be obtained in a reasonably constant condition and in a state of sufficient purity free from iron and other objectionable elements. Feldspars are represented by the formula $R_2O \cdot Al_2O_3 \cdot 6SiO_2$, in which R_2O represents an alkali oxide, K_2O in microline and Na_2O in albite. Two other feldspathic minerals are widely used as sources of alumina: nepheline syenite and aplite.

Alumina hydrate ($Al_2O_3 \cdot 3H_2O$). Alumina hydrate is a manufactured product prepared by a precipitation process from minerals, such as bauxite, dissolved as sodium aluminate.

Kaolin ($Al_2O_3 \cdot 2SiO_2 \cdot 2H_2O$). It is feasible to use some of the purer grades of kaolin as a source of alumina.

Kyanite and sillimanite ($Al_2O_3 \cdot SiO_2$). These minerals do not have the low iron content of feldspars, but have higher alumina content and may offer some cost advantage.

Slag. Blast furnace slags, beneficiated and milled to a condition suitable for use in glass batches as a source of alumina, silica, lime, and magnesia, offer a readily soluble source of Al_2O_3. Raw slags are not very suitable in the production of glass from either a physical or chemical quality standpoint.

Antimony oxide (Sb_2O_3). The principal mineral source is stibnite (Sb_2S_3). Antimony trioxide is formed when the metallic antimony is heated in the air or treated with nitric acid. Its role in glass composition is not clearly understood, but it is used in concentrations up to 1%, and it appears to make the glass more easily meltable and workable.

Arsenious oxide (As_2O_3). Arsenious oxide is a by-product of copper metallurgy. It collects as a fume in smelter flues and is refined by sublimation. It volatilizes at red heat without melting and condenses in fine crystals. It acts as an oxidizing agent toward ferrous oxide and also toward sulfur and carbon.

Barium oxide (BaO). Barium oxide is usually furnished by the compound barium carbonate ($BaCO_3$). It occurs naturally as the mineral witherite, but rarely pure enough for use in the glass batch.

Boric oxide (B_2O_3). Boron, or more specifically boric acid, is supplied to the glass batch in a variety of forms.

Lime: calcium oxide and magnesium oxide, CaO and MgO, respectively. This oxide is readily obtained from calcium carbonate ($CaCO_3$), available as quite pure high-calcium carbonates (whiting). Calcium carbonate is found widely distributed and in varying degrees of purity and consolidation in marble and calcareous earths or marls. Magnesia is used in glass in the absence of burnt lime.

Burnt lime, formed by the complete calcination of limestone in lime kilns, is another source of calcium for the glass batch. Burnt lime offers a few advantages. It is already calcined, reacts quickly with silica with the evolution of heat, contains no carbonaceous matter, and presents a minimum of weight and bulk.

Hydrated lime or slaked lime is obtained by mixing burnt lime with water in the proper proportions. Slaked lime contains about 75% CaO, and is preferred by some glass makers as a source of CaO.

Gypsum ($CaSO_4 \cdot 2H_2O$). Gypsum, or anhydrite or dead-burnt gypsum ($CaSO_4$), offers possibilities as a substitute for salt cake in glass melting and, at the same time, furnishes some CaO. In the melting process, calcium sulfate must react with sodium carbonate and yield lime and sodium sulfate. Gypsum is stored quite easily.

Iron oxide (FeO and Fe_2O_3). Iron oxide is the principal source of iron for some green glasses and for glasses used to absorb ultraviolet radiation.

Lead oxide (PbO). Lead oxide may be supplied to the glass batch in a number of ways. An early and common method was to use red lead (Pb_3O_4) and many still prefer its use as a raw material.

Potassium oxide (K_2O). Potassium oxide is often supplied by the hydrated carbonate of potash ($K_2CO_3 \cdot 1\frac{1}{2}H_2O$) known as pearl-ash. It is manufactured from potassium chloride.

Potassium carbonate (K_2CO_3). Potassium carbonate, also known as calcined carbonate of potash, is a material preferred by glass makers. It is somewhat dusty when freshly calcined and is apt to become lumpy from hydrating in a humid atmosphere.

Saltpeter, niter (KNo_3). Saltpeter is used when the oxidizing action of a nitrate is desired.

Silica (SiO_2). Silica, also called silicon dioxide, is the most important constituent in common commercial glasses because it is the principal glass-forming oxide. The chief source of silica sand, commonly called glass sand, consists essentially of quartz granules.

Sodium oxide (Na_2O). Sodium carbonate (soda ash) is the most common source of sodium oxide in the glass batch. It is readily available as a natural mineral or a manufactured product.

Fluorine (F_2). Fluorine is unique among the active elements in that it forms no stable oxides. It occurs in nature and in glasses, glazes, and enamels as fluorides.

Phosphate compounds. Apatite is a phosphate compound occuring as natural phosphate rock.

Titania (TiO_2). Titania is refined from rutile and ilmenite ores. It is marketed as a fine, white powder.

Lithia (Li_2O). Lithium carbonate (Li_2CO_3), manufactured from silicate minerals containing lithia, is the principal means of introducing Li_2O directly.

Zinc oxide (ZnO). This oxide is produced by direct oxidation of zinc and subsequent purification by sublimation. It is prepared in several grades based on color, particle size, and smoothness.

Zirconium oxide (ZrO_2). This material is usually introduced as zirconium oxide (ZrO_2). As a component of soda-lime glasses, it tends to increase viscosity, refractive index, and resistance to weathering. Its application is quite limited.

Types of Glasses

Pure silica glass

SiO_2: 99.5%+

Chiefly used for its low thermal expansion, high service temperature, and, when very pure, for its transparency to a wide range of wavelengths in the electromagnetic spectrum and to sound waves. It also has good chemical, electrical, and dielectric resistance. Its disadvantage is the very high temperature needed for manufacture, although it can be made by hydrolysis of $SiCl_4$; in either case, it is expensive. It is used for lightweight mirrors for satellite-borne telescopes, laser beam reflectors, special crucibles for the manufacture of pure single crystals of silica for transistors, and as a molecular sieve that lets hydrogen and helium through.

96% silica glass

SiO_2: 96%
B_2O_3: 3%

Made by forming an article, larger than the required size, from a special borosilicate glass, leaching out the nonsilicate ingredients with acid, and treating at high temperature to shrink the article and close the pores. Good thermal properties; service temperature higher and expansion coefficient lower than any other glass except pure silica. It is more expensive than borosilicate glass. Used for missile nose cones, windows of space vehicles, and some laboratory glassware where exceptional heat resistance is needed.

Soda-lime-silica glass

SiO_2: 70%
Na_2O: 15%
CaO: 10%

The addition of soda (Na_2O) and sometimes potash (K_2O) to silica lowers the softening point by 1472–1652°F (800–900°C). Lime (CaO) and sometimes magnesia (MgO) and alumina (Al_2O_3) are added to improve chemical resistance. Electrical properties can vary quite widely with composition. This is the most common of all glasses, used in quantities for plate and sheet (including windows), containers, and light bulbs. "Crown" glass is of this type, although modern optical crown glass usually contains barium oxide instead of lime.

Lead-alkali-silicate glass

SiO_2:	30–70%
PbO:	18–65%
Na_2O and/or K_2O:	5–20%

Lead oxide reduces the softening point even more than lime and also increases the refractive index and dispersive power. "Flint" glass for optical purposes and "crystal" glass for tableware are both lead glasses. They are also used for thermometer tubes, parts of electric lamps, and neon sign tubes. Compositions vary widely; a glass of high electrical resistance contains about 25% PbO and 6 or 7% each of Na_2O and K_2O; for a high refractive index the PbO content may be as much as 65%.

Borosilicate glass

SiO_2:	60–80%
B_2O_3:	10–25%
Al_2O_3:	1–4%

Has low thermal expansion, about one-third that of soda-lime glass, can be made with good chemical resistance and high dielectric strength, and is used where combinations of these are needed. Its high softening temperature makes it harder to work than soda-lime or lead glasses. Used for laboratory glassware, industrial piping, high-temperature thermometers, large telescope mirrors, household cooking ware, such as "Pyrex," enclosures for very hot lamps, and electronic tubes of high wattage.

Aluminosilicate glass

SiO_2:	5–60%
Al_2O_3:	20–40%
CaO:	5–50%
B_2O_3:	0–10%

Another low-expansion, chemically resistant glass that has a higher service temperature than borosilicate glass but is correspondingly harder to fabricate. It is used for high-performance military power tubes, traveling wave tubes, and many applications similar to those of borosilicate glass. Aluminosilicate glass without any boron is especially resistant to alkalis. Nearly all laboratory glassware is made from borosilicate glass, aluminosilicate glass, or a glass called aluminoborosilicate, which contains roughly equal amounts of Al_2O_3 and B_2O_3. The choice depends on the application; thus, aluminosilicate glass is used for high-temperature applications or alkali-resistant glass. Aluminoborosilicate glass is slightly better than borosilicate for chemical resistance but has slightly greater thermal expansion.

Optical glasses

The following list gives the approximate weight percentage compositions of a few optical glasses. The refractive index n_D and constringence V, defined in Chapter 9, are also given. The V-values are in descending order because, for a particular type of glass, they usually decrease as n_D increases.

Light barium crown glass (n_D = 1.54–1.55; V = 63–59)

SiO_2:	45–50%
B_2O_3:	3–5%
Na_2:	1%
K_2O:	7%
BaO:	20–30%
ZnO:	10–15%
PbO:	0–5%

Dense barium crown glass (n_D = 1.58–1.66; V = 60–50)

SiO_2:	30–40%
B_2O_3:	10–15%
BaO:	10–15%
ZnO:	0–10%
Al_2O_3:	10%

Very light flint glass (n_D = 1.54–1.55; V = 47–45)

SiO_2:	60%
K_2O:	8%
PbO:	27%

Very dense flint glass (n_D = 1.6–1.9; V = 34–20)

SiO_2:	20–40%
K_2O:	0–10%
PbO:	50–80%

Special glasses

The approximate weight percentage compositions of some less common inorganic glasses are given. Some have been developed for special applications. Others are of interest because of their unusual ingredients.

Silicon-free glass for sodium vapor discharge lamps

B_2O_3: 36%
Al_2O_3: 27%
BaO: 27%
MgO: 10%

Phosphate glass with high resistance to HF

P_2O_5: 72%
Al_2O_3: 18%
ZnO: 10%

"Soft-solder" glass with transformation temperature below 752°F (400°C)

SiO_2: 5%
B_2O_3: 15%
PbO: 64%
ZnO: 16%

"Lindemann" glass with low X-ray absorption. All metal atoms have low atomic numbers.

B_2O_3: 83%
BeO: 2%
Li_2O: 15%

Neutron-absorbing glass with high cadmium content

SiO_2: 26%
Al_2O_3: 2%
CdO: 64%
CaF_2: 8%

High-lead-content glass for absorbing gamma rays or X-rays; also, a very dense flint glass

SiO_2: 20%
PbO: 80%

Tellurite glass of very high refractive index (about 2.2) and dielectric constant (static value about 25)

TeO_2: 80%
PbO: 14%
BaO: 6%

Optical glass with high refraction and low dispersion (n_D = 1.68; V = 58)

La_2O_3: 20%
B_2O_3: 40%
Tm_2O_3: 20%
BaO: 20%

Semiconducting vanadate glass

V_2O_5: 85%
P_2O_5: 10%
BaO: 5%

Semiconducting chalcogenide glass transparent to infrared

As: 44%
Te: 24%
I: 32%

Two-dielectric chalcogenide glass transparent to infrared

(1) As_2S_3: 100%
(2) As: 40%
Ti: 20%
S: 40%

Elemental glass (consisting of a pure element)

S: 100%

Photosensitive gold ruby glass

SiO_2: 72%
Na_2O: 17%
CaO: 11%
Au: 0.02%
Se: 0.04%

Photochromic glass

SiO_2:	60%
Na_2O:	10%
Al_2O_3:	10%
B_2O_3:	20%
Ag:	0.6%
Cl:	0.3%
LiO_2:	0.9%

Glass Formers

A surprisingly large number of materials can be obtained as glasses under special conditions. It is even possible to obtain water in the form of glass, if it is cooled sufficiently rapidly, although the experiment is not easy to carry out. If the water has a large amount of some other material in solution, it is easier to obtain the glassy form of water with rapid cooling, since the atoms or ions of solute interfere with the crystallization process, and allow the temperature to fall to the point where the viscosity of the water becomes so high that crystal (ice) formation is difficult. In this connection it is known that simple living organisms (tadpoles, for example) will survive temperatures as low as −130°F (−90°C) if they are brought to this temperature rapidly, so that the formation of ice crystals in the tissue is avoided. In these cases, it may be that the water in the organisms enters the glassy state without the discontinuous and harmful change of physical properties entailed in ice formation.

The great bulk of common glasses are based on silicon dioxide (SiO_2) as the parent glass former. Silicon dioxide occurs abundantly in nature in pure crystalline form as quartz and cristobalite and as a component of numerous silicate minerals. Most common sand contains a high proportion of SiO_2. The pure crystalline forms of silicon dioxide have melting points around 3092°F (1700°C), so that pure SiO_2 glass is supercooled by almost 3092°F (1700°C).

The formula SiO_2 hardly conveys the nature of silicon dioxide. Actually, each silicon atom is bonded to four oxygen atoms, and each oxygen atom is shared between two silicon atoms. The resulting structure for one form of quartz is shown in the diagram in Figure 3.1.

Each silicon atom is shown at the center of a tetrahedron. The angle among the four bonds in which each silicon atom participates is fixed (108°) in both crystalline silica and silica glass. The angle between the two bonds of each oxygen atom is not as critical, howev-

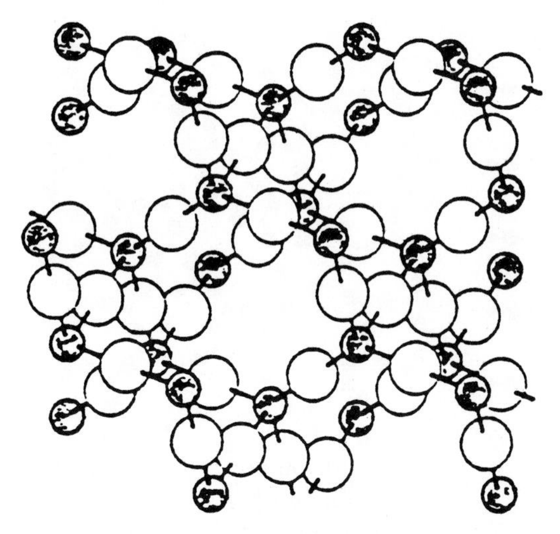

Figure 3.1 Regular crystalline assembly of tetrahedron in one form of quartz. Dark: One silicon atom. White: Four oxygen atoms.

er, so that the oxygen atoms joining two tetrahedrons provide the flexibility necessary for the amorphous structure of glass.

It is difficult to represent the nature of the structure possible with these SiO_2 tetrahedrons, and similar figures, in two-dimensional diagrams, as the reader will probably be prepared to grant from inspecting the diagram given. We can use the device proposed by Warren and Zachariasen and consider an imaginary two-dimensional element G (for glass former), that forms an oxide G_2O_3 only in two dimensions. The SiO_2 tetrahedrons, with oxygen atoms at the four corners, are replaced in our two-dimensional systems by oxygen triangles, with oxygen at the corners and the atom of our imaginary element G in the center of the triangle. The reader then must imagine the argument extended to three dimensions. The regular crystalline form of our two-dimensional oxide G_2O_3 and the amorphous disordered form found in the liquid (and thus, in the glass) are both shown in Figure 3.2. Note that, in either form, the number of bonds per atom is the same and that the structure is quite an "open" one with a good deal of unoccupied space. In the amorphous, or glassy form, the bond angles are slightly distorted from the 60° and 180° angles seen in the hypothetical crystalline form.

Pure SiO_2 glass, known as "fused quartz" or "fused silica," is used for critical applications, but has too high a softening temperature for general purposes. At the temperature necessary to melt quartz sand,

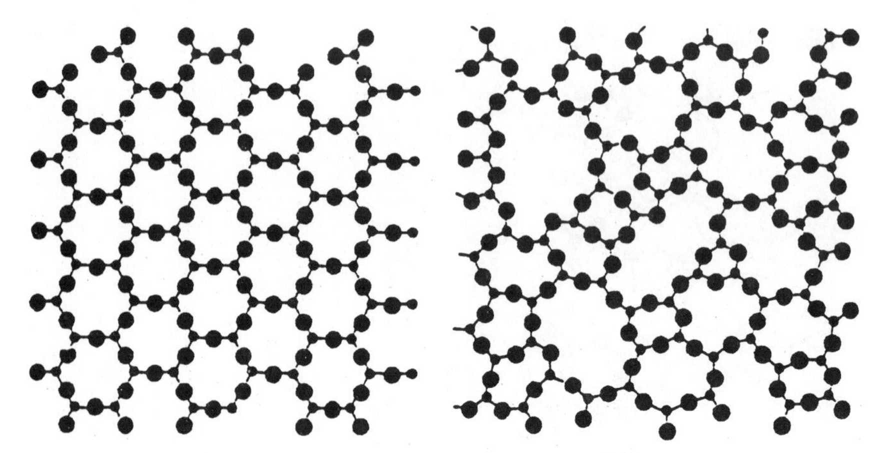

Figure 3.2 Two-dimensional glass. Left: Regular crystalline form of imaginary two-dimensional oxide G_2O_3. Right: Amorphous or glassy form of G_2O_3.

3092°F (1700°C), the liquid SiO_2 is quite viscous, and by the time the liquid SiO_2 has been supercooled to 2372°F (1300°C), the viscosity is already in the neighborhood of 10^{12} poises, far too stiff for convenient shaping by blowing or drawing. The addition of certain metallic oxides, such as soda (Na_2O) and lime (CaO), to the SiO_3 lowers the viscosity (and the temperature required to melt the materials initially) to more practical levels. Glass is normally worked at viscosities of 10^3 to 10^6 poises, and, for commercial soda-lime-silica glass, the required temperature is only 1112–1292°F (600–700°C).

The reason that the addition of the metallic oxides lowers the viscosity can be seen in the diagram, where we again use our imaginary two-dimensional glass, G_2O_3, now with the addition of a metallic ion, M, introduced as an oxide, M_20 (shown as circles in Figure 3.3). The metal ion occupies the space in the formerly open G_2O_3 network, and it would appear that this would stiffen the structure. However, each metallic ion removes one of the bonds of an oxygen atom from the basic G_2O_3 network or, in real glass, from the SiO_2 network. Furthermore, the bond between the metallic ion and the oxygen atom is essentially nondirectional. The result is that the structure is less well braced with the metallic ions present and, therefore, is less stiff or viscous.

Silicon dioxide exemplifies the primary characteristics a glass former must possess. First, the fact that each oxygen atom is shared between two silicon atoms and each silicon atom between four oxygen atoms allows the formation of complex three-dimensional networks. Second, the silicon-oxygen bond is a very strong bond. The result is

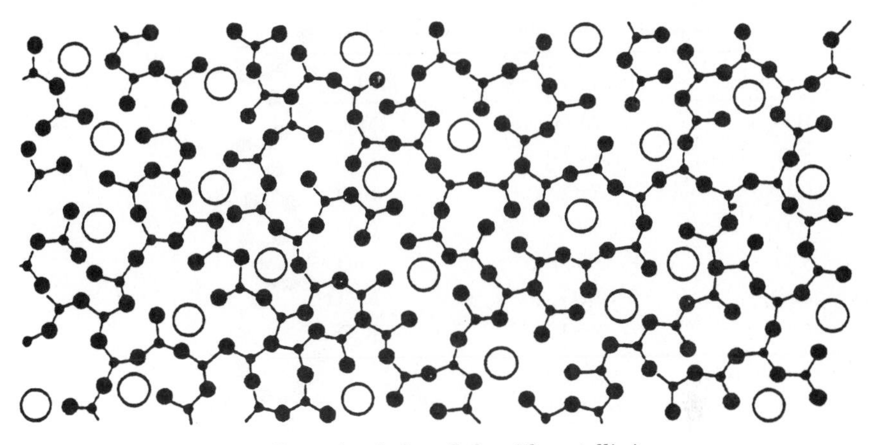

Figure 3.3 Imaginary two-dimensional glass G_2O_3 with metallic ions.

that it is very difficult to disentangle the network in its amorphous liquid form to permit the formation of crystals. The diagram of the structure of quartz shows an orderly crystalline arrangement of SiO_2.

If the reader will consider the diagram of our imaginary two-dimensional glass G_20_3, in Figure 3.2, he or she will be able to imagine how complex the structure of SiO_2 is in the liquid or glassy form.

In addition to silicon, there are a number of other elements whose oxides, sulfides, tellurides, and selenides form inorganic glasses. Boron, germanium, phosphorus, vanadium, arsenic, and zirconium are examples. Beyond these primary glass formers there are the so-called intermediates, titanium, zinc, lead, aluminum, thorium, beryllium, and cadmium, whose oxides do not form glasses alone but may take part in a glass network with one of the primary glass formers. Both the primary glass formers and the intermediates form strongly directional bonds that stiffen the glass structure. Some of these common oxides are listed in Table 3.1.

Beyond these are a number of modifiers, such as the Na_2O and CaO mentioned previously, whose presence does not contribute to the network strength, as such, and serves usually to lower the viscosity of glass and contributes desirable chemical, optical, or, indeed, economic properties to the basic glass. The bond strength decreases regularly as one passes from primary glass formers through intermediates to modifiers.

The oxide glasses are usually opaque to radiation in the infrared portion of the electromagnetic spectrum, and, for infrared work, the sulfides, tellurides, and selenides of arsenic have become quite impor-

TABLE 3.1 Functional Classification of Common Oxides

Glass formers	Intermediates	Modifiers
B_2O_3	Al_2O	MgO
SiO_2	Sb_2O_3	Li_2O
GeO_2	ZrO_2	BaO
P_2O_5	TiO_2	CaO
V_2O_5	PbO	SnO
AsO_3	BeO	Na_2O
	ZnO	K_2O

tant. Arsenic trisulfide, for example, forms a glass opaque to visible light but transparent in the infrared region. Pure SiO_2 transmits light in the ultraviolet region, but the addition of almost any of the common modifiers renders the glass opaque to ultraviolet light, and, because of the high melting point of SiO_2, the formation of pure SiO_2 glass for ultraviolet lenses is a matter of considerable difficulty.

At least one elemental glass, that is, a glass formed of a single element rather than a compound, is known. Liquid sulfur, if cooled very rapidly, becomes a rubbery, elastic "glass" at room temperature. This "glass" is only moderately stable and devitrifies fairly rapidly, however. Among organic compounds, the carbon-to-carbon linkage furnishes the basis for network formation, and many of the common plastics, particularly the transparent plastics, such as Lucite or Perspex, are, technically speaking, glasses. In this book, however, we will follow the common usage of the term glass and treat only the inorganic glasses, particularly the oxide glasses based on silicon.

The Transformation Temperature

We said that although glass does not crystallize (or if it did, it would not formally be glass), it undergoes a peculiar type of freezing. This peculiar type of freezing is of great importance in glass technology and in the theoretical understanding of glass. The effect of this "freezing" is not to alter the structure of the glass but to change the way it expands and contracts with changes in temperature.

We have seen that the heat energy in a liquid or a solid appears as motion of the atoms or molecules of the material. In a crystalline solid, the molecules occupy fixed positions relative to one another, and the molecular motions are vibratory motions around this fixed position. In a liquid, these vibratory motions occur, but there is some random movement of the molecules so that some intermolecular bonds are constantly being broken and new bonds formed. In both liquids and solids, part of the vibratory motion because of heat is motion along the axis of the interatomic bonds in the material, so that the distance between two neighboring atoms varies slightly as these atoms vibrate. In general, the higher the temperature, the greater the amplitude of the vibration.

Interatomic bonds may be formed in a number of different ways but always involve a balance of attractive and repulsive forces. At the normal interatomic spacing, these forces balance and the net force between two atomic nuclei is zero. It is almost always true, however, that, at this normal interatomic spacing, it is easier to pull two atoms a little further apart than to compress them further together. The result is that the thermal vibration has the effect of slightly increasing the interatomic spacing. Therefore, most materials expand slightly as the temperature rises and, correspondingly, contract as it falls. This normal thermal expansion involves no rearrangement of interatomic or intermolecular bonds but merely changes the effective average spacing between atoms.

Glass undergoes normal thermal expansion and contraction, but, above a temperature called the transformation temperature, it undergoes a second kind of expansion or contraction as the temperature changes. Glass has an "open" structure; the network of silicon and oxygen atoms (in a common glass) is not densely packed. As the temperature decreases, however, the structure becomes more dense. Consider a ring of eight silicon atoms alternating with eight oxygen atoms. Each of the silicon atoms in this ring is bonded, through oxygen atoms, to two silicon atoms not in the ring. At a high temperature, this configuration may be stable. At a lower temperature, two of the silicon atoms in the ring may release one of their bonds outside the ring, and, with an intervening oxygen atom, form a bridge across the center of the ring. The entire ring will now have a less open structure and occupy less space. This description of configurational shrinkage is greatly oversimplified; the important point is that configurational expansion or contraction involves rearrangement of interatomic bonds and, therefore, is quite different from the normal thermal expansion previously described.

The change in the volume of a given weight of glass, as the glass is cooled, is shown in Figure 3.4. Beginning at A, with the glass as a normal liquid (that is, above the melting point of the glass-forming materials), we cool the liquid to B, the theoretical freezing point. If the material crystallizes, there is a sharp decrease in volume to C, after which the crystalline material continues to shrink as temperature falls, but at a slower rate (per degree of cooling), to D, the vol-

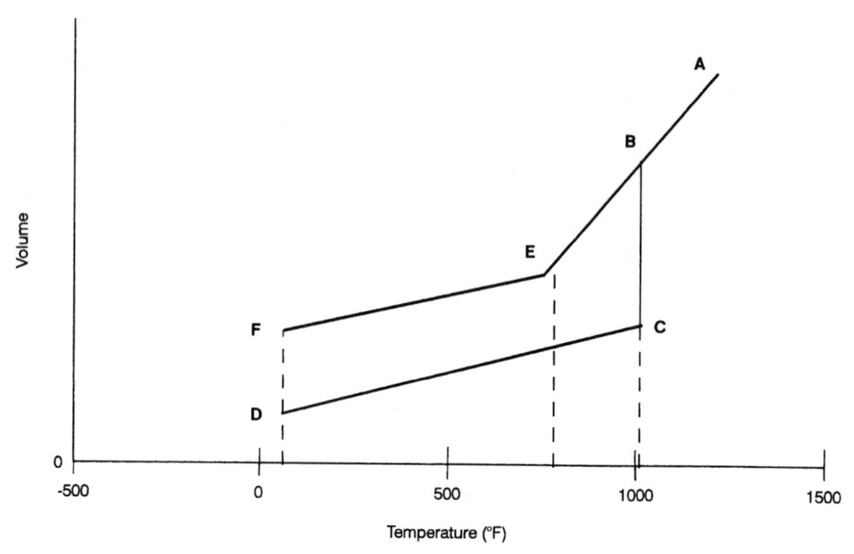

Figure 3.4 Changes in glass volume on cooling.

ume at room temperature. From A to B, the liquid has contracted by two processes, normal thermal contraction and configurational contraction because of a less open structure. From C to D, the crystal, which undergoes no configurational change, shrinks only by normal thermal contraction. (We should mention here that most materials shrink when crystallizing. The increase in volume when ice is formed from liquid water is the exception, not the rule.)

If we avoid crystallization and are able to supercool the liquid, the supercooled liquid continues to shrink at the same rate per degree of temperature decrease, as did the normal liquid. Both kinds of shrinkage are still occurring from A to E. At a particular point, however, the rate of contraction slows and, below that point, continues from E to F at the rate previously discussed for the crystal. Apparently, the configurational shrinkage has stopped at E, and, from E to F, the glass, although still a liquid, is shrinking only by normal thermal contraction. The temperature at which this occurs (the transformation temperature) is not a sharply defined point, but rather a range of about 932°F (500°C), as the bend in the curve suggests.

At the transformation temperature, the glass has become so viscous that the configurational changes necessary for the denser structure no longer have time to occur. These changes involve rearrangement of the intermolecular bonds, and, when the glass has reached a particular viscosity, the necessary changes can no longer keep pace with the rate at which the temperature is falling. The interesting point is that the viscosity at which this occurs, for practical cooling rates, is roughly the same for all glasses, 10^{13} poises, although the temperature involved may range from −130°F (−89°C) for glycerin to over 1000°C for pure silica glass. The fact that the configurational shrinkage ceases at this viscosity is the reason for making a viscosity of 10^{13} poises the formal borderline between a glass and a simple supercooled liquid.

The term "transformation temperature" is a bit of a misnomer, in that the glass undergoes no transformation at this temperature; for practical purposes, it ceases to undergo the internal transformations appropriate to its actual temperature. Further, the configurational changes do not actually cease. They simply continue at the rate set by the viscosity of the glass. The point is that, when this viscosity has reached the region of 10^{13} poises, the cooling rate involved is so slow that it causes the configurational changes to fall behind the temperature change. When the glass reaches room temperature and a viscosity of 10^{20} poise, for practical purposes, the change has stopped altogether.

In fact, there are two properties of a glass that change at the transformation temperature. The first is the rate of thermal expansion, as already discussed. The second, reasonably enough, is the specific heat of the glass, that is, the amount of heat energy required to raise the temperature of 1 gram of glass 1°C. Below the transformation temperature, all the heat supplied goes only into the molecular vibrations, as no appreciable configurational change is occurring; above the transformation temperature, the individual molecules can also move about somewhat in the material, and some of the heat energy goes into this motion and into the vibratory motion of the molecules. Accordingly, the specific heat may increase appreciably as the transformation temperature is exceeded.

The temperature at which a particular piece of glass effectively ceases to undergo configurational change, when cooling, is sometimes called the fictive temperature of the glass. This is the temperature at which the glass is thermodynamically stable, that is, has no spontaneous tendency to change to a denser or less dense structure. The fictive temperature of common glass is in the neighborhood of 932°F (500°C), and it follows that the glass is not in thermodynamic equilibrium at room temperature. In principle, common glass is unstable at room temperature. In fact, it is slowly shrinking, because configurational change will continue until the fictive temperature reaches the actual temperature. For glass at room temperature, however, the process goes forward so slowly that millions of years would be required.

Because rates (of cooling and of configurational shrinkage) are involved, it would seem that the temperature at which this transformation occurs should depend on the rate of cooling. This is, indeed, the case. As shown in Figure 3.4, if the cooling rate is high, the configurational change falls behind at a higher temperature; if it is very low, the transformation temperature is correspondingly lower. Thus, we may obtain glass of the same composition, but of different densities because of different fictive temperatures according to the rate at which the glass is cooled. Also, within the same piece of glass, the density may vary if the exterior part of the glass has cooled more rapidly than the interior. Finally, the glass may change in dimensions after it has cooled, because of continued configurational shrinkage. The effect is ordinarily negligible, but, in certain applications, such as precision thermometers, it can be a nuisance. We will refer later to the use of different densities obtained by different cooling rates within the same piece of glass in the manufacture of "toughened" glass.

Now, we can restate our definition of a glass in somewhat more formal terms. A glass is a material obtained by supercooling a liquid to a temperature, where the viscosity of the liquid exceeds 10^{13} poises, does not incur discontinuous changes in viscosity or structure, and retains the organization and internal structure of a liquid. As the preceding discussion has implied, cooling beyond the temperature where this viscosity is reached will have little effect on the internal structure of the material, except to increase the viscosity further; the viscosity of common glass at room temperature exceeds 10^{20} poises, and, over the time periods of interest to people, the material behaves as an ideal elastic solid.

Conclusion

Glasses can be designed to meet a set of specifications by using well-established and proven formulas. The glass design must be tested to correlate the formula with laboratory data.

In commercial practice, the final result will be very close to predicted performance.

References

1. Maloney, F. J. T., *Glass in the Modern World,* Doubleday and Company, Aldus Books, Ltd., London, 1967.
2. Tooley, F. V., *The Handbook of Glass Manufacture,* 3rd ed., Books for the Glass Industry Division, Ashlee Publishing Co., Inc., New York, 1984.

Chapter

4

Float Glass, Plate Glass, Soda Lime, and Colored

Joseph S. Amstock

President, Professional Adhesive and Sealant Systems

Introduction

In the 1950s, an ingenious new method of making relatively inexpensive flat glass of high quality was developed in England by Alistair Pilkington of the Pilkington Glass Co.[1] In the float process, a continuous strip of glass from the melting furnaces floats onto the surface of a molten metal, usually tin, at a carefully controlled temperature. The flat surface of the molten metal gives the glass a smooth, undistorted surface as it cools. After sufficient cooling, the glass becomes rigid and can be handled on rollers without damaging the surface finish. This method was introduced in the United States in the early 1960s. Four U.S. patents were very instrumental in the evolution of the float glass process in the United States.[2,3,4,5] The development of the float process took several years of massive effort and expenditure. Most of the problems encountered were primarily engineering and chemical in nature.

In 1965, the process was deemed capable of overtaking plate glass and, by 1970, was also clearly capable of supplanting sheet glass. As sizing techniques were developed for the manufacture of ⅛-in (3.0-mm) and then $^{3}/_{32}$-in (2.0-mm) glass, it became clear that the inferiority of sheet glass technologies to the float process would soon result in their demise.

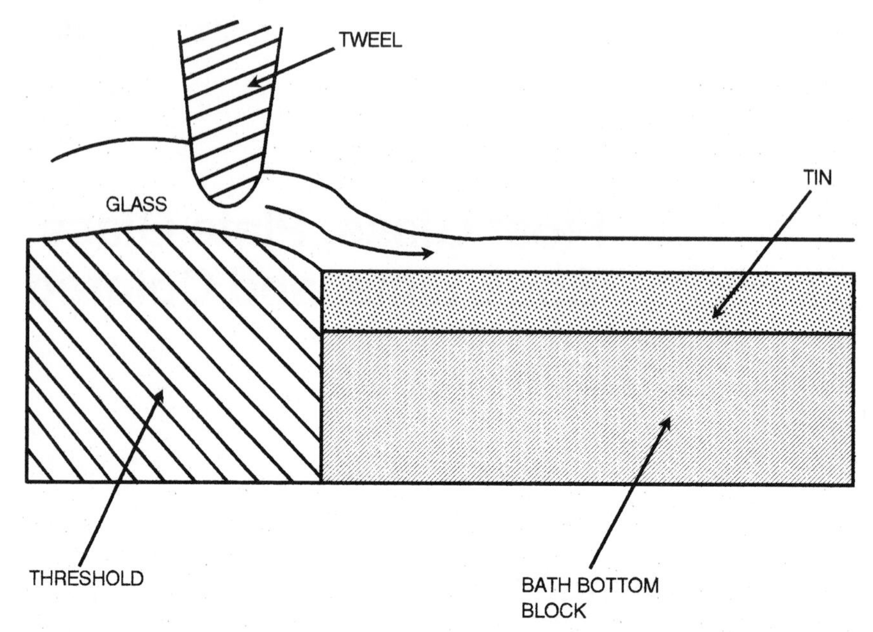

Figure 4.1 Typical process delivery system.

The Float Process

In the float process, glass from a tank flows under a tweel and over a lip (or spout) onto the tin bath. The temperature at this point is approximately 2000°F (1093°C). Figure 4.1 illustrates a typical process delivery system.

The temperature may vary slightly, depending on tonnage and other considerations, but it is always well above the liquid temperature. The glass follows a complicated flow pattern in this region of the bath. While the bulk of the glass is flowing forward and laterally to form what is called the "onion," the glass that was in contact with the lip refractory flows in the reverse direction to the "wetback" and then outwardly and forward to be in the outer edges of the ribbon. It is this wetback flow phenomenon which is at the heart of the float process.

Following are the most important factors in consistent glass manufacturing:

1. Controlled temperature to melt the raw materials.
2. Controlled temperature to cool the glass with proper stress control during the annealing process.
3. Consistent tons of pull to help temperature control.

TABLE 4.1 Typical Float Glass Compositions*

	Weight %			
	I	II	III	IV
SiO	7.08	73.11	72.98	72.65
Al_2O	0.10	0.10	0.12	1.15
CaO	8.0	8.80	8.40	
MgO	3.86	3.95	3.91	3.94
Na_2	13.83	13.90	13.76	13.04
MC†	0.29	0.37	0.43	0.82

*Representative compositions of four major float glass companies.
†Minor constituents, purposely added or incidental to raw materials used.

Batch house

Seven raw materials are used (five majors, two minor) to manufacture a soda-lime-silica float glass.[6] This is 11 hopper cars per day for two furnaces (tanks). Typical raw materials are listed in Table 4.1.

Unloading and storage of raw materials

Because of the vast tonnages required, raw materials are delivered to the processing plants via rail or truck. Each raw material is transported to a holding silo and stored before weighing. However, before each material is unloaded, it is sampled and checked to be sure that it is within the specification; this is accomplished by the Quality Control (QC) Department. Other on-site QC checks are made to assure that the raw material is put in the correct silo. A slip up here would take days to resolve and cause poor quality glass.

Weighing, mixing, and conveying to the furnace

Through a computerized process, the exact amount of raw material from each silo is put on the scales until the batch makeup is completed. After weighing, the materials are taken to a mixer where a precise mixing time is set and activated. Water is added to the batch to keep segregation to a minimum (similar to a large concrete mixer). As the raw material mixture is conveyed to the furnace, cullet is added on top of the batch. The cullet is clean ground glass used for economy because of decreased energy consumption, improved melting characteristics, improved air emissions, and increased production yields. The batch and cullet are sent through a metal detector to remove any staples, nails, or other traces of metal that may accidentally get into the batch. From the metal detecting area, the batch is placed into the storage silos over the furnace and is continually fed in the furnace to hold the glass level constant.

Furnace

Glass melting requires

1. high temperatures to obtain molten glass from the raw materials, a temperature range of approximately 2900°F (1593°C) at the melting end of the furnace and 2000°F (1093°C) at the exit from the furnace to the tin float bath, and
2. suitable refractories in the furnace to endure the high temperatures.

The several types of furnaces used throughout the world are generally side-port, regenerative types. A typical furnace 175 feet (53.34 m) long, 30 feet (9.144 m) wide, and four feet (1.219 m) deep holds in excess of 1600 tons of molten glass. The fuel is usually natural gas, sometimes with electric boosting; on occasion, fuel oil replaces gas as the main energy source.

Tin bath

Overall quality is very good. The few defects (stones and seeds) that are of tank origin occur less than once every 200 square feet (18.6 m^2). The batch continuously enters through the canal from the working end of the furnace to the tin bath. The problem of tin containment was solved by means of constructing a tin bath, which is the key to modern float glass technology. It holds approximately 150 tons of molten tin (value of $1.5 million). The tin is 2 to 3 in (5.28 to 7.82 mm) deep and the atmosphere inside is 94 percent nitrogen with 6% hydrogen to prevent oxidation of the tin and staining of the glass. The tin bath is 156 feet (47.55 m) long by 20 feet (6.10 m) wide at the front and 14 feet (4.267 m) wide at the end. Glass enters at 1900°F (1038°C), spreads to a wide ribbon, and floats on much denser and colder tin. The glass is cooled, receives a natural polish finish, and is precisely controlled to the specified thickness. The top roll machines control the glass ribbon through the tin bath. A schematic of the float line is seen in Figure 4.2.[7]

The glass is openly visible for the first time as lift rollers guide the ribbon from the bath into the lehr. The lehr is the annealing (cooling) apparatus using electrically heated air to cool the glass uniformly. The glass enters at 1125°F (607°C) and exits at 540°F (282°C). The lehr is approximately 381 feet (106 m) long.

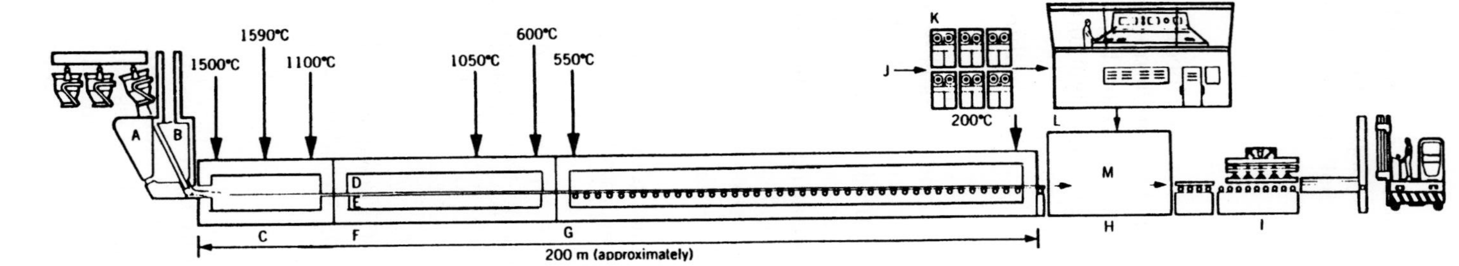

Figure 4.2 A typical float process. A. Raw material mix. B. Cullet. C. Oil-fired melting furnace. D. Controlled atmosphere. E. Molten tin. F. Float bath. G. Annealing lehr. H. Automatic warehouse (not shown). I. Automatic stacking. J. Orders. K. Computers. L. Control point. M. Computers govern the cutting processes, matching complex orders to the continuous ribbon of glass and directing the cut glass to the appropriate part of the warehouse for stacking and dispatch; the automatic warehouse stands by itself as an important advance in the flat glass technology. This illustration is not to scale.

Mass air

This portion of the lehr uses outside air to further cool the glass. Glass speed is 300 to 800 in/minute depending on thickness and production rates. The glass leaves the hot end and goes to the cold end of the plant.

Inspection booth

The inspection booth sits directly over the ribbon and permits a close inspection of the glass for defects. A xenon lamp source is directed at the glass and defects show up as shadows on the white paper below. The technician can mark defects or use a computer control model to have defective sections cut out.

Cutters

The cutters score the glass with carbide wheels in an X (across) and Y (along) direction for a particular customer or production runs. Over a dozen size changes are typically made per day. The cutters are computer controlled to adjust to the speed of the glass. On the X cut, the cutter moves at an angle compensating for the movement of the glass to wind up a with straight perpendicular cut. The first two "bridges" are Y cutters and the next four are X cutters. Approximately seven technicians per shift work on the line and in the quality control laboratory.

The snap rollers break the glass along the score lines by applying upward pressure. Takeout conveyors/rollers then separate the sheets or lites of glass by faster speeds and angling lites apart. Edge trim (this glass is imprinted from the top roll machined in the tin bath) glass is cut away and dropped into the hopper to be ground up and returned to the furnace.

Interleaving

Interleaving is the application of a powder spray of fine silicon beads, applied to eliminate scratching the glass surfaces when glass is packed together.

Glass handling/packing

Because of weight and specifically large sizes, this operation is accomplished by robotics. Machines with vacuum cups pick the glass off the line, rotate up the track, and place the glass on an A-frame. The glass is removed from the A-frame in a similar manner and placed into crates.

Plate Glass

Plate glass is defined as transparent, flat glass having plane polished surfaces and showing no distortion of vision when objects are viewed through it.[8] This process, in which the molten glass is made by a rolled process and, after cooling and solidifying, is mechanically ground and polished in the finished form, was previously used to produce all higher-quality glass for large areas of glazing. This process was developed to a high degree with on-line grinding and polishing of both sides of the continuous ribbon, but it is expensive and wasteful and, now, has been superseded by the float process, with the single exception of wired glass.

Again, because of the long usage of the term, much currently made float glass, particularly in higher thicknesses, is still incorrectly referred to as plate glass, and many standards, specifications, and other documentation still refer to plate glass. In the early days of its production, polished plate glass was used mainly in the manufacture of mirrors. Today, its expanded uses include automotive, furniture, and some limited construction. The methods used to produce polished plate glass and those required to make window glass are quite different after melting and fining processes have been completed. It is the rolling, grinding, and polishing operations that distinguish plate glass from sheet or window glass. Because of the surface contact with the base or table on one side and the roller on the other, the surfaces become roughened.

In 1688, St. Gobain of France established a plant to furnish commercial products. They desired to produce larger and thicker sheets than were available through the blowing process and developed methods of pouring molten glass on a flat base and of rolling it thereon into a plate.[9]

It was not until 1850 in Massachusetts that the first attempts were made to produce flat glass in the United States.

Early manufacturing

To restore the smooth surfaces and transparency, mechanical grinding and polishing operations are required. Plate glass factories are made up essentially of two parts:

1. the rough glass manufacturing operation, and
2. the grinding and polishing operations

The first stage in the manufacture of plate glass consists of melting the raw materials in a pot or tank furnace and casting or rolling the molten glass into a rough plate blank for annealing. The second stage of manufacture is the finishing or grinding and polishing of the rough plates. In the process of grinding and polishing, the plate glass acquires perfectly plane and parallel surfaces so that, in contrast to most sheet glass, there is no perceptible distortion of objects viewed through it.

To meet the varied requirements of modern plate glass usage, many thicknesses, sizes, compositions, and colors must be produced. This has necessitated three distinct methods of manufacture:

1. pot casting,
2. ladling, and
3. continuous tank rolling (this method is used for all large volume production units).

Except for the preliminary operations of melting and fining, the three processes are essentially the same. The pot casting and ladling systems are used primarily for special formulas and colors in relatively small batches and/or for extra large sheets or thicknesses. The continuous tank method is employed for the great quantities of regular glazing sizes, and for compositions ranging from $^{7}/_{64}$ in (2.78 mm) to $1^{1}/_{4}$ in (31.76 mm) in thickness.

At the beginning of the twentieth century, two important innovations in the casting process were introduced. The first and most significant was the continuous annealing lehr, which consisted of five staggered preliminary ovens and a straight tunnel runway 300 feet (91.44 m) to 400 feet (121.92 m) long. Its use permitted accomplishing the entire process of annealing and cooling in about three hours, compared to four or more days with individual ovens.

Until about 1922, plate glass was produced in the United States and Europe only by the pot casting method. The fundamental methods of producing plate glass changed very little. The glass, in a single, one-piece ribbon, now travels continuously by mechanical power, whereas in the old process, the plates were individually melted, rolled, annealed, ground, and polished. They were moved from one stage to another manually.

In 1939, Pilkington succeeded in developing its own twin grinding and polishing operations, thus, eliminating grinding and polishing tables by substituting two-side grinding and polishing machi-nes.

Enormous strides were made by Edward B. Ford Plate Glass Company and PPG Industries to further enhance the plate glass business and change the economic picture radically. A comparison of plant facilities for two eras provides a measure of the improvement. The investment for buildings, tools, and equipment averaged about $2600 per person in the 1900s compared to $45,000 per job in 1957. Prior to the introduction of machinery and aside from labor, the principal expenditure was for fuel.

The increased demand for glass by the construction and automotive industries and the development of a laminated safety glass market led to a search for improved glass-making processes. The older method of table casting began to be replaced by modified intermittent and continuous casting processes.

Bicheroux semicontinuous process

Bicheroux and Showers made some significant advances in the semicontinuous casting process. In its successful form, the Bicheroux process employs two rollers of the same diameter, as illustrated in Figure 4.3. The pot of glass is cast into a metal receiver or scoop and, when tipped, this receiver feeds the molten glass at the optimum rate between the paired forming rolls. Table sections receive the rolled plate and carry it away at the peripheral speed of the rolling machine.

A cutting device severs the plate into sections at the table junctions for immediate transfer to the annealing lehr positioned at right angles to the direct casting. Because of the more accurate control of the Bicheroux process, the rough plate thickness could be appreciably decreased. Also, the casting of rough plate for grinding to the desired $\frac{7}{64}$-in (2.778-mm) thickness for laminated glass became practical. Pots capable of holding about 28 cubic feet of glass were able to roll

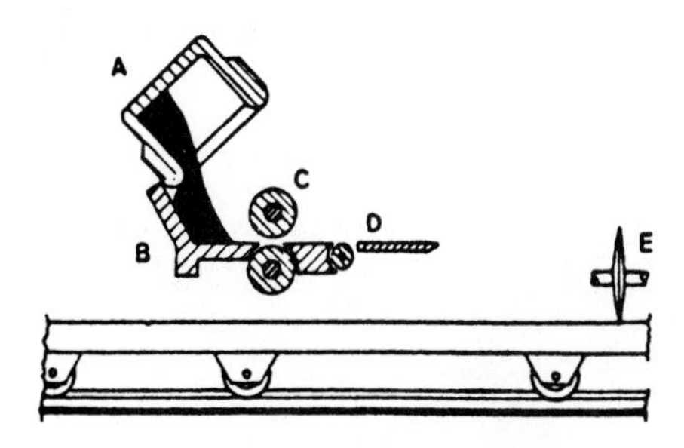

Figure 4.3 Bicheroux semicontinuous rough plate rolling machine. A. Glass pot in casting position. B. Glass receiver. C. Paired forming rolls. D. Tilting slip table. E. Hot glass cutter. F. Movable casting table sections.

plate 12 feet (3.658 m) wide and from 40 to 100 feet (12.192 to 30.48 m) long. In addition, greater accuracy was available with the continuous ribbon.

PPG ring roll process

Although similar to the Bicheroux process in that the glass is melted in individual pots and cast between rollers, this process differs mainly in the type of rolls and in the method used to deliver the glass to and remove it from the rolling machine. Molten glass is cast from the pot on the upper surface of a large hollow cylinder, 12 feet (3.657 m) in diameter (see Figure 4.4). The hollow cylinder is equivalent to the flat casting table curved to a cylindrical shape. Because of the complications involved in different rolling speeds for different glass thicknesses, different plate glass lengths, and occasional pouring delays, the front sections of the lehr must have a range of feeds different from that of the annealing lehr proper.

Continuous casting process

From the beginning, the continuous tank ribbon processes were confronted with two difficult problems. First, tank-made glass was considered inferior to pot-melted glass. Second, mechanical and other difficulties were responsible for keeping ribbon widths and, consequently, the maximum size of plates too low to satisfy a considerable portion of the polished plate glass market. The Bicheroux and

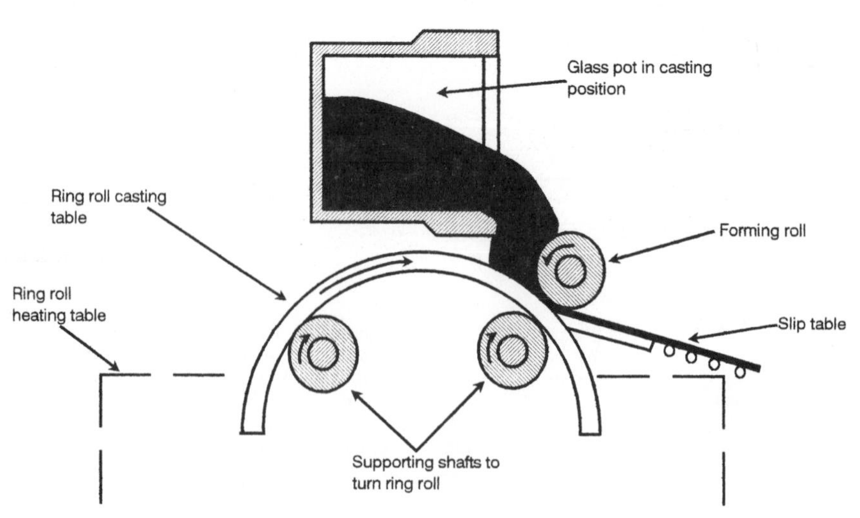

Figure 4.4 PPG-Showers ring roll.

PPG-Showers ring roll processes solved both of these problems almost from the start and permitted continuous grinding and polishing. The width of tank-melted, continuous ribbon-rolled glass has increased many times from 3 to 11 feet (0.914 to 3.353 m).

Libbey–Owens continuous rolling

Following the notable success of the Colburn sheet-drawing process, Libbey–Owens Glass Company modified the drawing machine, in 1925, to manufacture rough drawn plate blanks to help meet the unprecedented demand from the automotive industry. To overcome the slow drawing rate for thick glass, paired, knurled, water-cooled rolls were introduced as illustrated in Figure 4.5. These rolls pushed the sheet upward without slippage and greatly accelerated the normal drawing rate. This vertical rolling process became a regular source of plate glass blanks and supplemented the production of the faster horizontal rolling process for a number of years.

Continuous horizontal rolling process

Today's process for producing plate glass blanks by continuous rolling consists of

1. flowing the molten glass over a weir or through a refractory slot to give it a preliminary shape and, then, before the glass has had time to chill,

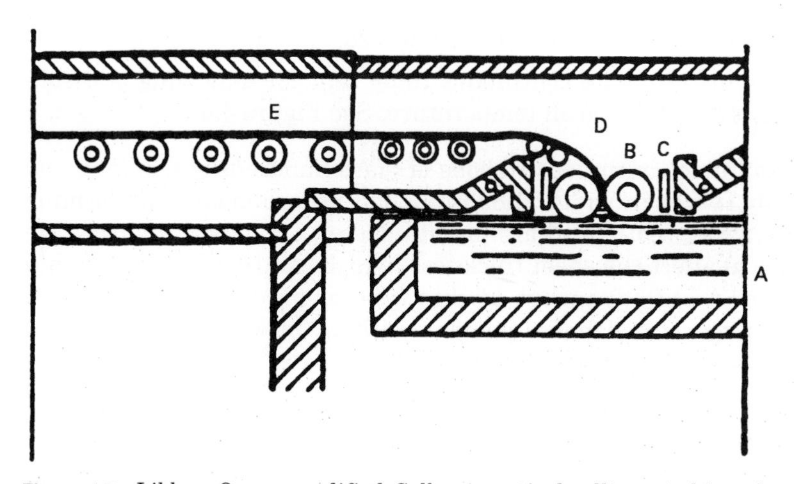

Figure 4.5 Libbey–Owens modified Colburn vertical rolling machine. A. Shallow-heated kiln or drawpot. B. Paired forming rolls. C. Sheet coolers. D. Apron roll conveyor. E. Horizontal lehr.

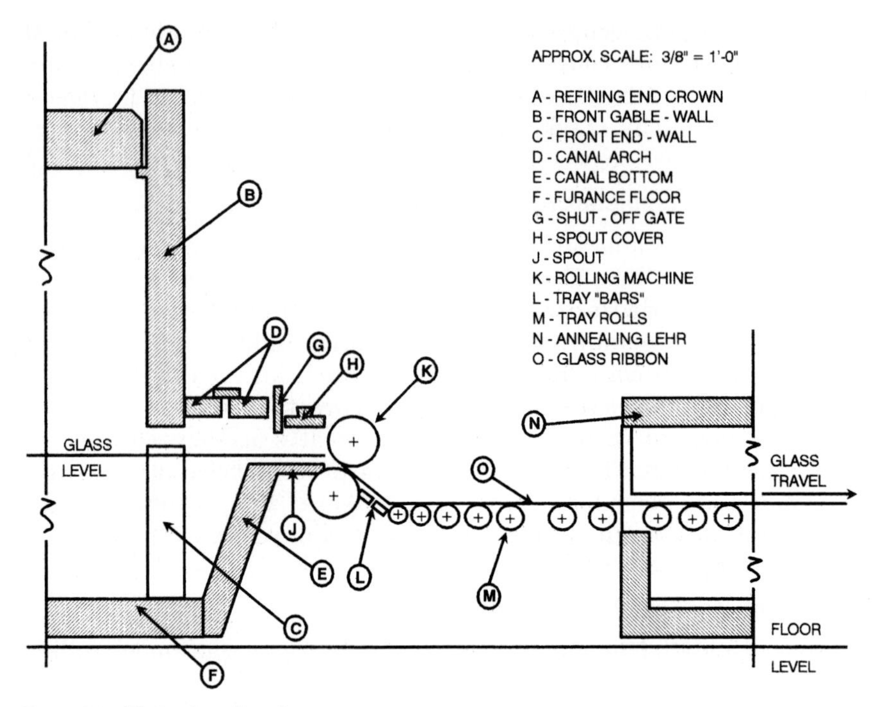

Figure 4.6 Plate glass forming process.

2. passing it between a pair of water-cooled rolls to give it accurate thickness and width,
3. stretching it slightly to improve flatness, and, finally,
4. passing it through a continuous roller lehr for annealing and subsequent cooling to room temperature. See Figure 4.6.

In this manner, continuous ribbons of glass hundreds of feet long and from 6 to 11 feet in width (1.828 to 3.353 m) are made. The forming rolls are water cooled and are maintained at a constant temperature so as not to affect the sheet thickness, which can be controlled within $\pm$ 0.01 in. A big advantage of this type of production is that the machines operate 24 hours per day without interruption except for occasional roll or lip changes required.

Continuous grinding and polishing

Both the Ford and Pittsburgh processes had tables mounted on flanged wheels that rolled on a precision track, and each car was separately laid with glass before being coupled to the train. In the

Pilkington system, grooved tables slid on guides very much like a lathe bed and were coupled together to give a continuous bed on which glass could be laid regardless of the location of table joints between units. The track had to be laid perfectly level and in a straight line. The lengths of this track line range from 800 to 1200 feet (243.84 to 356.76 m).

Soda-lime glass

The addition of soda (Na_2O) and sometimes potash (K_2O) to silica lowers the softening point by 1475 to 1651°F (800 to 900°C).[10] Lime and sometimes magnesia (MgO) and alumina (Al_2O_3) are added to improve the chemical resistance. Electrical properties can vary quite widely with composition. This is the most common of all glasses, used in huge quantities for plate and sheet glass, including window glass. In addition, it is used for containers and lamp bulbs. "Crown" glass is of this type, although modern optical crown glass usually contains barium oxide instead of lime.

Colored, Stained, and On-Line-Coated Glass

Body-colored[11] and tinted glasses are produced by varying the basic composition in the melting tank and can be produced by any of the processes mentioned earlier. Because changing the basic composition of a glass tank is a lengthy and large-scale operation, glasses with modified properties are being produced from basic clear glass by surface modification or surface coating applied either during manufacture or subsequently. These modifications to the basic glass that are applied during manufacture, that is, either on the float bath or while the glass is still in the annealing lehr, may still be considered basic products. They include solar control glasses in which a colored layer of metal ions is produced by injecting into one surface of the (clear) glass, while it is still in a semimolten state in the float bath, or by chemical vapor deposition in the float bath. There are also surface coatings, either for solar control or for reduced emissivity, which are sometimes called pyrolitic coatings because they are applied to the hot glass during its passage through the annealing lehr and involve pyrolitic decomposition of materials sprayed onto the glass to form a layer that fuses to the surface.

Stained glass is glass that has been treated with a material (silver nitrate, for example), which, when the glass is reheated to a point near softening, actually penetrates the glass and unites chemically with it to form a colored glass, usually yellow-orange. The term, however, is usually used to refer to all forms of body-colored or tinted glasses and clear glass on which translucent ceramic paints or even unfired (oil-based) paints have been used. These glasses are principally used for decorative purposes in relatively small pieces, assembled within a matrix of lead sections to make up a leaded light or window.

Conclusion

Float glass has been one of the most advantageous improvements seen in the glass industry during the past 50 years. More than 90 percent of the flat glass produced in the Western world is made by the float glass process. In this method, the molten glass is poured continuously from a furnace onto a bed of molten tin. The glass floats on the tin, spreading out and forming a level surface.

The properties of float glass vary only slightly among manufacturers. However, the manufacturer should be consulted for exact figures.

References

1. *Grolier Interactive, Inc.,* Grolier Electronic Publishing, Version 7.0, New York, 1995.
2. Lombardi, L., U.S. Patent 661,250.
3. Neal, W., U.S. Patent 710,350.
4. Hitchcock, H. K., U.S. Patent 789,911.
5. Pilkington, L. A. B., U.S. Patent 3,220,816.
6. *Processing Procedure,* courtesy of AFG Industries, Inc. Kingsport, TN, 1945.
7. Wilkes, J. A., *Encyclopedia of Architecture Design,* Engineering & Construction, Vol. 2, John Wiley & Sons, New York, 1984.
8. Tooley, F. V., *The Handbook of Glass Manufacture,* 3rd Ed., Vol. 11, Ashlee Publishing Co., Inc., New York, 1984.
9. Maloney, F. J. T., *Glass in the Modern World,* Doubleday & Company, London, 1967.
10. Maloney, F. J. T., *Glass in the Modern World,* Doubleday & Company, London, 1967.
11. Wilkes, J. A., *Encyclopedia of Architecture Design,* Engineering & Construction, Vol. 2, John Wiley & Sons, New York, 1984.

Chapter

5

Borosilicates

Joseph S. Amstock

President, Professional Adhesive and Sealant Systems

Introduction

Silicate glasses containing boric oxide compose another major group, borosilicate glasses. These glasses have a higher percentage of SiO_2 (70 to 80 percent) compared to lead crystal and soda lime. The rest of the composition is as follows: 7 to 13 percent boric acid (B_2O_3), 4 to 8 percent Na_2O and K_2O, and 2 to 7 percent aluminum oxide (Al_2O_3). Compared to most other types of glasses, such as soda-lime glasses, borosilicates have a very low alkali content. This allows them to retain excellent chemical and electrical properties and the low coefficient of thermal expansion inherent in the silica itself.

At the same time, the boric oxide fluxing agent in borosilicate provides good processibility, a characteristic unobtainable in high silica content glasses, such as 96 percent silica glasses. These properties have helped establish borosilicate glasses as the workhorse glasses for industrial applications. They are widely used for such products as laboratory glassware, boiler gages, process piping and drain lines, centrifugal pumps, impellers, seals to low-expansion metals, large telescope mirrors, enclosures for incandescent lamps, and electronic tubes of high wattage. Borosilicate glasses are also used to make baking and cooking dishes.

Technical Data[1]

Chemical properties

Hydrolytic resistance according to ISO 719-HGB

Hydrolytic resistance according to ISO 720-HGA

Acid resistance according to ISO 1776

Alkali resistance according to ISO 695-A

Borosilicate flat glass is highly resistant to water, neutral, acidic, and saline solutions and to chlorine, bromine, iodine, and organic substances. Even over long periods of time and at temperatures above 212°F (100°C), the glass exceeds the chemical resistance of most metal and other materials.

Exposure of the glass to water and acid results in leaching out only small amounts of mostly monovalent ions from the glass. In this process, a very thin, pore-free silica coating forms on the outer surface of the glass, slowing down additional attacks.

Mechanical properties

Density at 77°F (25°C)	2.2 g/cm^2
Modulus of elasticity	63 kN/mm^2
Knoop hardness HK 0.1/20 (according to E DIN/ISO 9385)	480
Poisson's ratio	0.2

Physical impact

The resistance of borosilicate glass to physical impact depends on the type of installation, the size and thickness of the panel, and other parameters. Therefore, data describing the physical impact can be supplied only on the actual application, along with information about the impact requirements for the specific application.

Optical properties

Refractive index, n_d	1.472
Dispersion ($n_f - n_c$)	71.9×10^{-4}

Thermal properties

Linear thermal coefficient of expansion, 68/572°F (20/300°C)	3.25×10^{-6}/K
Transformation temperature, Tg	986°F (530°C)
Annealing point (10^{13} dPa × s)	1040°F (560°C)

Thermal conductivity at 194°F (90°C)	1.12 KW/ (m × K)
Mean specific thermal capacity, Cp, 68–212°F (20–100°C)	0.83 kJ (kg × K)
Maximum operating temperature (in consideration of RTD*)	
short term	932°F (500°C)
long term	842°F (450°C)
Resistance to temperature differences (RTD*)	
short term exposure	
(1 h)	RTD = 110 K
(1–100 h)	RTD = 90K
long term exposure	
(>100 h)	RTD = 80K
Resistance to thermal shock (RTS**)	
Thickness up to 4 mm	RTS = 175K
Thickness 4–6 mm	RTS = 160K
Thickness 6–15 mm	RTS = 150K

Many compositions of borosilicate glasses have been developed to meet specific types of applications. In general, these materials can be grouped into six basic types. The family of borosilicate glasses is extraordinarily broad, depending upon how the boron compounds within the glass melt interact with the other constituents.

Types and Forms

Low-expansion borosilicate glass

Low-expansion borosilicate glass has one of the lowest coefficients of expansion 0 to 570°F (−18 to 299°C: 18×10^{-7} in/in°F) of the commonly used glasses. Another outstanding property of this glass is its excellent chemical resistance, which is much better than that of any other borosilicate glass. Because of its excellent thermal and chemical stability, low-expansion borosilicate glass is popular for chemical process piping where lines are commonly flushed with steam, hot water, or hot detergent solutions.

Low-electrical-loss borosilicate glass

Low-electrical-loss glass, which has the highest boric oxide content of the borosilicate glasses, has a dielectric loss factor of only 0.025 at mc and 68°F (20°C). Other than fused silica and some 96 percent silica

*RTD—Resistance to temperature differences.[2]
**RTS—Resistance to thermal shock.

glasses, no other glass has an equally low value. The low-electrical-loss glass is available as blown and pressed ware, centered shapes, and tubing and rod for various electrical applications, such as insulators, electronic tube components, and resistors.

Sealing borosilicate glass

Sealing borosilicate glasses, including tungsten, Kovar, and series sealing materials, are widely used for electrical applications, particularly lamps.

Ultraviolet-transmitting glass

Ultraviolet-transmitting glasses are special glass compositions having more than 50 percent transmittance at about 230 nm. These glasses, generally produced as tubing, rod, blown ware and panels, are useful for ultraviolet-sensitive photomultiplier tubes and low-pressure mercury germicidal and ozone-producing lamps.

Laboratory apparatus glass

Laboratory apparatus glasses are aluminoborosilicate glasses with relatively high alumina content. They are characterized by high softening temperatures, relatively low coefficients of thermal expansion, and good corrosion resistance.

Optical grade glass

Optical grade glasses, which are referred to as borosilicate crowns, are special composition glasses characterized by excellent homogeneity (freedom from cord and striae), high transmission, and good corrosion resistance.

Borosilicate glasses are available in the usual forms, except for rolled sheet and large castings. Depending upon the exact glass composition, the following five common forms are available.

Pressed glass. Pressed glassware was one of the first forms of this family of glasses. Today, with the exception of a few sealing type materials, all the borosilicates are available in the form of pressed glassware. Compared to other types of glass, the borosilicates have higher processing viscosities; consequently, they are slightly more difficult to press mold.

Blown ware. Nearly all the borosilicate glasses are available as blown ware. The low-expansion borosilicate types can sometimes be blown to closer tolerances than standard listed values.

Drawn glass. Drawn sheet glass can be prepared in a wide range of sizes, such as microsheet, which is available in 0.002- to 0.024-in thicknesses and 14-in width. Ribbon glass can be prepared in 0.0014- to 0.018-in thicknesses and can be made as narrow as 0.019 in or as wide as 1.5 in. Lengths of microsheet and ribbon are standardized for convenience at 100 and 1600 feet. Rods and tubing are also available in a wide range of diameters and lengths.

Spun ware. Even though the borosilicate glasses have some interesting properties applicable to spun ware uses, this type of glass has never been commercially available as spun ware.

Rolled and cast ware. Borosilicate glasses are available as rolled and cast sheets but not in sizes as large as soda-lime glass. Rolled glass is produced in sheets up to 2.5 feet × 18 feet (0.76 × 5.5 m).

On November 4, 1993, the joint venture at Jena Glasverke between Schott Glaswerke and Asahi Glass Co., Ltd. produced the world's first borosilicate glass using microfloat technology. Now, the flawless quality and transparency that are synonymous with the float method are combined with the superior thermal and chemical resistance properties of borosilicate glass. The revolutionary new product is known as Borofloat™.[3]

Engineering Properties

The standard practice in selecting glasses is to specify the type of application for which the material is intended and also to indicate the property requirements that the finished form must meet. Glass engineering is a highly refined art, and it is common practice for the engineer to rely on his material supplier to tailor a glass for specific requirements. Because properties such as ultraviolet transmission and dielectric loss are affected by even minute variations in minor constituents, it is wise to work directly with the material supplier.

Effect of viscosity on properties

Glass has been defined by ASTM as "an inorganic product of fusion which has been cooled to a rigid condition without crystallization." This definition emphasizes the fact that glasses have a noncrystalline, amorphous structure. The end properties of glasses are closely allied to their original state. Thus, a knowledge of the viscosity characteristics of glasses is important in studying their properties in the vitreous state. Knowledge of the viscosity of the glass, for example, is helpful to the engineer in determining the maximum permissible operating temperature for the material. Other properties, such as annealing and fabricating temperatures, are also affected by viscosity.

Softening point

The softening point of glasses with an original viscosity of 10^5 to 10^8 poises ranges from 1300 to 1500°F (704 to 816°C). Thus, borosilicate glasses have a fairly high distortion temperature.

Annealing point

The annealing point of glasses with an original viscosity of 10^{13} poises ranges from about 900 to 1050°F (482 to 566°C), which is about the average range for glasses. The annealing point represents the temperature at which the internal strains in the glass are reduced to an acceptable commercial limit in 15 minutes, as defined by ASTM C 336-54T.

Strain point

The strain point (the temperature below which permanent internal stresses cannot be introduced) of borosilicates with a viscosity of 10^{14} poise is in the intermediate range for glasses: 800°F and 1000°F (427 to 538°C).

Because the temperature gap between the softening point and the flow point of borosilicate glasses is small compared to most other glasses, they have a short working range; this makes fabricating operations somewhat more difficult.

Strength, elasticity, and fatigue

Like other glass materials, borosilicate glasses fail abruptly without any previous yielding. Consequently, they are classified as nonductile or brittle materials. All glass failures result from tensile stresses acting on surface defects, never from pure compression.

Mechanical strength

Even though glasses have extremely high inherent strength (possibly as much as 3 million psi), their design strength is quite low. The mechanical strength of glass is related to the small imperfections or flaws found at the surface which act as stress concentration points. The size of a glass product has a direct bearing on its strength. For example, borosilicate fibers with a breaking stress of about 500,000 psi have been drawn; however, pressed, borosilicate ware has a design value of only 2000 psi.

When an adequate safety factor is provided, the prolonged working stress of annealed borosilicate glass, as for other types of glass, is taken as 1000 psi. For tempered or thermally strengthened borosilicate glass, the design stress is 2000 to 4000 psi, depending on the piece in question.

The composition of the glass has no practical effect on its strength, although, compared to other glasses, most borosilicate glasses generally resist scratching better and have a better mechanical surface.

Elastic properties

In general, the rigidity of borosilicate glass is intermediate to that of other types of glass. The ultraviolet-transmitting borosilicates have a low Young's modulus (7.2×10^6 psi) whereas the sealing type borosilicates have a high value (9.2×10^6 psi). Aluminosilicate glasses are more rigid and lead alkali glasses are less rigid than the borosilicates. The Poisson ratio of borosilicate glasses ranges from 0.20 for the sealing types to 0.23 for the ultraviolet-transmitting types.

Although Young's modulus and the modulus of rigidity for most borosilicates decrease with an increase in temperature, the reverse is true for the low-expansion borosilicate glasses. See Figure 5.1.

Fatigue strength

Because the strength of glass is directly related to surface imperfections, such as cracks, it follows that the fatigue strength of glass is related to crack propagation. However, it is interesting that if the maximum stress and time duration are the same, there is little difference in the stress-time curves of glass loading with complete stress reversal. A comparison of the static and dynamic fatigue of an annealed borosilicate glass in bending is shown in Figure 5.2.

Hardness and wear properties

The hardness of any glass material, including the borosilicates, cannot be measured by conventional Brinell and Rockwell tests. However, many hardness tests have been developed for glasses. The most common are:

1. scratch hardness,
2. grinding or abrasion hardness, and
3. penetration hardness.

Although some correlation exists between the hardness values obtained from these tests, the order of hardness for a series of glasses

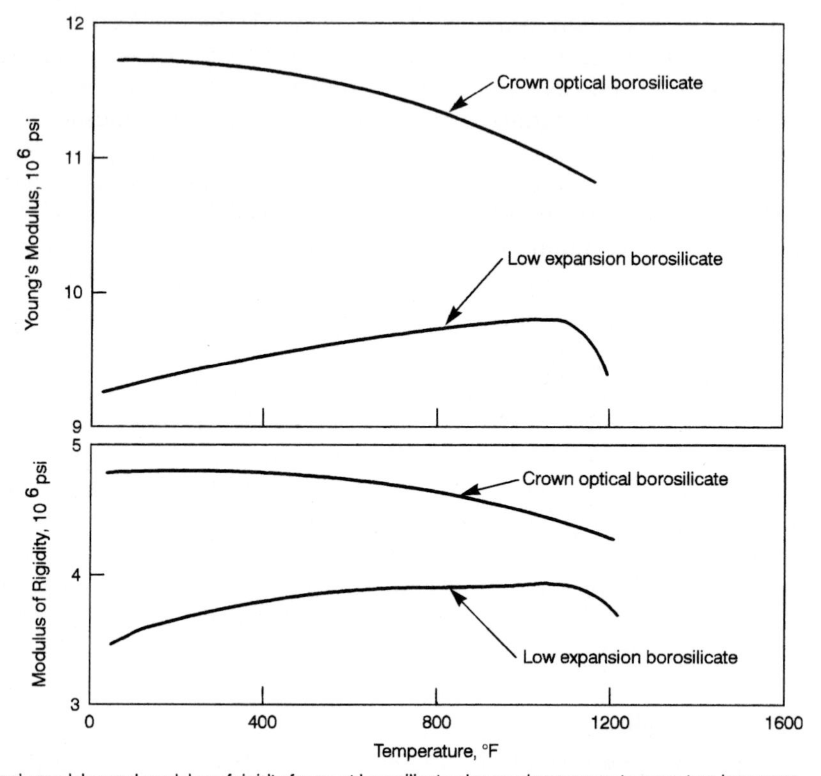

Young's modulus and modulus of rigidty for most borosilicate glasses decrease as temperature increases.

Figure 5.1 Young's modulus and modulus of rigidity.

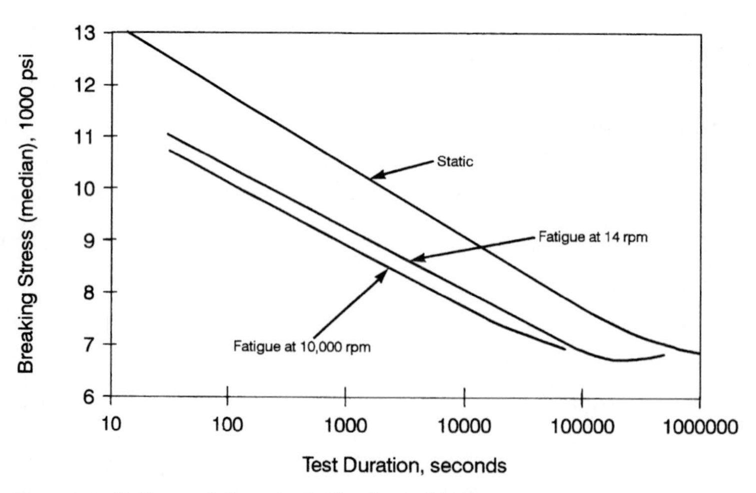

Figure 5.2 Fatigue of glass, including borosilicates.

may differ materially among the different methods. Consequently, care must be taken to select the test methods that best represent the actual service conditions.

Scratch hardness of glasses on the Moh scale generally lies between apatite and quartz. Some of the common materials hard enough to scratch glass include agate, sand, silicon carbide, hard steel, and emery. On the other hand, glasses are generally harder than mica, mild steel, copper, aluminum, and marble.

The grinding and abrasive hardness of glasses is a measure of the rate of removal of volume from a unit area of glass using carbide grit (No. 220) as an abrasive. The range of mechanical hardness of this glass is about 1.47 (low-loss-electrical type) to 1.52 (low-expansion type). By comparison, soda-lime plate glass has a value of 1.00 and aluminosilicate glass a value of 1.36.

Indentation hardness of the borosilicate glasses, a measure of penetration by a small diamond pyramid with less than a 50-g load, is quite high. Low-expansion borosilicate has a value of 630; only silica glass has an appreciably higher value: 780 to 800.

Although the smooth surface of glasses might indicate that the materials have a low coefficient of friction, the static coefficient of friction of glass under normal conditions is about the same as that of metals. The static coefficient of all glasses ranges from about 0.16 (glass on chromium) to about 0.29 (glass on tin plate). When lubricating materials (such as water-dispersible liquids which are allied to the silicones) are coated on the glass, however, the static coefficient may be greatly reduced to values as low as 0.005.

Thermal Properties

The thermal properties of primary interest include specific heat, thermal conductivity and diffusivity, and coefficient of thermal expansion. Values of the true and mean specific heat of low-expansion borosilicate glasses are shown in Figure 5.3.

The true value of specific heat is the specific heat measured at constant pressure (c_p) at a single temperature, whereas the mean specific heat (c_m) is the average specific heat taken over a temperature range. In general, except for soda-lime materials, the borosilicates have higher specific heat values than other types of glasses.

Just as for other types of glasses, the thermal conductivity of borosilicate glasses drops steadily as temperature decreases and reaches very low values in the neighborhood of absolute zero (see Figure 5.4). At temperatures above 750°F (399°C), radiation conductivity in glass becomes effective. Heat transfer by radiation conductivity is more effective in thick sections of glass than in thin sections.

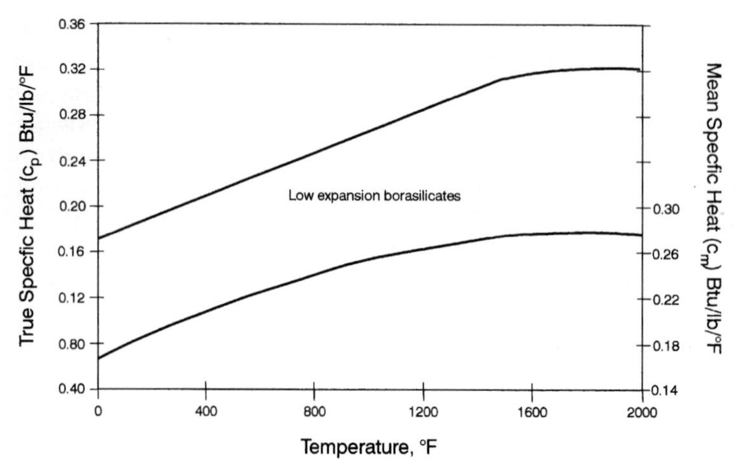

Figure 5.3 True specific heat vs. temperature.

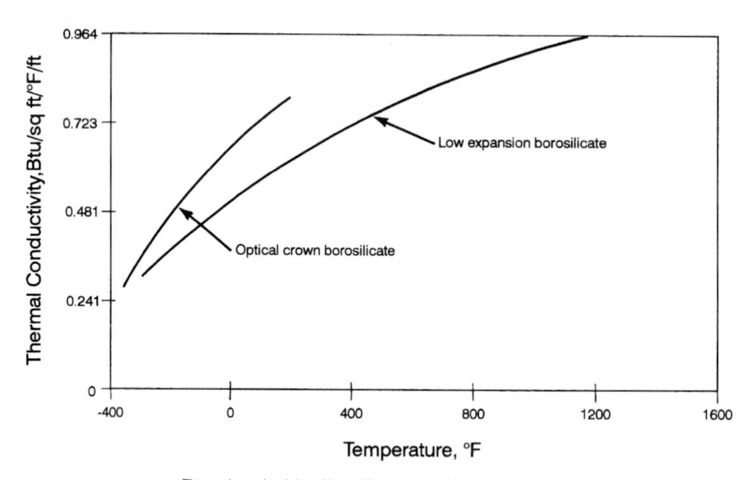

Figure 5.4 Thermal conductivity vs. temperature.

The thermal diffusivity (thermal conductivity/density × true specific heat) of low-expansion borosilicate glass at room temperatures is 0.007 cm²/s. At a temperature of about 750°F (399°C), diffusivity values decrease to 8 to 10 percent but they increase very rapidly at still higher temperatures.

The thermal expansion of other glasses, in general, is less than that of most other engineering materials and quite low compared to the other types of glasses. Furthermore, a wide range of coefficients of expansion is available in borosilicate glasses. For example, at temperatures from 32°F (0°C) to about 570°F (299°C), low-loss-electrical borosilicates have a coefficient of expansion of 17.7×10^{-7} in/in/°F, compared to Kovar sealing type borosilicate with a value of 28.3×10^{-7} in/in/°F. Thermal expansion of borosilicate glasses increases materially above their strain point of 800°F (427°C) (see Figure 5.5). Within the annealing range, values can be as much as two to three times those measured below 570°F (299°C).

Heat resistance

The terms soft and hard are used by the glass industry to indicate low and high softening temperatures of materials. In general, the borosilicates are classified as hard glasses.

The upper working temperatures (for mechanical considerations only) of annealed borosilicates range from about 400°F (204°C) in normal service to about 900°F (482°C), the extreme limit. Corresponding values for tempered borosilicates are 410 to 550°F (210 to 288°C).

If suddenly heated, a glass surface develops compressive stresses which, for all practical considerations, will not produce failure. However, sudden cooling of the surface will produce tensile stresses which will greatly increase the probability of failure.

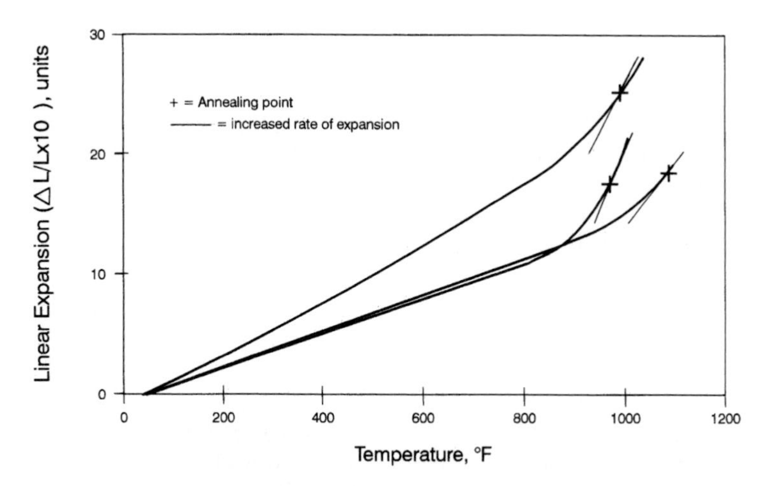

Figure 5.5 Linear expansion vs. temperature.

Tests for determining the thermal endurance of glass are often conducted by heating the part in an oven or heated bath and then plunging it into water. The temperature difference between the oven and water temperatures required to produce failure is recorded as the measure of thermal endurance. Data obtained from annealed borosilicate plates ranging in thickness from ⅛ in to ½ in shows that the borosilicates have a thermal shock resistance range of about 218 to 319°F (107 to 159°C) for thin sections and about 128 to 178°F (53.5 to 82°C) for thicker cross sections.

Thermal endurance is also related to the coefficient of expansion of the glass and to the shape of the glass article. Complex shapes usually have high stress and, consequently, lower thermal endurance. The medium used for chilling the surface also has an effect on the severity of the test. Oils, other fluids, and air generally have a less severe effect than water.

Electrical Properties

Borosilicate glasses are extensively used for electrical products. For example, the sealing type borosilicates are used to seal a number of different metals for incandescent lamps, gaseous discharge devices, and electronic tubes. In addition, although soda-lime glass is usually used for general-purpose lamps, low-expansion borosilicate glasses are often used for higher wattage, for smaller bulb sizes of a given wattage, or for outdoor applications. The low-expansion borosilicate glasses are also used for radio transmitting tubes and rectifiers where higher temperatures, thermal shock, or special sealing metals are involved.

The advantages of borosilicate for electrical applications include high dielectric strength, high volume resistivity, high surface resistivity, a good range of dielectric constants, and a low power factor.

Because the dielectric strength of borosilicate glasses is very high (>9000 kv/cm), this property is relatively unimportant compared with the problem of designing for the true breakdown value. The dielectric breakdown of borosilicate glasses, like many other insulating materials, can occur in two different ways:

1. an electronic breakdown that occurs when the true dielectric strength of the material is exceeded, and
2. thermal breakdowns.

The range of intrinsic (theoretical) and true breakdown voltages of various thicknesses of a low-expansion borosilicate glass is shown in Figure 5.6. Note that for a low-expansion borosilicate glass immersed in oil, its actual measured dielectric breakdown may be as much as 10

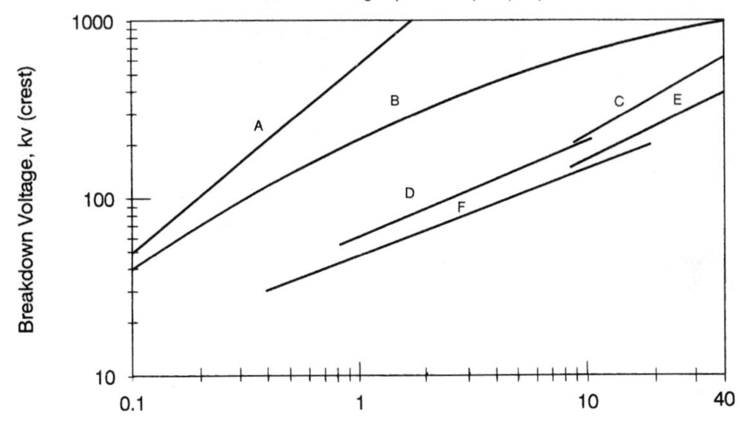

Figure 5.6 Breakdown voltage vs. section thickness.

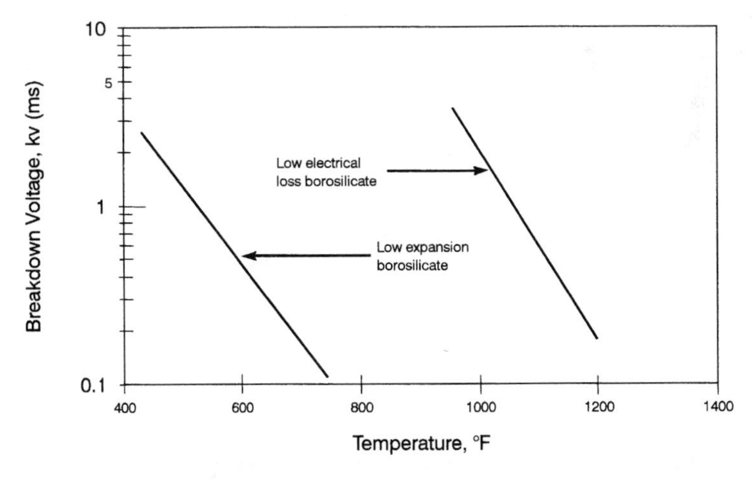

Figure 5.7 Breakdown voltage vs. temperature.

percent less than its intrinsic value. Bear in mind that the dielectric breakdown voltages decrease with an increase in frequency.

The breakdown voltage of borosilicate glass as a function of temperature is shown in Figure 5.7. At elevated temperatures, a breakdown is governed mainly by the resistivity of the glasses at those temperatures.

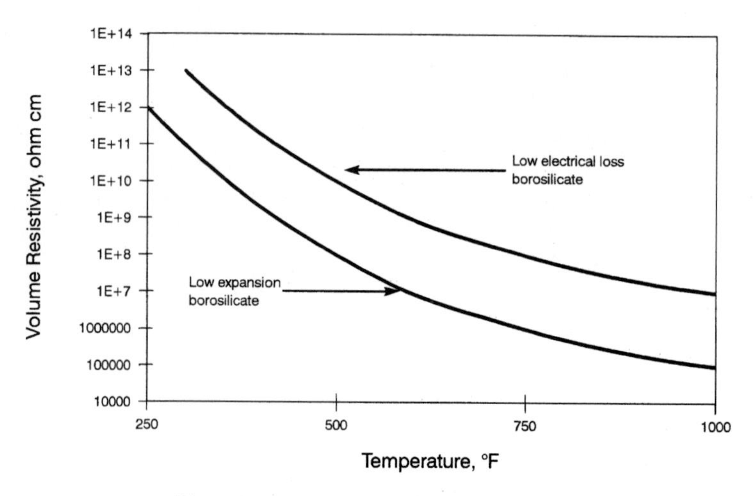

Figure 5.8 Volume resistivity vs. temperature.

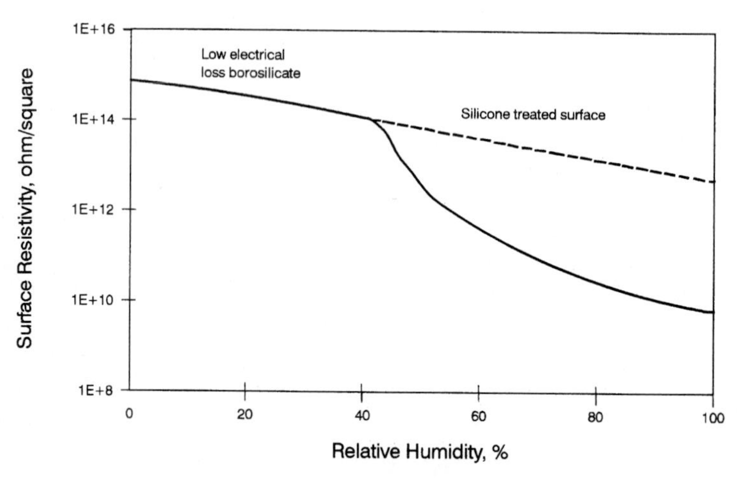

Figure 5.9 Surface resistivity vs. relative humidity.

They also have a high range of volume resistivity. At room temperature, low-electrical-loss borosilicate has a volume resistivity of 10^{17} ohm-cm, and the low-expansion type has a value of 10^{15} ohm-cm. The variation of volume resistivity with temperature is shown in Figure 5.8.

The surface resistivity of low-electrical-loss borosolicate is also extremely high (see Figure 5.9). The low alkalinity of this type of borosilicate glass means that its surface conductivity (which is a function of the amount of alkali present at the surface and the atmospheric conditions) is low.

Pure silica glass has the lowest dielectric constant found in commercial glasses—3.7 to 3.8. However, the borosilicate glasses also have a very low range of dielectric constants—4.1 to 5.1. Furthermore, their dielectric constant values are relatively insensitive to temperature changes ranging from 32°F (0°C) to about 750°F (399°C).

Low-electrical-loss borosilicate glass has one of the lowest power factors (−0.06) of any glass material. The other types of borosilicates, however, have power factor ranges from −0.2 to 0.50 at 1 mHz and 68°F (20°C).

Corrosion Resistance

Low-expansion borosilicate glass is extremely corrosion resistant, particularly in acid or neutral solutions. However, the other types of borosilicates are generally less resistant to corrosion than other types of glasses. Since it is the inert nature of silica that provides glasses with their chemical durability, the high percentage of silica (80.5 percent) in low-expansion borosilicate glass makes it practically insoluble in water except at high temperatures. Although the laboratory apparatus type of aluminoborosilicate glass does not have as high a silica content (74.7 percent) as the low-expansion material, the addition of alumina (5.6 percent) gives it increased durability.

Water and neutral solutions may increase alkalinity when alkalis are extracted from the glass. Thus, their rate of attack on the glass may be otherwise. The change in the pH value of the corrosive solution depends on its original volume and also on the chemical properties of the glass. The effect of solution pH on the relative rate of attack for a low-expansion borosilicate glass is shown in Figure 5.10. As is the case with other types of glass, the rate of attack on borosilicates by all reagents increases rapidly with temperature (Figure 5.11). In general, the rate of attack increases on the order of two and a half times for each increase of 18°F (−7.8°C).

Weathering corrosion from atmospheric moisture, like other forms of corrosive attack, increases as a function of the alkaline content of the glass. Consequently, due to their low alkali content, the low-expansion borosilicate and aluminoborosilicate glasses have high weathering durability. However, glasses with high boric oxide content, usually in excess of 18 percent (such as the low-electrical-loss and the tungsten sealing borosilicate glasses), are subject to attack by weathering. The boric oxide at the surface is leached out leaving a weather-susceptible surface high in silica content.

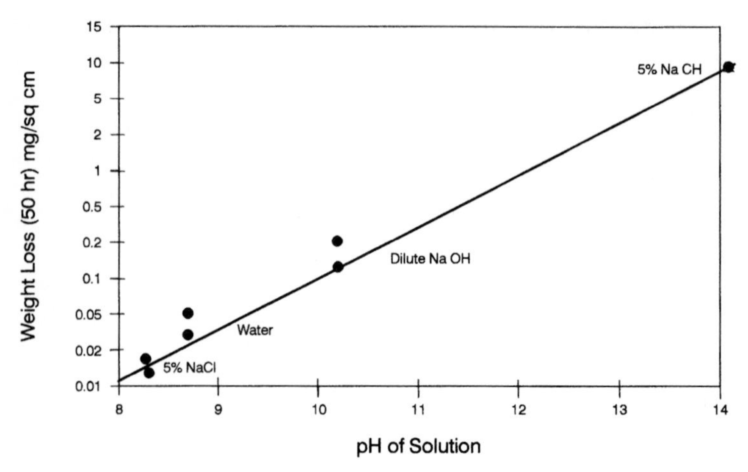

Figure 5.10 Weight loss vs. pH of solution.

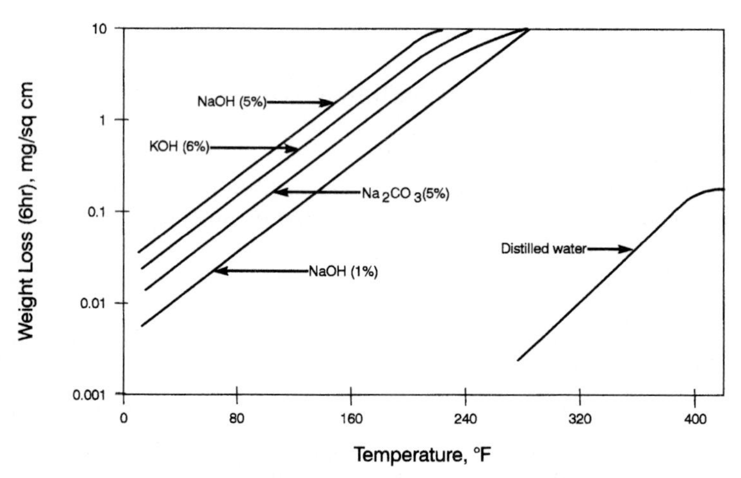

Figure 5.11 Weight loss vs. temperature.

Optical Properties

The refractive index of borosilicate type glasses ranges from about 1.469 (low-loss-electrical type) to 1.487 (sealing type). These materials transmit about 90 percent of visible light and virtually no ultraviolet or infrared light (see Figure 5.12). However, special ultraviolet-transmitting borosilicates are available which transmit both 185 nm and 254 nm. These special glasses are useful for low-pressure mer-

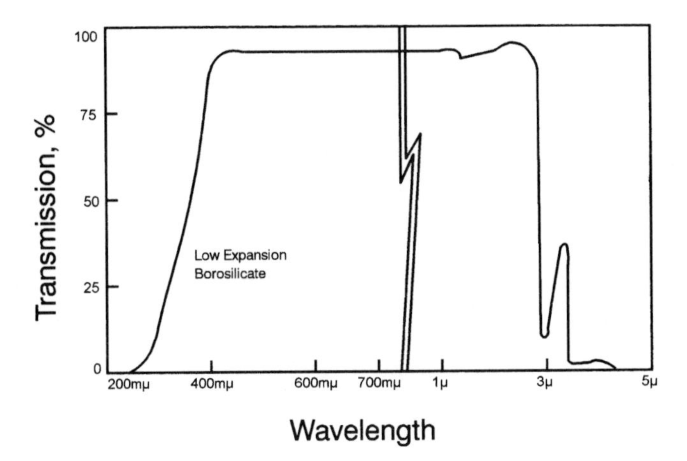

Borosilicates in general, transmit about 90 % visible light and virtually no ultraviolet or infrared light. Special colored glasses are available that precisely control transmittance in the visible range.

Figure 5.12 Borosilicates in visible light.

cury germicidal and ozone-producing lamps. Colored borosilicate glasses with precise control of transmittance in the visible range are available for filtering, signaling, and illumination.

The effects of stress, i.e., fire resistance, on borosilicate glasses are more noticeable than in other types of glass. The stress-optical coefficient ranges from about 3.8 Brewsters (low-expansion type) to about 4.8 Brewsters (low-electrical-loss type).

Processing and Fabrication

Heat treating

Internal stresses and strains are induced when a glass object gradually cools from its processing and fabricating temperatures to room temperature. The magnitude of these residual stresses and strains is determined by the rate of cooling within the transformation range and the material's coefficient of expansion and section thickness.

Stresses and strains can be reduced by cooling the glass object gradually in an annealing operation. Thus, annealing of borosilicate glasses accomplishes the same function as annealing of metals, i.e., reduction of residual stresses and, to a small degree, modification of structure. Naturally, the operating temperature of borosilicate glasses and other glasses must be kept below their annealing temperature.

Nearly all borosilicates can be tempered to increase mechanical strength. Tempering, which is accomplished by rapidly chilling the glass object to high temperature, also increases thermal shock resis-

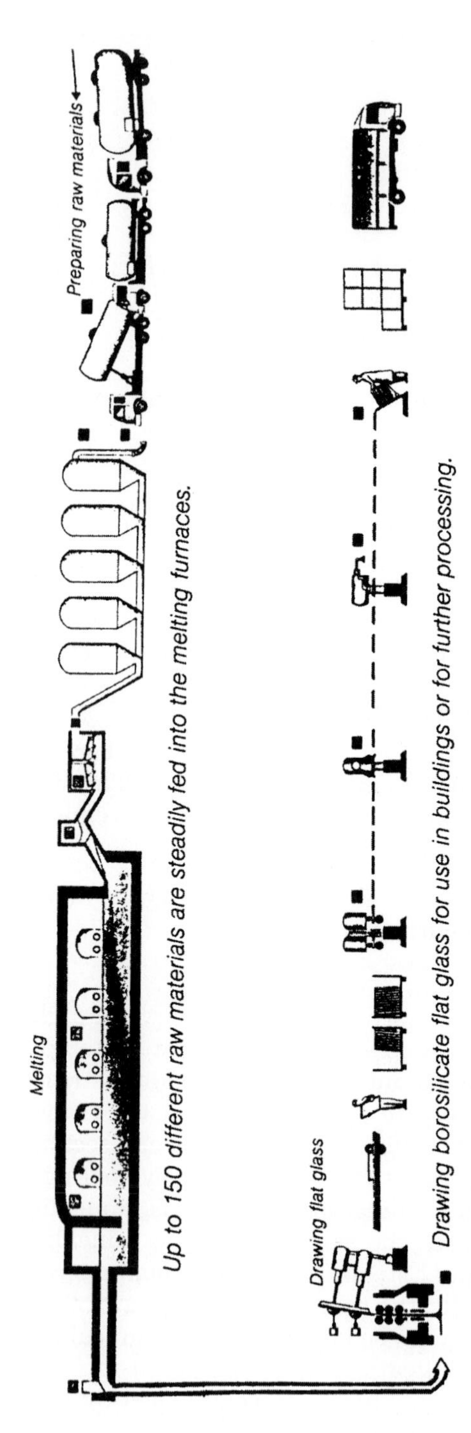

Figure 5.13 Borosilicate flat glass line.

tance. In Figure 5.13, a complete line for producing borosilicate glass is illustrated.

Annealed glasses subjected to sustained loading (1000 h), may fracture under stresses for prolonged periods without failure. Furthermore, fully tempered glass can withstand from three to four times the temperature differences for thermal shock than annealed glass can take.

It is difficult to fully temper this type of glass or, for that matter, most other types of glass which are less than ¼ in thick. In addition, because the tempering process leaves the glass with high compressive stresses on the surface, all cutting and internal tensile stress and all grinding or shaping must be done before the glass is tempered.

Joining

Borosilicate glasses can be joined with adhesives provided that

1. the adhesive sets with little or no dimensional change, and
2. the adhesive sets other than by solvent loss, polymerization, for instance.

The thermal coefficient of expansion should also be taken into account. This is especially true with low-expansion borosilicate glasses. Adhesives that remain soft after setting are very useful for glass. However, silicone, one of the softest groups of adhesives, does not possess high bond strength.

Where maximum bond strength is required, a relatively hard adhesive, such as an epoxy, should be used. Cement thickness and joint area should be made as small as possible.

Borosilicate glasses, especially the sealing type, can be sealed directly to metal substrates where the thermal expansion rate of the glass is compatible with that of metal. Glass-to-metal seals are produced by heating the glass until it softens, bringing it in contact with the metal substrate, and then allowing the seal to cool. A primary requirement of a good glass-to-metal seal is that a tightly adhering oxide layer is formed on the metal so as to promote wetting by the molten glass.

Good glass-to-metal seals also can be obtained with components having widely different thermal expansions. Considerable expansion differences can be tolerated by specially designing the joint so that it exerts only compressive stress on the glass. For example, large differences in expansion can be effectively bridged by using a graded seal consisting of a number of separate glass sections with different coefficients of thermal expansion, which are fused together.

Finishing and coating

Borosilicate glasses can be finished by grinding, polishing, drilling, and cutting. Grinding is often employed when dimensional tolerances are required closer than those obtained after fabrication. Polishing is used, especially with the crown glasses, to closely control surface conditions. Polished lens surfaces can be held to within one wavelength of light. Several methods of drilling, including ultrasonic drilling, are used with borosilicates. Cutting operations can be done either mechanically or thermally.

Typical surface treatments that can be applied to borosilicate glasses include staining, metallizing (silver or platinum), metal oxidation (tin oxide), lacquering, etching, abrading, and engraving. Metal oxidation is especially suited for the low-expansion borosilicate glasses in such applications as resistors and heat reflectors. A special metallic oxide coating on the surface of the borosilicate increases heat reflectivity from about 5 to about 50 percent. This type of coated glass is used as a heat shield.

Ceramic decorative finishes on these glasses are limited to small areas because of differences in their coefficients of thermal expansion.

Design Capabilities

Borosilicate glass products are produced in a tremendous variety of shapes, sizes, and tolerances. There is no distinctive difference between the design tolerances of borosilicate glasses and other types of glasses.

As with other types of glass materials, the form, rather than the type of material, controls the design limitations of borosilicate glasses. The design limitations of a piece of glass are controlled by thickness, size, complexity, strength, level of precision, and solidity or hollowness. Consequently, for information on the design capabilities of borosilicate glasses, the engineer is referred to product managers who can match capabilities to needs.

References

1. Schott Corporation, *Borofloat™, A World First from Jena,* Yonkers, NY, 1994.
2. *Test Procedures,* Schott Glaswerke, Mainz, Germany.
3. Borofloat is a trademark of a Schott Group Company.

Chapter

6

Crown Glass, Cylinder Glass, and Bent Glass

Joseph S. Amstock

President, Professional Adhesive and Sealant Systems

Introduction

The evolution of glass, which was a precious commodity for much of America's early history, influenced the size, appearance, and often the number of windows in a building. Until the nineteenth century, most of the window glass was crown or cylinder glass, as it was known, and came from western Europe, principally England.[1]

Crown and cylinder glass

Crown glass had fewer imperfections than cylinder glass but was available in more limited sizes. To make crown glass, the glassblower would twirl a heated bulb of blown glass repeatedly, permitting centrifugal force to flatten it into a large disk (Figure 6.1). After reaching a diameter of about four feet, individual panes of different sizes would be cut from the disk, using established cutting grids. Crown glass can be identified today by its concentric ripples. A varying thickness may also be noticeable, because the glass closest to the center of the disk would have been thicker than the edges.

A hand-blown glass cylinder was produced by blowing a cylindrical tube, slitting it lengthwise, and then reheating and unrolling it into a flat sheet. Because cylinder glass provided larger sheets than crown glass, it was often referred to as broad glass. Cylinder glass is characterized by parallel ripples and a more uniform thickness than crown glass.

Figure 6.1 Manufacturing crown glass.

Hand-blown cylinder glass continued to be produced into the nineteenth century, and the size of the tubes was increased up to seven feet in length and 18 inches in diameter by blowing and swinging the cylinder over elevated platforms or pits dug into the floor of the glasshouse.

The glass industry continued to advance throughout the nineteenth and into the twentieth century. In response to the growing demand for large display windows, plate glass was introduced in America around the 1830s. Plate glass offered advantages over early cylinder glass by providing larger sizes and an optically truer surface. It also was much stronger and could sustain large wind loads.

In the twentieth century, the manufacture of plate glass was improved by the use of rolling machines that squeezed a stream of molten glass between rollers. Then, the surface of the glass was ground and polished to a smooth finish.

Cylinder glass was not abandoned, however. In 1903, this early manufacturing method was further improved when John Lubbers introduced a mechanical means to blow and draw cylinders approximately 40 feet long and 30 inches in diameter. By producing larger sheets of glass in a shorter time using less labor, Lubber's process represented a major step in the mechanization of the window glass industry.

Nonetheless, continued research by the window industry soon supplanted cylinder glass production. Machines were created that would draw molten glass in a continuous sheet, producing a flat product in quantity without requiring the numerous steps involved in making cylinder glass. This method continues to be used today.

Bent Glass

Bent glass, also called curved glass, slumped glass, or bowed glass, is not new and has been around for years. Along with blown glass and the early antique window glass achieved by casting or flattening blown glass, bent glass was in use by medieval cultures for special furniture pieces and adornments of specialized types. Primitive pit kilns and clay molds were utilized in creating shapes to enhance the furnishings of the nobles and gentry. The up-scale uses of bent glass as an element of major construction came to Europe in the nineteenth century with the creation of Victorian "greenhouse structures" like the Kew Gardens Conservatory in England.[2]

The primary appeal of bent glass has always been a mixture of the practical and the aesthetic, enabling designers to extend their creations to encompass difficult options where flat glass products would not, or could not, measure up.

Although glass products, fuels, refractories, and mold products have improved greatly over the centuries, the basic techniques for custom architectural bent glass have changed only in subtle ways, leaving the methods generally in the "craft" areas that are often operator-sensitive and lacking in the application of highly mechanized approaches. High-speed lines are the avenues of mass production for bent glass items, such as automotive glass and high-volume greenhouse eaves.

Architectural uses

There are many different uses for curved or bent glass other than the traditional architectural elements of corners and transition areas. The use of bent glass for interior applications creates exciting shapes for stair-rail balustrades and landings, elevator cabs and enclosures, and interoffice partitions. Other common applications include revolving door enclosures, skylights, barrel vaults, store-front entrances, and fixtures such as display cases, sneeze guards, and canopies.

Bent safety-laminated glass can be used for high-security applications, including bullet-resistant glass, acoustical control, and overhead glazing.

Methods of manufacture

Bent glass is formed, basically, by introducing two elements to a piece of float glass: heat and gravity. These are the foundations of the process. When set up and fired properly, it is a relatively straightforward process. Therefore, the shapes that lend themselves to bending are likewise straightforward. Quality and cost are directly related to the complexity of the shape and the size of the bend required.

Bent glass has certain limitations in size, shape, and type of glass specified. The height (or a straight line dimension) is normally not of great importance other than in terms of windloading, job-site handling, product availability, and, of course, the bending capability of a particular source. The bent dimension or arc length is of much more concern to the bender and his costs.

In most construction uses, an arc of 90° or less is generally required. In these cases, the standard arc length of around 96 inches is considered the upper end of feasibility when using ¼-inch glass. If a given arc exceeds this general dimension, usually the lite is split in sections horizontally, thicker material is used, or more exotic bending techniques are brought into play.

The incorporation of straight legs or tangents is generally discouraged except where they are short relative to the length of the curve. Slight deviations from the straight are inherent in these types of bends and can be difficult to glaze. It is common for an architect to design lots of extreme legs stuck onto a curved segment, not realizing all the added cost and glazing problems accompanying such a design.

Most fabrication techniques are applicable to bent glass. Lamination, insulation, edge work, holes, pattern cutting, and the like are available from a complete bent glass source. It is always advisable to request data from the fabricator on any aspect of bending that seems questionable to the designer or glazier.

Generally, most types of commercial glass can be bent routinely. These include float glass up to and including one inch thick, wire glass, tints, patterns, cathedral and opalescent, and others. At the present time, the use of reflective glass is limited to those types of pyrolytic glass, and these are currently rated at the medium-performance level.

When dealing with bent glass, we are really dealing with circles. When discussing bent glass, the terms used to describe the circle and its parts become very important, particularly when trying to convey an idea by phone to a bent glass estimator.

Bent glass terminology

Curved glass is a whole new world and it takes a bit of adjusting to think in three dimensions.[3] All circles have a center which is a point equidistant from all points on the circle. See Figure 6.2. The circle can

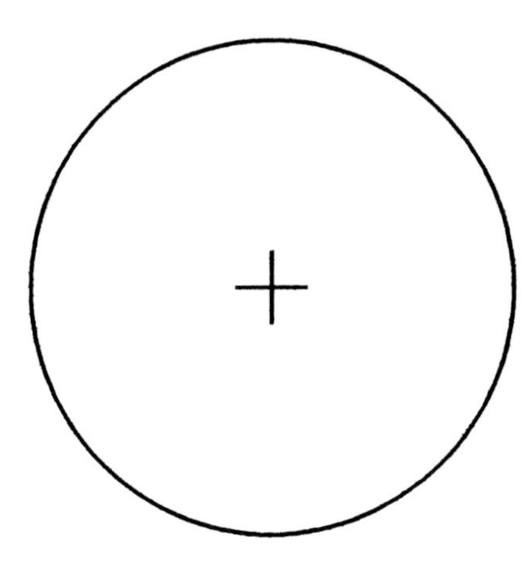

Figure 6.2 The center and circumference.

be divided into halves, quarters, eighths, and so on. The complete distance around the circle is called the *circumference.*

Even when the blue-line drawing shows only a small segment of the circle, it is best to try visualizing the entire circle that this segment would scribe if it were complete. This incomplete segment is called an *arc.* It can be further qualified by connecting a straight line from the center of the circle to the outside edge of the arc, describing the *radius* of the arc and its circle. The plural of radius is radii, and all radii are equal within the same circle.

Then, if we draw two radii on either end of the arc to the middle (looking like a piece of pie), the angle created where the lines intersect at the middle describes the *degrees* of the arc, which can be measured with a protractor. There are 360° in a circle. A quarter of a circle equals 90°, an eighth equals 45°, etc. See Figure 6.3.

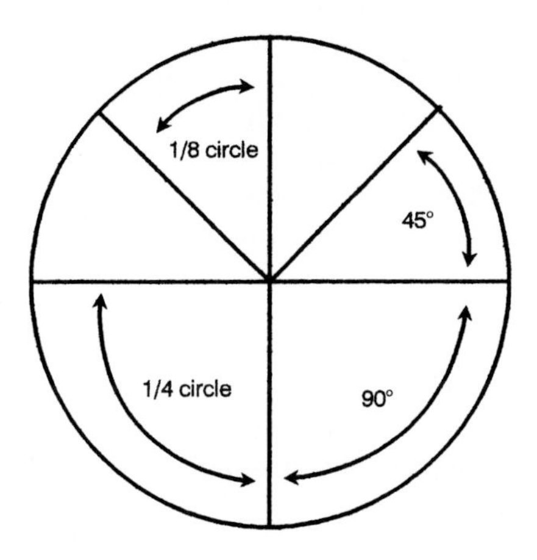

Figure 6.3 Geometry of a circle.

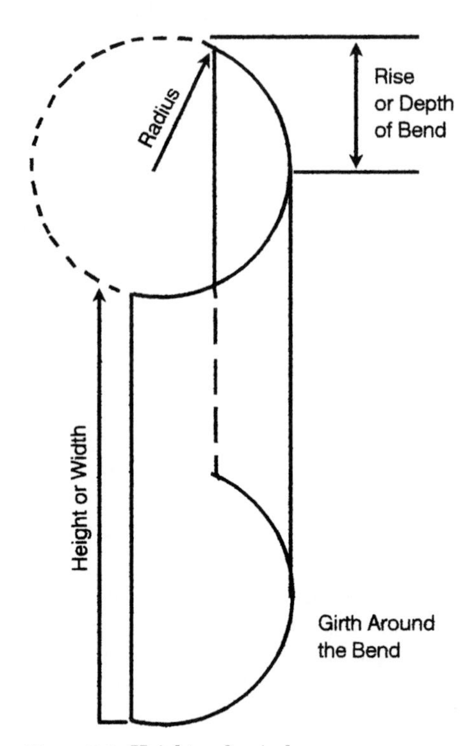

Figure 6.4 Heights of a circle.

A *chord* is a line drawn across the curve connecting the ends of an arc. The distance between the middle of the chord and the arc, measured at right angles to the chord, is called the *rise of the bend.* See Figure 6.4.

Any straight line drawn from the arc at a right angle to the radius is known as a *tangent* or a *straight* leg and is best drawn with a square where one leg is laid out as the radius and the other is the tangent. A true tangent must start at the arc at right angles to the radius and extend straight for a specified distance. If the radii are visualized as the spokes of a wheel, when the lowest spoke is exactly vertical, the ground forms a tangent to it. See Figure 6.5 for other parts of a circle.

The other necessary measurement that comes in to play when measuring for bent glass is the *reversed curve.* Use a flexible tape to obtain the girth or arc measurement. See Figure 6.6. This represents the distance around the bend.[4] The measurement must be taken on either the concave or convex side. If an exact measurement cannot be taken, give the chord and depth measurements.

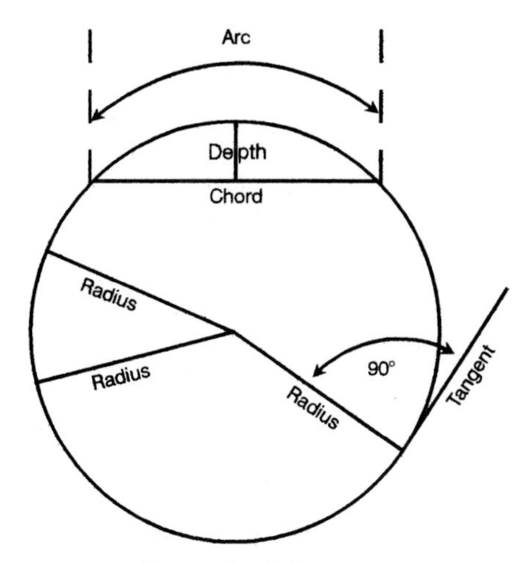

Figure 6.5 Parts of a circle.

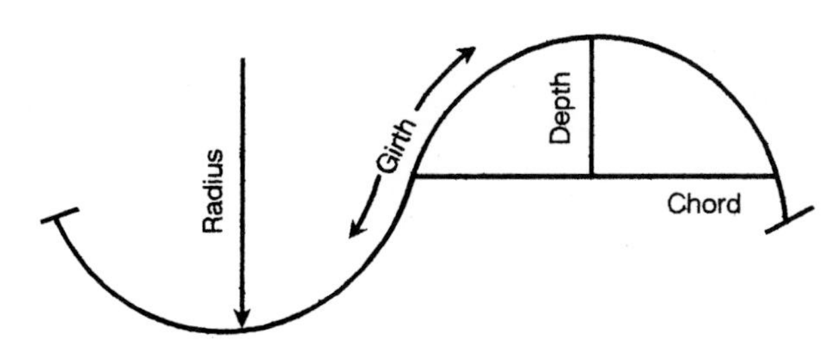

Figure 6.6 Reverse curve.

Glazing

The handling and glazing of bent glass employ techniques similar to those used with flat glass. The setting blocks are the same and the spacers at the corners are similar, but a center block is used to carry the bent section of glass. Be certain not to create pressure points on the glass. Gunnable sealants, such as polyurethane, polysulfide or silicone, are preferable. Coordination with the metal-forming source can be helpful, but it is not necessary.

References

1. *The Construction Specifier,* July 1991.
2. Green, K., *U.S. Glass, Metal and Glazing,* November 12, 1985.
3. Dlubak, F. C., "Everything You Ever Wanted to Know About Working with Curved Glass," *Fenestration* September 10, 1992.
4. Block, V., "Laminated Glass," *Glass Magazine,* November 1986.

Chapter

7

Tempered, Laminated, Heat-Treated, and Heat-Strengthened Glass

Joseph S. Amstock

President, Professional Adhesive and Sealant Systems

Introduction

Tempered glass, also known as toughened glass, is produced by heating annealed glass to approximately 1200°F (650°C), at which point it begins to soften. Its outer surfaces are then cooled rapidly, creating high compression in them. Its strength is usually increased by a factor of four or five times that of annealed glass, embodying safety and strength, as well as beauty. Tempering also imparts a unique safety characteristic, so that when fully tempered glass breaks, it fractures into relatively harmless fragments. This phenomenon, referred to as "dicing," significantly reduces the likelihood of injury from broken glass because there are no jagged edges or shards. Tempered glass also resists breakage by moving objects that could break ordinary glass. It can also resist temperature differences of up to 350°F (177°C) which would cause ordinary glass to crack. The heat treatment process shown in Figure 7.1 introduces surface compressions in heat-strengthened and fully tempered glass.

Despite its highly desirable characteristics, tempered glass does not fit everywhere. It should not be used where building codes require fire-resistant glazing, such as wired glass, or where a primary requirement is security. Because of its reputation for strength, many mistakenly believe that it will withstand careless handling and

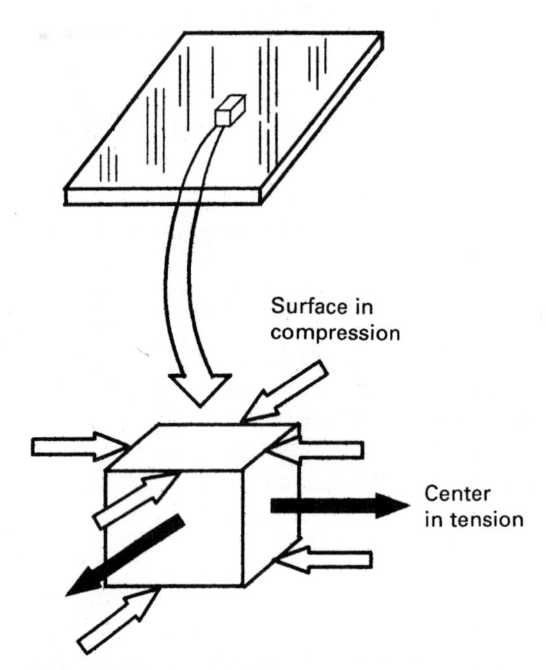

Figure 7.1 Heat-treated glass.

improper installation. This is not the case, however. Tempered glass must be handled and installed with the same care as ordinary glasses to prevent damage to the edges or surfaces.

Tempered or toughened glass is a safety glazing material and is recognized as such under the various codes. Most fully tempered glasses are capable of attaining the highest classification in these building codes.

Flat Glass Tempering Systems

There are two basic methods of producing air-tempered glasses:

1. The glass is held vertically by tongs along its top edge and is subjected to heating and an air blast. Figure 7.2[1] shows a schematic

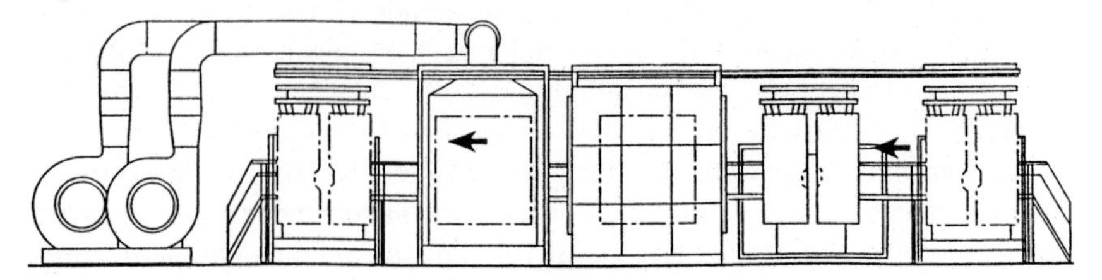

Figure 7.2 Tempering line for flat glass.

consisting of a loading station, generally one or more in-line heating chambers, a tempering zone, and the unloading station. The production of small-size glass sheets can be carried out on special frames without the use of clips. All operating lines can be controlled by a distribution control system consisting of microprocessor instrumentation interfaced to a personal computer, with the necessary software designed to meet process data control and supervisory requirements.

2. The glass is supported horizontally on rollers. The plant is installed at floor level and consists of a loading station, a heating cell, a tempering zone, and an unloading station. The glass sheets are rested on conveyor rolls and follow the zones described in sequence with a continuous unidirectional movement. The rolls are provided with an oscillating movement in the heating and tempering zones.

This is one of the most sophisticated types of plants for producing tempered flat glass sheets. It has the advantage of shorter processing times over the vertical tempering plant. Hence, a greater production output is achieved with no clip marks. See Figure 7.3.[2]

Each method produces some degree of bow and warp. The first method noted above inevitably causes "tong marks" along the edge where it is gripped by the supporting tongs. All float and sheet glass can be tempered or toughened, as can many rolled glasses. However, wire glass cannot. It is also possible to coat one surface with an opaque ceramic material, which then is fired onto the glass during the tempering process, producing a durable, clad material often referred to as spandrel glass or enameled glass. Table 7.1 shows the limitations and maximum sizes imposed by the tempering process.

Heat-treated glass

Heat-treated glass is produced by subjecting annealed glass to a heating process, a process similar to that used for tempered glass. The glass is heated to a temperature of approximately 1150°F (621°C), lower than used for fully tempered glass, and the cooling process is slower. The cooling process locks the surfaces of the glass in a state of high compression and the central core in compensating tension. The strength developed is about twice that of the equivalent annealed glass, bow and warp are generally less, and the product has many uses, but it does not usually meet the criteria for a safe glazing material because its breakage pattern more nearly resembles that of an annealed glass than that of a fully toughened glass. As shown in Figure 7.4, the factor which relates the average failure pressures of heat-strengthened versus annealed glass is 2.0, whereas the factor which relates failure pressures at the 0.008 failure rate is 3.2.[3]

The color, clarity, chemical composition, and light transmission

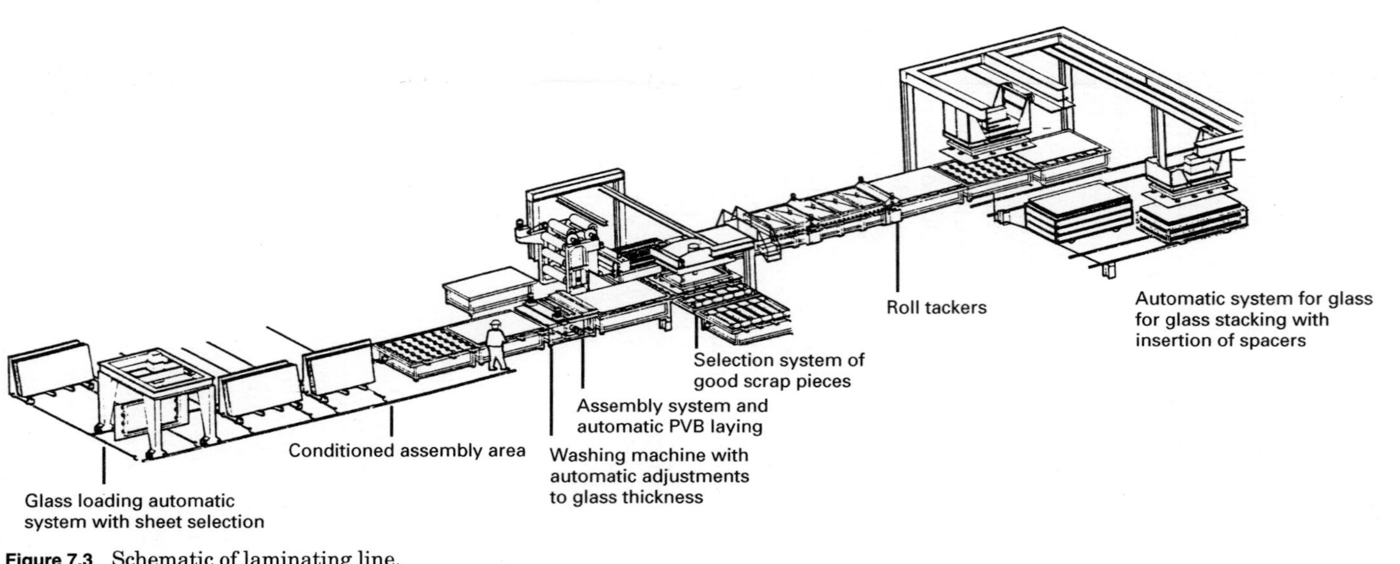

Figure 7.3 Schematic of laminating line.

TABLE 7.1. Limitations on Maximum Size Imposed by the Tempering Process

Glass thickness, in (mm)	Maximum size, in (mm)
$\frac{5}{32}$ (4)	90 × 51 (2100 × 1300)
$\frac{3}{16}$ (5)	102 × 78 (2600 × 2000)
$\frac{1}{4}$ (6)	165 × 78 (4200 × 2000)
$\frac{3}{8}$ (10)	165 × 78 (4200 × 2000)
$\frac{1}{2}$ (12)	165 × 78 (4200 × 2000)
$\frac{5}{8}$ (15)	158 × 71 (4000 × 1800)
$\frac{3}{4}$ (19)	158 × 71 (4000 × 1800)

characteristics remain unchanged. Compressive strength, hardness, specific gravity, coefficient of expansion, softening point, thermal conductivity, solar transmittance, and stiffness also remain unchanged. The only physical property that changes is bending strength. Under uniform loading, heat-treated glass is stronger than annealed glass of the same size and thickness and thus more resistant to thermally induced stresses, cyclic wind loading, and impacts by windborne objects and hail.

Heat-treated glass consists of two products, heat-strengthened and fully tempered glass, by definition of the degree of residual surface compression or edge compression. Most furnaces can produce both. A furnace and its quench must be adjusted by its operator for one or the other during a product run. The adjustments may include changes in the furnace temperature, exit temperature of the glass, residual time in the furnace, and the volume and pressure of the quench.

Production

There are two basic methods of producing air-quenched, heat-treated glass. In one method, the glass moves through the furnace and is quenched in a vertical position; in the other it moves on rollers of stainless steel to high-strength ceramic in a horizontal position.

Tong-held glass, the vertical process, may exhibit a long arc or "S" curve plus some minor distortion at the tong points. Horizontally heat-treated glass will have characteristic waves or corrugations caused by the support rollers. Tolerances are permitted under ASTM C 1048.

Limitations

Recommended maximum service temperature for the heat-treated glass is approximately 500°F (260°C). Heat-treated glass is not classi-

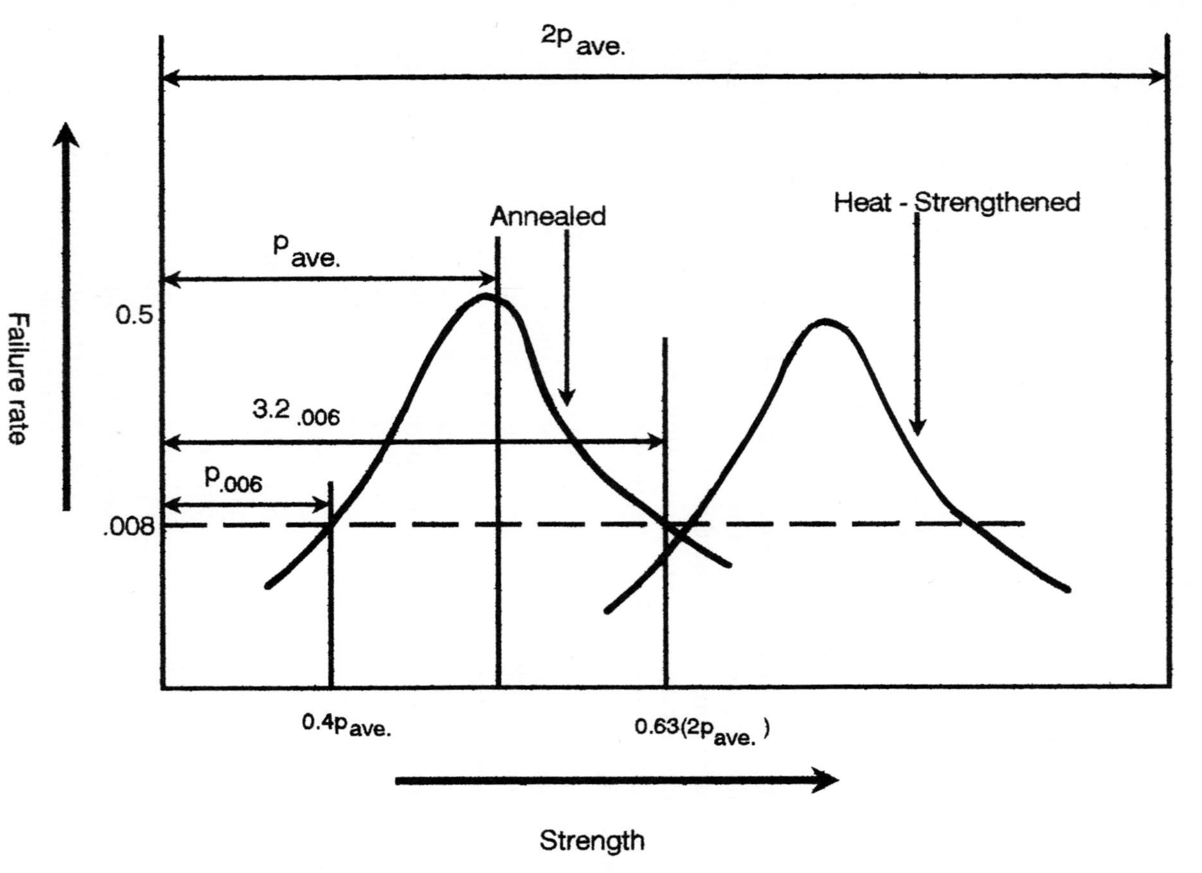

Figure 7.4 Relative strengths for annealed and heat-strengthened glass.

fied as a fire-retardant material. Tempered glass, although four times stronger than annealed, and heat-strengthened glass, twice as strong as annealed glass, should not be selected to meet a given wind load simply because annealed glass of the same size and thickness is sufficient. The stiffness of annealed glass and heat-treated glass is the same; deflection under a given uniform load (wind load) will be identical for glass of the same size and thickness. Excessive deflection can cause glazing seal failure, glass breakage by contact with the framing, and psychological discomfort to the occupant.

Heat-treated glass cannot be cut, drilled, or edged after being heat-treated. It should be sandblasted or acid-etched with caution. Either of these processes may reduce the thickness of the compression layer somewhat, thus, reducing the strength of the lite. Some deep patterns of rolled glass cannot be heat-strengthened, or, if tempered, will not break in the same manner prescribed by CPSC 16, CFR 1201, or ANSI Z 97.1 for safety glazing.

Iridescence, spots, or splotches may be visible at times, especially on tempered glass, when viewed in certain types of reflected light or through polarizing glasses. The intensity will vary with lighting conditions and viewing angles. This is caused by the strain pattern induced during the cooling stage and inherently is not a cause for rejection.

Heat-Strengthened Glass

Heat-strengthened glass is defined in ASTM C 1048 as having residual surface compression greater than 3500 psi and less than 10,000 psi or edge compression greater than 5000 psi and less than 9700 psi. When broken, the appearance of the break pattern will vary widely depending on the amount of compression and the surface quality. At the lower portion of the surface compression range, there will be a large, open pattern reminiscent of annealed glass but generally without the sharp-pointed shards. In the midrange of 4500 to 7000 psi, the pattern becomes increasingly smaller, or more closed, with more fragments. At the high end above 7000 psi, it becomes smaller and smaller until, above 9000 psi, it is often difficult to determine if the lite was heat-strengthened or tempered.

Heat strengthening in the 4000- to 7000-psi range is probably the most desirable for most uses. The break pattern is relatively large, the fragments tend to lock together and remain in the glazing, and the strength is sufficient for most purposes and readily produced on horizontal equipment. It is difficult to produce ⅛-in and ¼-in thicknesses.

All heat-treated glass will break when the compression layer is penetrated. Surface or edge chips, scratches, gouges, or smalls which do not completely penetrate the compression layer can be slowly propagated by heat or wind cycling and result in breakage from no apparent cause days or months after the damage has occurred, leading to the mysterious breakage.

Chemically Strengthened Glass

Chemically strengthened glass has no accepted standard for surface compression, for edge compression, or for properties not defined in ASTM C 1036. The surface compression can vary widely, even within the same sheet. The Glass Tempering Association's (GTA) *Engineering Standards Manual* states the following regarding chemically strengthened glass.

Chemical strengthening of glass is brought about through a process known as ion exchange. Glass is submersed in a molten salt bath at a temperature below the annealing range of the glass. In the case of soda-lime float or soda-lime sheet glass, the salt bath consists of potassium nitrate. During the submersion cycle, the larger alkali potassium ions exchange places with the smaller sodium ions on the surface of the glass. The larger potassium ions "wedge" their way into voids in the surface created by the vacating smaller sodium ions. This "strengthened " surface may penetrate to a depth of only a few thousands of an inch. The compression strength of a chemically strengthened glass can reach as high as 10,000 psi.

This level can be drastically reduced due to surface flaws. Much data and specifications are published as an average psi. This obviously means that there were glass samples with high psi readings and samples with low psi readings. It seems that chemically strengthened parts from the same salt bath have a wide range of psi measurements. When chemically strengthened glass is broken, it does not dice but breaks similarly to annealed glass. Therefore, it should not be used where safety tempered glass is needed.

Some glass technologists and researchers claim that the ion exchange is actually only a few molecules in depth—a few millionths of an inch—not "a few thousands of an inch" as stated in the Glass Tempering Association Manual. Although chemically strengthened glass can be cut after treatment, the cutting process causes total loss of the added strength for an inch or so on either side of the cut. It simply reverts to annealed glass. Where and when it is scratched also causes a similar loss of strength. Chemically strengthened glass is used mainly in the opthalmalic and aeronautical industries where glass less than ⅛ inch thick is required to have properties stronger than annealed glass. It has also gained some favor in the security glazing field as a protective covering over polycarbonate.

Laminated Glass

Laminated glass consists of two or more lites of glass bonded together by a polyvinyl butyral (PVB) plastic interlayer, such as Saflex® interlayer.[4] The glass plies may be equal or unequal in thickness and may be the same or different in heat treatment. Further, laminated glass may be used as a single lite or both lites of an insulating glass unit. Therefore, a wide range of glazing products using laminated glass can be manufactured to accommodate the varied structural design conditions which may be encountered.[5]

Research has demonstrated that the same common laminated glass products (same ply thickness, same glass type) behave as if the unit is monolithic, except for two unusual loading situations. Hence, most design procedures will treat laminated glass products as if they are monolithic lites with the same nominal total thickness.

Manufacture of Laminated Glass

General

To eliminate contamination within the finished laminates, the storage and layup of cut blanks is an area that requires close attention. Being a labor-intensive part of the operation, placing the underlayer on the glass in an open environment always has the potential of contamination from hair, clothing lint, airborne dust, and foreign contaminates from workers' hands. Additionally, the cleanliness of the incoming glass pairs can significantly affect the level of contamination and the performance properties of the final product, To successfully produce high-quality laminated glass products, it is imperative that the storage and layup of the underlayer be performed in as clean an environment as possible with consistent glass surface quality.

Layup practices also set the stage for the laminating process. Wrinkles, misalignment of the tint band, and "short vinyl" can have their root cause in the conditions and practices found at layup. Hot glass and poor trimming techniques can cause severe problems during deairing, tacking, and autoclaving. A routine monitoring program for layup conditions and practices can help ensure a high-quality product.[6]

Glass preparation. Glass temperature and cleanliness are important variables for successful laminating. Whether the process involves glass washing prior to layup or it involves "no wash after bending," the cleanliness of the glass surfaces will affect visual quality and laminate performance. A multitude of foreign substances can interfere with glass–polyvinyl butyral (PVB) adhesion when located at the bonding interface. Spot delaminations and low adhesion are frequently caused by poor or variable glass cleanliness.

In the processes that wash the glass just prior to lamination, the goal of the washing step is to remove particulate contamination plus any surface residue from the glass pairs. Once particulates have been removed or residues dissolved, the detergent solution must be swept away in a final rinsing step. Final rinse water quality and the quality of the water blow-off system have a large effect on adhesion level and adhesion variability. The presence of water spots on the washed glass layup is a good indicator that the glass is still dirty. Laminators should contact their glass-washing machine equipment manufacturer and their glass supplier to recommend detergents, washer operating conditions, and blow-off systems. Refer to Chapter 21 for more specific procedures on glass-washing techniques and handling of glass.

The salt condition of the final rinse water will affect the ultimate level of the glass-PVB construction. As can be seen from Figure 7.5, water hardness (expressed as ppm of salt content) lowers the adhesion level by interfering with the glass–PVB bonding mechanism. Water hardness usually varies with the water supply used: well water, river water, etc. Because many municipalities use different sources of water depending on the demand and the season, the salt content can vary significantly throughout the year. It is recommended that each lami-

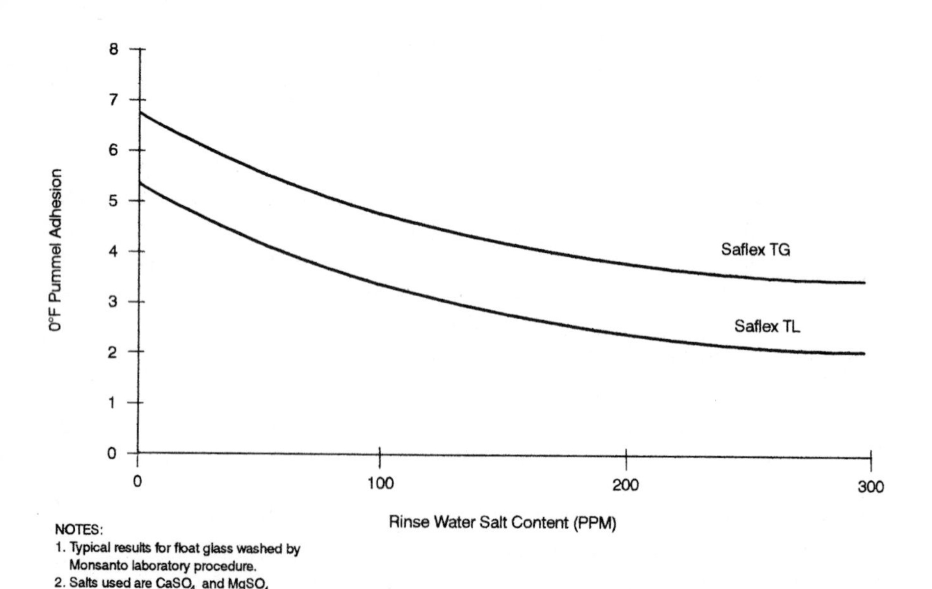

Figure 7.5 Salt content vs. adhesion.

nator keep abreast of water quality and water sourcing changes to anticipate hardness variations that will affect laminate quality. A way of correcting this problem is to change to demineralized or deionized water. Commercial units are available that will remove minerals or hard water salts from the water source used for final rinsing.

In bent laminate constructions that utilize a "no bending after washing" process, the glass is washed prior to applying a separating or parting agent. Upon removal from the bending furnaces, bent glass pairs are fed into the laminating line without further washing. Depending on the amount of separating agent applied, removal of the agent may or may not be necessary. Under normal circumstances, the application of the separating agent suspensions is not a major factor in the final laminate quality unless the application is highly variable. However, in applications that require the removal of excess separating agent after bending, the method of agent removal can affect performance properties. It is strongly suggested that separating agents be removed with a vacuum system. Brushing or hand wiping with rags can result in surface contamination and nonuniform removal.

Polyvinyl butyral preparation. Upon cutting the rectangular blanks of PVB for layup, a small amount of oversize may need to be added to the length to compensate for any shrinkage that occurs in the stacks. When using refrigerated vinyl, shrinkage will depend on the inherent vinyl shrinkage level plus the stresses applied during unwinding. Experience and product type will dictate the level of overlength needed. To allow for relaxation of stresses and temperature equilibrium, stacks should be allowed to sit overnight at a temperature of 55 to 65°F (13 to 18°C) and a relative humidity matched to the product moisture desired. For 0.40 percent moisture in the final product, relative humidity of 21.5 to 24.5 percent is required. The controlled storage temperature will result in a good temperature match with the incoming glass while avoiding sticking in the stacks. The controlled humidity level will prevent changes in the moisture level of the sheet that will ultimately affect adhesion. During the storage period, stacks should be covered with protective film (i.e., virgin polyethylene) to avoid contamination of the top layer with airborne dust. A virgin grade of polyethylene is recommended to avoid the transfer of anti-stick/antistatic agents to the PVB that may alter adhesion. Stack heights of 4 to 6 in are recommended.

White room conditions. Environmental control is critical for the assembly of clean laminates at the proper processing conditions. The following recommendations apply to white room installation and procedures:

1. double-door entrance and exit so that the inner door closes before the outer door opens.
2. positive air pressure supplied with filtered air.
3. dust collection mats and rugs at doorways.
4. controlled air temperature at 55 to 65°F (13 to 18°C).
5. controlled relative humidity matched to the desired final moisture with control accuracy of ± 2%.
6. limited access by visitors.
7. special workers' clothing: lint-free coveralls, hair nets, lint-free gloves.
8. routine cleaning of floor and overhead equipment.

In practice. The glass surface temperature at layup is normally slightly elevated to allow for some sticking between the glass and the PVB interlayer. This slight tackiness maintains the interlayer position without the top piece of glass sliding as the assembly goes down the line. A glass temperature in the range of 70 to 105°F (21 to 41°C) is typical in the industry when using 0.030-in-thick (0.76-mm) interlayer. A temperature higher than 105°F (41°C) can cause shrinkage of the interlayer at layup and result in wrinkling or "short vinyl." Also, a high glass temperature may cause premature edge sealing and/or difficulty in repositioning the plastic. Premature sealing is a primary cause of trapped air in the final laminate. For thinner 0.015-in-thick (0.38-mm) interlayers tackiness at layup is more severe. Layup temperatures below 95°F (35°C) are recommended.

The last procedure before deairing is trimming of excess plastic from the edges of the laminates. Care should be taken not to stretch the plastic during trimming because this can result in reduced thickness, snapback, and blow-ins during autoclaving. Experience will dictate the closeness of the edge trim for each type of process to avoid "short vinyl" problems.

Design loads

When load durations are short (i.e., wind gusts) or when laminated glass temperatures are low (i.e., snow load conditions), designs can assume that laminated glass acts monolithically. There are only two design situations in which laminating glass will not act monolithically:

1. When laminated glass temperatures are high, greater than 120°F (48.9°C), at the time the design wind load occurs

TABLE 7.2. Approach to Design of Laminated Glass

Single Laminated Lite
See Table 7.3
1. Design requirements (from building code)
2. Trial design (designer)
3. Basic lite strength (see Figure 7.6)
4. Strength modifications (see Table 7.3)
5. Compare strength with design requirement (step 1)
6. Modify design and repeat

2. When laminated glass at room temperature, 75°F (23.9°C), is loaded for long periods.

The first situation occurs when a laminated spandrel panel heated by the sun has not cooled when the wind gust arrives. The second situation can occur to the interior lite in an insulating glass unit under snow load, when the interior space below is heated and the temperature of the laminated lite is raised to and maintained at 75°F (23.9°C).

Design loads are determined from building codes. Strengths of "trial" designs are determined from figures and multiplying factors, shown in Tables 7.2 and 7.3. Modifications of "trial" designs and reanalysis will bring the design into conformance with loading requirements, helping to assure selection of an economical laminated glass product.

Glass strengths obtained from figures are based on a failure rate of eight lites per 1000. This failure rate is employed in design procedures for both vertical and sloped glass. Although sloped glass is more susceptible to impact from falling objects, laminated glass minimizes debris falling from the opening in the event of breakage. Designs employing laminated glass do not require special attention to minimize potential hazards.

Structural design

This section outlines procedures that can be followed by designers or architects to determine required thicknesses of laminated glass lites used singly or in insulating glass units for interior or exterior installations. The procedures consider uniform lateral pressures s (dead load, wind, snow), glass types (annealed, heat-strengthened, fully tempered), and environmental conditions (temperature).

TABLE 7.3. Laminated Single Lite Example

Design a laminated glass lite supported on four sides in a 60 × 80-in opening, 30° from horizontal with a snow load of 50 psf.

1. Design requirement: snow load 50 psf (long-term load, acting vertically; from building code)

$$w_s = 50 \text{ psf}$$

2. Trial design: ⅜-in laminated glass, symmetrical plies, both annealed
3. Basic lite strength: from Figure 7.6 (⅜-in)

$$p = 45 \text{ psf}$$

4. Strength modification factors: heat treatment factor and long-term load factor

$$p = 45 \times 1.0 \times 0.6 = 27.0 \text{ psf}$$

5. Compare with design requirement: design pressure = dead load* + snow load

$$s = w_d \cos\theta + w_s \cos^2\theta = \text{design requirement}$$

$$s = 5 \cos 30° + 50 \cos {}^2 30° = 41.8 \text{ psf}$$

6. Design requirement not met by trial design; modify design and repeat process:

   Try laminated ¼-in laminated glass, symmetrical plies, both heat-strengthened (HS).

   From Figure 7.6 (¼-in) and strength modification factors,

$$p = 27 \times 2 \times 0.8 = 43.2 \text{ psf}$$

   Design requirement: design pressure = dead load* + snow load

$$s = 3 \cos 30° + 50 \cos^2 30° + 40.1 \text{ psf}$$

   Modified design acceptable

*A ⅜-in lite weighs 5 psf; a ¼-in lite weighs 3 psf.

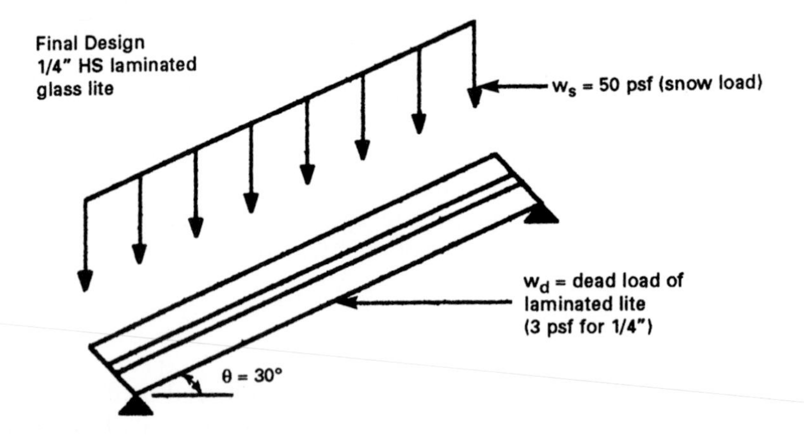

Laminated glass may be used singly or in combination with monolithic glass plate or another laminated lite to form an insulating glass unit. The design methodologies for these uses are outlined. The designer begins with a design pressure s (obtained from dead, wind, and/or snow load calculations which consider site conditions and building geometry) and an opening dimension ($h \times w$).

For sloped glazing, uniform loads are modified to generate a calculated design pressure:

$$s = w \cos^2 \theta + w_d \cos \theta \text{ \{snow\}}$$

$$s = \pm w_w + w_d \cos \theta \text{ \{wind\}}$$

The designer should be guided by the provisions of the applicable building codes. With this information, the designer must establish a minimum thickness t for a laminated glass lite that will carry the design pressure. Because glass strength figures provide relationships between pressure and the lite dimensions for specific annealed glass thicknesses (see Figures 7.6 and 7.7), the design process becomes one of trial and error.

1. Try a glass thickness by selecting a specific glass strength figure.
2. Enter the glass strength figure with lite dimensions and obtain strength of lite p.
3. Modify the lite strength by heat treatment factors and load duration factors, if applicable.
4. Compare modified lite strength p_m with design pressure s; accept minimum thickness t, or move to a thicker or thinner t, as appropriate, to "close" on a proper design.

Insulating glass

1. Determine design pressure s from building code design loads.
2. Establish "trial" design: lite thickness t and heat treatment, annealed glass, heat-strengthened, or fully tempered.
3. Then determine the strength p of one lite from Figure 7.6.
4. Multiply strength p by 1 for annealed glass, 2 for heat-strengthened, and 4 for fully tempered. For long-term loads, multiply strength p by 0.06 for annealed, 0.8 for heat-strengthened, and 0.95 for fully tempered.
5. Multiply modified strength p_m from above by 1.8.
6. Compare 1.8 p_m with design pressure s from step 1.

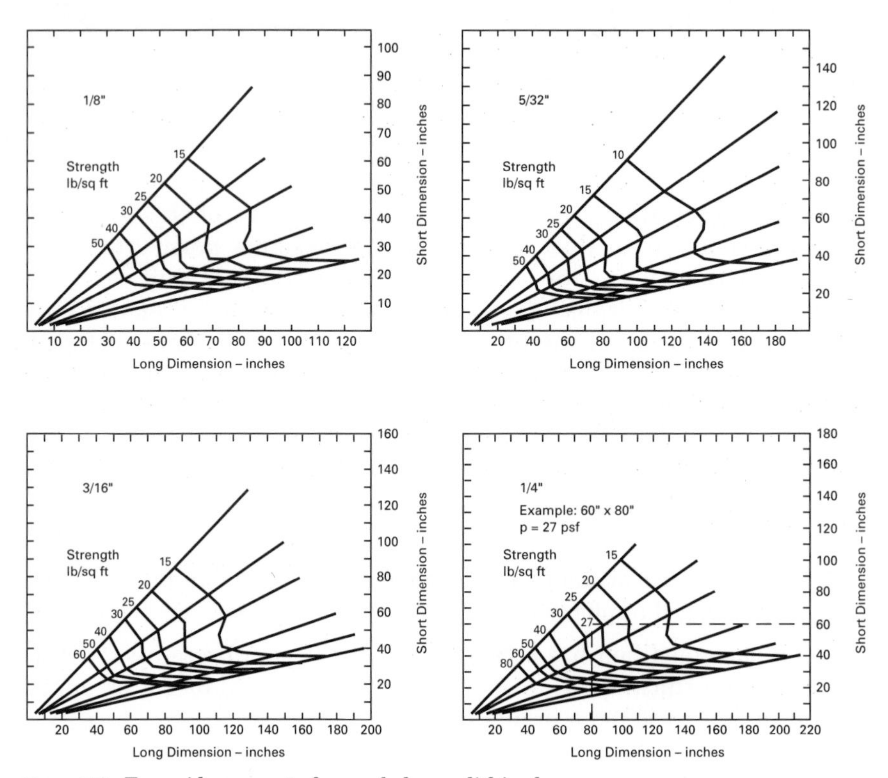

Figure 7.6 Four-side support of annealed monolithic glass.

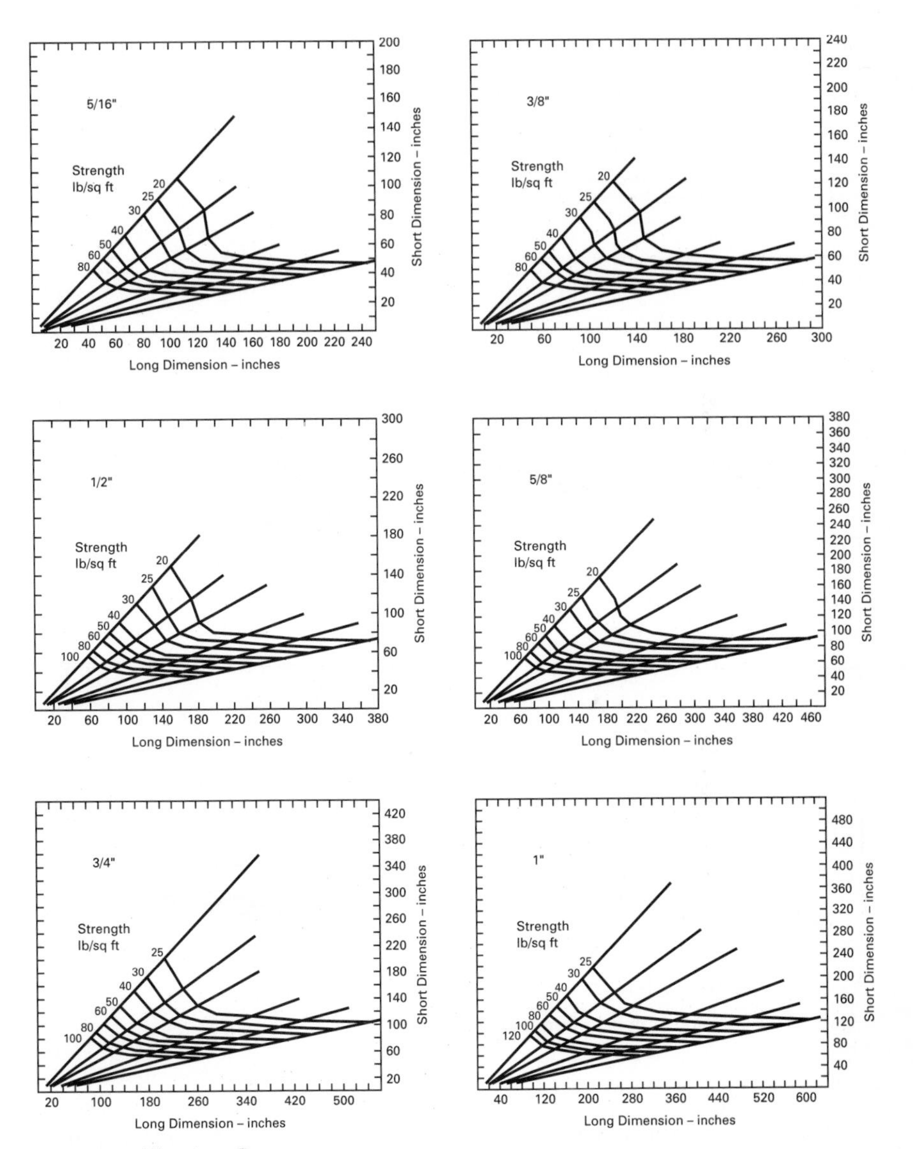

Figure 7.6 *(Continued)*

4-Side Support

To use:

a. Assume "trial" thickness, locate proper figure.
b. Enter this figure with lite dimensions.
c. Find glass strength (See 1/4" Chart).
d. If design load exceeds glass strength, move to figure for larger thickness and repeat from Step b.

Note: Due to manufacturing constraints, some glass configurations may be practical. Please consult the glass supplier.

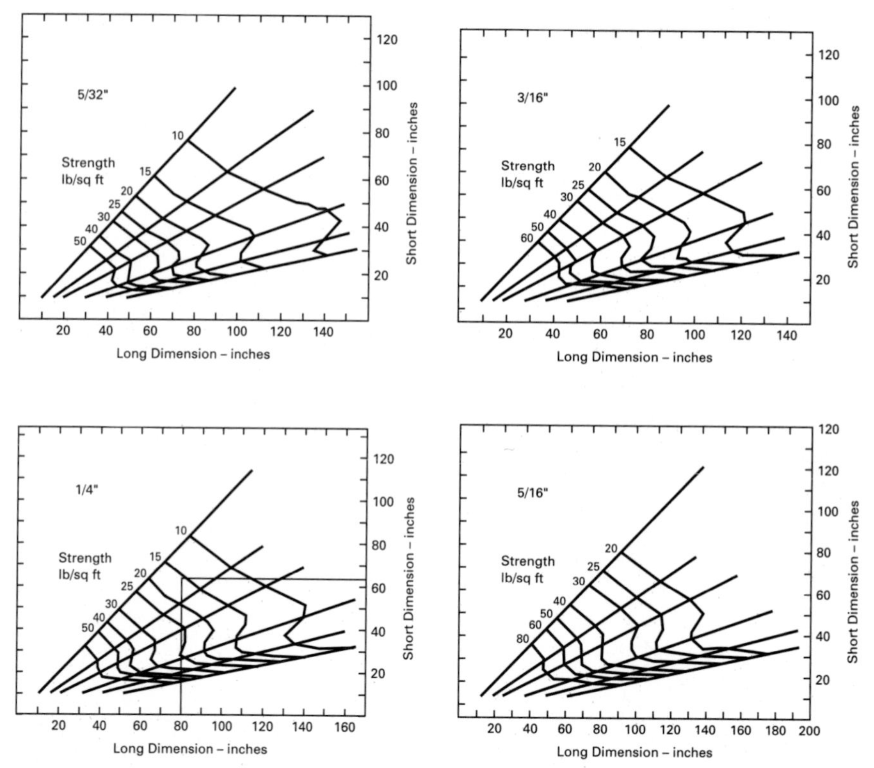

Figure 7.7 Four-side support of annealed layer.

7. Revise design, if necessary, and repeat the procedure to close on an appropriate design. Figure 7.8 shows a variety of load-sharing conditions.

References

1. Courtesy of EMIITALIA, Milano, Italy.
2. Courtesy of Boune Elettromeccanica, Italy.
3. Minor, J. E., Basic Glass Strength Factors, *Glass Digest,* August 15, 1990.
4. Saflex® is a trademark of Monsanto Chemical Company, St. Louis, MO.
5. *A Guide to Structural Performance of Laminated Architectural Glass with Saflex® Plastic Underlayer,* Monsanto Chemical Company, St. Louis, MO., MCC-8-181, 1988.
6. Courtesy of Monsanto Chemical Company, St. Louis, MO.

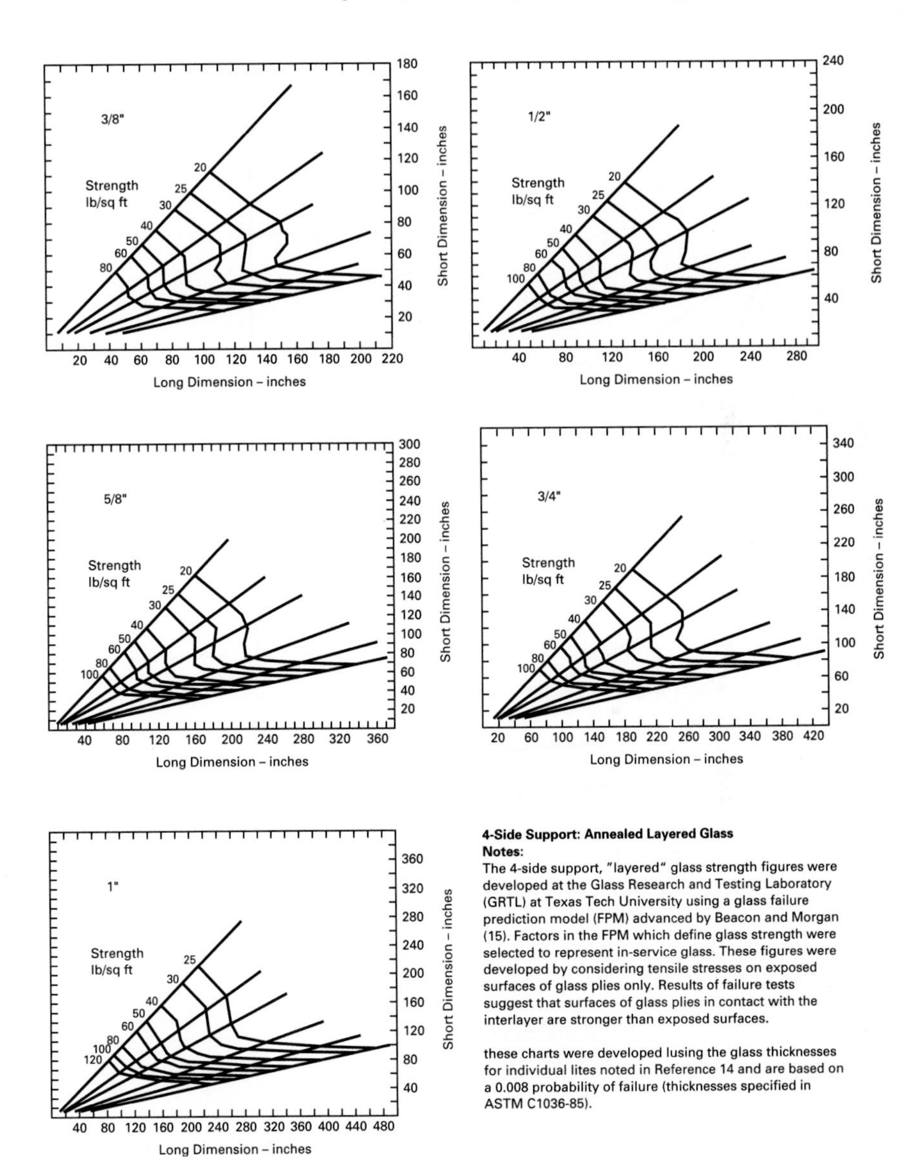

4-Side Support: Annealed Layered Glass
Notes:
The 4-side support, "layered" glass strength figures were developed at the Glass Research and Testing Laboratory (GRTL) at Texas Tech University using a glass failure prediction model (FPM) advanced by Beacon and Morgan (15). Factors in the FPM which define glass strength were selected to represent in-service glass. These figures were developed by considering tensile stresses on exposed surfaces of glass plies only. Results of failure tests suggest that surfaces of glass plies in contact with the interlayer are stronger than exposed surfaces.

these charts were developed lusing the glass thicknesses for individual lites noted in Reference 14 and are based on a 0.008 probability of failure (thicknesses specified in ASTM C1036-85).

Figure 7.7 *(Continued)*

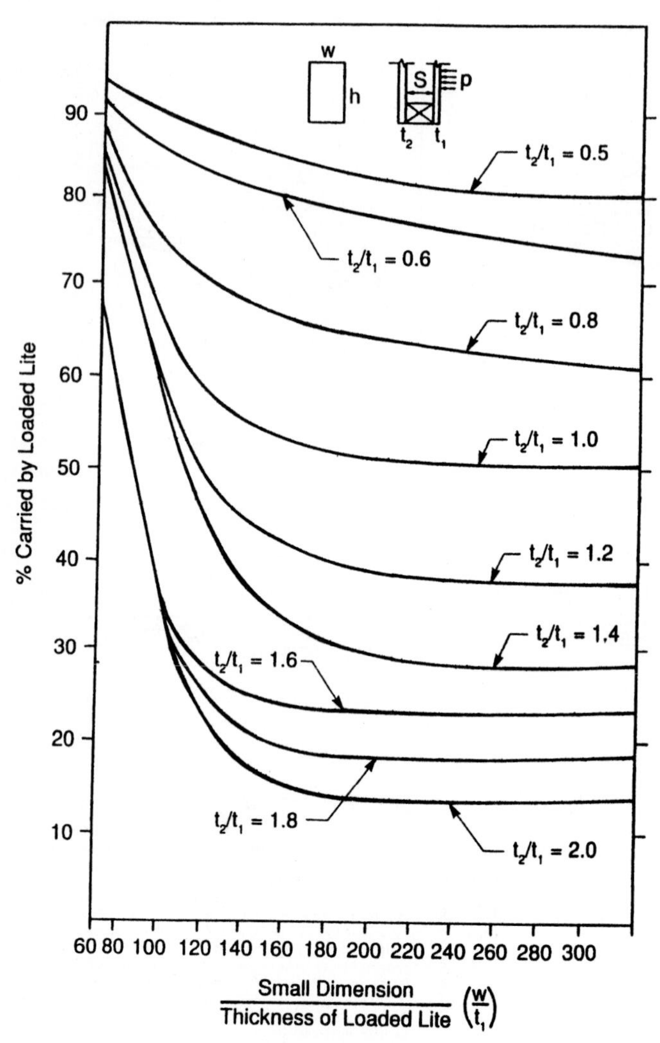

Figure 7.8 Insulating glass load sharing.

Chapter

8

Fire-Rated, Sound Control, X-ray Shielding, and Bullet-Resistant Glass

Joseph S. Amstock

President, Professional Adhesive and Sealant Systems

Fire-Rated Glass

Introduction

Fire protection became an issue in the United States in the 1940s, following a rash of fire-related fatalities. At that time, the groundwork was laid for more effective regulation of fire safety through the use of building codes. The key to fire protection is compartmentalization, that is, keeping the fire and smoke contained in an area so that further damage is prevented and the fire is easier to extinguish. Consequently, all materials used in fire-rated locations, including glazing, must withstand extremely high temperature.

In the architectural process, areas of a building are designed according to building codes for fire ratings. These areas, or openings from the areas, are assigned 20-, 45-, 60-, or 90-minute or two-, three-, or four-hour ratings depending on the degree of fire protection required. The window assemblies constructed in these locations must carry an equal or greater fire rating.[1]

Testing

Building codes in the United States require that all products placed in a fire-rated window/glass assembly pass code conformance requirements, when tested by Underwriters Laboratories, Inc., or equal and bear certification labels by the agencies. During the tests, glazing products are installed in frames or windows subjected on one side to a furnace fire that increases to 1640°F (893°C) for a 45-minute rating, 1700°F (926°C) for a 60-minute rating, and 1925°F (1052°C) for a three-hour rating.

Glazing choices

Historically, wire glass has been the only glazing product available for fire-rated locations in commercial buildings. There are a few glazing products able to achieve a fire rating in the United States:

1. *Wire glass.* Wire glass randomly cracks two or three minutes into the fire. However, the wire acts as a webbing that holds the cracked glass into place. Unrated clear, annealed or tempered glass will usually fail after two or three minutes, depending on the location, and pieces of the broken glass will fall from the building. Wire glass is still widely used, primarily because it is inexpensive and readily available.
2. *Contraflam™, a unique insulating glass unit.*[2] This is an insulating glass, consisting of two or more lites of tempered glass and specified as a fire-rated wall system. The product's design requires a special framing.

   In the event of a fire, a clear gel between the glass lites turns to an opaque, white barrier and retards the radiant heat.
3. *FireLite™, a transparent, wireless glass ceramic.* This product has a low coefficient of expansion that keeps it from breaking down when cooled rapidly.
4. *Pyran™, a 6.5 thermally heated flat glass.*[3] This product is a wireless, clear, transparent glass which recently passed the UL test at 20 minutes. However, tests indicate that the glass is capable of withstanding two hours of heat exposure. At that time, the glass begins to sag. It is a prestressed, heat-resisting glass with low thermal expansion. Thus, it does not shatter during the sudden rise in temperature which occurs during the early stages of a fire. Of importance, during the first 20 minutes of the fire, the glass remains transparent making it easier for the occupants of a building and rescuers to find their way. Because of the prestressing nature of the product, it can disintegrate into small pieces in an emergency. Pyran™ also has a ninety percent light transmission providing architects and builders with an open and transparent design in buildings, and, at the same time, giving the necessary fire protection.
5. *Pyrobel™ is a multi-ply sheet glass with three interlayers.*[4] Following the outbreak of a fire, the inner interlayers of this unique product expand protecting fire-trapped victims from smoke, heat, and flames. The face not exposed to the fire heats only slightly, thus pre-

venting the fire from spreading by heat radiation or convection. As the intensity of the fire increases and the interlayers expand, the glass becomes opaque. Flames are obscured and the danger of panic is reduced during evacuation. Transparency and noise protection are additional qualities.

6. *Superlite™.* This system is a safety-rated insulating glass unit,[5] has a 45-minute fire rating, and provides full, clear, wireless vision. The unit's makeup is designed to retain structural integrity under fire conditions because of its three-layer design. The two layers of tempered glass encapsulate a colorless polymer gel. When exposed to fire, the gel absorbs heat through evaporative cooling to form an insulating layer providing protection from radiant heat. Figure 8.1 illustrates two profiles for different areas of application.

Fire versus fire-rated

Requirements for fire-rating are not fully understood because wire glass held a monopoly on the fire-rated market for decades, and it became accepted as a generic product. According to the National Glass Association Guide to Federal Model Safety Codes 1990, many building inspectors and most fire prevention officials consider wire glass a safety product. As the National Glass Association (NGA) states, "Nothing could be further from the truth." Wire glass is one-half as strong as plain, unwired, annealed glass. It breaks more readily than plain plate glasses.

When considering which product to specify, it is generally considered by code that the fire requirement overrides the safety requirement when the two are in conflict. The proper product to use is always a fire-rated one, permanently labeled to indicate certification.

Sizes

The maximum allowable sizes for lites of fire-rated glazing material are generally controlled by two factors:

1. the maximum size for which the glazing material has been tested and passed, and
2. the local building code.

Often, these factors go unobserved and are frequently misunderstood by industry professionals. With such stringent testing procedures in place for fire-rated glass, it is ironic that they are not followed up with strict enforcement at installation. Many of the problems and much of the confusion stems from the lack of consistency in building codes and their interpretation.

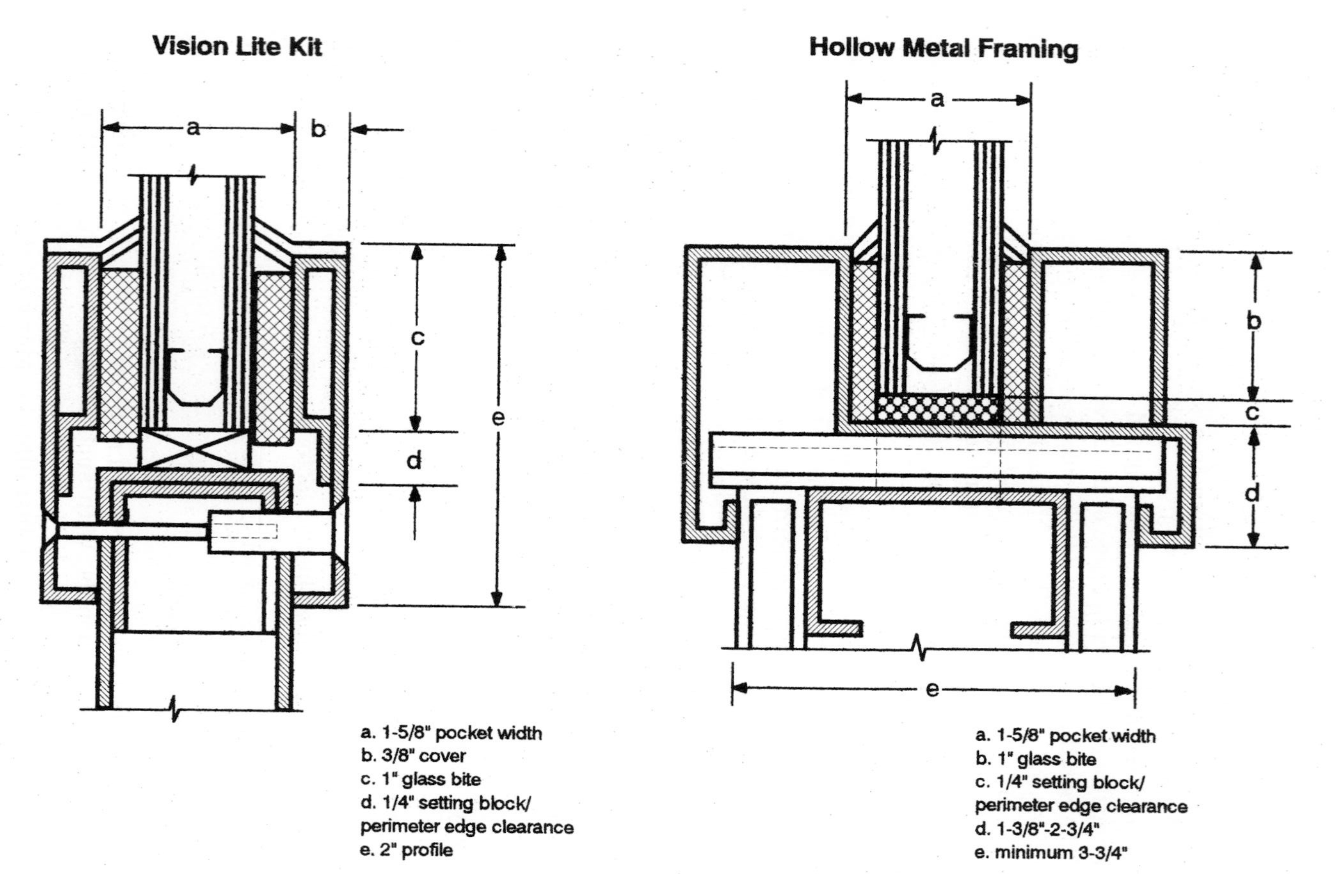

Figure 8.1 Cross section of fire and safety units. (*Courtesy of SIFTI division of O'Keefe's.*)

Codes

There are different types of codes, resulting in various levels of compliance and optional involvement in the process by those affected by them. Some codes are enforceable, and some are not. There are codes written as standards for legal enactment, others that are nothing but standards, and a few that are models made available for adoption by legal jurisdictions.

Building codes have been in existence within civilized societies for thousands of years, but like most everything else in our world, they lack universal standardization and are subject to interpretation by some authority.

In many instances, fire-rated glazing is installed improperly because the professional may not actually be aware of the code requirements. Glazing, in this instance, is not the only thing involved. It is all the components, including doors, frame sealants, etc.

Standards

The National Fire Protection Association's NFPA 80 Standard for Windows and Fire Doors is one of the installation standards that conform to UL's testing results. NFPA 80, which is reflected in all major national and regional model building codes, offers criteria for testing procedures used by UL and other recognized testing agencies. Final authority for developing and enforcing building codes and fire regulations, however, rests with local building code officials, whose guidelines, in fact, can override or reverse NFPA standards.

The Future

New products always bring change to the fire-rated glazing industry. As these developments occur, questions regarding testing, installation, and specifications of fire-rated glazing will continue to arise.

Increased awareness and recognition of fire-rating requirements will help simplify specifiers' purchase decisions for fire-rating glazing. With litigation a common threat in the United States, lack of awareness is not acceptable or affordable.

Sound Control Glass

Developers, specifiers, and others in the building industry must keep abreast of new laws and realize that they may have been hastily developed and may bring about additional expense without necessarily solving a problem.

During this decade, sound control products will become as popular as those designed to save energy. These products will ultimately be judged by their performance. In 1990, the new ASTM standard was published, and, as the decade progresses, additional information will be available to assist the specifier through this highly technical fenestration application. It is crucial to keep in touch with this technology as it evolves.

The logarithmic system of sound characterization can be weighted to best describe the pattern for sorting information. The A-weighted scale is deemed closest to the operation of the human ear. All sounds are not alike. The human ear picks up sounds from approximately 20 hertz (Hz) through 15 kilohertz (kHZ). Figure 8.2 illustrates the audible range, and loudness ranges of speech and music. Note the threshold of hearing and that of feeling or pain.

A problem that is increasingly common in these days of high-decibel living is noise control.[6] Architects have succeeded in designing windows which let in natural light, but help is needed in preventing unwanted sound from entering buildings. Sound control laminated glass is highly effective in reducing sound transmission. Sound control is an important factor in building design considering today's construction market. Unfortunately, the best available sites are often potentially the noisiest.[7] The sites can be measured by a level of sound known as a decibel (dB).

Windows can be the weak link in the transmission of unwanted sound into a building. Just as a window lets in light, it also can trans-

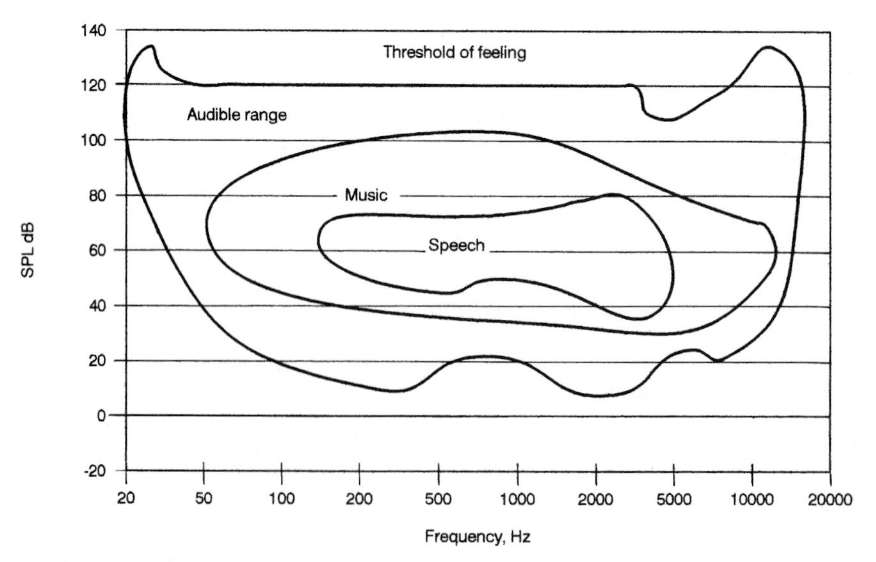

Figure 8.2 Loudness and frequency ranges.

mit unwanted noise. To establish the desirable acoustical environment, the noise control requirements and the sound-reducing properties of glazing materials should be considered an integral part of total space design.

Laminated glass is highly effective in reducing unwanted sound transmission, and laminated glass can be used in standard window design. Figure 8.3 is a cross section of laminated glass incorporated into a sound control unit.

Laminated architectural glass also effectively reduces the coincident dip in the 1000- to 2000-Hz range common to glass products. The panel becomes highly transparent to sound in this frequency range. Coincidence impairs the overall acoustical performance of glazing systems because general environmental noise sources, such as traffic and aircraft, have significant amounts of sound energy in this frequency range.

Frequency

Frequency indicates the number of vibrations per second and is measured in hertz (Hz); the higher the frequency, the higher the pitch. Perception of sound by the human ear depends on, among other things, the frequency. Table 8.1 defines this statement.

Tables 8.2 and 8.3 illustrate the excellent noise reduction performance of laminated glass with a plastic interlayer. The sound transmission loss (TL) of a panel is the difference in decibels (dB) between the sound energy incident on a panel and the sound energy transmitted through the panel.

The more effective the sound isolation provided by the material, the greater the TL.

Sound transmission class

To provide a single number rating for describing the sound isolation performance of a material, the ASTM has developed the sound transmission class, or STC rating.

TABLE 8.1 Sound Range

Definition	Infrasonic	Normal sound	Ultrasonic
Frequencies	0	20	20,000 hertz
Perceptibility	Perceived but not in audible range for man	Audible to humans	Not in audible range for man but audible to some animals

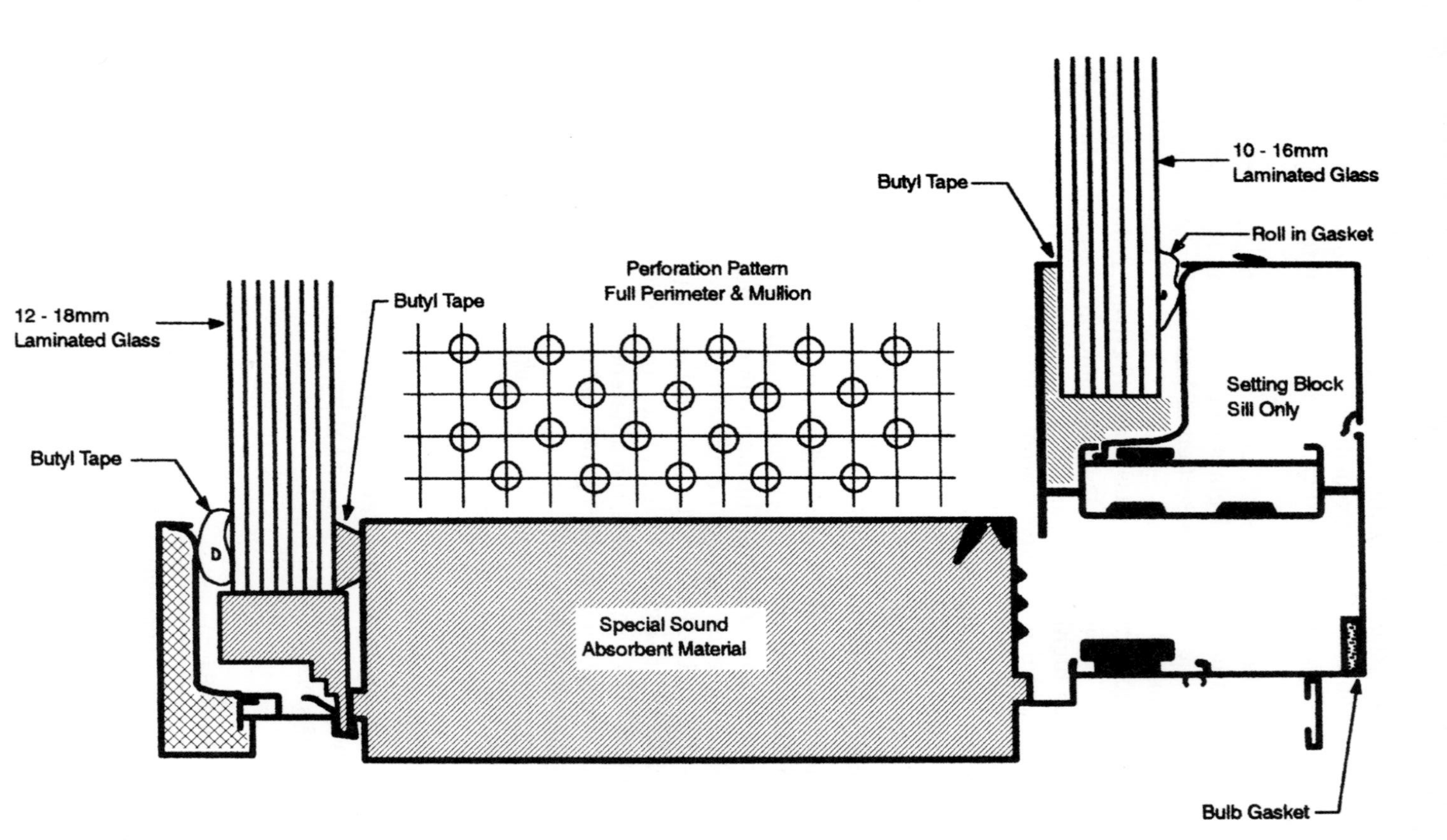

Figure 8.3 Sound control unit.

TABLE 8.2 Acoustical Performance of Glass Fabrications

	Thickness, in	Thickness, mm	Inside, in	Construction space, in	Outside, in	STC value
Laminated glass	1/4	7.24	1/8	0.030 PVB	1/8	35
	3/8	9.53	1/4	0.030 PVB	1/8	36
	3/8	10.5	1/4	0.060 PVB	1/8	37
	1/2	12.1	1/4	0.030 PVB	1/4	38
	9/16	12.9	1/4	0.060 PVB	1/4	39
	5/8	16.2	3/8	0.030 PVB	1/4	40
	3/4	19.9	1/2	0.060 PVB	1/4	41
Laminated insulating glass	1	26.1	1/4 laminated	1/2 airspace	1/4	39
	15/16	24.6	1/4 laminated	1/2 airspace	3/16	39
	1 1/8	29.3	3/8 laminated	1/2 airspace	1/4	40
	1 7/8	37.3	1/4 laminated	1 airspace	3/16	42
Double-laminated insulating glass	1	27.9	1/4 laminated	1/2 airspace	1/4 laminated	42
	1 3/4	45.9	1/2 laminated	1 airspace	5/16 laminated	46
Non-laminated insulating glass	1/2	14.5	1/8	1/4 airspace	1/8	28
	1	27.9	1/4	1/2 airspace	1/4	35
	1 1/2	40.6	1/4	1 airspace	1/4	37
Monolithic glass	1/4	5.59	1/4	—	—	31
	1/2	12.4	1/2	—	—	36

STC is a single-number rating derived from individual transmission losses at specific test frequencies (ASTM-E-90, ASTM-E-413). It allows a preliminary comparison of TL performance at all frequencies of the actual sound spectrum to be controlled.

Sound diffusion

To be diffused, sound requires a medium capable of retransmitting vibrations. The media may include solids, gases, and liquids. Depending on the medium of propagation, the distinctions detailed in Figure 8.4 apply. Airborne waves striking a solid surface can take different directions, depending on the type of wall, as in Figure 8.5.

TABLE 8.3 Transmission Loss (TL) Measurements*

		1–3 Octave band sound transmission loss (TL), dB																		
	Glazing configuration, in	STC	100	125	160	200	250	315	400	500	630	800	1000	1250	1600	2000	2500	3150	4000	5000
Single-laminated glass	$\frac{1}{8} \times 0.030 \times \frac{1}{8}$	35	24	26	28	28	29	30	30	32	34	35	35	36	36	35	35	38	43	45
	$\frac{3}{16} \times 0.030 \times \frac{3}{16}$	36	27	27	27	30	31	31	33	34	35	36	36	35	34	37	41	45	49	52
	$\frac{1}{4} \times 0.030 \times \frac{1}{4}$	38	25	29	28	30	33	33	34	36	37	37	37	36	37	41	45	48	51	53
	$\frac{1}{4} \times 0.060 \times \frac{1}{4}$	39	26	29	28	30	33	33	35	36	37	38	38	37	38	41	44	47	51	54
	$\frac{1}{2} \times 0.060 \times \frac{1}{4}$	41	29	32	29	32	34	35	36	39	38	37	37	41	44	47	50	53	56	57
Laminated insulating glass	$\frac{1}{4}$ laminate × $\frac{1}{2}$ air × $\frac{3}{16}$	39	26	23	25	23	27	31	34	36	38	39	41	43	45	46	43	49	55	55
	$\frac{1}{4}$ laminate × $\frac{1}{2}$ air × $\frac{1}{4}$	39	28	20	29	24	26	30	34	36	39	42	43	44	44	41	40	47	52	56
	$\frac{1}{4}$ laminate × 1 air × $\frac{3}{16}$	42	22	27	27	28	31	35	38	41	42	43	44	45	47	47	45	50	58	61
	$\frac{1}{4}$ laminate × 2 air × $\frac{3}{16}$	45	24	25	34	33	34	40	41	44	44	46	47	47	48	48	46	50	55	56
	$\frac{1}{4}$ laminate × 4 air × $\frac{3}{16}$	48	26	36	34	37	37	43	44	48	49	51	51	50	51	50	47	51	58	60
	$\frac{1}{4}$ laminate × $\frac{1}{2}$ air × $\frac{1}{4}$ laminate	42	26	21	29	28	30	34	36	40	42	44	44	44	45	46	47	52	57	58
	$\frac{1}{2}$ laminate × 4 air × $\frac{1}{4}$ laminate	51	34	38	34	40	41	45	47	51	52	53	53	51	52	55	58	60	62	64
Air-spaced glass	$\frac{1}{8} \times \frac{1}{4}$ air $\times \frac{1}{8}$	28	26	21	23	23	26	21	19	24	27	30	33	36	40	44	46	39	34	45
	$\frac{1}{4} \times \frac{1}{2}$ air $\times \frac{1}{4}$	35	29	22	26	18	25	25	31	32	34	36	39	40	39	35	36	46	52	58
	$\frac{1}{4} \times 1$ air $\times \frac{1}{4}$	37	22	19	27	23	31	30	35	35	36	39	41	42	41	36	37	46	51	56
Monolithic glass	$\frac{1}{4}$	31	23	25	25	24	28	26	29	31	33	34	34	35	34	30	27	32	37	41
	$\frac{1}{2}$	36	26	30	26	30	33	33	34	36	37	35	32	32	36	40	43	46	50	51

*Tested under ASTM E90

Airborne Noise

Vocal chords or resonating bodies of musical instruments causing air molecules to vibrate

Mechanical Noise

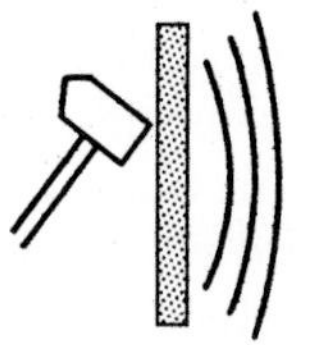

Direct stimulation, for instance knocking at the door, causing the vibration of solids

Footfall Noise

Noise produced when walking on floors or ceiling

Figure 8.4 Different types of sound.

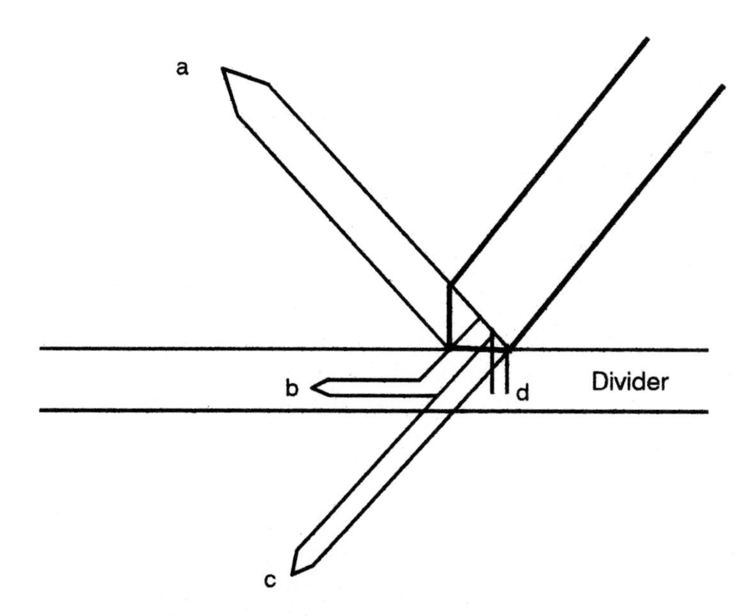

a=noise is deflected (by reflection), therefore the angle of incidence equals that of emergence.
b=the wall made to vibrate retransmitts part of the sound as material sound
c=the vibrating wall stimulates the surrounding air, thereby letting through part of the sound energy.
d=part of the sound energy is absorbed by the wall and transformed into heat

Figure 8.5 Incident noise.

Infills

Different glazing infills will achieve different transmission losses. For example, a window designed to reduce automotive traffic noise should achieve the greatest transmission loss from 80 to 400 Hz. Generally speaking, the heavier the infill (mass), the better suited it is to low frequencies. An infill mass of 8 psf (typical of ½-in thick glass) would be significantly better at decreasing the rumbling noise of traffic than one of half that weight, but would, on the other hand, still pass some high-frequency sounds, such as the whine of sirens or diesel turbochargers. Therefore, the mass usually must be enhanced by using two pieces of glass—one 4 psf and one 3 psf laminated together with a vinyl or resinous cement. This asymmetrical makeup achieves better transmission loss over a broader spectrum of frequencies, even though it has approximately the same mass as a single piece of glass.

This window could be further refined by using two pieces of glass of slightly different mass, separated by an air space or gas-filled space. Air spaces should be as wide as possible—4-in spaces are much better than 3-in spaces. Small air spaces of ¼ to 1 in will actually diminish the product's performance because resonating sound transfer may occur.

So, to dispel a myth, insulated glass, as commonly manufactured with a symmetrical makeup and narrow air space, makes a poor sound control infill. But this product's characteristics change when its small air space is filled with such gases as argon, SF_6, and helium. Certain mixtures produce excellent high-frequency (550 to 4,000 Hz) transmission loss. Unfortunately low-frequency transmission loss is then degraded.

Because the atomic weights of these gases differ so radically from that of air, optical quality could be distorted. A heavy gas will be forced to expand at the bottom of a sealed unit, making it pear-shaped; gasses lighter than air, such as hydrogen and helium, will expand at the unit's top.

The Complete Package

The construction specifier must know the performance rating of a composite window tested as a unit, not just the rating of the window or glazing infill tested separately. The key to success is in the assembly of the product to achieve airtightness, positive latching, good joinery, and all-around quality. The most sophisticated glazing products are rendered nearly useless if not correctly installed in a fenestral unit able to sustain the same transmission losses as the infill. Some guidelines follow.

1. Specify products that have been tested not only with the glazing infill, but as a complete windows with weep and weatherstripping,

installed according to factory guidelines. All fenestration products lose 2 to 12 dB or more, depending on their construction. Do not simply use laminated glass to treat sound control unless you know the transmission losses of the glass.

2. Ensure tight installation and proper sealing and flashing of all sound control products to eliminate any air infiltration that would destroy the acoustic properties of the wall composition.

3. Supply a test report of the acoustic product to your client, disclosing all the one-third octave, band-transmission losses.

4. Inspect each product before installation to confirm that all weatherstripping, gaskets, and seals have been properly installed by the manufacturer, ensuring acoustic and airtightness.

5. Create a fenestration product that can pass generic, ASTM, CAWM, AAMA, and federal specifications for fenestration products. Does the product have proper weep holes? Has drainage been allowed to treat condensation? What steps have been taken to hinder condensation in sealed products? Were the proper sealants and gasketing used to avoid contamination of the glazing infills? Was the correct paint used to avoid outgassing in sealed products?

Remember, an acoustic product must be a good fenestration product first, as shown in Figure 8.6.

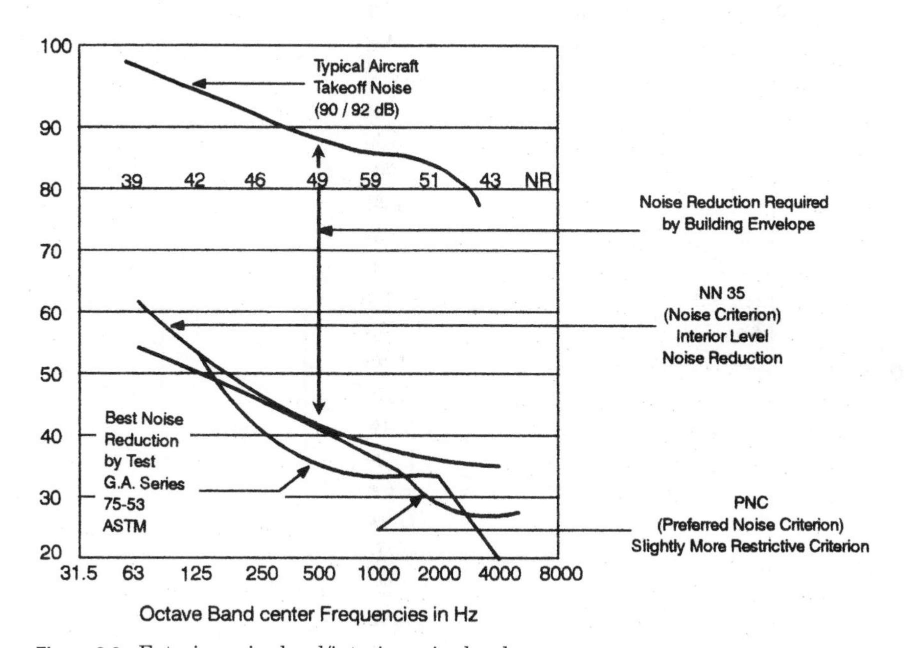

Figure 8.6 Exterior noise level/interior noise level.

STC validity

For years, architects and specifiers have been using the infamous STC (Sound Transmission Class) single-number rating system, ASTM specification E 413. This single-number system has proven effective when applied to interior wall partitions, for which it was specifically developed. But, in fact, the STC rating system specifically excludes exterior noise penetration from its scope. The first paragraph of the standard states that STC was designed to correlate with subjective impressions of the insulation provided against the sounds of speech, radio, television, music, and similar sources of noise in offices and dwellings. Outside this specification's scope are applications involving sources of noise spectra that differ markedly from those described above. Thus excluded, for example, would be noise generated by most machinery, certain industrial processes, bowling alleys, power transformers, and the like. As shown in Figure 8.7, air leakage plays an important role in the STC rating.

And finally, the exterior walls of buildings, whose noise problems are most likely to involve motor vehicles or aircraft, are particularly excluded. Imagine sitting on the witness stand, trying to explain to a plaintiff's attorney why you used an STC number to control exterior traffic noises, when the standard specifically excludes that application in the first paragraph.

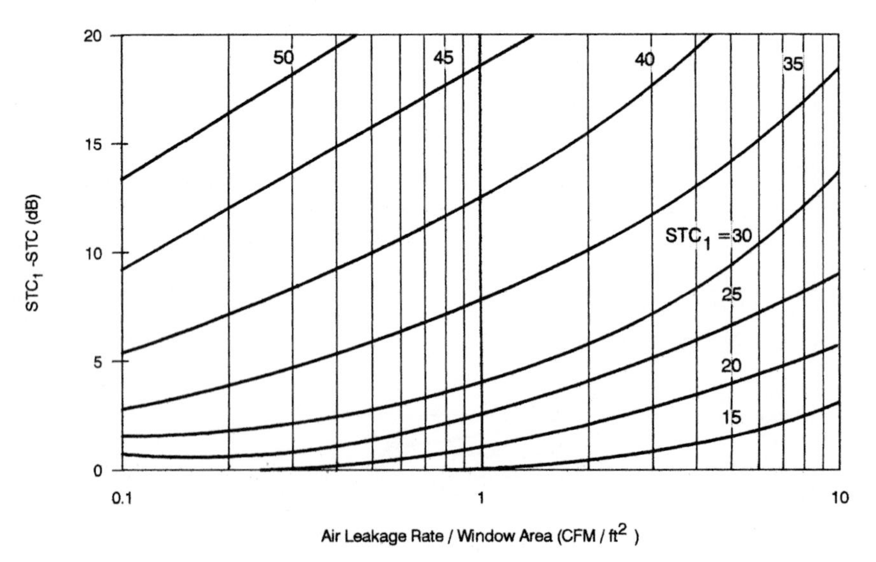

Figure 8.7 Sound transmissivity.

New standards

New standards are being developed to remedy this situation. The first is a replacement standard for STC, to be used for exterior noise penetration involving transportation sounds. This new, single-number rating system, being developed by ASTM task group E33.32 (Fenestration Ratings), is soon to be crowned ASTM E13.32, Outdoor Indoor Transmission Class (OITC). It will be based on an A-weighted rating methodology in frequencies best suited to exterior noise.

The new system was developed on data gathered by E90 testing procedures, the same database used for STC. However, some lower frequencies associated with transportation noise have been incorporated. Although this document is not currently available, a sound control engineer can explain the merit of incorporating frequencies lower than those currently used by STC into a single-number rating system to better depict noise generated by transportation.

The other new standard, the ASTM Standard, Specification for Rating the Acoustical Performance of Exterior Windows and Doors, for the first time, will provide a classification system allowing one to specify windows according to a predetermined classification based on manufacturers' product testing. Simply stated, the manufacturer will test its product, using the E90 data collection process, and data will then be reduced to a single-number rating pursuant to OITC E13.32, referenced above. That number will correspond with a type, numbered on a scale of 1 through 7. Therefore, in judging manufacturers of fenestration control products, it will be held, generally, that two manufacturers offering windows with the same rating have equivalent products.

Specific sounds

Any single-number rating system—the new standard included—leaves much to be desired when a specific noise problem must be remedied. A specific noise might be a factory whistle for signalling workers or siren noises near the receiving area of a hospital. When specific noises are identified, it is best to treat just those frequencies involved, instead of taking the general remedy provided by a single-number system.

Future possibilities

Right now, state legislatures are writing and passing new laws aimed at controlling sound pollution. As an example, California's Title 25 (the Housing Law Earthquake Protection Bill), Section 6.3 (Subtitle T25–28) states that

> The purpose of this section is to establish uniform minimum noise isolation performance standards to protect persons within new hotels, motels, apartment houses, and dwellings other than detached single-family dwellings from the effects of excessive noise including but not limited to hearing loss or impairment and interference with speech and sleep.

Soon, we will see such legislation expanded beyond individual residential, unconnected units. Building inspectors will monitor fenestration products to ensure that they meet the transmission losses required for particular developments relative to adjacent highways, airports, and other sources of noise pollution.

X-ray Leaded Glass

Personnel working in examination rooms, operating rooms, intensive care or cardiac care units, and mammographic and nuclear medicine suites require protection against secondary radiation. Historically, X-ray departments have been designed with conventional X-ray shielding products, such as heavy, bulky, lead paneling and small lead-glass windows. These had a tendency to create a "closed-in" environment with a strong negative effect on patients. But medical facilities planners had little else to choose from, until the introduction of lightweight, lead plastic shielding.

Since its introduction more than 10 years ago, this type of shielding has become the number one choice for radiation shielding. Its enormous design flexibility and versatility are unmatched by other conventional X-ray shielding products. Now, for the first time, architects, specifiers, and hospital/medical facilities planners truly have a choice.

Today's competitive environment has created the need for state-of-the-art products that will enable medical facilities to create modern, attractive, spacious facilities. Whether new construction or renovation, considerable dollars are being invested in radiology departments. This type of panel will help to highlight this investment in equipment. It will dramatically enhance the appearance of the facilities and create a comfortable and relaxed environment for the patient.

Shielding opens up a whole new world of design opportunities for this type of product. They should be shatter-resistant, transparent, lead-plastic materials impregnated with 30 percent lead by weight. They should provide effective radiation protection and the ability to create modern, attractive, space-saving X-ray rooms with panoramic viewing areas that allow the technologist and patient to see each other, giving both a greater feeling of security.

Usage

In the hospital, Clear Pb[8] shielding design flexibility is unmatched for shielding requirements in radiology suites, special procedures rooms, nuclear medicine areas, cardiac cath suites, specialty operating rooms, and CT rooms.

Outside the hospital, lightweight lead shielding is perfect for private radiology groups, mobile radiology/CT units, veterinary hospitals, and dental operations. Pb shielding is being used wherever X-ray shielding is required.

Properties

The clear lead shielding is a lead-containing plastic sheet developed for use as an optically transparent radiation shield. Lead is introduced into the resin as an organolead salt producing a 30 percent by weight lead-acrylic copolymer. Lead equivalencies of 0.3, 0.5, 0.8, 1.0, 1.5, and 2 mm are provided by changing the thickness of the sheet. Its lightly tinted brown color has an eye-ease effect, and its low reflective and refractive indices result in less distortion and reflection when compared to lead glass.

High resistance to shattering contributes to its ease of handling and use in large continuous panels, allowing unobstructed views of patients and equipment. When exposed to ^{60}Co gamma radiation of 1×10^4 R, the clear lead barrier shows no deterioration of physical properties. Table 8.4 shows the thickness and lead equivalents to obtain the necessary protection, and Table 8.5 outlines the mechanical and thermal properties.

Additional tables are provided covering optical constants, Table 8.6, the energy coefficients, Table 8.7, and the source of linear coefficients, Table 8.8. All this data is advantageous when writing a specification for the use of X-ray panels.

TABLE 8.4 Thickness and Lead Equivalents

(Density = 1.6 g/cm^3)		
Thickness, mm	Lead equivalence, mm Pb	Weight, lb/ft^2
7 ± 0.7	0.3	2.3
12 ± 1	0.5	3.9
18 ± 1	0.8	5.9
22^{-0}_{+3}	1.0	7.2
35 ± 2	1.5	11.5
46^{-2}_{+3}	2.0	15

TABLE 8.5 Typical Mechanical and Thermal Properties of Lead Glass

at 23°C, 65% rh			
Properties	Unit	Lead glass	Conventional acrylic
Specific gravity		1.6	1.2
Tensile strength	kg/cm^2	490	780
Elongation	%	18	6
Tensile modulus	kg/cm^2	2.4×10^4	3.3×10^4
Flexural strengths	kg/cm^2	710	1200
Flexural modulus	kg/cm^2	2.1×10^4	3.3×10^4
Izod impact strength (notched)	kg-cm/cm	3.0	2.2
Falling ball test	cm	>200	>200
Minimum breakage ht.		Lead glass	
Rockwell hardness	—	M-73	M-100
Heat deflection temperature	°C	80	100
Maximum stress: 18.6 kg/cm^2			
Coefficient of linear thermal expansion (below 35°C)	$cm/cm^{-°C}$	1×10^{-4}	7×10^{-5}

TABLE 8.6 Optical Constants

Properties	Clear Pb	Lead glass	Plate glass	Conventional acrylic
Refractive index n_D	1.54	1.98	1.52	1.49
Refractive index p*	0.0454	0.1081	0.0426	0.0387

*$p = (n_D - 1)^2/(n_D + 1)^2$ at 23°C.

TABLE 8.7 Energy Linear Coefficients

	68.8		43.0	
X-ray, kV	μ, cm^{-1}	μ/ρ, cm^2/g	μ, cm^{-1}	μ/ρ, cm^2/g
Conventional acrylic	0.208	0.175	0.245	0.206
Lead glass*	8.04	1.84	—	—
Plate glass	0.563	0.227	1.07	0.431

μ: Linear attenuation coefficient, cm^{-1}.
ρ: Absorber density, $g \cdot cm^{-3}$.
μ/ρ: Mass attenuation coefficient, $cm^2 \cdot g^{-1}$.
* Lead content 51%.

TABLE 8.8 Source Linear Coefficients

	^{60}Co		^{137}Cs	
γ-ray source	μ, cm^{-1}	μ/ρ, cm^2/g	μ, cm^{-1}	μ/ρ, cm^2/g
Clear Pb	0.103	0.0646	0.152	0.0948
Lead glass*	0.25	0.057	0.38	0.087
	−0.28	−0.056	−0.47	−0.094
Lead	0.685	0.0606	1.25	0.111

μ: Linear attenuation coefficient, cm^{-1}.
ρ: Absorber density, $g \cdot cm^{-3}$.
μ/ρ: Mass attenuation coefficient, $cm^2 \cdot g^{-1}$.
* Varied with lead content.

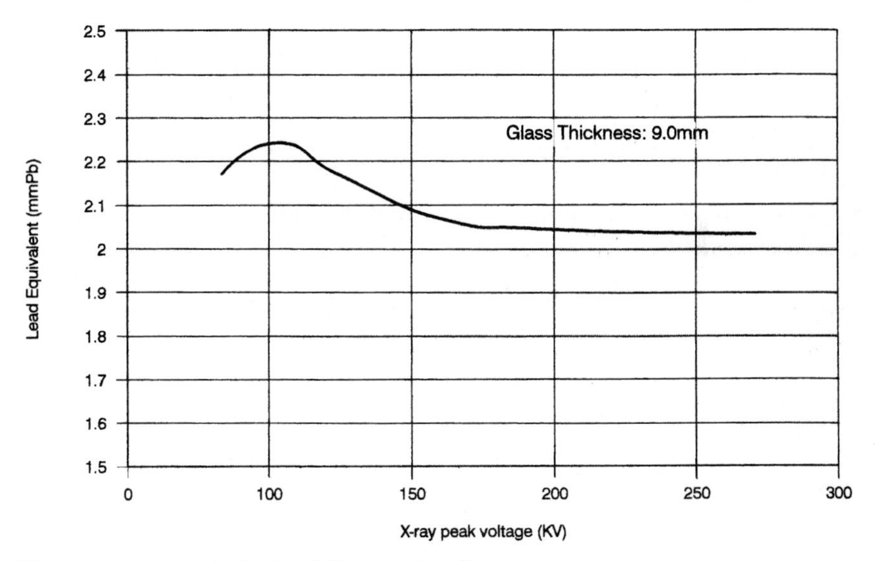

Figure 8.8 Lead equivalent and X-ray peak voltage.

Shielding panels are shatter-resistant and distortion-resistant and are lead-free composites. The following graphs, Figures 8.8 and 8.9, illustrate the lead equivalency at peak voltage and lead equivalency versus glass thickness, respectively. The data is for an X-ray source potential of 150 kV and 200 kV using a $^{23}/_{64}$-in (9-mm)-thick glass. The lead equivalent slightly decreases when the potential exceeds 150 kV but remains constant at 200 kV and higher.

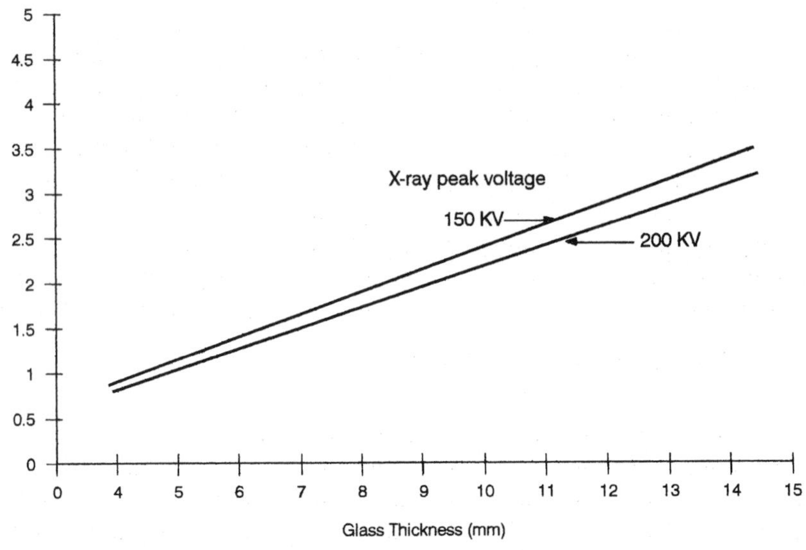

Figure 8.9 Glass thickness and lead equivalent.

Installation

Because hospital environments are involved, all work on permanent installations should be under the supervision of the manufacturer. Be certain that all the panels are supported by the floor, wall, or ceiling brackets. Figure 8.10 illustrates the proper hanging method for mounting a unit on the ceiling. Because of the potential dangers involved with X-ray, exercise extreme caution when specifying, handling, and installing this type of panel.

Bullet-Resistant Glass

Threats to facilities in the form of burglaries, forced entries, ballistic attacks, and bomb blasts can be met with various configurations of laminated glass. Large storefront glass windows at street levels are one of the best ways retailers, small service businesses, and restaurants have to showcase their merchandise or services. But these areas are increasingly at risk, particularly from "smash and grab" vandalism. According to the FBI's Unified Crime Report, almost 70 percent of all burglaries involved forced entry through windows or doors. The most critical step in a burglary is entry, and that entry most commonly involves a window. Standard, laminated, bullet-resistant glass can provide a significant improvement over monolithic glass products in its ability to withstand handheld weapons, such as hammers, bricks, and handguns. Although the burglar may muster

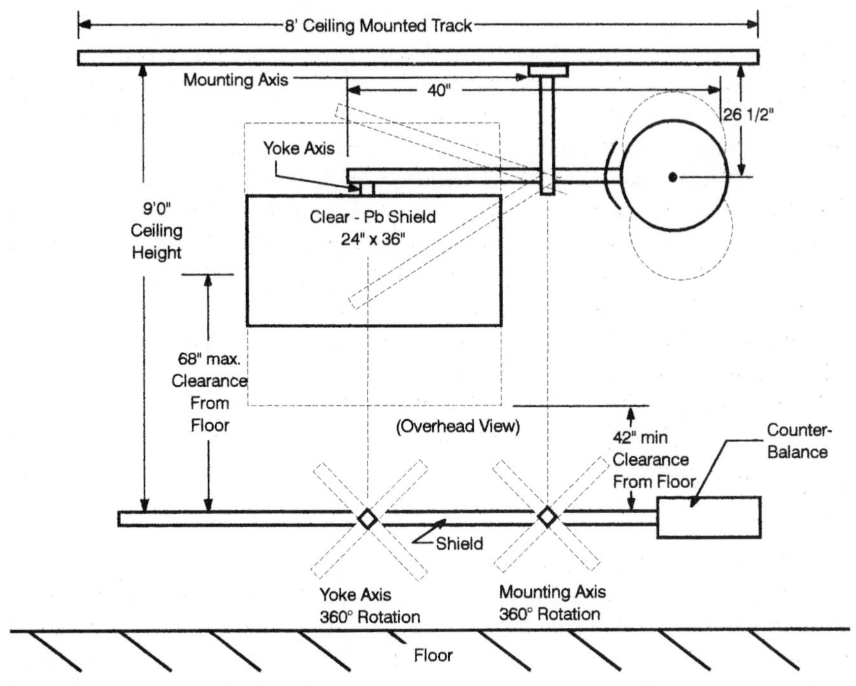

Figure 8.10 Ceiling mount.

enough force to shatter the glass, it requires repeated and attention-getting blows to break through the plastic interlayer. Even quiet glass cutting tools are ineffective, because laminated glass needs to be cut on both sides of the interlayer. Laminated bullet-resistant glass should be considered for display cases for jewelry, electronic, and other high-ticket items. Some of the world's greatest art treasures and documents are protected by bullet-resistant glass, including the Mona Lisa and the Declaration Of Independence.

High-Security Glazing Applications

In the tragic 1995 bombing explosion at the Federal Building in Oklahoma City, more than 75 percent of the injuries sustained by those who survived in the area surrounding the building were due to flying shards of glass. Glass experts who visited the site in the days after the blast saw repeated examples of failed window glass in an area that extended several miles in each direction from ground zero. The property damage to the interior of the buildings was costly, as shattered and fallen glass left dozens of buildings unprotected from the weather and looters. In the few instances where laminated glass was used in buildings near the Federal Building, the glass had cracked but remained in the frame, protecting the interior of the building.

In investigating the glass breakage in the nearby buildings, it was also discovered that adhesive "security films" performed poorly. The glass shards did not adhere to this type of film but left the film and became flying missiles. These findings were consistent with the testing done at the Glass Research Laboratory at Texas Tech University. Although a blast as severe as the Oklahoma City event is uncommon, bomb explosions are not.

Laminated bullet-resistant glass deflects in an explosion, enabling it to adsorb the blast pressure and some of the energy generated. When laminated glass remains in its frames, it protects the interior of the building from the blast wave, which causes the majority of the damage to the building's interior.

In high-risk buildings, such as banks, prisons, or other facilities that could come under gun attack, bullet-resisting security glazing can successfully prevent penetration of bullets and impact-induced spall (fragmentation from the protected side of the glazing).

Bullet-resistant glass is generally a multi-ply, annealed, glass laminate. There are normally four levels of bullet-resistant glasses listed with Underwriters Laboratories, Inc. (UL). Bullet-resistant glass is manufactured to resist penetration by medium- and superpowered small arms and high-powered rifles. Although no product manufactured today is bullet-proof, bullet-resistant glasses are designed to afford enhanced protection as a safety shield against direct bullet injury. Table 8.9 contains data showing four levels of bullet resistance.[9]

Underwriters Laboratories (UL) tests for low-, medium-, and high-powered rifles require glass from $1^{13}/_{16}$ to 2 in thick. The spalling encountered on the inner side of the bullet-resistant glass can be prevented by using combinations of plastic and glass. The term "bullet-resisting" as used here signifies protection against complete penetration, passage of projectile fragments, or spalling (fragmentation) of the protective material to the degree that a person standing directly behind the bullet resistant barrier would be injured. It is used in conjunction with the rating of bullet-resisting materials.

Polycarbonate laminates

Lexgard laminate, manufactured by General Electric Co., is a composite of Lexan sheet bonded together with a patented GE interlayer. For a higher level of protection from ballistic and physical attack, plastic laminates can be used (in combination with a glass component system) which remain intact when the glass breaks. GE is the only

TABLE 8.9 Bullet-Resistant Glass

Product						Tolerances	
Ballistic threat	Level	Certified	Thickness, in, and fabrication required	Approx. wt/psf	Visible light* transmitted	Thickness, in	Dimensional, in
.38 Super automatic MPSA	Level 1	UL752	$1\frac{3}{16}$ 2 Lts $\frac{1}{8}$ + 1 Lt $\frac{3}{8}$ + 1 Lt $\frac{1}{2}$ + 3 pcs .015 AG	16	68%	$+\frac{1}{16}$ $-\frac{1}{8}$	$+\frac{1}{8}$ $-\frac{1}{8}$
.257 Magnum HPSA	Level 2	UL752	$1\frac{1}{2}$ 2 Lts $\frac{1}{8}$ + 1 Lt $\frac{3}{4}$ + 2 Lts $\frac{1}{2}$ + 4 pcs .015 AG	21	65%	$+\frac{3}{16}$ $-\frac{1}{16}$	$+\frac{1}{8}$ $-\frac{1}{8}$
.44 Magnum SPSA	Level 3	UL752	$1\frac{3}{4}$ 1 Lts $\frac{1}{4}$ + 2 Lts $\frac{3}{8}$ + 1 Lt $\frac{1}{2}$ + 1 Lt $\frac{1}{8}$ + 4 pcs .015 AG	25	62%	$+\frac{3}{16}$ $-\frac{1}{16}$	$+\frac{1}{8}$ $-\frac{1}{8}$
.30-06 Rifle HPR	Level 4	UL752	2 2 Lts $\frac{1}{8}$ + 1 Lt $\frac{1}{4}$ + 3 Lts $\frac{1}{2}$ + 5 pcs .015 AG	28	60%	$+\frac{3}{16}$ $-\frac{1}{16}$	$+\frac{1}{8}$ $-\frac{1}{8}$

*Based on clear glass.

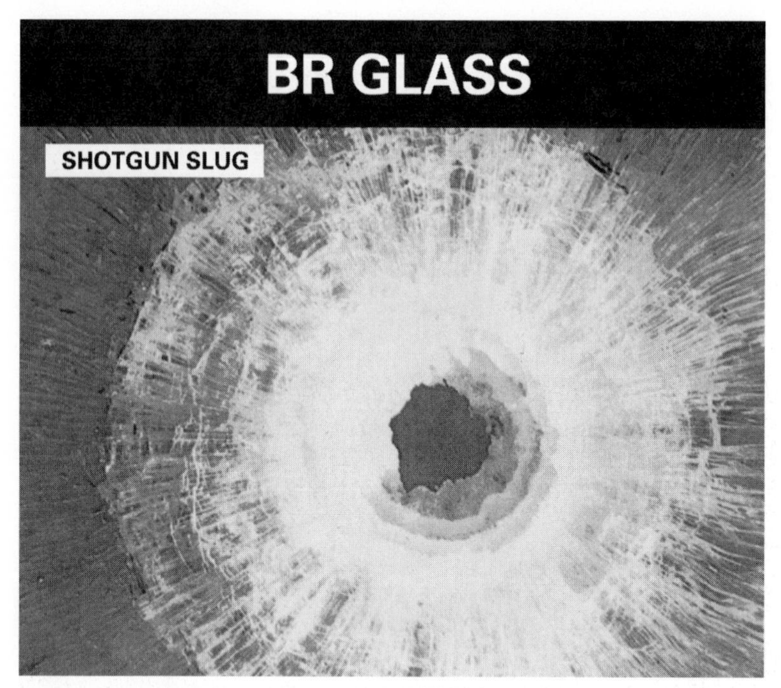

Figure 8.11 Bullet holes. (*Courtesy of General Electric Co.*)

TABLE 8.10 Performance vs. Weight and Thickness

	UL Listing		12-Gauge shotgun
	Medium-power small arms (MP)	Super-power small arms (SP)	Rifled slug*
	.38 Super 475 ft · lb 1280 ft/s	.44 Magnum 1150 ft · lb 1470 ft/s	$\frac{7}{8}$ and $1\frac{1}{8}$ oz slugs 2175 and 2750 ft · lb 1400–1600 ft/s
Lexgard® laminates			
Thickness	1 in	$1\frac{1}{4}$ in	$1\frac{1}{4}$ in
Weight, sq. ft.	6.5 lb	8 lb	8 lb
BR glass			
Thickness	$1\frac{3}{16}$ in	$1\frac{3}{4}$ in	Not available
Weight, sq. ft.	14.9 lb	23 lb	Not available
BR acrylic			
Thickness	$1\frac{1}{4}$ in	$1\frac{3}{4}$ in	Not available
Weight, sq. ft.	7.7 lb	10.8	Not available

*Lexgard laminates data is based on tests by Underwriters' Laboratories using Remington Model SP12R5 and Brenneke 491 Grain 12-gauge slugs. Three and two (close spacing) shot patterns were used at −25 and 120°F. A copy of the UL report is available on request from General Electric.

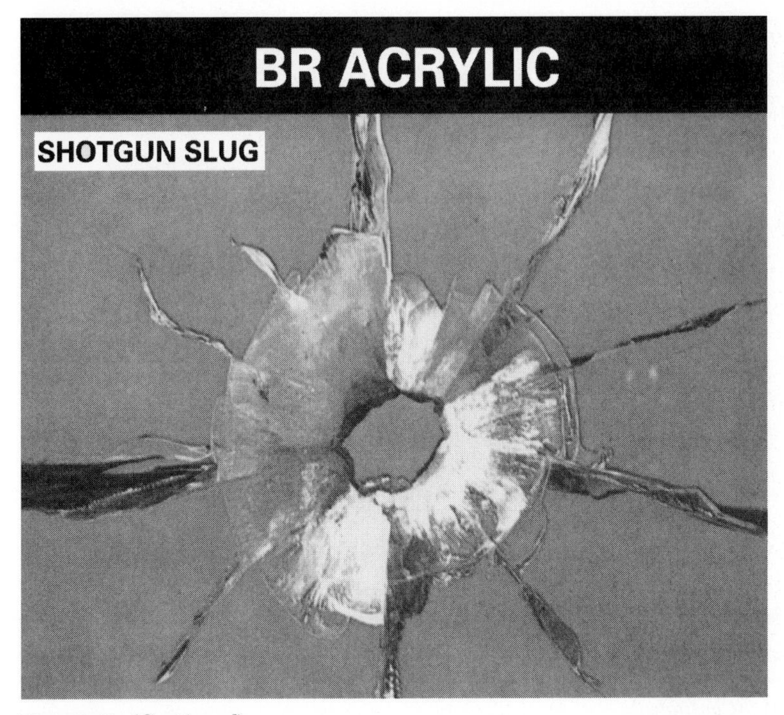

Figure 8.11 *(Continued).*

manufacturer of an all-plastic laminate.[9] Table 8.10 shows the performance versus weight and thickness and Figure 8.11 graphically illustrates the configuration of bullet holes in three different materials.

Conclusion

There are a variety of building materials that can be specified for use in preventing flame spread, preventing forced entry, and, lastly, in attempting to keep our homes and workplaces quiet.

References

1. Razwick, J., In the line of fire, *The Construction Specifier,* April 1995.
2. Contraflam is a trademark of Eich Corp., Los Angeles, CA.
3. Pyran is a trademark of Schott America, Yonkers, NY.
4. Pyrobel is a trademark of Glaverbel, Belgium.
5. Superlite is a trademark of SAFTI, Division of O'Keefe's, San Francisco, CA.
6. Sound Control Windows a Specialty, *Los Angeles Times,* 1984.
7. Hardman, B. G., Fenestration for Sound Control, *The Construction Specifier,* April 1990.
8. Clear Pb is the registered trademark of Nuclear Associates, Carle Park, NY.
9. Lexgard is the registered trademark of General Electric Company, Waterford, NY.

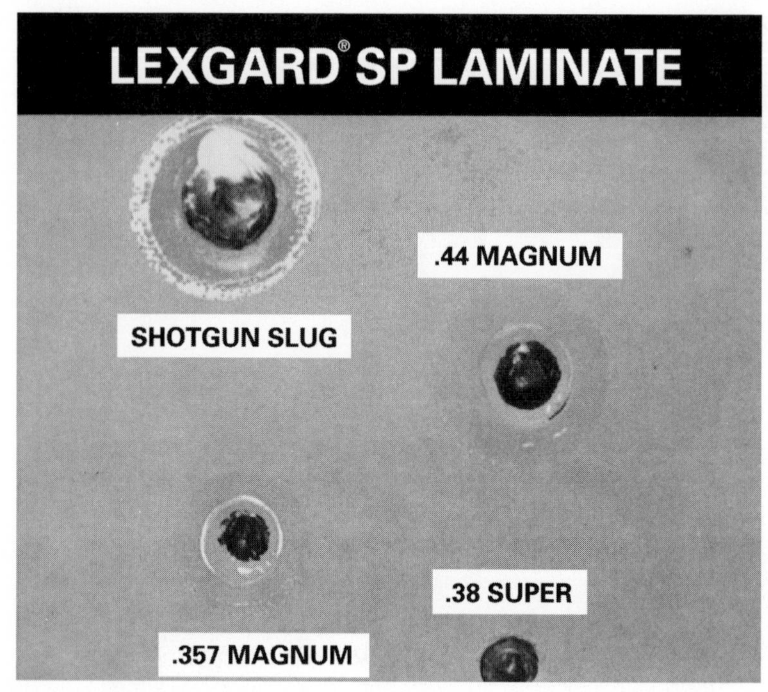

Figure 8.11 *(Continued).*

Chapter

9

Properties of Glass

Joseph S. Amstock

President, Professional Adhesive and Sealant Systems

Introduction

Although glasses may be formed from a number of different materials with correspondingly different properties, many of the most common properties of glass are direct consequences of the glassy state. In this chapter, we will see how such properties arise from that state.

Mechanical Properties

On a practical human time scale, stresses are usually applied and results measured at a rate far too fast for the flow properties of glass at room temperature to show. Under these circumstances, common glass behaves like an elastic solid. In bulk, glass appears almost completely rigid, but in thin sheets or fibers, it is quite pliable, provided that the radius of curvature is large compared to the thickness of the glass.[1]

We saw that the silicon–oxygen bond is very strong, so that glass should be a very strong material. In principle, this is true. Newly formed, fine glass fibers will support loads greater than 70,000 kg/cm^2 (1 kilogram per square centimeter is roughly 14 pounds to the square inch). This figure is five times that obtained by our best steels. More important, it is twice as much as steel could support even in theory. The practical strength of materials, however, is greatly affected by defects and irregularities. Such defects, particularly surface defects that lead to cracks, make the actual strength of ordinary glass less than one-hundredth of the theoretical value calculated from the strength of the interatomic bonds.

Because glass is effectively rigid, an applied force will be concentrated, by the leverage involved, on the few intermolecular bonds at the base of a small surface crack. Although these bonds are very strong, the rigidity of the glass allows macroscopic forces to be applied to a few submicroscopic structures. The bonds, strong or weak, must give way because a glass has the homogeneous structure of a liquid, and a crack, once begun, encounters no internal boundaries or discontinuities to interrupt its progress. Further, as the crack deepens, the effective leverage increases, and the fracturing of further bonds becomes that much easier. The crack, therefore, spreads rapidly across the entire piece of glass.

The crack sensitivity of glass is such that glass is weak in tension, although very strong in compression, where stresses do not lead to cracks. To increase the useful strength of glass, at least three courses are open. One may attempt to prevent surface defects that allow cracks to start, one may try to avoid subjecting the glass surface to tension or, finally, one may attempt to prevent a crack from spreading. All these techniques have been used. It is not very profitable to try to prevent minute cracks in all exposed glass surfaces, but one may protect the glass surface with a tough skin of another material.

The troublesome cracks originate from surface defects and spread to the interior if the glass surface is placed in tension. The second method of strengthening glass mentioned above is to avoid placing the surface in tension. Although it seems a bit magical, there are at least two ways to do this: by thermal and chemical "toughening." In thermal toughening, the piece to be toughened is heated above its transformation temperature, and then, the surface is rapidly chilled. The configurational shrinkage of the surface is thus arrested, but the interior, not yet cooled, undergoes further configurational shrinkage. When the entire piece is cool, the interior is permanently in tension, and the surface permanently in compression. A similar result can be obtained by placing the hot glass article in contact with a salt that will exchange large metallic ions for smaller ones from the surface of the glass. For example, if a glass containing lithium (Li^+) is brought into contact with sodium chloride (NaCl) under appropriate conditions, some of the Li^+ ions near the surface of the glass will be replaced by Na^+ ions, on a one-on-one basis. The Na^+ ions are larger than the Li^+ ions they replace, so that when the glass cools, the surface cannot shrink as much as the interior. Before a crack can begin in toughened glass, the piece must be bent far enough to overcome the surface compression and place the surface in tension. Such glasses may have impact strengths four to ten times as great as that of untoughened glass.

The final technique for strengthening glass is to provide a way of stopping a crack from spreading. If glass fibers are embedded in a matrix of suitable resins that, though weaker than the glass, can deform to distribute a stress, a crack across a single fiber will not seriously weaken the material. The load is transferred to the remaining fibers by the matrix and, indeed, to the portions of the broken fiber remaining intact. The crack that destroys only a small part of one fiber in the fiber resin system would have spread over the entire piece in bulk glass. Crystals are deliberately formed in the glass; that is, it is partially devitrified to provide discontinuities and therefore barriers to crack propagation.

In discussing the mechanical properties of glass, we must point out that the drawing (and related techniques, such as blowing) of glass depends on its behavior as a proper theoretical liquid. Strain (deformation) of a viscous liquid is directly proportional to the applied stress. When glass is drawn, it does not "neck" or develop weak spots, for example, as metals do.

Transparency

The reader will not be surprised to learn that the transparency of glass derives from its basic liquid structure. It is common for liquids to be transparent, whereas transparency is relatively rare in solids. Not all liquids, and indeed not all glasses, are transparent, however, and we must show how transparent glasses differ from opaque liquids. To explain this point, we must say a little about the nature of light.

We may think of light as a particulate (that is, composed of a stream of discrete packets of energy, called photons) or as a wave phenomenon. In modern theory, the two descriptions are completely equivalent. For the moment, we are interested in light as a stream of photons. There is a particular amount of energy carried by each photon. When light falls on a bulk metal, the energy of a trapped photon is absorbed and promptly reradiated by "free" electrons within the metal. These electrons are free in the sense that they can absorb or release energy over a wide energy range, in particular over the range required to absorb and then reradiate light. Therefore, the metal reflects light falling on its surface. In typical glasses, electrons are not free in this sense; they are bound quite rigidly to particular energy levels, and the metallic reflection does not occur.

But glass absorbs light of particular wavelengths, that is, particular colors of light. In fact, common glass is opaque to wavelengths at the infrared and ultraviolet ends of the spectrum. Molecular vibrations that do not involve any permanent change of molecular position are possible in a material. These vibrating systems are "tuned" systems; that is, they can vibrate only at specific frequencies or modes of vibration, although each system may have several of these allowable frequencies. There is a definite amount of energy per photon increasing as the wavelength of the light decreases; that is, photons of violet light contain more energy than photons of red light. It may be that a photon of light of a particular color may have just the right amount of energy to excite one of these vibrating systems from one mode of vibration to another, more energetic mode. If so, photons of that frequency will be absorbed by the material, and the energy absorbed will not necessarily be reradiated immediately or at the same frequency. The material will then absorb the particular color of light concerned.

It happens that the common glasses do not have energy systems that can accept the range of energies represented by photons of visible light, although they do absorb both infrared and ultraviolet radiation. The fact that common glass does not absorb light is not because of the inherent character of glass, but because of careful selection of materials. Very small amounts of certain impurities will lead to tinted or virtually opaque glass. Iron oxide, yielding deep green and brown colors, is, in practice, the most troublesome impurity of this kind. Of course, appropriate "impurities" may be added deliberately when a tinted glass is specifically required.

Common glass, then, does not reflect light in the metallic sense nor does it absorb visible light. Except insofar as bulk metals are unlikely to form glasses (the type of interatomic bond that permits the free electrons in metal is not the type required to form glass), neither precondition for transparency is a necessary consequence of either the liquid or the glassy state. But there is more coming: The structure of a glass or of any liquid is irregular, and the individual molecules are much smaller than the wavelength of visible light. Thus, there are, in fact, no structures large enough to obstruct the passage of a light wave. The wavelength of visible light is roughly from 4000 to 7000 angstrom units (one angstrom unit, that is, 1 Å is 10^{-8} or 0.00000001 cm). Individual glass molecules are roughly 2 to 3 Å in size. A light wave is no more obstructed by a single one of these molecules or by an irregularly spaced set of such structures than an ocean wave is obstructed by an individual pebble on a beach. To the light, the nonreflecting, nonabsorbing glass is simply "another kind of space."

Glass is distinctly another kind of space to the light, however. It is a space filled with charged particles, the electrons and protons that make up the individual atoms. Light is, after all, an electromagnetic disturbance in space, and the disturbance (that is, the light wave) does not travel as fast through a space filled with charged particles as through a vacuum. The ratio of the speed of light in a vacuum to its speed in glass (or any other transparent medium) is called the refractive index for the medium, and ranges from 1.5 to 1.9 for different kinds of common glass. Whenever light passes the boundary between media of different refractive indexes, as from air to glass, a certain amount of light is reflected at the boundary. The amount reflected depends on the difference in the refractive indexes of the two media; for common glass about 4 percent of the incident light is reflected at each air–glass boundary. The reflection is quite different from the metallic reflection discussed previously.

Now we come to the crux of our argument. A liquid, including a glass, has no internal boundaries or discontinuities. We have spoken loosely of individual glass molecules, but we could perfectly well view the whole of a piece of glass as one large molecule. Accordingly, as light passes completely through a piece of glass, it encounters only two optical boundaries, the first one on entering the glass and the second on leaving it. Eight to ten percent of the light may be lost by reflection at these two boundaries, but almost all the rest passes through. It is rare for crystalline solids in bulk to possess this kind of internal homogeneity. A single perfect crystal may be quite transparent, but bulk solids are usually made up of millions of single crystals. As light passes the boundaries of these crystals, some light is lost by reflection at each boundary, and the material is effectively opaque. Thus, the internal homogeneity of glass, which is typical of liquids but not of solids, is the basis for the transparency of glass.

Refraction

We introduced the refractive index of a transparent medium as the speed of light in a vacuum (or the speed in air, which is practically the same) divided by its speed in the medium concerned. When a ray of light passes between two media in which it travels at different speeds, it undergoes a change in direction unless it encounters the boundary exactly at right angles. The effect of glass in bending (refracting) light was well known before it was realized that the relative speeds of light in glass and in air were involved.

In discussing refraction, the direction of the ray of light is measured from a line called the normal which is perpendicular to the air–glass surface. The angle so measured of the ray falling on the glass is called the angle of incidence (I); the angle made by the refracted ray within the glass is called the angle of refraction (r). It was found empirically that, for a given kind of glass, the ratio $\sin I : \sin r$ was the same for all values of I, and this ratio was called the refractive index (n) of the glass. See Figure 9.1. Refraction is the result of the difference between the velocity of light in a vacuum (V_v) and the velocity in glass (V_g). The direction of the ray is measured from the perpendicular to the vacuum–glass boundary, as shown at the left side of Figure 9.1. Since the frequency of light is fixed (for a particular color), the wavelength—the

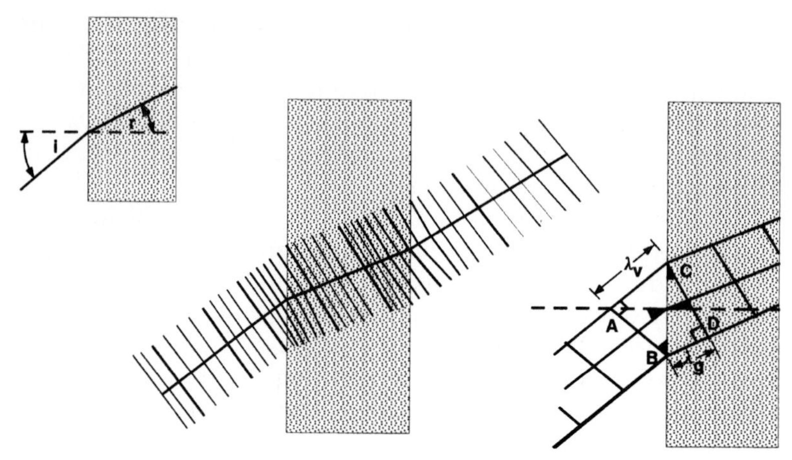

Figure 9.1 Refractive index.

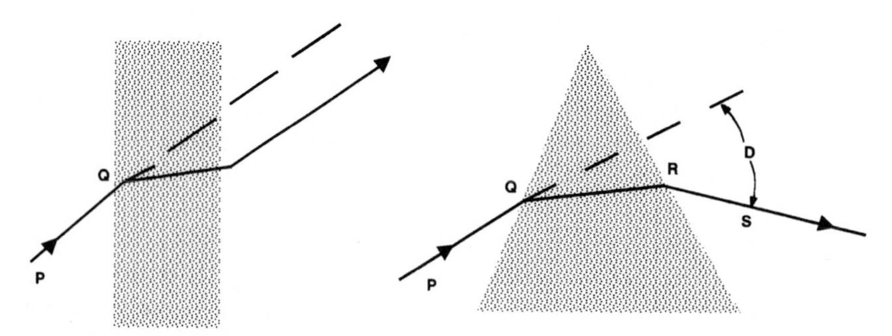

Figure 9.2 Laws of refraction.

distance the light travels during one cycle—depends directly on its velocity and the wavelength in a vacuum. Referring to Figure 9.1, AC is greater than that in glass, BD. Because the wave fronts AB and CD are perpendicular to the ray, the line BC is the common hypotenuse of the right triangles ABC and BCD; hence the ratio sin I/sin $r = \lambda_V/\lambda_g = V_v/V_g$ is the refractive index:

$$n = \sin I/\sin r$$

Figure 9.2 demonstrates the laws of refraction. A ray of light passing through a parallel-sided block emerges parallel to its original path but displaced laterally from it. If the sides of the block are not parallel, the ray is deviated through the angle D, called the angle of deviation. The path of a ray is reversible; that is, if a ray were traveling from right to left along SR, it would emerge traveling along QP.

The reader may be satisfied by referring to the diagram that n, as just defined, is the same as the ratio of the speed of light in air to its speed in glass.

When a ray of light passes out into the air again, its speed increases to its original value and its direction changes again, in a sense opposite to the change that occurred when it entered. If the sides of the glass slab are parallel to each other, the final direction of the light is parallel to its initial direction; the ray has been displaced but not deviated, or more precisely, the deviations suffered at each surface cancel each other out. If the glass does not have parallel sides, the ray will suffer a net deviation, as shown in Figure 9.2. This diagram also illustrates the fact that the path of a light ray through a system is reversible in the sense that it can be traced backward and still be consistent with the laws of refraction.

When a ray emerges from glass into air, it is bent away from the normal. It follows that, as the angle r increases, the angle I will eventually reach 90° so that the emergent ray is traveling parallel to the surface. For any larger value of r, refraction is impossible, and the ray is simply reflected inside the glass. This phenomenon is called total internal reflection ("total" as opposed to the partial rejection that always accompanies refraction). The value of I for which the changeover occurs is called the critical angle, and because $\sin 90° = 1$, its value must be given by $\sin r = 1/n$. If $n = 1.50$, the critical value of r is about 42°. Figure 9.3 illustrates how total internal reflection functions.

Total internal reflection: Light emerging from glass into air is reflected away from the normal (dotted line), and at the same time partial internal reflection occurs (left). When the angle r is at its critical value, the emergent ray is parallel to the surface (center), and for values of r larger than this, total internal reflection occurs (right).

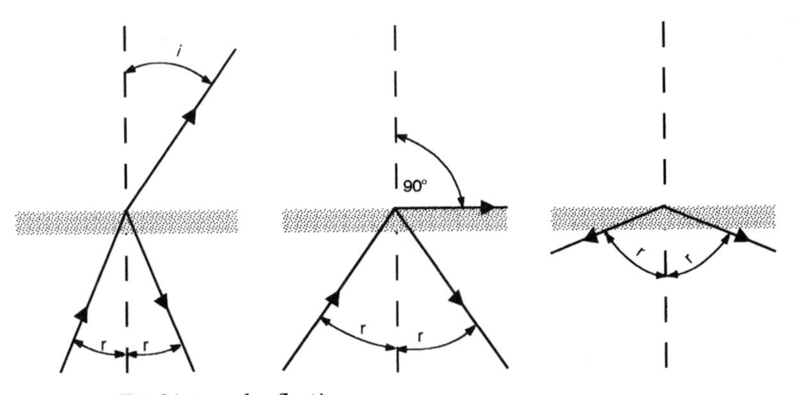

Figure 9.3 Total internal reflection.

(The letters I and r are used here to refer to the rays in air and glass, respectively, and no longer stand for the words "incident" and "refracted." The equation $\sin I = n \sin r$ still holds true.)

Dispersion

The fact that a prism of glass, or of any transparent material, will disperse white light into its constituent colors was first discussed in detail by Isaac Newton. The reason for this dispersion, which is illustrated in Figure 9.4, is that refractive index varies with wavelength, being greater for the short wavelengths at the blue end of the spectrum than for the longer wavelengths at the red end. The value quoted for the refractive index of a glass has no precise meaning, therefore, unless the wavelength of the light to which it refers is also given. When a single value is quoted, it is usually n_d, the value measured for the yellow light emitted by helium, whose wavelength is 5876 Å, or the mean value for the double yellow line of sodium at 5893 Å. Yellow is chosen because it is near the center of the visible spectrum, and the value for such a color is sometimes called the mean refractive index. Refractive index depends on color, so that a glass prism splits a ray of white light into its components. The angle between the extreme rays of the spectrum d is the angle of dispersion.

The difference in refractive index for two arbitrarily chosen wavelengths near the ends of the visible spectrum (red and blue light) is called the mean dispersion of the glass. Using n_F, the refractive index for the red line of hydrogen at 6563 Å, and n_C, the refractive index for the blue line of hydrogen at 4861 Å, to represent the ends of the spectrum,

$$\text{Mean dispersion} = n_F - n_C$$

For a light borosilicate crown glass with n_d given in a glassmaker's catalog as 1.5097, the mean dispersion is 0.0079. A particular double

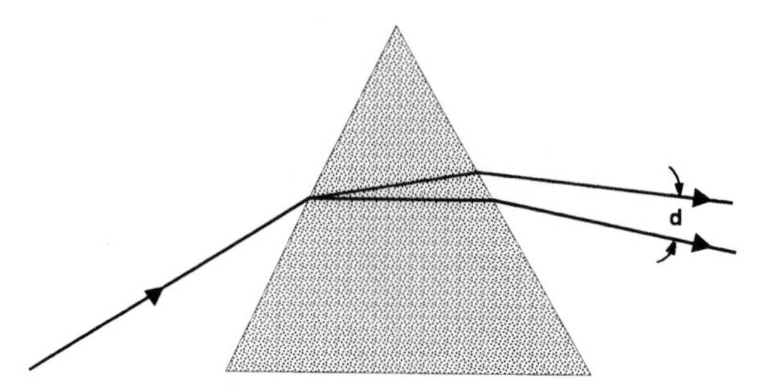

Figure 9.4 The dispersion of light.

extradense flint glass with $n_d = 1.9271$ has a mean dispersion of 0.0441, more than five times as great as the lighter glass. Generally, mean dispersion increases rather quickly as the refractive index goes up, but there is no simple rule relating these two quantities.

We have said that light can be considered as a form of wave motion. Without going into details, the simplest kind of light wave to picture (though not the most common in nature) is that of plane-polarized light, whose waves vibrate in a plane at right angles to the direction of travel, like a string whose particles move up and down when a wave travels along it. Normal glass is "optically inactive," which means that it does not preferentially transmit light with any particular plane of vibration, nor does it rotate this plane. This is what we expect from a liquid with its random internal structure. However, if glass is under compression in one direction, the molecules will be pushed together in that direction. A wave traveling with the plane of vibration parallel to the force will see the glass as denser and so having a higher refractive index than a wave whose plane is perpendicular to the force.

Chemical Stability

There is nothing in the glassy state as such that guarantees chemical stability, and many otherwise potentially useful glasses fail on this account. Soda-silica glass, for example, is quite soluble in water, and this "water glass" has very limited applications. Nonetheless, by proper choice of materials, glasses of very great chemical durability can be prepared. For applications where corrosion resistance at ordinary or at moderately high temperatures is required, glass is usually the material of choice.

Pure silica is attacked by alkaline solutions, however, and thus common (soda-lime-silica) glasses contain the seeds of their own chemical destruction. When these glasses are exposed to water, the water dissolves sodium ions from the surface of the glass to form alkaline sodium hydroxide. This in turn attacks the silica. Newly formed or freshly broken glass surfaces are rapidly attacked in this fashion. Workers of glass surfaces have put this effect to good use by wetting a glass before scribing it for "cutting." The chemical attack on the surfaces exposed by the scriber initiates regular cracks that assist in getting a clean break.

Common glass exposed to water develops a silica-rich (i.e., sodium-poor) surface layer that slows the attack to the point where it has no further consequences for most practical purposes. Considerable damage is done, however, and the fragility of common glass is largely caused by the numerous minute surface cracks and pits developed by this alkaline corrosion, using the glass' own alkali. We mentioned earlier that newly formed glass will support stresses of as much as 70,000 kg/cm^2, but for practical engineering purposes, a tensile strength of only 350 to 700 kg/cm^2 is assumed. Newly formed glass articles, or those kept in a controlled dry atmosphere, are much stronger than glass that has been exposed to a normal atmosphere, even for a few hours.

Glass "fatigues" under tension; that is, under static load, it will ultimately break at stresses lower than those that even aged glass would initially support, and it is the corrosion process just described that is responsible. The initial tension enlarges surface cracks, not enough to start a fracture, but enough to expose new surfaces to chemical attack. As the corrosion proceeds, the cracks open further and ultimately a crack that was progressing at the rate of fractions of a millimeter per hour by chemical erosion becomes a mechanical crack spreading at the rate of several thousand meters per second. This fatigue does not develop if the glass is tested in a vacuum or in an inert atmosphere.

For resistance to alkali, the use of boric and aluminum oxides in place of soda and lime is quite successful, and this aluminoborosilicate glass is used in critical applications, such as laboratory and industrial glassware, and for packaging delicate pharmaceuticals. It is fair to say that a special glass can be developed to meet almost any corrosion problem, but often only by sacrificing other desirable properties. The reverse is also true. Many of the most interesting and sophisticated optical glasses have very poor corrosion resistance and must be used in sealed cells where they are protected against atmospheric attack. As we should expect, the corrosion resistance of glasses decreases as the temperature rises, and it is particularly important to anneal glass in an atmosphere free from any dangerous contaminants.

Thermal Properties

The linear coefficient of thermal expansion is the fraction of original length at O°C by which material increases per degree centigrade rise in temperature. For glass, values range from 5.5×10^{-7} (5.5 parts per 10 million) for pure silica to as much as 125×10^{-7} for certain commercial glasses. The normal range is 60 to 90×10^{-7} for most glass. At temperatures over the transformation temperature of glass, the coefficients may increase as much as 50 percent, because configurational changes and normal thermal expansion occur.

Glass is a poor conductor of heat, and glass exposed to sudden changes in temperature may develop dangerous stresses, leading to fracture resulting from the temperature difference between the surface and the interior. For fairly obvious reasons, sudden heating, which puts the glass surface in compression, is less dangerous than sudden cooling, which puts the surface in tension.

For applications where thermal shock is anticipated, toughened glass offers some advantage, because the surface is already under compression. The problem is usually attacked more directly, however, by using glasses with low coefficients of thermal expansion. Pure silica is ordinarily too expensive (because it is so difficult to fabricate), but 96 percent silica glass has a coefficient of thermal expansion of only 8×10^{-7}, compared to 5.5×10^{-7} for pure silica, but, say, 90×10^{-7} for ordinary glass. For less demanding service, borosilicate glasses with a coefficient of 32×10^{-7} are satisfactory, and these are the basis of the well-known Pyrex brand of heat-resistant glassware. Where excellent corrosion resistance is to be combined with thermal shock resistance, aluminoborosilicate glass (42×10^{-7}) is used.

Electrical Properties

The materials that form glass readily are characterized by strongly directional, interatomic bonds. The outer electrons of the individual atoms are restricted to these bonds and thus are not free to conduct an electric current, as the bonding electrons in bulk metal do. Glasses, therefore, have very high electrical resistance, and for practical purposes are electrical insulators. On the other hand, a glass is a liquid, and common glasses contain metallic ions (Na^+, for example) that, if free to move, can carry a current. Common glass, then, is an insulator only insofar as its viscosity is so high that the metallic ions are bound in place in the glass and not free to move. The result is that the electrical resistance of glass decreases as the viscosity decreases, that is, as the temperature rises. At sufficiently elevated temperatures, a glass may carry an appreciable electrical current.

An immediate consequence of the conductivity of glass at high temperatures is that it is possible to heat glass by passing current through the molten material. Electrical melting of glass-forming materials may be used where the greatest care must be taken to avoid contamination and to control the process closely. The production of high-grade optical glass is a case in point.

The electrical properties of glass are of particular interest, of course, where glass is used in electrical devices that operate at temperatures above room temperature. Electron tubes, where the internal temperature may be several hundred degrees above the ambient temperatures, are an example. We cannot go into all the details that assume importance in this kind of application, but we will mention one point that bears on our general remarks about the nature of glass. As we have seen, the electrical conductivity of glass depends on the movement of metallic ions within the glass structure. Clearly, the closeness of this structure depends on the degree to which the configurational shrinkage previously discussed has proceeded. In making "pinches," which are the small bits of glass used to space the wires in an electron tube, the conductivity of the glass, which operates at an elevated temperature, assumes some importance. It is found that this conductivity may be reduced by a factor of 3 if the pinches are carefully annealed; the closeness of the carefully annealed structure decreases the mobility of the ions.

Conclusion

Understanding the various theories of the glassy state is critical for those individuals involved in developing new products for specific applications.

References

1. Maloney, E. J. T., *Glass in the Modern World,* Doubleday & Company, Aldus Books Limited, London, 1967.

Chapter

10

Mirrors

Joseph S. Amstock

President, Professional Adhesive and Sealant Systems

History and Introduction

The evolution of mirror use has been slow until our lifetime. The ancient Egyptians used highly polished bronze, gold, and silver as mirrors as early as 2500 B.C. The ancient Greeks also had metal mirrors, as did the Etruscans, by the sixth century B.C., and they were both copied by the Romans. At that early time, mirrors were not used to decorate a wall or a ceiling, but were usually small hand mirrors prized for their utility.[1]

The wall mirror, protected by a frame and used as a decoration, appeared in Europe in the fifteenth century, rather late in the history of mirrors. The manufacture of mirrors made of glass with a backing of tin and mercury amalgam began in Venice in the sixteenth century.

The Venetians were the chief suppliers of mirrors to Europe for several centuries. Originally, glass mirrors were called "looking glasses" to distinguish them from the already familiar metal "mirrors."

France led the way in making cast glass after the method was invented by Abbe Thevart in 1688. In England, glass casting was done to a limited extent as early as 1735, but large scale production was not introduced until 1773.

In America, only English blown plate glass was available until the late eighteenth century. After the American Revolution, cast glass was extensively imported from France and, to an increasing degree, from England. No plate glass was made in the United States by either blowing or casting until the mid-nineteenth century. The first silver plated mirrors are believed to have been developed by Professor

Justin Von Liebig of Germany, using a cold process, and by Thomas Drayton of England, using a hot process. Mirrors were originally made one at a time, being transferred by hand from one process table to another. As time went on, these process tables were heated and other improvements in production management took place.

Actually, the first major advance in mirror manufacturing came around the fourteenth century, when a "tin-mercury" process was used to bond thin tinfoil to glass with mercury. These were known as mercury mirrors.

You only have to go to an antique store to see beautiful examples of these mirrors which are still prized as lovely additions to any home. But today we have expanded from limited decorator use and the utilitarian application of mirrors to many uses with entire walls and ceilings mirrored in a variety of tints and colors. Today's mirror manufacturers have spent time and money to conquer many problems both in the mirror manufacturing process and the results of the finished product. These efforts have produced quality mirrors for a variety of uses.

Mirrors with a safety backing and acrylic mirrors are two of the more recent additions by the industry. Changes have brought about pleasing touches for every decor and architectural design with new ideas such as mirror-on-mirror combinations that can enhance commercial and home decoration.

The general public does not realize the work and research that have produced the mirror we know today. It was not until 1940 that new, faster-setting silver solutions were developed, using strong reducing agents which plated faster and made it possible to advance into today's more automated types of manufacturing. Along with mirrors, there are many new decorative glass items, such as reflective mirrors, spandrel glass, and highly polished ceramics which give us a new range of products for commercial decorating inside and out.

Support industries have gone through an evolution of their own. This has been true for manufacturers of basic glass, mirror-backing products, adhesives, hardware, sealants, and finishing tools.

All of the innovations in these areas have necessitated the use of more educational materials and seminars within the industry directed to architects, specification writers, designer/decorators, and contractors.

Acrylic Mirrors

Acrylic mirrors are being used in many ceiling applications because they are lightweight and will not break. But there are some guidelines you must follow when installing acrylic mirrors.

First, be sure that the adhesive used is compatible with acrylic mirrors. Many adhesives are not and will cause almost immediate damage to the mirror.

Another factor that should be considered with an acrylic mirror is that there are two types of acrylic mirrors. One is extruded and has stress strains that may cause the mirror to warp away from an adhesive or to curl before the adhesive can set. It is usually necessary to use mechanical fasteners with this product.

The other type of acrylic mirror is cell-cast acrylic. It has no stress strains, stays in the plane as applied with no curl, and has less distortion. When buying acrylic mirrors to be applied with an adhesive, be sure to specify a cell-cast mirror.

The application of the adhesive is another area that can cause problems with the new products used for substrates and the new variations being used in mirror installation with strips and overlays, just to name two. The adhesive manufacturer's directions must be followed for best results.

Just because an application has been done a certain way for the last 25 years does not mean it will necessarily continue to be done that way. The philosophy that when all else fails, read the directions is not the best policy. Read the directions first and then if you have reservations or questions, call the manufacturer.

Most important, use an adhesive that is formulated by technical research for mirror application. There are many products that will hold a mirror on the wall, but the length of time they will hold and the amount of damage they cause to the mirror are important considerations. Failure in this respect can occur in a few days or months, or it may be several years before failure occurs.

There are products being used by nonprofessionals that are not formulated for mirror installation or have not been properly tested, and they are causing almost immediate damage to certain types of mirrors. In other instances, it may take as long as seven or eight years before damage appears. This is the case where clear sealants believed to be safe have been used only to have mirrors ruined years later.

The industry needs to exercise care that mirrors are installed to accentuate their beauty and to provide trouble-free installation for a lifetime. After all, the public has been taught to use mirrors, and now we must be sure that installers use the best methods to apply them.

Method of manufacture

Mirrors are made through a process of successive layering of various materials to the back side of a sheet of glass. What makes a good mirror is high-quality materials and properly controlled application techniques which assure strong interlayer adhesion.[2]

Initially, the glass is scrubbed with a polishing compound to remove any particles or contaminants, generally a Lucor packing powder (adipic acid and Lucite beads). After a brief exposure to this solution, a 125-psi, high-pressure spray is used to rinse off the surface. The final cleaning stage is a precleaner. Then the glass is sensitized with a dilute solution of tin chloride.

Silver is applied to the glass by a chemical process which causes the precipitation of silver by mixing silver nitrate with a reducing agent. This ultrathin silver foil adheres to the glass.

Then the first metal layer is covered by a second metal layer of copper. This constitutes the mirror's first layer of protection. Copper can be applied in two different ways:

Chemically: By precipitating a solution of a copper salt mixed with a reducing solution as copper metal. This is basically the same process as described for silver. The copper metal produced by precipitation adheres securely to the layer underneath.

Galvanically: By spraying a solution of a copper salt mixed with a suspension of metal powder (usually zinc or iron) onto the silver layer that acts a conductor. Galvanic current develops between the silver layer and the metal powder, generating copper metal which adheres to the underlying silver layer. Excess chemicals and metal powders then are thoroughly rinsed from the glass with deionized water.

The glass is now dried and preheated and moved into the painting section. Painting may be done with a single or a double coat.

Double-coating: A first layer of paint is curtain coated over the copper.[2] The paint has the following characteristics:

1. It adheres firmly to the underlying copper layer.
2. It has strong resistance to corrosion.
3. It remains constantly elastic, a prerequisite for absorbing the consequences of cutting and the vibrations of the various processing operations.

Following the first application of paint, the glass is thoroughly dried in infrared ovens. A second paint then is curtain-coated or roller-coated onto the glass, and has the following characteristics and functions:

1. Good covering capability, required to fully close all pores and to insulate the underlying layers from the atmosphere.
2. Hard and able to resist scratches, mechanical aggression, and abrasion

The second coat of paint then is fully dried and cured.

Single-coating: Paint is selected which combines the characteristics of the two paints above. The paint is applied by a curtain or a roller coater, fully dried, and cured.[3]

The front side then is cleaned to eliminate smears and drops of either silver or copper. Finally, the mirror is washed, dried, and inspected for imperfections in the glass or the reflective coating.

Fabrication and handling

It is important to emphasize that care is taken during every step of fabrication to maintain the integrity of the back and edges of a mirror. Any major damage to these two areas will result in a useless product. Equally important, however, is cleanliness in the fabrication shop. Dirt, grit, solvents, and other contaminants can lead to damage to the surface and to the backing.

Always use gloves when handling mirrors. This protection works two ways. Hands are protected from sharp edges, and the edges and backing of the mirror are protected from body salts and chemicals.

Vacuum or sweep the cutting tables with a stiff brush regularly to keep dust down and to eliminate glass grit and particles which could scratch mirrors.

When grinding and polishing edges with a belt sander, always cool with an adequate flow of water. Grind slowly enough to keep the heat generated to a minimum.

When machine grinding, diamond wheels should always be dressed and maintained in good cutting condition. Set wheels so as not to grind excessively on the paint side, and edge grind in only one direction.

Diamond wheels should also be used with clean water or coolant as a lubricant. If coolant is used, it should be clean, pure water with a pH between 6 and 8.

Operators of diamond grinding equipment should be familiar with the risks of thermal, mechanical, and chemical aggression. Be sure that mirrors, especially backs and edges, are thoroughly washed after fabrication. Use only clean water for washing. If a glass-washing machine is used, employ a recommended mild detergent. No commercial glass cleaner is recommended because most contain ammonia or other strong chemicals which can damage the edges of the mirror.

After fabrication and cleaning of the edges, the glass shop should apply an approved sealant to all edges. This will provide additional protection against moisture, other degrading chemicals, or atmospheric penetration of the backing. Be sure to allow adequate drying time before rehandling.

Attaching the mirror

Mirror adhesives, now truly embraced by the mirror industry, have had to adapt to the new types of mirrors and new ideas in applying decorative mirrors, such as appliqués, overlays, and beveled strips which require close tolerances.

Some applications, for example, around hot tubs and swimming pools, require new methods of installation to prevent moisture from attacking the mirror or affecting the installation. Furniture, bifold doors, shower enclosures, and other mass production items are requiring faster-setting but mirror-safe adhesives.

There are four problem areas with mirror installations most often dealt with by the adhesive manufacturers. They are substrates, ceiling applications, the use of acrylic mirrors, and adhesive application.

Probably the most complicated of the four is the substrate. There are many new products being used as substrates, creating problems that must be addressed, and there are substrates used for years about which many installers still have questions.

Substrates made of wood products, which are manufactured in a variety of forms, can cause several problems. They are susceptible to taking on moisture and causing damage to the mirror backing, and they are treated with fire retardants that contain chemicals or salts that may also damage the mirror backing.

Wood products can also warp, causing adhesives to loose adhesion from the surface and allowing the mirror to fall. Keeping these facts in mind, if a wood product must be used, it should be sealed on both sides to provide a properly stable surface for the adhesive and to prevent moisture and chemical damage to the mirror, The product used in sealing should be compatible with the adhesive that is being used. An alternative, and by far the best substrate to use for a mirror installation, is moisture-resistant drywall, also called green board.

Finished wall covering products, such as wallpaper, may not be compatible with the mirror adhesive used, and the manufacturer should be consulted. It is usually best to remove the wall covering and seal the wall with a primer before applying the adhesive. Mirrors should not be applied directly to a painted surface because the bond of the adhesive will be no better than the bond of the paint.

Masonry walls, such as concrete block or brick, should be sealed to provide a good surface for the adhesive and to prevent alkalies, salts, or moisture from damaging the minor. Again, the product used to seal the wall should be compatible with the adhesive.

Vitreous brick, glazed tiles, or glass are easier surfaces to deal with, as is metal. These surfaces must be clean and dry but do not need to be sealed.

In recent years, ceiling applications have been controversial. Many installers will not even discuss the project or offer any warranties. There are different ways to install ceiling mirrors that are safe, produce an end result that the architect, decorator, and owner are looking for, and still have the installer feel comfortable with the application. This can be accomplished with an adhesive specified for a given installation. The key to this kind of installation is proper application of the adhesive and the correct substrate preparation. Another method would be a combination of mechanical fasteners with an adhesive, and a final option would be using premounted mirrors or panels in a track system or dropped into a grid system.

Again, the installer must be comfortable with what has been done from a safety standpoint. If a safety film is on the back of the mirror, use an adhesive that will not damage or dissolve the safety film.

Transportation, receiving, and storage

Every time a mirror crate or an open mirror is moved, there is potential for damage. Therefore, the key to successful handling is to keep movements to a minimum. Plan your storage in an efficient manner. Use proper handling techniques and equipment.[3]

Before unloading the truck, verify the weight of the cases, and confirm that the handling equipment is adequate. Check your shipments on arrival. If there appears to be moisture present, the mirrors should be unpacked and allowed to dry using a separating technique. Do not allow mirrors to remain in contact with wet protective pads. Be sure that your mirror storage areas are in dry, adequately ventilated spaces and do not store mirrors in areas of high humidity, where exposed to chemical fumes, or near high heat, such as steam or water pipes. These conditions can cause staining or deterioration of the mirror edges.

Do not store mirrors outdoors or in unheated areas. The mirror can be affected by the moisture prevalent under these conditions and by excessive expansion and contraction caused by cyclic temperatures. Block mirror cases up off the floor to prevent any water damage to the bottoms of the cases. Also, do not store crates or mirrors on uneven surfaces. This can lead to stresses and strains on the mirrors which can lead to cracks on the surface or in the backing.

Store mirrors vertically, not horizontally. Even minor movements of a flat stack of mirrors will cause abrasion if not breakage. Never slide a mirror across the one next to it.

Do not move or reship partially unpacked mirror cases without proper repacking. Movement within the case can cause damage or breakage. If mirrors are transported in an open or exposed condition and become spattered or come in contact with foreign elements such as road salt, they should be washed with warm water and dried with a soft rag.

Thermal aggression in edge grinding

Three decades ago, glass was cut, shaped, and fabricated prior to silvering, sparing the silver and paint from the tortures of edge grinding and providing the finished edges of each piece with a protective coat of paint. Today's mass production techniques, while lowering the cost of mirror production, have reversed this otherwise favorable method. Glass is now silvered in sheets which are later cut, leaving the silver's edge exposed and vulnerable to numerous forms of destructive aggression.

During edge grinding, the generation of excessive frictional heat is a major cause of edge failure or "black edge." When heated, the silver and copper layers expand at a faster rate than the glass, eventually detaching from the glass. Although not always immediately visible, this weakened bond will facilitate oxidation of the silver and a blackening of the mirror's edge.

Excessive heat buildup in grinding can be caused by any of the following factors:

1. Low concentration of diamond in the grinding wheels
2. Excessively hard binder in the grinding wheel
3. Incorrect peripheral speed of the grinding wheel
4. Excessive grinding pressure
5. Excessive feed speed (bringing the glass to the wheel at a faster rate than the wheel can cut)
6. Coolant water too hot to cool adequately (usually resulting from too small a holding tank or too low a coolant concentration)
7. Inadequate or inaccurate delivery of the coolant to the point of friction

Mechanical aggression in edge grinding

Other than excessive heat buildup, improper grinding techniques may result in mechanical aggression, equally dangerous to the mirror's edge. When portions of the metal layer are exposed even in microscopic amounts, the silver begins to oxidize, initiating the gradual process of black edge "creep." Too coarse a diamond grain is a common occurrence; sometimes coarse grain, rather than a higher diamond concentration, is purposely used to increase the capability of the wheel to remove stock. Coarse grain does not "cut" but rather "tears" off metals (silver and copper), thereby producing chips on the edge of the sheet.

Imbalances of the wheel (or of the spindle) impart infinite incisions on the mirror edge. These repeated incisions, impacting head-on onto the superposed layers, can cause the silver to become detached from the glass or the paint to become detached from the copper and are a result of an excessively deep groove of a peripheral wheel or improper inclination of a cup wheel. The more metal which remains unprotected, the more contact with the atmosphere will promote deterioration. In other words, the edging process should reveal the metal layers as little as possible.

Chemical aggression

Although today's mirror backing paints are engineered to resist many forms of chemical attack, there is still a limit to what they can endure. Furthermore, it must not be forgotten that the edge itself is not protected by the backing paint. Following are some of the more typical chemicals likely to attack mirrors:

1. Ground glass raises the pH of coolant, and therefore, recirculated coolant should be frequently changed (pH should be maintained in a range from 6 to 8).
2. Adhesives and mastics typically contain solvents, most of which can harm mirror backing. Care must be taken to select adhesives with noncorrosive solvents and to apply them so as to allow the solvent to escape.
3. Adhesives can also leach out harmful chemicals from substrates, such as gypsum board and particle board. All substrates should be sealed prior to mirror installation.
4. Glass cleaners may contain ammonia or other chemicals harmful to the silver. They should never be allowed to spill or run over onto the mirror's edge.
5. Airborne acids, alkali, and solvents are frequently present on construction sites. Care should be taken to avoid them, and mirrored rooms should remain well ventilated.

Cleaning

The "final touch" on any outstanding mirror installation is proper cleaning. The techniques described here are good practices to be followed and passed on to the end user to maintain the mirror.

The best and safest cleaner for a mirror is clean warm water applied with a soft cloth. Be careful not to allow the edges of the mirror to become or remain wet over a period of time and do not use any acid or alkali cleaners for mirror cleanups after installation. They can attack both the surface and edges as well as the backing of the mirror. Never use an abrasive cleaner on the surface.

Remove surface marks or stubborn dirt with oil-free steel wool, not solvents which can attack and damage the edges and backing.

Never spray any cleaner directly onto a mirror. Instead, apply the cleaner to a soft cloth and then wipe the mirror, as this will also prevent "puddling" at the mirror edge where the cleaner can attack the backing.

Be sure to dry all joints and edges thoroughly so that no cleaner contacts the edge and backing.

Be sure to let your building owners know that routine cleaning maintenance can be accomplished simply and effectively by washing, rinsing, and drying the mirror.

Installation

The best mirror installation is one that is striking in appearance and trouble-free during installation. Proper techniques, carefully and professionally employed, can virtually guarantee this result. Always use gloves when handling any mirror to prevent damage to the face or backing from skinborne salts or chemicals.

Never install mirrors on new plaster, new masonry, or on a freshly painted wall without proper sealing. Also, do not install in any new construction area where airborne solvents or heavy-duty cleaners or chemicals are in the air.

Never install mirrors outdoors without additional engineered protection for the backing of the mirror. The mirrors should have breathing space behind them when installed to prevent moisture entrapment during and after installation. Never install a mirror in contact with a splash board or sink back. Insist on at least $^{25}/_{64}$ in (10 mm) of space between the bottom edge of a mirror and other surfaces to prevent moisture entrapment and permit drainage.

Whenever possible, use mechanical means of installation, such as J-mouldings, clips and screws, or framing, in preference to tapes, adhesives, mastics, etc. J-mouldings should have weep holes, and mirrors should always have $^{1}/_{8}$-in (3-mm) neoprene setting pads between the mirror and clip or moulding used. Figure 10.1 illustrates the proper installation of a mirror with mastic and J-Bars. Figure 10.2 covers the installation with mastic and L-Bars.

If mechanical means are not acceptable or practical, choose an adhesive system. Be sure materials selected are compatible with the mirror backing and avoid using materials containing strong, harsh, or corrosive solvents or acids. Acetone, toluol, methylene chloride, acetic acid, or acids of any nature can severely damage mirror backings.

When adhesives are used, do not apply dollops of material; when placed against a wall, they flatten to a larger, pancake-size diameter. Perimeters dry relatively fast. This normally traps solvent, curing additives, and potential chemical attackers of the mirror centrally, which can detract from overall adhesive strength.

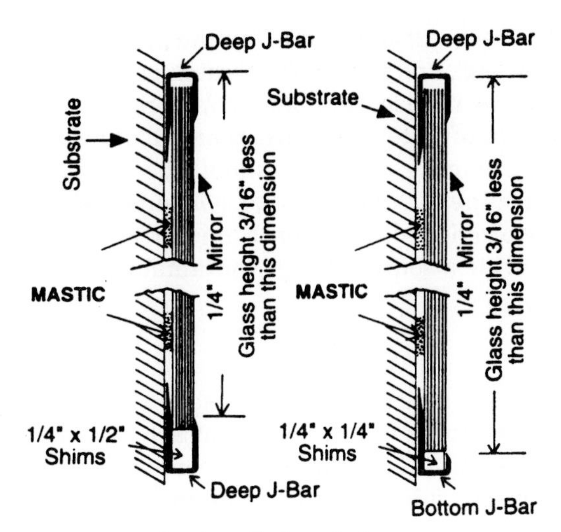

Figure 10.1 J-Bar installation of mirror.

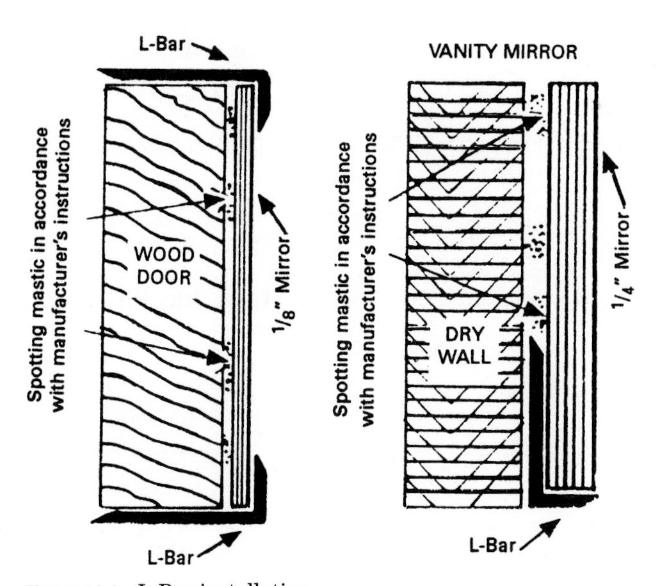

Figure 10.2 L-Bar installation.

Adhesives should be applied in a straight line with ⅛- to ½-in (10- to 13-mm) width beads. Beads should be vertical when installed. Do not loop or crisscross beads which create entrapped areas that prevent venting. Vertical application assures that heavier-than-air or lighter-than-air fumes escape. Caulking gun application is ideal. It allows adhesive application without mechanically contacting the mirror backing with trowels, putty knives, etc., and minimizes scratch potential.

Cleaning of installed mirrors is also very important. Solid soils, such as paint, excess edge sealants, felt buttons, tape, or adhesives should be razor-bladed off. Cleaning solutions should be mild and preferably free of ammonia, vinegar, bleaches, or solvents. Do not spray directly on mirrors. Spraying directly on a mirror can allow material to run down and provide a source of edge contamination.

Conclusion

Working with mirrors is always a delicate task because of the nature of the product. It is one of the features in buildings that are most visible. Always maintain extreme care and caution when handling and installing mirrors. Eventually, the mirrors will end up in showrooms, shops, and homes around the world.

References

1. Palmer–Ball, S., Adhesive Use with Today's Mirrors, *U.S. Glass, Metal, and Glazing,* October 1990.
2. Czopek, J., The Mechanics of Making a Good Mirrors, *Glass Magazine,* December 1995.
3. Courtesy of North America Association of Mirror Manufacturers, Potomac, MD, 1996.

Trademarks

Sunbelt™—A registered trademark of PPG Industries, Inc. Pittsburgh, PA.

Chapter

11

Fabrication Equipment

Joseph S. Amstock

President, Professional Adhesive and Sealant Systems

Introduction

Perhaps the most notable development in the equipment market in recent years has been the response of the small glass shop to the demands of a more discerning and critical public. The corner glass shop has found that customers seek much more than replacement windows. They are looking for table tops, beveled mirrors, mirrored walls, and more complicated cuts in glass.[1]

Responding to this interest in decorative glass and glass that is functional and, more important, economical, the glass entrepreneurs have sought out equipment for custom and production needs. Manufacturers have scaled down some of their larger equipment and have begun more vigorous promotion of the smaller variety.

Data Availability

The glass industry, trade journals, and the many associations offer monthly publications with a wealth of information for the glass supplier. By constantly checking past and present publications, you are in a position to compare. Once that is accomplished, you are in a position to contact the equipment supplier and solicit quotations. The equipment and machinery distributors can also present new insight to the business venture with discussions on space allocation and placement of equipment for improved production and future needs. It is most important that the machinery supplier visit the intended facility to understand your goals.

Types of Equipment

Bevelers and edgers

The rush to provide a medium-size small shop beveler has not been uniformly successful. It is a delicate piece of equipment and does not lend itself well to size reduction. Medium-sized machines designed for specific types of work but not necessarily for high production output have been successful. Other types of bevelers are shape bevelers used for specific shapes, double edge bevelers, and custom fabricated machines. Table 11.1 offers a list of several manufacturers of edging and beveling machines.

Beveling machines generally use three different types of wheels to create a polished bevel on a glass surface:

1. Metal-bonded diamond wheels which cut and remove glass from the surface to achieve the desired bevel width and angle. A diamond coolant is used to reduce heat and add lubricity to this operation.
2. Resin-bonded diamond wheels, sometimes called polishing wheels, which remove coarse wheel marks left by the metal-bonded diamond wheels. A diamond coolant is also used in this operation. After the glass passes through this wheel section, it has an opaque or frosted surface where the bevel angle has been cut. Even though the frosted surface appears smooth, a microscope reveals many jagged protrusions on the glass.
3. Natural or synthetic felt wheels are used in the final section of the polishing operation and are the wheels that must polish the frosted surface to an optically clear finish. The finish is achieved by introducing a cerium oxide slurry, a fine optical abrasive powder mixed with water, onto the wheel where it contacts the beveled glass surface. Felt wheels cannot polish by themselves. They are merely efficient carriers of abrasive particles. As slurry is introduced, polishing compound is captured in the wheel's fibrous matrix. Under pressure, these wheels rotate between 1,000 and 4,000 rpm in carrying the cerium oxide across the beveled area. The resulting chemical and mechanical interaction polishes the bevel's surface.

Wheel categories

Solid wheels are harder and more durable but also more expensive. On machines that require a thin wheel wall, i.e., less than 25 mm of material between the wheel's outside and inside diameters, a solid wheel offers more dimensional stability than can be expected from the spiral type. Additionally, they polish more aggressively because of their hardness and ability to operate under greater pressures. Solid wheels can have a lower cost per polished inch of glass. One major drawback is

TABLE 11.1 Edging and Beveling Machinery

Name and address of company	Machine name	Min/Max thickness, in	Min/Max width/ length, in	Machine speed	Type edging	# spindles	Straight line or shapes	Size, in	Other
Alpha Glass Machinery, Inc. Woodbridge, Ont., Canada	ADA 1403—automatic straight line edger	$\frac{1}{8}$ to $\frac{1}{2}$ (3 to 12 mm)	Max. sheet size 98×98 (2.5×2.5 m)	$27\frac{1}{2}$ to 150 in/min (0.7 to 3.9 m/min)	Pencil	3	Straight line	$216\frac{1}{2}$ in long× $37\frac{1}{2}$ wide×98 in high	Weight 2,822 lb
	ADA 2401—automatic straight line edger	$\frac{1}{8}$ to $\frac{1}{2}$ (3 to 12 mm)	Max. sheet size 98×98 (2.5×2.5 m)	20 to 216 in/min (0.5 to 5.5 m/min)	Flat (with arrises)	4	Straight line	$216\frac{1}{2}$ in long× $37\frac{1}{2}$ wide×98 in high	Weight 2,756 lb
	ADA 2601—automatic straight line edger	$\frac{1}{8}$ to $\frac{1}{2}$ (3 to 12 mm)	Max. sheet size 118×118 (3×3 m)	20 to 196 in/min (0.5 to 5.0 m/min)	Flat (with arrises)	8	Straight line	$277\frac{1}{2}$ in long× $37\frac{1}{2}$ wide×$98\frac{1}{2}$ in high	Weight 4,630 lb
	ADA 2606—automatic straight line edger	$\frac{1}{8}$ to $\frac{3}{4}$ (3 to 20 mm)	Max. sheet size 118×118 (3×3 m)	20 to 196 in/min (0.5 to 5.0 m/min)	Flat (with arrises)	8	Straight line	277 long× 47 wide×102 in high	Weight 4,299 lb
	ADA 3000—automatic straight line edger	$\frac{1}{8}$ to $1\frac{3}{16}$ (3 to 30 mm)	Max. sheet size 118×118 (3×3 m)	20 to 196 in/min (0.5 to 5.0 m/min)	og, pencil	10	Straight line	299 long× 47 wide×102 in high	Weight 6,834 lb

TABLE 11.1 Edging and Beveling Machinery (*Continued*)

Name and address of company	Machine name	Min/Max thickness, in	Min/Max width/ length, in	Machine speed	Type edging	# spindles	Straight line or shapes	Size, in	Other
Alpha Glass Machinery, Inc. Woodbridge, Ont., Canada	ADA 7000—automatic straight line edger	$\frac{5}{32}$ to $\frac{3}{4}$ (4 to 20 mm)	Max. sheet size 118×118 (3×3 m)	20 to 196 in/min (0.5 to 5.0 m/min)	Flat and mitre bevel	8	Straight line	299 long× 47 wide× 102 in high	Weight 6,830 lb
	ADA 3503—straight line edger and mitre beveler	$\frac{1}{8}$ to 2 (3 to 50 mm)	Max. sheet size 118×118 (3×3 m)	16 to 197 in/min (0.4 to 5.0 m/min)	Flat and mitre bevel	12	Straight line	315 long× 53 wide× 104 in high	Weight 7,940 lb
	ADA 3504—straight line edger and mitre beveler	$\frac{1}{8}$ to $\frac{3}{8}$ (3 to 35 mm)	Max. sheet size 118×118 (3×3 m)	12 to 126 in/min (0.3 to 3.2 m/min)	Flat (with arrissed edges)	12	Straight line	335 long× 43 wide× 110 in high	Weight 8,380 lb
	ADA 4008—automatic straight line edger	$\frac{1}{8}$ to $\frac{3}{4}$ (3 to 20 mm)	Min. width 12 max. width 116	24 to 240 in/min	Beveled	10 on each side (total 20)	Straight line	197 long× 197 wide	Weight 9,260 lb
	ADA 5003—automatic straight line beveler	$\frac{1}{8}$ to $\frac{3}{4}$ (3 to 20 mm)	Max. sheet size 118×118 (3×3 m)	20 to 102 in/min	Flat, og, pencil, beveled	7	Straight line	238 long× 45 wide× 98 in high	Weight 7,050 lb

	ADA 2001—automatic shape edger	$1/8$ to $1/2$ (3 to 12 mm)	59 diameter (1.8-m diameter)	Variable	Pencil, flat (with arrises)	1 (3 stacked edging wheels)	Shape	110 long (fully opened)× 63 wide× 63 high	Weight 1,170 lb
	ADA 2003—automatic edger with template	$1/8$ to $3/4$ (3 to 20 mm)	Max. 71 diameter (1.8-m diameter)	39 to 394 in/min on periphery	Pencil, flat (with arrises)	2 (3 stacked edging wheels)	Shapes	110 long (fully opened)× 98 wide× 79 high	Automatic and manual operations
	Angoglass-corner edging machine	Up to 1 (26 mm)	Any size	Several seconds per pass	Pencil, flat (with arrises)	1 (3 stacked edging wheels)	Corner	80 long× 72 wide× 64 high	Weight 1,430 lb
BGM Diamond Tools, Inc. Greensboro, NC	MB-MBP double edger	All	N/A	0 to 400 in/min	Pencil, edge, flat	Variable	Straight line	N/A	
	TB/64 beveler/edger	$1/8$ to $5/8$ (3 to 15 mm)	N/A	0 to 200 in/min	Bevel/edge	6	Straight line	N/A	
	PR/88 edger	$1/8$ to $1\,1/4$ (3 to 31 mm)	N/A	0 to 240 in/min	Flat	8	Straight line	N/A	
	TR/87 edger	$1/8$ to $1\,1/4$ (3 to 31 mm)	N/A	0 to 280 in/min	og	8	Straight line	N/A	
	PR/54 edger	$1/8$ to $3/4$ (3 to 15 mm)	N/A	0 to 240 in/min	Flat	5	Straight line	N/A	

TABLE 11.1 Edging and Beveling Machinery (*Continued*)

Name and address of company	Machine name	Min/Max thickness, in	Min/Max width/ length, in	Machine speed	Type edging	# spindles	Straight line or shapes	Size, in	Other
BGM Diamond Tools, Inc.	Mini PG/ 15-TM edger	$\frac{1}{8}$ to $\frac{5}{8}$ (3 to 15 mm)	N/A	0 to 240 in/min	Pencil edge	3	Straight line	N/A	
Greensboro, NC	CAT 250/4M groover	$\frac{1}{8}$ to 1 (3 to 20 mm)	N/A	20 to 75 in/min	Plate groove	4	Straight line	N/A	
	CR/1111 edger	$\frac{1}{8}$ to $1\frac{1}{4}$ (3 to 31 mm)	N/A	0 to 200 in/min	Flat	11	Straight line	N/A	
	Max 50 TMP beveler	$\frac{1}{8}$ to 1 (3 to 25 mm)	N/A	Up to 160 in/min	Bevel	13	Straight line	N/A	
Bilco Manufacturing, Inc. Zelienople, PA	Straight line edge seaming machine	$\frac{1}{16}$ to $\frac{3}{4}$	18×18 in to 84×120 in	Size dependent	Seaming prior to tempering	4	Straight line	N/A	
Contract Developments & Projects (Leeds) Ltd. Leeds, West Yorkshire, England	Two-spindle grinding and polishing machine (round edge and polish or flat and two arrises)	2 mm to 10 mm	N/A	N/A	Flat or round edge and polish	2 (1 grinding, 1 polishing)	Straight line	N/A	
Downey Glass Co. Inc.	Bevelers	$\frac{1}{8}$ to 1 in	2×2 in to 72×100 in	N/A	Beveled	N/A	Straight line	N/A	
Los Angeles, CA	Flat polishers	$\frac{1}{8}$ to 1 in	2×2 in to 96×120 in	N/A	Flat polishers	N/A	Straight line	N/A	

	Pencil edgers	S.S. to $\frac{1}{2}$	2×2 in to 72×100 in	N/A	Pencil	N/A	Straight line	N/A	
	Shape edger and bevelers	$\frac{3}{16}$ to 1 in	20- to 88-in diameter	N/A	Flat, og, pencil, bevel	N/A	Shape	N/A	
Glassline Corp. Perryburg, OH	CNC-shaped glass grinding machines RS 8.14	1.7 to 12 mm	8×12 in up to 52×84 in	Up to 400 in/min	Penciled, seamed	1 or 2	N/A	N/A	Complete digitizing system included; digitizer and complete processor are available on model
	Rotary edge grinder	1.7 to 12 mm	2×4 in up to max. of 88-in diagonal	400 in/min	Seamed and pencil edge	1 to 4 spindles	N/A	4 different models	Used in automotive, appliance, and furniture industries
	Vertical edge grinding machine	2 to 19 mm	Depends on model	Depends on model and thickness of glass	Flat, arris, pencil	Depends on model	N/A	N/A	Glassline can supply vertical washers
	KR302 round and oval grinder	3 to 19 mm	8 to 60 in	Depends on shape and thickness	Pencil edge	1	N/A	N/A	For grinding furniture glass in rounds and ovals
	Double edge grinders	3 to 25 mm	3×3 in up to 140 × 140 in	1000 in/min	Seam, pencil, flat, arris	2 to 10	N/A	12, 40, 60, 80, 100, 120, 140 in	Various models available

TABLE 11.1 Edging and Beveling Machinery (*Continued*)

Name and address of company	Machine name	Min/Max thickness, in	Min/Max width/ length, in	Machine speed	Type edging	# spindles	Straight line or shapes	Size, in	Other
C.R. Laurence Co., Inc. Los Angles, CA	EP 3000 two-spindle edger	$\frac{1}{8}$ to $\frac{1}{2}$ in	$4\frac{3}{4}\times4\frac{3}{4}$ in to 60×100 in	20 to 132 in/min	Pencil and flat with arris sides	2	N/A	162 in long× 48 in wide× 84 in high	Warranty program, installed by service techs
	AL 4 four-spindle edger machine	$\frac{1}{8}$ to $\frac{1}{2}$ in; as an exception up to $\frac{3}{4}$ in	Min. 2×6 in; max 84× 120 in	32 to 157 in/min standard; can be varied upon request	Flat with arris sides	4	N/A	209 in long $\times30\frac{5}{8}$ in wide× $86\frac{5}{8}$ in high	Warranty program, installed by service techs
	AL 5 five-spindle edging machine	$\frac{1}{8}$ to $\frac{1}{2}$ in; as an exception up to $\frac{3}{4}$ in	2×6 in	32 to 157 in/min; other on request	Flat with arris sides	5 (1 diamond, 2 seaming, 2 polishing)	N/A	217 in long× $39\frac{5}{8}$ in wide $\times86\frac{5}{8}$ in high	Warranty program, installed by service techs
	AL 6 six-spindle edger	$\frac{1}{8}$ to $\frac{3}{4}$ in	2×6 in	32 to 157 in/min; other on request	Flat with arris sides	5 (2 diamond, 2 seaming, 2 polishing)	N/A	229 in long× $39\frac{5}{8}$ in wide $\times86\frac{5}{8}$ in high	Warranty program, installed by service techs
	AL 8 eight-spindle edger	$\frac{1}{8}$ to 1 in; as an exception up to $1\frac{3}{16}$ in	2×6 in	32 to 157 in/min; other on request	Flat with arris sides	5 (3 diamond, 2 seaming, 3 polishing)	N/A	256 in long× $39\frac{5}{8}$ in wide $\times86\frac{5}{8}$ in high	Warranty program, installed by service techs
	Shape beveling machine	$\frac{1}{8}$ to 1 in	6×6 in min; up to 8×8 ft; circle diameter 16 in, max 102 in	N/A	Beveled, pencil, flat	1	Distance 8 ft		Manual and automatic models available

John De Gorter Inc. Roslyn, NY	Schiatti-FPS-10	$1/8$ to $3/8$ to $3/4$ in	4×6 in to 100×120 in	20 to 120 in/min	Flat	6	Straight line, no circles	N/A	
	Schiatti-BFT double edger	$1/8$ to $1/2$ in	Min. 4 in, max. 100 in	25 to 280 in/min	Flat or pencil	2 to 12	Straight line	N/A	
	Schiatti-FPS-15	$1/8$ to $5/8$ to $3/4$ in	4×6 in to 100×120 in	20 to 120 in/min	Flat	8	Straight line, no circles	N/A	
	Schiatti-FPS 20/4L	$1/8$ to 1 in	4×6 in to 120×140 in	20 to 120 in/min	Flat	11	Straight line, no circles	N/A	
	Schiatti-FPS-50	$1/8$ to 2 in	4×6 in to 120×140 in	20 to 120 in/min	Flat	13	Straight line, no circles	N/A	
Hilemn Machinery Sales Ltd. Greensboro, NC	Besana R6 og	$1/8$ to $3/4$ in	Min. 6 mm glass, 4×8 in; min. 19 mm glass, 8×8 in	20 to 180 in/min	og and pencil	6	Straight line	N/A	2 in diameter stainless steel spindles
	Besana R6 Cup	$1/8$ to $3/4$ in	Min. $4\frac{3}{4}$ in	21 to 106 in/min	Flat	6	Straight line	$19\frac{3}{4}$ ft long× 4 ft wide× $8\frac{1}{4}$ ft high (rack only)	Automatic self-adjusting spring conveyer
	Besana R8 Cup	$1/8$ to $3/4$ in	$4\frac{3}{4}$ in	26 to 132 in/min	Flat	8	Straight line	23 ft long× 4 ft wide× $8\frac{1}{4}$ ft high (rack only)	Designed for heavy glass production with a polished flat edge

TABLE 11.1 Edging and Beveling Machinery (*Continued*)

Name and address of company	Machine name	Min/Max thickness, in	Min/Max width/ length, in	Machine speed	Type edging	# spindles	Straight line or shapes	Size, in	Other
Hilemn Machinery Sales Ltd. Greensboro, NC	Besana R6V	$\frac{1}{8}$ to $\frac{3}{4}$ in	Min. 6 mm glass, 4×8 in; min. 19 mm glass, 8×8 in	20 to 180 in/min	Pencil and flat	8	Straight line	N/A	Versatile combination pencil and flat edger
	Besana R9	$\frac{1}{8}$ to $1\frac{1}{2}$ in	N/A	20 to 160 in/min	Flat	9	Straight line	N/A	Oscillating diamond wheel group for laminated and heavy glass
	Besana R11	$\frac{1}{8}$ to 1 in	$3\frac{1}{4}$ in	20 to 106 in/min	Flat	11	Straight line	30 ft long× 4 ft wide× $8\frac{1}{4}$ ft high (rack only)	Designed using two felt wheels with cerium oxide for a brilliant polished edge
	Besana R13/TS	3 to 30 mm	Min. 15 to 30 mm, 130 mm square; min. 3 to 10 mm, 100 mm square	N/A	Flat, seams, mitering	13	Straight line	10 m long× 150 m wide× 2.5 m high	Digital display
	Besana horizontal double edger	$\frac{1}{8}$ to $\frac{1}{2}$ in or $\frac{3}{4}$ in	Min. 8×8 in; max. 63 in×126 in	27 to 355 in/min	Flat or pencil	Varies with need	Straight line	N/A	Designed to meet customer's needs
	Lovati 1000	$\frac{3}{16}$ to 1 in	Min. 19 in diameter; max. 82 in diameter	N/A	Edging and beveling	1	Shape	53 in long× 67 in wide× 69 in high	Bevel angle 0 to 15 degrees; width 5 to 35 mm

Hilemn Machinery Sales Ltd.	Lovati 15/AF	$^3/_{16}$ to $^3/_4$ in	12- to 60-in diameter	N/A	Pencil	4	Shape	N/A	Most effective pencil edging $^1/_4$ to $^1/_2$ in glass
Greensboro, NC	Lovati 20/AF	$^3/_{16}$ to $1^5/_8$ in	20 to 88 in; length/width ratio 2:1	1 to 20 r/min	Pencil, flat, og	5	Shape	$6^1/_2$ ft long × $11^1/_2$ ft wide × $8^1/_2$ ft high	Fully automatic production machine
	Lovati 152AB	$^3/_{16}$ to $^1/_2$ in	13 to 64 in; length/width ratio 2:1	1 to 12 r/min	Beveling	5	Shape	144 in long × 80 in wide × 78 in high	Four beveling heads, one edging head
	Lovati 152ABF2	$^3/_{16}$ to $^5/_8$ in	13 to 64 in	1 to 12 r/min	Beveling	6	Shape	152 in long × 80 in wide × 78 in high	Four beveling heads, two edging heads
	Lovati Bilux A	$^3/_{16}$ to $^3/_4$ in	13 to 64 in	1 to 12 r/min	Bevel polishing	2	Shape	88 in long × 80 in wide × 78 in high	
	Lovati Bilux A4	$^3/_{16}$ to $^3/_4$ in	13 to 64 in	1 to 12 r/min	Bevel polishing	4	Shape	104 in long × 88 in wide × 78 in high	
	Lovati 202 AB	$^3/_{16}$ to $^3/_4$ in	19 to 88 in	1 to 12 r/min	Bevel	5	Shape	208 in long × 100 in wide × 80 in high	Four beveling heads, one edging head
	Lovati 202 AB6	$^3/_{16}$ to $^3/_4$ in	19 to 88 in	1 to 12 r/min	Bevel	6	Shape	240 in long × 100 in wide × 80 in high	Five beveling heads, one edging head
	Lovati Bilux SA	$^3/_{16}$ to $^3/_4$ in	19 to 88 in	1 to 12 r/min	Bevel polishing	2	Shape	120 in long × 76 in wide × 76 in high	

TABLE 11.1 Edging and Beveling Machinery (*Continued*)

Name and address of company	Machine name	Min/Max thickness, in	Min/Max width/ length, in	Machine speed	Type edging	# spindles	Straight line or shapes	Size, in	Other
Hilemn Machinery Sales Ltd. Greensboro, NC	Lovati 301/FB8	$\frac{1}{8}$ to $\frac{3}{4}$ in	4 to 16 in	12 to 40 in/min	Bevel	7	Shape	100 in long× 112 in wide× 84 in high	Completely automatic with eight positions
	Lovati 301 Lux/4	$\frac{1}{8}$ to $\frac{3}{4}$ in	4 to 16 in	12 to 40 in/min	Bevel polishing	3	Shape	88 in long× 104 in wide× 84 in high	Ideal complement for 301 FB8
	Maclav Alba	2 to 15 mm	Min. 20 mm square; max. 600×2000 mm	0 to 5 m/min	Bevel	9	Straight line	6400 mm long ×2000 mm wide× 2700 mm high	Patented vacuum transport system
	Bensana R3	$\frac{1}{8}$ to $\frac{3}{4}$ in	N/A	0 to 250 in/min	Pencil	3	Straight line	N/A	52 lb spindles
	TAIA B.S.– Coopmes	Max. cutting head travel 3.210 mm	Max. locator travel 2000 mm	N/A	N/A	N/A	N/A	N/A	For single and laminated sheets
	Microbo– Coopmes	Working cut 370×250 cm	Glass thickness 2 to 12 mm	N/A	N/A	N/A	N/A	300 cm long× 425 cm wide× 130 cm high including bridge	Hydraulic tilt, air flotation
Lo-Glass Machinery Co. Atlanta, GA	Lustre-Kwik I and II	$\frac{1}{8}$ to $\frac{3}{4}$ in	N/A (open top)	40 to 200 in/min	P.E. and F.E.	2	Straight line	160 in wide× 36 in deep× 78 in high	Now offering two-cup wheel edger with two arms

Orient Glass Inc. Clark, NJ	Bando B-100 beveler	Min. 1/8 in; max 3/4 in	Min 2×2 in; max weight up to 120 kg/m	Up to 5 m/min	Bevel width up to 1 5/8 in	10	Straight line	N/A
Powergrind Ltd. Buckinghamshire, England	MULTIRAD	2 to 12 mm	100 to 2000 mm	4 m/min	Flat and og	1	Straight line and shaped	N/A
	TEAMSTER–COLT	2 to 12 mm	100 to 2500 mm	0 to 10 m/min	Flat and og	8	Straight line	N/A
Salem Distributing Co. Inc. Winston–Salem, NC	Zanetti FP/10	1/8 to 3/4 in	3×8 in	2 to 19 ft/min	Flat and seam	10	Straight line	4 ft long×27 ft wide×9 ft high; 3980 lb

that their thickness is usually limited to below 40 mm. The manufacturing process required to make a thicker needle-punched sheet can be cost prohibitive.[3]

Cutting

With computerization, cutting tables have advanced with very accurate, programmable systems for both straight and shape cutting, along with optimizing programs which control the waste to a level which the operator and the accountant can determine. Loading and unloading devices are built into cutting table systems which can take most sizes of glass and scoop them quickly into predetermined positions. Table 11.2 lists the manufacturers of cutting equipment. The other significant advancement in cutting technology is the process of waterjet cutting.

The equipment for industrial applications was developed and patented by Dr. Norman Franz in 1968. That machine operated with a pressure of 10,000 psi (700 bars) and was designed to cut paper tubes. Since then, about 1500 machines have been installed worldwide.

The operation of a state-of-the-art waterjet cutting machine requires regular municipal water that is filtered and brought up to a pressure 58,000 psi (2800 to 4000 bar) by a "pressure intensifier," which is basically a sophisticated pump, driven by a 40-kW electric motor.

The pressurized water then travels through a thin, flexible stainless steel tube to the cutting head, where it passes through a sapphire nozzle to form a thin jet, 0.008 to 0.016 in (0.2 to 0.4 mm), flowing at about twice the speed of sound. Depending on the material to be cut, an abrasive is added to the water. Soft materials, such as fabric, leather, frozen food, and the like can be cut with pure water, and the resulting cutting gap is a maximum of ¼ in (0.2 to 0.4 mm). Abrasive must be added to cut hard materials, such as glass, ceramics, metal, etc., resulting in a cutting gap of approximately 1⁄16 in (1.5 mm).

Even though the jet of water is very thin and the smooth effortless movements of the waterjet appear almost like cutting through butter, the forces at work are significant, bordering on the violent. The stream of water meeting the workpiece is equivalent to the impact of a car hitting a wall at 55 mph.

The cutting table is usually designed according to the stationary workpieces, moving tool principle, and resembles in concept somewhat a single head XYZ glass-cutting table. The optimum cutting process is usually arrived at by experience. The best combination of water pressure, type and amount of abrasive, and cutting speed differs for the various materials and the workpiece thickness.

TABLE 11.2 Cutting Machinery

Name and address of company	Name of machine	Min/Max glass thickness	Min/Max glass width and length	Machine speed	Manual, semiautomatic, or fully automatic	One- or two-bridge system	Programmable	Size	Other
AGM Glass Machinery Inc. Charlotte, NC	Bottero 352 BCS Moduline shape-cutting table	2 to 19 mm	1000× 500 mm; 3650× 2750 mm	100 m/min	Fully automatic	One-bridge	Yes	Standard and jumbo size	
Bystronic, Inc. Hauppage, NY	XYZ-F-89 automatic shape-cutting machine	0.090 in ss to $\frac{3}{4}$ in	48×60 in to 130× 240 in	400 ft/min	Fully automatic	One-bridge	Yes	130×19 in to 155×256 in	Data input by direct keyboard entry or from optimization by on-line connections, $5\frac{1}{4}$-in floppy, $3\frac{1}{2}$-in compact disk, or perforated tape; remote lubrication and pressure adjustments; electronic control is made specifically for glass cutting by Bystronic.
C. R. Laurence Co. Inc. Los Angeles, CA	CRL Semi-automatic XY double-bridge cutting table	ss to $\frac{3}{4}$ in	72×96 in to 108× 144 in; custom sizes on request	25 s per complete cycle	Manual and semi-automatic	Two-bridge	No	96×12 in to 132×168 in	Options include air flotation, breaker bars, auto remote bridge operation, and free-fall racks.

TABLE 11.2 Cutting Machinery (*Continued*)

Name and address of company	Name of machine	Min/Max glass thickness	Min/Max glass width and length	Machine speed	Manual, semiautomatic, or fully automatic	One- or two-bridge system	Programmable	Size	Other
C. R. Laurence Co. Inc. Los Angeles, CA	CRL CNC glass cutting machine	ss to $\frac{3}{4}$ in	36×48 in	Up to 3600 in/min	Manual, semiautomatic, and fully automatic; semiautomatic are upgradeable to fully automatic	One-bridge	Yes—includes optimization labels, pattern print-out summary reports, and storage assignments locations	36×48 in to 130× 240 in	Includes precision cutting bridge and head with $\frac{1}{64}$-in cutting tolerances with repeatability of cut pattern plus 0.005 in
ESAB Automation Inc. Fort Collins, CO	GXA & GXB gantry shape cutters	$\frac{1}{8}$ in and up	130× 204 in	25 to 250 in/min	Manual	Two-bridge		Variable assorted sizes	Integrated with Flow Systems waterjet cutting system
The Flecher-Terry Co. Farmington, CT	Flecher 3000™ glass, plastic, matboard cutting machine	$\frac{1}{4}$ in	4 to 16 ft wide and any length	Depends on operator	Manual	N/A		48×60 in	
G.F. Technologies Cuyahoga Hts., OH	Opti-Glass system	ss to $\frac{1}{4}$ in	0 to 72× 84 in		Manual			28 ft× 10 ft 6 in	
Glass Equipment Development Inc. Twinsburg, OH	Series 1500 computerized glass cutting system	0 to $\frac{1}{4}$ in	48×60 in to 72× 96 in	2500 in/min	Fully automatic	One-bridge		30 ft 11 in ×14 ft $1\frac{1}{2}$ in	X-Y-Z cutting

Bilco Manufacturing Inc. Zelienope, PA	Versa-Cutter (for specialty glass and small automotive)	1 to 6 mm+	24×36 in or as required	3000 + ipm	As required	One-bridge	Yes	Varies	
	CNC-One/ 500/20000	0.062 to 75 in	65×84 in to 132×204 in	1500 to 3000 ipm	Automatic	One-bridge	Yes		
Dynacut, Inc. Springtown, PA	Model 1500/HD Precision cutoff machine (14- and 20-in models)	To 5 in	To 6 in wide, any length	Determined by blade size	Manual and semi-automatic	N/A	To order	24×36× 36 in	Custom accessories available
Glass Equipment Development Inc. Twinsburg, OH	Series 6000 computerized glass cutting system	0 to $^3/_8$ in	48×72 in to 108×144 in	3400 in/min	Fully automatic	One-bridge	Fully computer-controlled system	44 ft 10 in× 17 ft $3^1/_8$ in	X-Y-Z cutting out of square and contour cutting
Grenzebach Corp. Newnan, GA	Contour cutting machine, type SMK	1.5 to 6 mm	2400× 1500 mm to 500× 300 mm	85 m/min	Fully automatic	One- and two-bridge	Yes	3200× 2300 mm	Features cutting head with infinitely variable cutting force to optimize cutting speed and maintain contour dimension
Hilemn Machinery Sales Ltd. Greensboro, NC	Duraflex by Malav	6 to 25 mm	400 to 2500 mm	N/A	N/A	1 cutting head			Maximum width to length ratio 2:1

TABLE 11.2 Cutting Machinery (*Continued*)

Name and address of company	Name of machine	Min/Max glass thickness	Min/Max glass width and length	Machine speed	Manual, semiautomatic, or fully automatic	One- or two-bridge system	Programmable	Size	Other
Klopper Mashinentechik DmbH Co. Dortmund, Germany	K-MA 90	1.5 to 5 mm	200× 1100 mm to 590× 200 mm		Fully automatic	Two-bridge	CNC		
The Lockformer Co. Lisle, IL	Excalibur	0.020 to 1.00 in	60 to 130 in× 120 to 204 in	True cutting speed 1000 in/min	Fully automatic	One-bridge	Yes, you can score any shape	Special size made to order	
Pankoke Flachglasterchnil GmbH Luebeck, West Germany	DIG-semi-automatic cutting bench NC 2/3—fully automatic cutting bench also for curves and models	3 to 19 mm	6000× 3250 mm	60 to 100 m/min		DIG-4 longitudinal 1 cross-cutter NC 2/3-single head pivoting system	Yes		
Perfect Timing Corp. Lancaster, TX	Aculite 2000, Aculite 1000, Aculite 500	¾ in	130× 204 in	3000 in/min	Aculite 2000 fully automatic, Aculite 1000 and 500 semi-automatic	Aculite 2000 and 1000—one-bridge; Aeulite 500—two-bridge	Yes		Counterbalance and hydraulic tilt top manual cutting tables also available

Peter Lisec GmbH Hausenening, Austria	ESL-E, ESL-EF, ESL-SF, ESL-K	2 to 19 mm	Jumbo sizes	300 ft/min	Fully automatic	One-bridge, single head	Yes	Complete lines, single tables	Agent: Bud Shelton, Fayetteville, GA
W. J. Savage Co., Inc. Knoxville, TN	Tysamon/by Savage Model WJS-20	$\frac{1}{8}$ to 36 in	$\frac{1}{8}$ in to 20 ft $\times$ $\frac{1}{8}$ in to 40 ft	240 in/min	Manual, semi-automatic, and fully automatic	One- and two-bridge	Yes	Unlimited size; custom manufacture available	Cuts bullet-proof glass

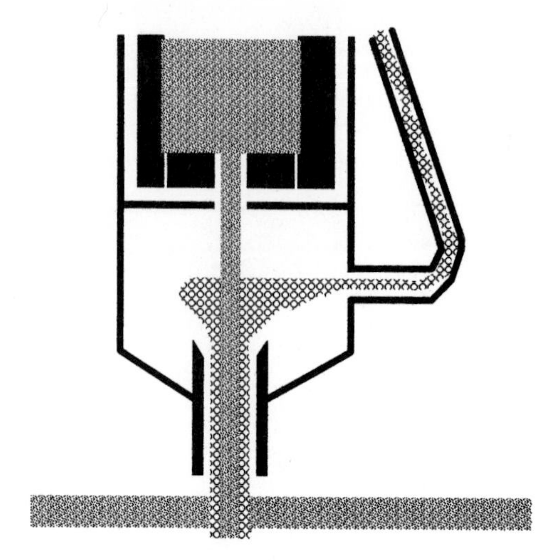

Figure 11.1 Waterjet cutting head.

The cutting speed of a waterjet machine is quite slow. It can be as slow as ¾ in/min (20 mm/min) if, for example, laminated glass 2½-in thick is to be cut. But to move to a positive note: Some machines may be equipped with two cutting heads, cutting two workpieces in parallel and thereby reducing the processing time in half.[3]

Figure 11.1 illustrates a waterjet cutting head for cutting hard materials such as glass.

Tempering lines

Tempering lines are now mostly horizontal, employing well-controlled technology. The oscillating roller hearth, whose design was largely pioneered by a domestic company, is particularly successful because of its compactness and efficiency. Table 11.3 lists the companies selling glass tempering machinery. Refer to Chapter 7 for additional data on tempering.

Insulating glass lines

Complete lines now feature totally automatic computer-controlled systems for sealing insulating glass units, including the construction and layup of the spacer units in accordance with variable glass dimensions. The cleaner automatic lines have enabled manufacturers to clean up some of the unsightly butyl-smeared plant areas. Most of the more automated, vertical insulating glass lines are manufactured

TABLE 11.3 Tempering Machinery

Name and address of company	Name of machine	Min/Max glass thickness	Min/Max glass width/ length	Furnace speed	Thru or chamber (batch) furnace	Size	Other
Casso-Solar Corp. Pomonal, CA	Horizontal flat glass tempering	3 to 15 mm	12 to 120 in× 12 to 240 in	As required by production rate	Continuous or batch type	Custom design for each customer	Provides high quality optics and flatness
Cattin Machines S.A. LaChaux De-Fonds, Switzerland	Flat tempering-type TPR bending and tempering-TBR (horizontal) and TBVE (vertical)	2.8 to 19 mm	75×170 mm to 2500×5200 mm		Both	Flat temp: 1250×2700 mm to 2500×5200 mm; bend temp: to 2200×2800 mm	
Glasstech, Inc. Perryburg, OH	Advanced tempering systems for architectural glass	$\frac{1}{8}$ to $\frac{3}{4}$ in	36- to 96-in widths; lengths up to 23 ft	Quoted on application	Both		

by German firms. Refer to Table 11.4 for particular types and configurations of equipment for applying butyls, tapes, warm-edge materials, and other sealants.

Washing machinery

Although a full chapter has been devoted to glass washing (Chapter 21), we have listed the specifics for the type of washers available to the industry. Table 11.5 lists some of the major manufacturers of this specialized equipment.

Conclusion

Overall, the demands and rewards for the growing retail market in flat glass and other fabricated glass have outweighed the prices being charged, especially in view of the dramatic increase of foreign equipment because of the devaluation of the dollar. The strong demand for good equipment underlines the fact that the glass sector is in a dynamic, expansive mode. Renewal and growth are still taking place in North America, and these fabricators are not about to miss out.

References

1. Bushell, S. B., Fabrication equipment: The trends, *Glass Magazine,* 9/1987.
2. Bally, M. A., Debunking myths about waterjet cutting, *Glass Digest,* 4/1994.
3. Brennan, M. P., Zeroing in on natural and synthetic felts for beveling machinery, *Glass Digest,* 9/1995.

TABLE 11.4 Insulating Glass Machinery

Name and address of company	Name of machine	Min/Max glass thickness	Min/Max glass width/length	Total number of units produced in eight hours	Manual, semi-automatic, or fully automatic	Other
Aztech Inc. Somerville, NJ	Model AGT automatic gunning system	$^3/_8$ to $1^1/_4$ in IG units	6 to 80 in	1000	Semiautomatic	
Besten Inc. Cleveland, OH	Butyl extruder		Any frame length and width	1200 frames	Semiautomatic	
	Ultrasonic soldering machine		No frame size limit	1 spacer/min	Semiautomatic	
	Heated roller press		36 to 96 in any length	400 to 3000 units	Semiautomatic	
	XY glazing table	6×8 in frame size to 48×78 in			Fully automatic	
	Automatic Swiggle Strip applicator		16×16 in to 61×61 in	750	Semiautomatic	
Bilco Manufacturing Inc. Zelienople, PA	Equipment for manufacturing IG (using Tremco Swiggle Strip)	0.062 to 0.75 in	Any	As required	All	

TABLE 11.4 Insulating Glass Machinery (*Continued*)

Name and address of company	Name of machine	Min/Max glass thickness	Min/Max glass width/length	Total number of units produced in eight hours	Manual, semi-automatic, or fully automatic	Other
FDR Designs Inc. Loretto, MN	In-line fill percentage tester	All thicknesses	All sizes	Can test as many as produced	Fully automatic	The in-line fill percentage tester tests one IG unit at a time prior to sealing; then circulates gas in the IG unit for reading; shows percentage of argon in unit
	RGS 15/30 high-speed gas filling machine	All thicknesses	All sizes	1200 IGs per shift	Fully automatic	RGS 15/30 is a patented high-speed gas filler with a computer-controlled flow rate of 15 to 30 L/min; interior IG pressure is monitored during filling process to eliminate implosion or explosion
Glass Equipment Development Inc. Twinsburg, OH	Series 2000 linear hot melt system	Unit thickness 0 to $2\frac{1}{2}$ in	12×12 in to 60 in wide $\times$ unlimited length	1000	Semiautomatic	Uses patented Linear U and Linear Skip sealant application method

Glass Equipment Development Inc. Twinsburg, OH	Series 3000 linear hot melt system	Unit thickness 0 to $2\frac{1}{2}$ in	N/A	2500	Semiautomatic	Uses patented Linear U and Linear Skip sealant application method
Graco/LTI/Pyles Wixom, MI	Unit 8900 Hydra Mate dispensers	N/A	60×84 in	N/A	Manual and automatic	
Lake County Sales Inc., Grand Rapids, MI	LCS 6084 heated platen press	3 in	7.875×13.750 in to 63×98.375 in	250	Manual	
Lenhardt/ Bystronic Inc. Hauppauge, NY	Combination hot melt dual seal production line	0.090 to 0.500 m	Up to jumbo sizes	500 to 1000	Semiautomatic	Vertical production line with integrated three-side linear hot melt coating system
Peter Lisec GmbH Hausmening, Austria	Insulating glass line	3 to 60 mm		Up to 2000	All types available	Agent: Bud Shelton, Fayetteville, GA

TABLE 11.5 Washing Machinery

Name and address of company	Name of machine	Min/Max glass thickness	Min/Max glass width/length	Type of glass	Number of blowers	Number of brushes	Horizontal or vertical	Operating speed	Size	Other
Besten Inc. Cleveland, OH	Glass washer	ss up to $\frac{1}{4}$ in	36 to 96 in wide, any length		1	4 or 6	Horizontal	15 to 30 fpm	36, 48, 60, 72, 84 and 96 in	
Bielefelder Union Bielefeld, Germany	Union 2070, 2076, 2077, 2070, 2185	0.5 to 25 mm	50×100 mm to 3.300× 6.000 mm	All types	1	As many as required	Both	Up to 25 m/min	Depends on max. glass size and no. of brushes	Speciality high-tech washing lines for high vacuum coating
Bilco Manufacturing Co. Zelienope, PA	Curved glass washer	Flat to 15 in	Width to 94 in	All types	Varies	None	Both	0 to 35 + fpm	Varies with glass size and line speed	
	Float glass washer	0.020 to 0.75 in	$1\frac{1}{2}$ to 144 in long	All types for secondary processes	1	As required	Horizontal	Standard range, 0 to 7	Varies	High pressure (brushless) washers with same parameters also available
Emmett Engineering Ovada, Italy	Mini Washing 45	0.4 to 3 mm	25×65 to 450×450 mm		1 turbo in the drying section	2 nylon pairs	Horizontal	0 to 3 m/min	2000×1200× 1150 mm	Stainless steel machine for cleaning small glass

Glass Equipment Development Inc. Twinsburg, OH	84-in wide 6-brush horizontal glass washing machine	0 to $\frac{1}{4}$ in	Minimum diagonal measurement $14\frac{1}{4}$ in	Any flat glass	1	6	Horizontal	0 to 22 ft/min	$207\frac{1}{2} \times 106\frac{3}{4}$ in	Flip-top design, stainless steel tanks, options: prewash spray bar, hydraulic cylinder assist for lift-top; static bar
	96-in wide 6-brush horizontal glass washing machine	0 to $\frac{1}{4}$ in	Minimum diagonal measurement $14\frac{1}{4}$ in	Any flat glass	1	6	Horizontal	0 to 22 ft/min	$207\frac{1}{2} \times 106\frac{3}{4}$ in	Flip-top design, stainless steel tanks, options: prewash spray bar, hydraulic cylinder assist for lift-top; static bar
Hilemn Machinery Sales Ltd. Greensboro, NC	Mod "S" 1300, 1600, 2000 by Taulzi	3 to 16 mm	Maximum 1300, 52 in; 1600, 61 in; 2000, 80 in	All	4	4	Horizontal	1 to 5 m/min		
Klopper Mashinentechik DmbH Co. Dortmund, Germany	HWM and VWM washer	0.4 to 40 mm	10×40 mm to 3300×6000 mm	Laminated plate and spec. coated	1 to 2	Customer request	Both	Individual		
Peter Lisec GmbH Hausmening, Austria	VHW (vertical) RTL (horizontal) washers	3 to 60 mm	Up to 118 in; any length	All types, even coated	1	6 or 8	Both	Up to 50 ft/min	Depends on type	Agent: Bud Shelton, Fayetteville, GA
Sommer & Maca Industries Inc. Chicago, IL	Washers from 12 to 144 in wide	ss to $\frac{3}{4}$ in	Varies with size of machine	All types	1	4, 6, and 8	Horizontal	Varies	Varies	Made in the U.S.

Chapter

12

Sealants

G. Bernard Lowe*

Technical Marketing, Morton International Inc.

Joseph S. Amstock

President, Professional Adhesive and Sealant Systems

Polyurethanes

Introduction

In recent years the use of polyurethane sealants in insulating glass has shown a dramatic increase. Their worldwide market share has grown from below 0.5 percent in 1980 to 12 percent in 1983 and an additional 10 percent through 1995 making the polyurethanes the fastest growing adhesive/sealant system in this industry. Figure 12.1 details the market share of polyurethane in 1995.

Since the mid-1980s, more than 9.3 million pounds of polyurethane insulating glass sealants have been consumed by manufacturers on sealed units.[1] On a worldwide basis, this narrow base of sealant manufacturers has increased by twofold, supplying more than 3000 manufacturers of insulating glass units.

Even more significant has been the increase of research and development by the insulating glass sealant manufacturers on polyurethane back-boned sealants in the past several years.[2] During the last 30 years, insulating glass manufacturers have approached the challenge of producing a hermetic seal between two separated pieces of glass in many ways. These have ranged from fusing glass to metal, wraps with

*Coauthor of the section on polysulfides.

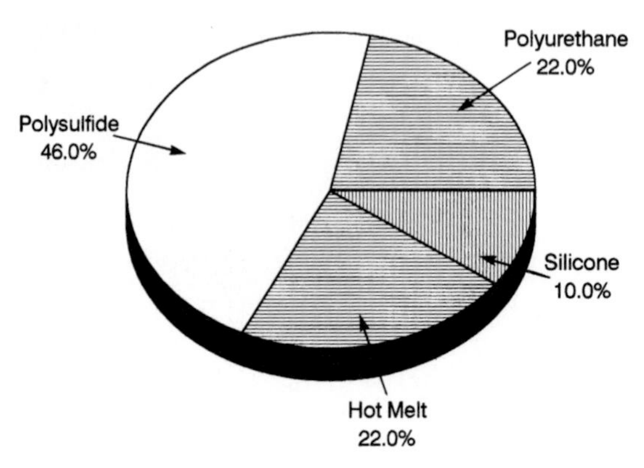

Figure 12.1 Sealant usage.

mastic, to the present standard of using organic sealants. Earlier manufacturing methods were abandoned because of poor product performance, limited insulating glass overall size, and thickness of the air space. Still others became obsolete purely on the basis of cost. Although some of the methods produced units that performed better than those made by current methods, the units were not competitive, and their seal longevity did not warrant the additional cost.[3]

Polysulfide sealants solved some of the problems, as did silicone and hot melt butyl sealants. Unfortunately, however, each of these sealants added problems, too, although most have now been corrected by the development of polyurethane sealants.

Like other industries, the insulating glass industry has seen the evolution of materials and methods. The technologies of material (sealant, spacer, desiccant, etc.) and application techniques have advanced the industry, along with the general recognition of the longevity requirements for insulating glass, to the current state of the art. Steel and aluminum spacers, Swiggle Strip™, Super Spacer™ and the new Intercept™ system now used to separate the glass, for example, have been developed to hold the unit together and to keep water vapor from getting to the desiccant inside the spacer.

Many other demands have been placed on the sealant by the manufacturers and by the climatic conditions where insulating glass units are being specified. The air inside a sealed unit typically expands or contracts with changes in barometric pressure and temperature. The expansion and contraction raise several questions about the performance of the sealant. Should the edge seal, then, be soft enough to accommodate every change in volume of the air space? No, because the moisture vapor transmission (MVT) area would be constantly changing. This observation raises another question: If keeping the MVT area largely constant is important, then shouldn't the sealant be rather hard? Again, no, because preventing movement would break the glass. Ideally, the sealant should resist opening without stressing the glass to the point of breakage. This property, known as tensile modulus, should show minimal change with temperature.

Furthermore, the sealant should have high elastic recovery so that, when the sealant flexes, it can return to its original position. When sealant elastic recovery is maximized, the permanent "opening" of the insulating glass unit due to the effects of weathering is maintained. Such weathering effects include heat, trapped water (from glazing problems), and ultraviolet light transmitted to the seal/glass interface, which are, undoubtedly, the source of most seal failures.

Seeking long-term solutions to these problems, many laboratories set out to develop an insulating glass sealant that would perform better than polysulfides, overcome their limitations, and still remain cost effective.

Poly BD®resins—Basic characteristics

In 1974, Bostik, Inc. began experimenting with a new resin produced by Arco Chemical Company. In their research, they tested many resin systems including hot melts and fluorinated elastomers. The result was a polyurethane-based sealant with distinct advantages over polysulfides. A starting formulation using Poly BD® R-45HT resin as the base for an insulating glass window sealant is shown in Table 12.1 and basic elastomeric properties are shown in Table 12.2.

TABLE 12.1 Insulating Glass Formula–Poly BD Based

Parts by weight		
Ingredients	Base	Accelerator
1. Poly BD™ Resin R-45HT (OH value = 0.85 MEG/G	100.0	
2. Insol 100 (equivalent weight = 104)	3.0	
3. Super-Pflex 200 (calcium carbonate)	85.0	
4. Ultra-Pflex (calcium carbonate)	40.0	
5. Sterling R (carbon black)	2.0	
6. Santicizer 278	75.0	
7. Cab-O-Sil N-70-TS	7.5	
8. Vanox NBC	2.0	
9. Organosilane ester A-187	2.0	
10. T-12 Catalyst	0.045	
11. Isonate 143-L (equivalent weight = 143.5)		(NCO/OH-0.95) 15.49 (NCO/OH-0.925) 15.10

TABLE 12.2 Elastomeric Physical Properties

	NCO/OH Ratio	0.926 NCO/OH Ratio
Tensile strength, psi	237	256
Elongation, %	383	473
Tensile set, %	6	9
Tear strength, pli	57	52
Hardness, Shore A	43	39
Modulus, 100%, psi;	131	105
300%, psi	211	179
Water vapor transmission, g/m²/24 h	8.29	—
Permeance, g/m²/24 h	0.20	—

TABLE 12.3 Application properties at NCO/OH-0.95

This mixing ratio (weight 100 parts base/4.74 parts of accelerator); the ratio is variable as part or all the Santicizer 278 plasticizer can be put on the isocyanate accelerator side.	
Work life, adjustable*	20–45
Tack-free time—adjustable*, h	3–8
Sag, channel at 70°F (21°C), in	0.15
Peel adhesion, aluminum to glass, pli	12–14
Lap shear, glass to glass, psi	>130 glass failure

*Can be adjusted by changing the T-12 content.
Aluminum channel, $\frac{3}{4}$ in wide, $\frac{3}{4}$ in deep × 6 in long; $\frac{1}{6}$ in wall, maximum
SIGMA Method Test A.3.B.
SIGMA Method Test P.7.A.
SIGMA Test Method P.6.A.

Based on the suggested formula outlined above, it is possible to formulate a sealant to achieve the properties listed in Table 12.3.

In retrospect, perhaps the advantages of polyurethanes could have been anticipated. Polyurethane-based sealants had found wide acceptance for high-performance products in many markets, particularly in OEM and construction waterproofing markets.

Poly BD® resins are liquid, hydroxyl terminated polymers of butadiene with an average molecular weight of 2800. The degree of polymerization is in the range of 50 and the microstructure of the product is similar to that prepared by emulsion polymerization. The predominant configuration is trans-1,4 (60 percent), with approximately 20 percent each cis-1,4 and vinyl-1,2 (Figure 12.2). The terminal hydrox-

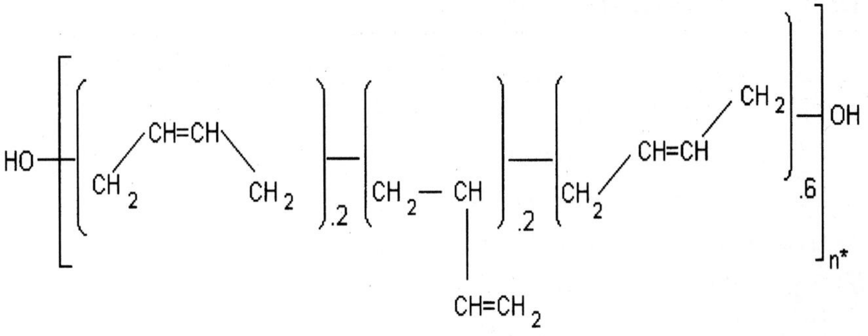

*=50 for R-45M and 45HT
*=25 for R-20LM

Predominant Microstructure

Cis	1,4 unsaturation	20%
Trans	1,4 unsaturation	60%
Vinyl	1,2 unsaturation	20%

Figure 12.2 Poly BD resins R-45HT.

TABLE 12.4 Typical Physical Properties of Poly BD™ Resin

	R-45H
Nonvolatile material,wt %	99.9
Viscosity, mPa s @ 72.5°F (23°C)	8000
Viscosity, mPa s @ 86°F (30°C)	5000
Hyroxyl number, mg KOH/g	48.2
Hydroxyl value, meq/g	0.85
Hydroxyl function	2.4–2.6
Molecular weight, M_n	26
Polydispersity, W_w/M_n	2.5
Water, wt %	0.03
Specific gravity at 72.5°F (23°C)	0.901
Iodine number, g/100g	400
Glass transition temperature (Tg), 32°F (0°C)	−75
Heat capacity (cp), kcal/kg°C	0.51

yl groups are primary (95 percent) and predominantly of the allylic type. Depending on the grade, the hydroxyl functionality varies from 2.2 to 2.6.

Poly BD® resins are characterized by low volatility, coupled with low viscosity and moisture levels. Some typical physical properties are shown in Table 12.4.

Poly BD® resin properties

1. Hydrolytic stability
2. Resistance to aqueous acids and bases
3. Low temperature flexibility
4. Low moisture vapor permeability
5. Low embedment stress
6. Thermal cycling stability
7. Electrical insulation properties

Systems based on these materials can be cured at ambient temperatures by reaction with di- or polyisocyanates. Pot life and cure time can be readily adjusted by using typical urethane catalysts. The most desired physical properties can be achieved by formulating with short chain diols, fillers, and plasticizers. Technically the cured elastomer is a polyurethane. However, the nonpolar, hydrocarbon butadiene backbone imparts several properties of significant advantage to the insulating glass industry. It is important to remember that all sealants change dramatically when subjected to weather. Table 12.5 compares a polysulfide-based formula to that of a Poly BD® resin system.

TABLE 12.5 Elastomer Formulation/Effects of Accerator Aging on Physical Properties

	Poly BD™ Resin-based starting formulation			Typical commercial polysulfide formula		
Test	NCO/OH—0.95					
Days in weatherometer*	0	10	30	0	10	30
Tensile strength, psi	237	276	288	164	178	160
Elongation, %	383	432	476	431	105	95
Tensile set, %	6	8	9	32	4	2
Hardness, Shore A	43	43	39	50	60	61
Modulus, 100%, psi	131	114	111	119	177	—
300%, psi	211	230	225	145	—	—
Peel adhesion, aluminum/glass,† lb	14	38.3	39.5	8.0	10.5	10.3
Permeance, metric perms	0.20	0.30				

*Weatherometer utilizes 6500-watt xenon arc, borosilicate glass filtered with a cycle of 102 minutes, 18 minutes water spray at 50% relative humidity, and a black body temperature of 140°F (60°C).

†At zero starting time, the Poly BD™ Resin elastomer is not fully cured to allow for longer pot life for laboratory testing. A higher catalyst concentration is expected to increase the initial peel strength value.

The relative advantages of a polyurethane insulating glass sealant, compared to a polysulfide benchmark, can be divided into three categories: application, performance, and cost. Over the years, polysulfides have been reformulated to improve wet application characteristics, the ability to go on a stack of lites of insulating glass more easily and smoothly. Once applied, the sealant should not run out or sag, especially at the corners of large air space units. Polysulfide formulas have improved markedly in this respect, but they are hampered by the inherently high viscosity of the base polymer, even with the addition of organic solvents. Urethane-based sealants are lower in viscosity and can be compounded without the addition of solvents with better flow properties and sag resistance.

Qualitative comparison of the viscosities of polyurethane and polysulfide-based sealants at different shear rates is shown in Figure 12.3. The polyurethane sealant has a higher viscosity (greater sag resistance) at lower shear rates, but it has a viscosity lower than a polysulfide when it is troweled or pumped.

The net result is a "whipped cream" consistency. The sealant is easily applied by gun or trowel with full penetration into the spacer profile, but with good resistance to sagging. In addition, the urethane-based sealant can be formulated to maintain its viscosity longer after mixing the base and activator components. Figure 12.4 shows a comparison of viscosity versus time after mixing the polysulfide and polyurethane sealants; both sealants have the same pot life and initial viscosity. The urethane stays lower in viscosity until the pot life limit is reached.

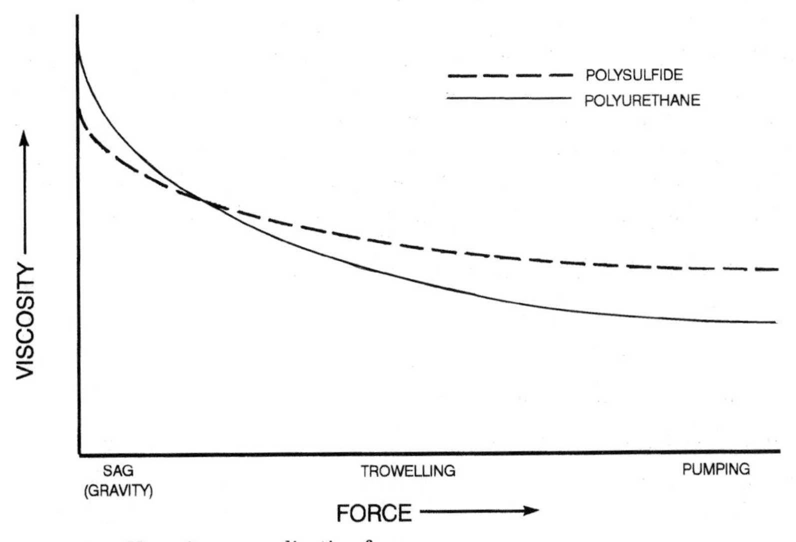

Figure 12.3 Viscosity vs. application force.

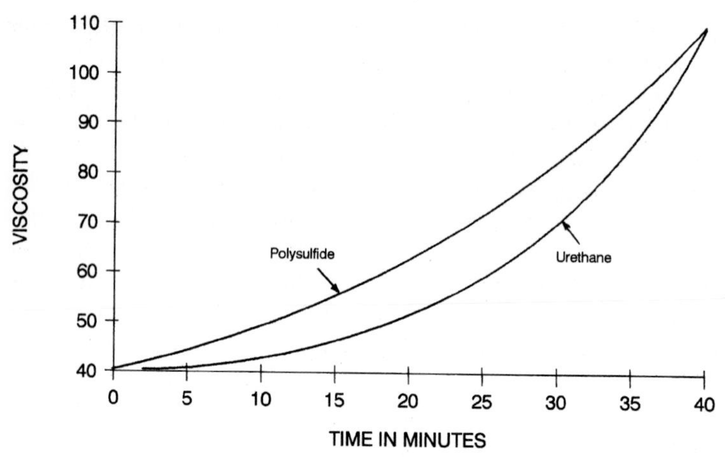

Figure 12.4 Viscosity vs. time after mixing.

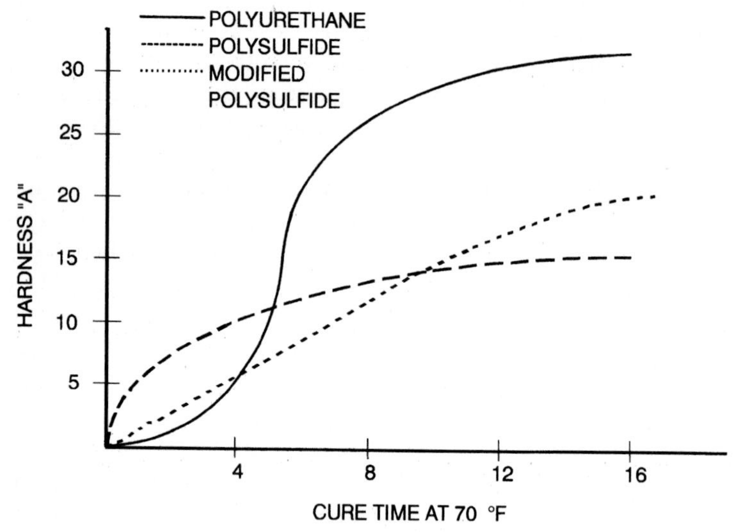

Figure 12.5 Hardness build of IG sealants.

Lower viscosity translates into easier application and better wetting of the glass spacer without stringing or pullout of the sealant. The lack of stringing is especially useful for automated lines, because it minimizes or eliminates the cleanup of units after robot gunning heads have applied the sealant.

Polyurethane-based sealants also cure faster than polysulfides. As shown in Figure 12.5, a plot of durometer hardness build at 70°F (21.1°C) versus time after mixing, the polyurethane reaches a manageable cure state for cutting stacks apart, handling units, etc., after three hours. The polysulfide and the modified polysulfide require six to eight hours. All had the same pot life, and because the polyurethane develops its adhesion to the glass and the spacer faster, there is no chance that the sealant has set up and not bonded to the glass.

The single most important advantage of a polyurethane-based sealant is that it does not contain solvent. Most commercially available polysulfides contain 4 to 12 percent solvent by weight. Because solvents have a low density material, after they evaporate, volume is reduced by seven to twenty percent. The lack of solvent in polyurethane-based sealants results in several improvements for the insulating glass manufacturer.

Improved properties

1. No sealant shrinkage, especially at the corners.
2. No reduction in design MVT path caused by shrinkage.
3. No special storage and handling. The solvents normally added to the polysulfide, such as toluene or methyl ethyl ketone, are flammable and give low flash points to the compounded sealant. Polyurethane insulating glass sealants have flash points greater than 200°F (93.3°C).
4. No loading of desiccant by adsorption of organic solvent vapors. With polysulfide edge seals, five to ten percent of the adsorption capacity of the desiccant can be consumed by solvent.
5. No shrinkage-induced glass stress. As solvents evaporate from the edge seal and the sealant maintains full contact with the glass, the stress induced by shrinkage becomes a stress on the glass, with the spacer acting as a fulcrum. The effects of the stress depend on the glass thickness, air-space size, and unit dimension and are most apparent on single strength glass and large spacer sizes. This stress can be relieved by sealant relaxation, which will increase the vapor transmission area of the unit, or it can remain as a residual stress causing glass deflection. The stress increases the probability that the lite will break during the first winter, as the glass deflects inward.
6. Reduction of in-factory glass breakage. Because solvent evaporation stresses the glass, if the stress becomes excessive (large stacks of single strength insulating glass units with medium to large air spaces), the glass can break. Breakage can occur in only one lite in a stack of thirty, or in a much smaller percentage when stacked lites are cut apart. When the edge seal is solvent-free polyurethane, dramatic reductions of in-factory glass breakage have been found independent of other glass deflection/breakage variables.
7. Reduction of "first winter" breakage. Reports from parts of the country with severe weather indicate that the use of solvent-free

polyurethane can reduce in-factory and "first winter" breakage from as much as 0.5 percent of production essentially down to zero.

8. Better performance of desiccants. Because they can be supplied without solvents, polyurethane insulating glass sealants provide an advantage in the selection of desiccants for the sealed unit. Most available desiccants are designed to absorb solvent vapors from the sealant because solvent can be present in the insulating glass unit as readily as water vapor.

Along with the ability to absorb solvents comes the ability to absorb nitrogen in the air space. The negative effect of nitrogen absorption causes glass deflection inward, which opens the unit (elongates the sealant), thereby increasing the MVT rate area; it also contributes to glass breakage. With a solvent-free edge seal, the desiccant can be designed only for adsorption of water vapor, which maximizes desiccant efficiency and essentially eliminates nitrogen absorption and resulting glass deflection.

Although polysulfides have done a good job of achieving a balance of properties, urethanes perform better. Figures 12.6 and 12.7 show modulus versus temperature characteristics and elastic recovery of the two sealant types.

The polyurethane displays a lower modulus at low temperature and a higher modulus at elevated temperatures. Also, the elastic recovery data at room temperature and high temperature favors the urethane.

Another interesting property of polyurethane insulating glass sealant is its ability to bond to contaminated surfaces. In laboratory tests, urethane has been applied to glass and spacers treated with various contaminants, including hand grease, silicone spray, and cutting oils. The

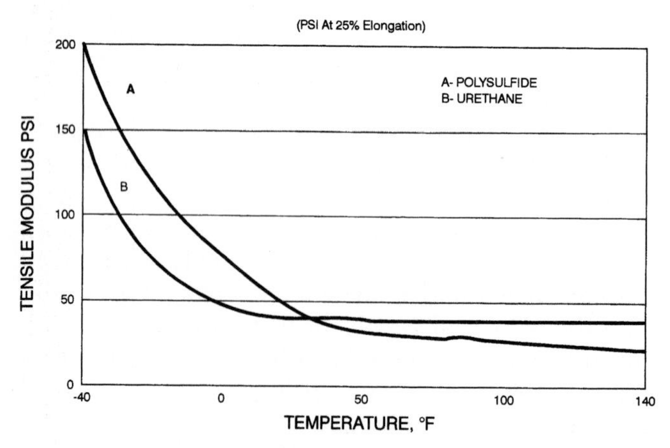

Figure 12.6 Tensile modulus vs. temperature.

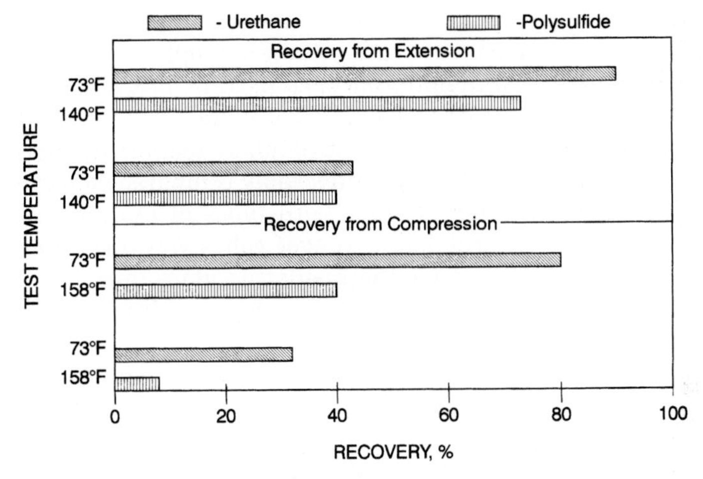

Figure 12.7 Elastic recovery properties. Extension, tensile: 25%, elongation at test temperature for 24 hours, then relaxed 1 hour at 73°F. Percent to original length. Compression: cylindrical specimen compressed to 50% of original height for 72 hours at test temperature, then relaxed 1 hour at 73°F. Percent return to original height.

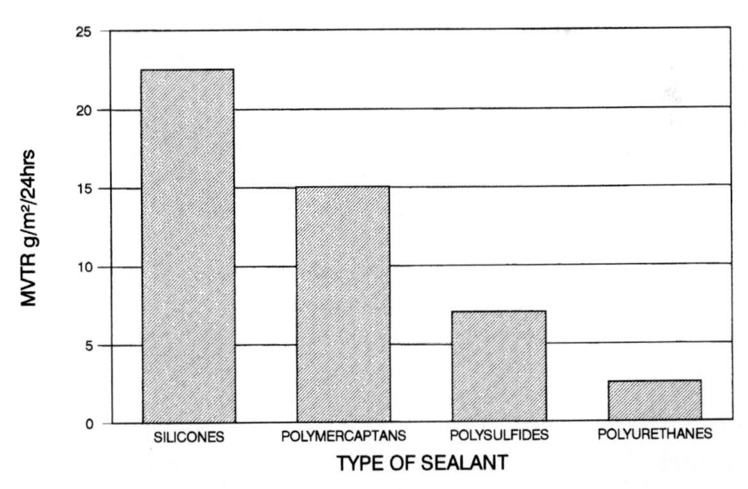

Figure 12.8 Moisture vapor transmission rates.

urethane bonds through most of these materials. This property gives the insulating glass manufacturer a safety margin in sealant adhesion.

Laboratory values for moisture vapor transmission indicate that the urethane sealant is slightly superior to polysulfide, but when the mechanical performance parameters (tensile modulus and elastic recovery) are combined with MVT, as in Fig. 12.8, the advantage of polyurethane is more apparent. This test simulates the performance of an insulating glass unit more accurately than a simple MVT rate. By this test, the urethane transmits only 42 percent of the water vapor that a polysulfide transmits.

Adhesion values for the urethane insulating glass sealant in various modes (tensile, shear, and peel) rank favorably compared to polysulfide. This may be misleading, because both types of sealant fail cohesively when tested to destruction. Stronger cohesively, urethane offers greater structural integrity for the insulating glass unit, especially for architectural lites and patio doors or when site line specifications are encountered.

Tests show that a urethane insulating glass sealant is equal to or surpasses the performance of polysulfide in nearly all aspects. When tested for hardness and modulus increase on long-term exposure to 160°F (71.1°C), urethane and polysulfide are about the same and both increase at similar rates. But, when tested long-term at 120°F (48.9°C), the urethane retains its original properties over the polysulfide. On continuous water immersion, the urethane maintains glass adhesion better, with the exception of lead-cured polysulfide which has specific disadvantages for heat and ultraviolet resistance.

The resistance to ultraviolet light (UV) of insulating glass sealants has also been tested in many ways. Direct simultaneous exposure of urethane to ultraviolet light and atmospheric oxygen causes its surface to check and erode. But this is a highly misleading test because ultraviolet light and oxygen are never available together in a glazed unit. Moreover, the supply of ultraviolet light at the glass/sealant interface is restricted by incident angle and the relative opacity of window glass to ultraviolet light.

Still, the most reasonable test for UV resistance must be the direct exposure of the sealant to UV transmission through glass. By these tests, polyurethane shows substantially better retention of glass adhesion (Fig. 12.9).

These tests were run on glass/sealant adhesion specimens, not on sealed units, so as to separate UV-induced effects from air-space expansion/contraction effects on the edge seal. After two years exposure (45° South, S.E. Pennsylvania), the polyurethane shows less "UV burn" than the polysulfides, which show surprising differences among types.

Glazing

The chemical resistance of polyurethane insulating glass sealants, for the purposes of compatibility with wet and dry glazing systems, compares with polysulfides. Oil-based caulks and some butyls, normally

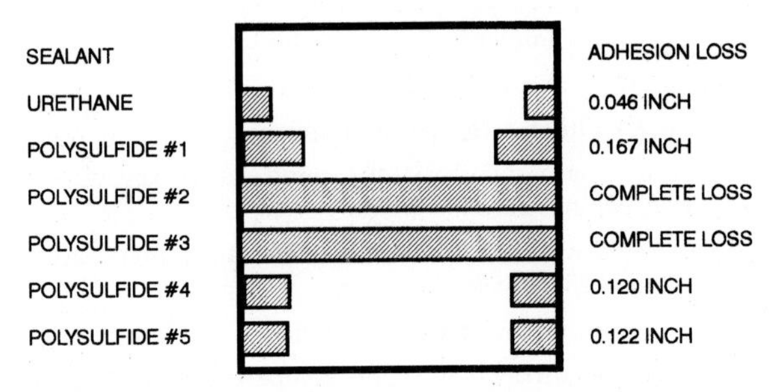

Figure 12.9 Glass adhesion loss ("UV burn"). 1.5-in wide specimens on detergent washed glass (3/32 in); adhesion loss is average distance from sealant edge where glass adhesive failure was evident.

containing slow evaporating solvents, should be avoided. Silicones, acrylics, and polysulfide caulks are generally compatible, as are all the vinyl wraps, setting blocks, and butyl/isobutylene tapes tested to date.

Polyurethane insulating glass sealant shows wide compatibility with reflective and laminated glass. Even some of the more difficult, sputter-coated, high-performance glasses can be sealed with polyurethane for long term-performance. Test results on Low-E and Hi-T glasses are very encouraging.

ASTM E 774 specification

U.S. ASTM E 774 specification is a modification of the durability test of the Sealed Insulating Glass Manufacturers Association (SIGMA). But, instead of SIGMA's single level, it has three levels, two of them more difficult than SIGMA's single requirement.

ASTM E 7744[4] requires that the insulating glass unit pass the ASTM durability test E 773. This consists of the cycles listed in Table 12.6. The test is conducted in a large, complicated test chamber which subjects the 14 by 20-in insulating glass units to cycles of cold followed by water spray and ultraviolet light at high temperatures. Failure occurs when the frost point is exceeded.

The frost point is determined by ASTM E 546. This employs a metal container with dry ice cooled to the test requirement of −20°F (−29°C) held against one of the lites of glass. Failure occurs when frost appears inside the air space of the sealed unit being tested.

Chemical fogging is generally a deposit of some volatile substance that volatilized out of the edge seal onto the inner lite of glass after

TABLE 12.6 One Cycle of Accelerated Weathering Test for Insulating Glass Units

Step No.	Time, Minutes	Test Procedure
1	65	Cool from ambient to −22°F (−30°C)
2	65	Maintain at −22°F (−30°C)
3	65	Turn on heat, temperature ambient
4	30*	Water spray and ultraviolet; temperature continues to rise; cabinet becomes humid
5	35*	Turn off water; continue ultraviolet until temperature rises to 134°F (57°C)
6	65	Maintain ultraviolet and humidity
7	65	Maintain ultraviolet; reduce temperature to ambient; after 65 minutes, turn off ultraviolet

*The 65 minutes of the two steps constitutes one cycle.

TABLE 12.7 Qualification Test Cycle for Sealed Insulation Glass Units

Classification	High Humidty	Accelerated Weather Cycle
C. First run	14 Days	<−26°F (−32°C)
B. Then run	14 Days	<−20°F (−29°C)
A. Then run	14 Days	<−20°F (−29°C)

exposure to ultraviolet. These oxidizable materials are solvents and plasticizer which can be generated when ultraviolet light and cold occur simultaneously. The method for determining chemical fog is described in ASTM E 773.[5] The test sequence is as follows:

1. 14 by 20-in sealed insulating glass unit is cycled for the first cycle shown in Table 12.6.
2. Then it is exposed to ultraviolet radiation for seven days with the glass temperature maintained at 65°F (18°C) by the ultraviolet lamp.
3. Then it is inspected for chemical fogging.

Fogging could also appear on the glass as a hazy film during the frost point test. This deposit is termed the chemical dew point. It would be the same as chemical fogging.

The specification delineates three classes. The requirements are shown in Table 12.7.[6] Each sample will be qualified, first as rating "C," then the same specimen for rating "B," and finally the highest rating "A." The test specimen must endure 42 days of high humidity and 252 cycles of accelerated weathering.

Cost factors

On a gallon per gallon basis, a urethane insulating glass sealant is normally cost competitive with polysulfide, but other savings are not so apparent. First, the nonsag characteristics described earlier provide for less sealant waste. At three large facilities manufacturing aluminum windows, wood windows, and commercial insulating glass, the manufacturers ran urethane for an extended period parallel to polysulfide. The three manufacturers reported material savings with urethane which ranged from 3 to 7 percent.

Furthermore, the reduction of in-factory glass breakage can represent a significant economic advantage. Calculations indicate that the sealant contributes about one-eighth of the manufactured cost of a single strength insulating glass unit, so reducing glass breakage has an eightfold effect on the sealant's real cost. For example, reducing sealant-related breakage from 0.35 percent to 0.1 percent of production, a net 0.25 percent change, is equivalent to a 2 percent savings on sealant. Of course, a reduction in "first winter" breakage is worth proportionately much more in replacement cost and customer satisfaction.

Other cost savings have been reported from users of urethane sealants in processing speed/labor, desiccant usage, and other factors. The total cost savings are equivalent to about 10 percent of the sealant cost.

Polysulfides

The first synthetic rubber was manufactured in the United States in 1929[7] and was known as a polysulfide. Its most interesting property was unusual inertness to solvents and hydrocarbon fuels in contrast to the easy swelling of natural rubbers. In early 1942, work began that led to the invention by J. C. Patrick and H. R. Ferguson[8] of a process of reductively cleaning, to a predetermined degree, a portion of polymeric polysulfide groups to a curable rubber (—SH polymer chain terminals). Subsequently, these studies led to a wide range of liquid polymers, ranging from below 1000 molecular weight to viscous liquids of about 7000 molecular weight. LP™ became the trademark for these liquid polymers. Polysufides are polymers of bis (ethylene oxy) methane containing disulfide linkage. The reactive terminal groups are mercaptans (general structure is HS).[9]

Mercaptan-terminated liquid polymers

Conventional mercaptan-terminated liquid polymers (HS—R—SH) are available in wide ranges of molecular weight with cross-linking densities ranging from 0.05 mol% to 2.0 mol%.[10]

TABLE 12.8 LP™ Physical Properties Polysulfide Polymer

Typical Properties	LP-2
Color—MPQC-29	70 max
Viscosity, poises at 77°F (25°C)	410–525
Moisture content, %	0.3 max
Mercaptan content, %	1.7–2.20
General Properties*	
Average molecular weight	4000
Pour point, °F (°C)	45 (7)
Flash point (PMCC), °F (°C)	>350 (177)
% Cross-linking agent	2.0
Specific gravity at 77°F (25°C)	1.29
Average viscosity, poises at 4°F (−16°C)	3800
Average viscosity, poises at 150°F (65.6°C)	65
Low temperature flexibility, 10,000 psi, 708 kg/cm², °F (°C)	−65°F (−54°C)

*Cured sealant.

$$HS[—CH_2—CH_2—OCH_2—CH_2—S_x]_4—CH_2CH_2—OCH_2—OCH_2—CH_2—SH_{x\ Average} = 2.25$$

Table 12.8 lists the physical properties of the liquid polysulfide polymer, which allows the formulator a choice in selection.

LP™ polysulfide basic characteristics. Polymerization of liquid polysulfide polymers to a high molecular weight elastomer is normally accomplished by oxidizing the thiol (SH) terminals to disulfide —S—S bonds.

$$2—RSH + (O) \rightarrow —RSSR— + H_2O$$

$$2\sim RSH + MnO_2 \rightarrow \sim RSH\sim + MnO_2 \pm H_2O$$

The curing agents most commonly used as oxygen-donating materials are manganese dioxide, lead peroxide, calcium peroxide, cumene peroxide, and *p*-quinone dioxime. Lower valence metallic oxides, other organic oxides, metallic paint dryers, and aldehydes can also function as curatives. However, the most widely used are manganese dioxide, lead dioxide, and selected dichromates (protected by U.S. Patents 2,787,608 and 2,964,503). As with most sealant formulations, the selection of ingredients, such as fillers, pigments, and plasticizer, is governed by the end use. This is particularly true with the curing mechanism for sealants to be used as insulating glass compounds. Compared with other elastomeric sealants, polysulfides have some of the most extensive history as sealants. Polysulfide sealants were introduced into the construction market in the early 1950s. They enjoyed increasing popularity for the next 10 years and then began sharing the market with solvent-based acrylics and polyurethanes through the 1970s. In so far as polysulfide-based sealants for insulating glass are concerned, the first version of an organic sealant was originally used as a fuel tank sealant in aircraft. This material was a two-component, lead peroxide cured sealant. Unfortunately, after a few minor UV exposure tests, it was learned that this type compound would burn at the glass interface when exposed to ultraviolet light. The next improvement resulted from extensive experimentation with manganese dioxide based catalyst systems, still used in the industry today.

The insulating glass industry is a large market for sheet glass and the amount used is in excess of 10×10^7 m^2 worldwide. The edges of insulating glass units are sealed with a polymer-based sealing material, to hold the two or more lites of glass together and to provide a barrier against permeation of water into the unit. It must also be resistant to glazing compounds and paints used in the installation of the sealed unit. Polysulfides have been a dominant polymer in this application accounting for about 76.0 percent of the worldwide market, along with polyurethane, polyisobutene, and silicone.[11] Insulating glass units have anticipated lives of more than 10 years, but premature failure can occur. The manifestation of failure is misting or condensation of the internal surfaces of the sealed unit. The major cause of failure is adhesion loss between the sealant and the glass.

A typical insulating glass unit consists of two panes of double strength ($\frac{1}{16}$ in [4 mm in Europe]) glass separated by a hollow spacer. The spacer is generally aluminum or steel filled with a desiccant, silica gel, molecular sieve, or a combination of the two. Minute holes on the inner surface of the spacer allow water vapor in the unit access to the desiccant. The effect of water on polysulfide sealants, such as those used in insulating glass units, has been studied extensively. This data is used to assess the durability of the sealed units.

Materials. Polysufide sealants consists of two components, a base and a curing agent and/or catalyst. The properties of a 100 percent solids system are listed in Table 12.9.

The components are mixed together by adding the calcium carbonates and the titanium dioxide to the liquid polysulfide polymer in a high-shear mixer and mixing for 15 to 20 minutes to form a smooth viscous paste. The thixotrope and adhesion promoters are added, and the mixing is continued for an additional 20 minutes. Finally, the plasticizer is added and the mix is adjusted to a final workable viscosity.

TABLE 12.9 Typical Cured Properties of 100% Solids

Composition Base (Part A) pigments Accelerator (Part B)	LP® Liquid polysulfide polymer, fillers, and metallic oxides and plasticizer
Physical characteristics (typical values)	
Base (Part A)	
Weight/gal	15 ± 0.4 lb
Total nonvolatile	100.0% ± 1.0
Viscosity 30 psi, 0.125 in orifice 175-g charge	200 ± 50 s
Slump	0.10 in, max.
Accelerator (Part B)	
Weight/gal	14.4 ± 0.4 lb
Viscosity	2000–5000 poise
Particle size	4 mil max.
Sealant (Part A & Part B)*	
Stick life	1–2 h, adjustable
Tack-free time	1.0–4.0 h, adjustable
Slump	0.05 in, max.
Hardness, Shore A 24 h 72 h	 20–40 30–50
Peel adhesion, piw 24 h 72 h	 10 min 10 min
Shear strength 24 h 72 h	 50 min 60 min
Mix ratio	Base/accelerator
By weight	13.0/1.0
By volume	12.5/1.0

*Tests were performed under standard laboratory conditions in accordance with ASTM E 773/774.

The curing agent is a paste consisting of manganese dioxide, plasticizer, and tetramethyl thiuram disulfide plus filler.

Fillers are added to the polysulfide polymer to reduce cost and to modify rheological properties. Precipitated calcium carbonates have much smaller particles than ground counterparts. About 90 percent of the particles are smaller than 1 μm, in comparison to the ground calcium carbonate, which has a distribution in the range of 2 to 10 μm. A coat of stearic acid on the calcium carbonate particles increases the

TABLE 12.10 Typical Cured Properties

Properties	Physical data	
100% modulus, psi (kg/cm^2)		
300% modulus, psi (kg/cm^2)	350 (24.6)	
Tensile strength, psi (kg/cm^2)	410 (28.8)	
Elongation, %	510	
Hardness, Shore A, 2 s	50	
Thermal conductivity (ohm-cm)	3.3×10^{-4} cal/s cm^2/^{0}C/cm	
Heat of combustion	5284 cal/g	
Specific heat	0.42 cal/g	
Coefficient of expansion	2.2–2.4 $\times$ 10^{-4}cm/cm/^{0}C	
Peel adhesion, pli, aluminum to glass		
	24 h	10.0
	72 h	15.0
Shear strength, piw	24 h	30.0
Glass to glass, piw	72 h	40.0

viscosity of the base and introduces thixotropy. Rutile titanium dioxide is added as a whitening pigment to assist in visualizing complete mixing of the base and catalyst. Plasticizer is used to lower the viscosity of the base compound. Generally, bentonite is added as a thixotropic agent and 3-glycidoxpropyltrimethoxysilane is incorporated as an adhesion promoter.

Manganese dioxides used for this purpose are not pure compounds but are activated by the addition of alkali metal salts or hydroxides. Manganese dioxide is activated by sodium hydroxide.[12] The role of the tetramethyl thiuram disulfide is to accelerate the reaction. The process involves oxidative coupling of two —SH end groups on the polysulfide molecules to form an —S—S—, resulting in chain extension. The reaction is complex but proceeds by a free radical mechanism.[13,14] Table 12.10 presents the typical cured properties of a polysulfide insulating glass compound.

Polysulfide insulating glass sealants. This segment is devoted to polysulfide-based insulating glass sealants. However, it should be noted that the final selection of a curing agent is based on its overall performance with respect to a number of requirements, including cost, stability, controllable cure rate, heat stability of cured composition, elastomeric properties, toxicity, etc. The following are some of the differences obtained with each curing agent. Certain polymer selections require a smaller amount of curative because of the molecular weight of the polysulfide polymer.

With the LP-2 polymer, PbO_2 provides an easily controllable polysulfide polymer curing rate. The amount required per 100 parts of polymer is 7.5 to 10.0 parts. An active grade of MnO_2 provides a polysulfide polymer-based composition with improved heat resistance, using a ratio of 7.5 to 10.0 parts of catalyst to 100 parts of polymer. CaO requires moisture for activation and is usually used in systems that require a white color and not in an insulating glass sealant formulation. Generally 10.0 to 12.0 parts are required to cure the polymer. ZnO is also used in white formulations and is slow to react. However, it does provide improved heat resistance, using 10.0 parts to 100 of the base polymer. There are other curing agents, such as cumene hydroperoxide and *p*-quinonedioxime, that can be used, depending on the application.

Durability and physical characteristics. The factors which influence the life of an insulating glass unit are numerous and complex. Although the durability and effectiveness of the sealant systems are certainly important, the adhesion to the glass and spacer material is fundamental. Short-term adhesion is relatively easy to obtain. Long-term adhesion is more difficult[15] because of the dynamic nature of the sealant and its interaction with the environment. As an example, water and water vapor enter and exit the sealant, plasticizers migrate to the surface, and chemical changes, such as curing and degradation, continually take place in the sealant.

At the same time, glazing compounds, wood preservatives, and cleaning fluids are brought into contact with the sealant. Thus, the sealant is also subjected to a variety of continuous mechanical stresses. The critical factors in determining the durability of the insulating glass sealant and, hence, the insulating glass unit itself are polymer type and polymer content. LP polysulfide-based sealants are the predominant product worldwide. The observations made in this section on polysufides refer only to this type of formulated sealant.

Liquid polysulfide polymer-based sealants display excellent resistance to a number of oils and solvents. The electrical properties are very good and can be improved even further by experimenting with other types of fillers and plasticizer. The cured sealant displays excellent resistance to ageing, ozone, oxidation, sunlight, and weathering. Long term "Weather-o-meter" tests and actual usage indicate that these materials have excellent resistance to daily exposure in varying climatic conditions.[16] The service temperature range for cured formulations is −65°F (−54°C) to 225°F (107°C). For intermittent service, temperatures as high as 350°F (177°C) are feasible. Excellent adhesion to most substrates, coupled with good flexibility, is obtained with a properly formulated composition. See Table 12.10 for adhesion values to some substrates.

Mechanical edge stress. It is difficult to determine the precise distortions occurring at the sealant edge but the shear and tensile stresses and strains due to different movements can be estimated. Figures 12.10 and 12.11 illustrate two types of movement to be considered. Figure 12.10 shows the effect of the differential thermal expansion/contraction at the unit edges; movement occurs in both the x and y directions. Figure 12.11 shows the pumping action created by temperature fluctuations and external pressure changes. These tensile and angular stresses are in addition to the shear stresses indicated in Figure 12.10. Moreover, stresses induced by wind, snow (in horizontal glazing applications), etc., must also be considered.

To withstand the forces applied, it is essential that the edge seal has good long-term adhesion, even when subjected to high humidity at elevated temperatures or even to water immersion, a condition often found in modern window glazing. In addition, the sealant must retain these properties at very low temperatures.

Interaction with water. The role of water and water vapor is complex and requires a thorough knowledge of permeation and also of solubility, diffusion, and interfacial hydrolyzability. Because all the mecha-

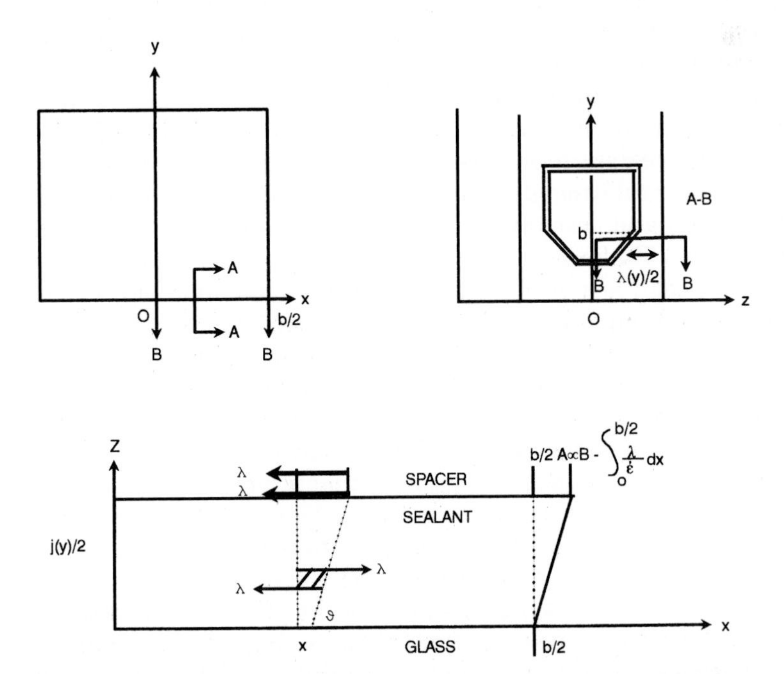

Figure 12.10 Changes because of shear action on unit edge resulting from thermal changes.

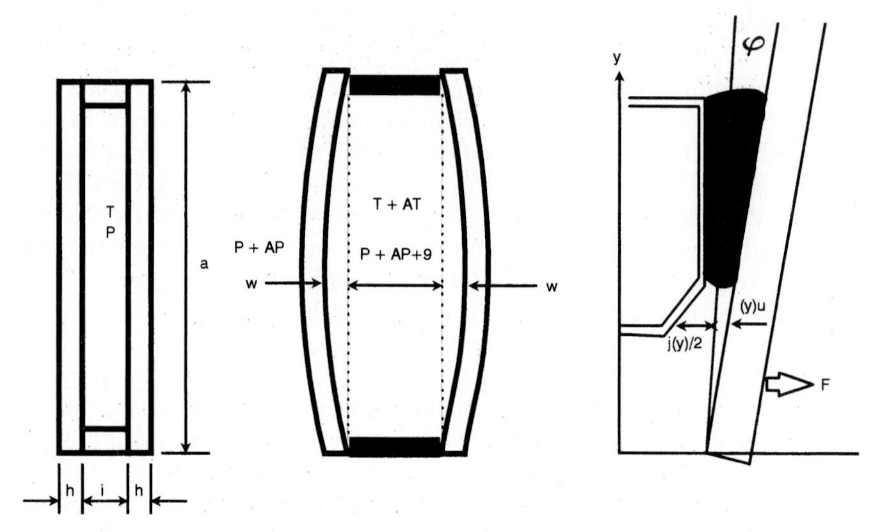

Figure 12.11 Pumping action because of temperature fluctuations and pressure changes.

nisms act simultaneously, the sealant formulators have to find a careful balance between them and understand that improving one property may have an adverse effect on the other properties. It will be shown that the best polysulfide sealants have a polymer content in the range of 30 to 40 percent. Great emphasis is placed on MVTR in insulating glass sealant literature. However, if this were the only factor, most of the insulating glass units in the market would endure for at least 50 years and possibly for as long as 300 years.[17] It is possible to reduce the MVTR for a specific polymer type by using low-polarity plasitcizers at high levels. However, if excessive amounts of a plasticizer with finite compatibility are used, adhesion to glass can be lost prematurely. Thus, it is wrong to base the sealant formula solely on minimizing the MVTR. In polysulfide sealants, the water diffusion in the system is related to the polymer content, the type of curing agent used, and the water mass uptake. For the last property, it has been shown, for ideal compounds, that the optimum polysulfide polymer level in the sealant (by weight) lies between 35 and 40 percent, based on low diffusion coefficients with long times to reach equilibrium.[18] It must be remembered that a low water mass uptake is not in itself an indication of service life. If the weakness of the bond is related to the fractional mass uptake, the optimum polymer concentration would be 43 percent, based on the time to reach equilibrium at a reasonably low saturation equilibrium (see Table 12.11).

TABLE 12.11 Time to Reach Equilibrium versus Polymer Content in Water Mass Uptake*

Experiments	
Polymer content, %	Time to equilibrum, days
54	45
51	63
48	144
45	169
43	210
41	196
38	169
37	196
27	81

*Test conditions at 140°F (60°C).

Water uptake. Four sheets of various sealant formulations were conditioned at 140°F (60°C) and 95 percent relative humidity on shelves of stainless steel wires so that both sides were exposed. The samples were periodically removed for weighing. The measured weight increases of the cured sheets of sealants exposed to the 140°F (60°C) and 95 percent rh were plotted in the form of mass uptake against the square root of time. The diffusion coefficient D was obtained from the mass uptake required to reach half the equilibrium value $t_{(1/2)}$ using equilibrium, where L is the thickness of the sample:

$$D = 0.049\ L^2/t_{(1/2)}$$

The behavior of the joints on ageing cannot be accounted for by water diffusing through the sealant to the interface and weakening it, because the rate of water diffusion is low. This point is illustrated by the information in Table 12.12, which presents the calculated water

TABLE 12.12 Calculations on the Amount of Water at the Edge of the Interfacial Failure Zone for Joints

Ageing time, mo	Dt/L^2	Distance of zone edge, mm	C/C_1 (concentration)	Concentration, %
2	8.6×10^{-3}	0.5	2.3×10^{-3}	0.12
3	1.3×10^{-2}	2.0	3.5×10^{-3}	0.16
4	1.7×10^{-2}	3.5	4.6×10^{-3}	0.21
6	2.6×10^{-2}	5.0	6.9×10^{-3}	0.31

concentration at the inner edge of the interfacial failure zone for joints of a given sealant. The table includes the value of Dt/L^2 for the joint, the value C/C_1 at the edge of the zone, and the actual water concentration at this point. If water diffusion were an important factor in debonding, we would expect that the concentration at that edge of the zone would be constant and that the actual concentration would be much higher. In fact, it would be expected to be a significant proportion of the 45 percent water absorbed by the polysulfide sealant at equilibrium. Values of C/C_1 were calculated from Dt/L^2 using the following equation which gives concentrations C within a lamina of thickness $2L$ immersed in water vapor or liquid so that the surface concentration diffusant C_1 remains constant; t is time. The origin of coordinates is at the center of the lamina.

$$C/C_1 = -(4/\pi) \sum_{n=0}^{\infty} [(-1)^n/(2n + 1)] \exp[-D(2n + 1)^2\pi^2 t/4L^2] \cos [(2n + 1)\pi/2L]$$

It might have been (but is not the case) that, in the polysulfide–glass joints, there is a critical water concentration above which the interface is significantly weakened. This has been demonstrated in some other systems. When adhesive joints with metallic adherents are exposed to air of high humidity (e.g., 80 to 100 percent), they weaken in time; this fact is widely known and has been reported in literature and covered in review.

In contrast, it has been observed that joints can withstand exposure at lower humidities (50 percent rh or less) for long periods without weakening. This refers to some epoxide–aluminum joints which showed no loss of strength after exposure to laboratory humidity for up to 11 years. Butt joints with an epoxide sealant exposed at 68°F (20°C) and 55 percent rh for 2500 hours did not weaken. Because polysulfides adsorb about 10 times more water than epoxides, it might be expected that the critical water concentration for sealants would be about 10 to 15 percent, much more than the very low values calculated in Table 12.12.

Dry and wet adhesion of the system are clearly important factors but are often misused. From thermodynamic study, it can be shown that both wet and dry adhesion are mostly independent of polymer content, so a high polymer content is not necessary for good initial adhesion. More important, it has been shown that a sealant with good initial adhesion does not necessarily have a reliable service life; although the dry work of adhesion is positive, that for wet adhesion can easily be negative. See Table 12.13.

Plasticizer migration. The rate at which the plasticizer migrates toward the adhesion surface depends on the plasticizer level and its

TABLE 12.13 Changes in Work of Adhesion (Dry) vs. Polymer Content

Polymer content, %	Work of adhesion
54	$99 \pm 12 \cdot$ psi
51	$82 \pm 21 \cdot$ psi
48	$98 \pm 22 \cdot$ psi
45	$76 \pm 15 \cdot$ psi
41	$83 \pm 15 \cdot$ psi
38	$80 \pm 18 \cdot$ psi
37	$71 \pm 16 \cdot$ psi

TABLE 12.14 Mass Uptake of Plasticizer Compared to Initial Concentration of Plasticizer in Sealant

Plasticizer in sealant, phr	Mass uptake of plasticizer, %
20	1.7
40	4.5
50	3.75
60	1.8

compatibility and temperature. Compatibility may change for the worse as the polysulfide-based sealant ages or as it adsorbs water. Early studies[19] found, in some cases, that adhesion failure had been accompanied by an increase in plasticizer concentration at the interface. Long-term durability studies have been carried out, showing that there is a maximum plasticizer to polysulfide polymer ratio beyond which premature adhesion loss occurs in a high-humidity environment. In essence, the sealant should contain a minimum of 30 percent polymer and a maximum of 15 percent plasticizer. See Table 12.14.

Sealants having glass transition temperatures within their normal working temperature range of −15°F to 140°F (−26°C to 60°C) will undergo massive stress changes as the temperature decreases. It has been shown[19] that the glass transition temperature of a polysulfide insulating glass sealant decreases as the polysulfide content increases. Thus, resistance to low-temperature stress is helped by a high polymer content (Fig. 12.12).

Examination of failure surfaces. The glass surface from the ruptured bond with a sealant, which contained a high proportion of plasticizer, was rinsed with diethyl ether, and the material removed was exam-

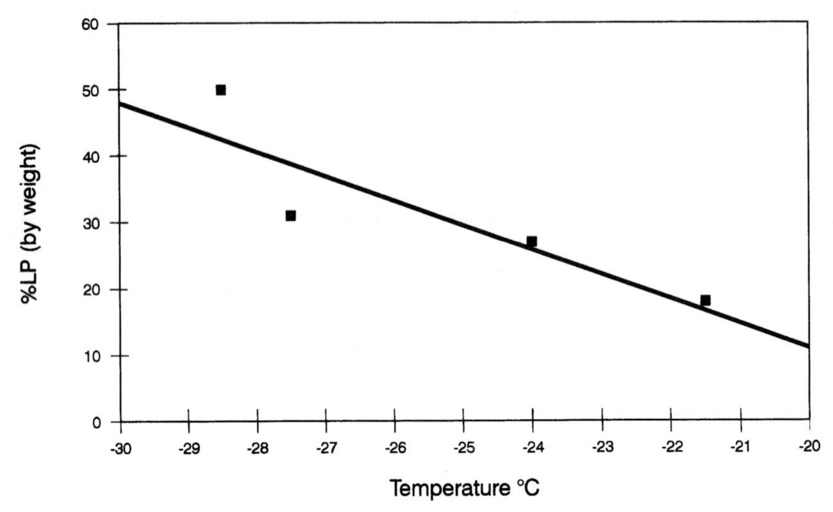

Figure 12.12 Percentage of LP polymer vs. glass transition temperature.

TABLE 12.15 Elemental Analysis (Atomic %) by XPS on Glass Surfaces from Joints

		Ageing time, mo.		
Element	Peak	1	2	10
C	1s	72	67	62
O	1s	20	20	22
S	2p	5.3	3.0	0.0
Mg	A	2.1	4.6	3.3
Ca	2p	0.9	1.3	1.6
Si	2p	0.0	3.0	7.7
Na	A	0.0	1.1	3.0

ined by infrared spectrophotometry. The spectrum obtained was identical with that of the plasticizer. Weakening of the interface could be due to the plasticizer diffusing at the interface. Indeed, joints with sealants, which contained the greatest proportion of plasticizer, all failed catastrophically on humidity ageing. X-ray photoelectron spectra were obtained on glass surfaces taken from joints after various periods of ageing and are shown in Table 12.15. Ignoring hydrogen, which is not detected by XPS, the atomic composition of the polysulfide is 53.7 percent C, 32.8 percent S, and 13.4 percent O, and the plasticizer composition is 81.1 percent C and 18.2 percent O. The composition of the matrix of a sealant would be 60.8 percent C, 24.6 percent S, and 14.6 percent O. The analysis of surfaces for the aged samples show an increase in oxygen and a reduction in sulfur, which is consistent with plasticizer displacing polymer at the interface. As the amounts of the elements Mg, Ca, Si, and Na (which are present in glass) generally increase on ageing, the layer of organic substances left on the glass becomes thinner. Interfacial sulfur decreases, but oxygen increases on ageing, which also indicates that the amount of plasticizer at the failure surface increases. It is possible, however, that the migration of plasticizer to the interface occurs after the sealant has failed.

Thermal analysis. Figure 12.13 shows the DMTA (dynamic mechanical thermal analysis) for a sealant after wet ageing. The unaged sealants all show a single peak in the region of −10 to −20°F (−23 to −29°C) which is the glass transition temperature of the unplasticized sealant. This shows that, although termed a plasticizer, the material used does not plasticize in the usual manner, that is, by lowering the transition temperature.

On wet ageing, the position of the peak did not change, but a second peak appeared in the region of 38 to 42°F (3.3 to 5.6°C). The latter peak results from the melting of the water and was absent in the samples that had been wet-aged and then dried. The data shows that water does not plasticize the polysulfide sealant and that most is isolated as droplets. Samples which had been wet-aged showed a prominent peak slightly above 32°F (0°C), which results from the adsorbed water.

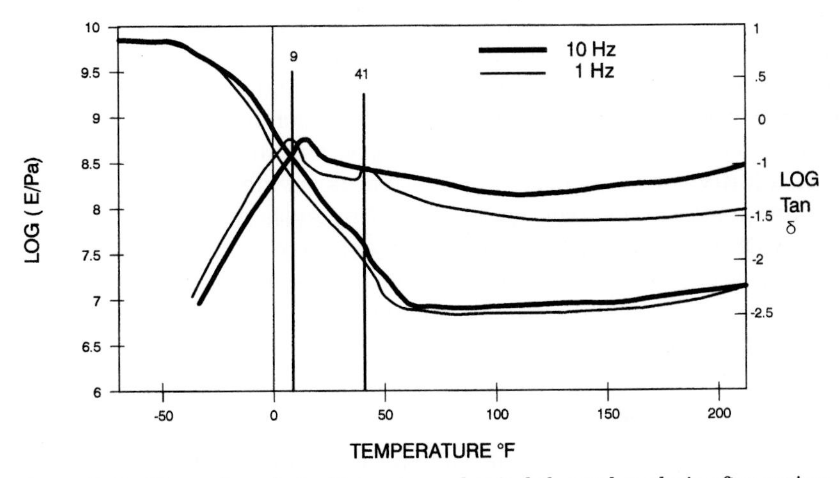

Figure 12.13 Thermal analysis. Dynamic mechanical thermal analysis after ageing at 140°F and 95% RH at 1 Hz and 10 Hz.

Displacement by plasticizer. The surface of a liquid is the sum of dispersive and polar components (Eq. 12.1) and the interfacial tension ($\gamma_{1,2}$) between two liquids (1 and 2) is given by Eq. 12.2.

$$\gamma = \gamma_L^D + \gamma_L^D \qquad (12.1)$$

$$\gamma_{1,2} = \gamma_1 + \gamma_2 - 2(\gamma_1^D \gamma_2^D)^{1/2} - 2(\gamma_1^P \gamma_2^P)^{1/2} \qquad (12.2)$$

The interfacial tension measured between a silicone fluid and plasticizer was 4.0 mNm^{-1}. The surface tension of silicone was 18.5 mNm^{-1} and that of the plasticizer was 37.0 mNm^{-1}. By making the assumption that the polar component of surface tension for the silicone is zero, Eq. 12.2 gives a value of the dispersive component for the plasticizer of 35.8 mNm^{-1}. Hence, the polar component is 1.2 mNm^{-1}. The work of adhesion between polymer and glass in the presence of plasticizer, calculated from these values, is 26 ± 14 mNm^{-1}, showing that plasticizer cannot displace the polymer from glass.

Hot Melt Systems

Hot melt sealants are known by many chemical definitions or classifications and, essentially, are used as the primary sealant in a dual-seal system for edge sealing of insulating glass units. They combine the speed and convenience, along with the durability, of liquid polyurethane or polysulfide sealants. Hot melt edge sealants provide many diverse benefits, including excellent heat resistance, water and solvent resistance, absence of solvents, and relatively low application temperatures, below 250°F (121°C), which provide immediate handling bonds and the ease of one-sided application.[21]

Butyl and sometimes certain polyisobutylene grades are used as the base or binder in a wide range of hot melt sealants for the insulating glass industry. Sealants are made from isobutylene polymer and are usually compounded with plasticizers, suitable inert fillers, and other additives designed for application as insulating glass sealers. Sealants used for construction of insulating glass must possess excellent weathering, ageing, and durability characteristics but must have a low moisture vapor transmission rate and be nonfogging. The butyl polymers fulfill all these requirements and have been successfully used in insulating glass sealants for years.

Types of hot melt sealants

It is essential to compare the various hot melts and their specific properties:

Ethylene vinyl acetate (PSA)

Butyl rubber hot melt

Polyester hot melt

Pressure-sensitive adhesive hot melt

Chemical backbone

Structure of butyl rubber molecule[22]

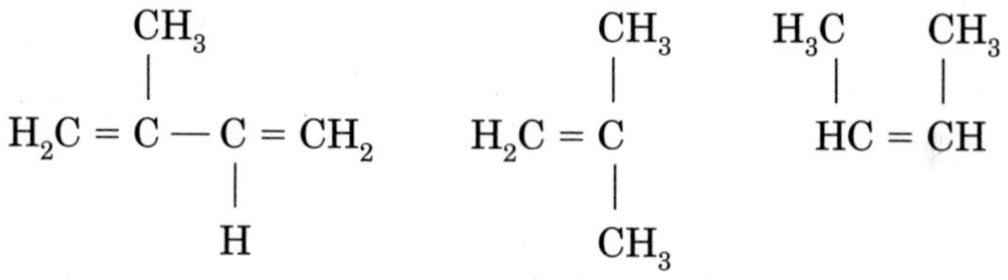

$$\underset{\text{Isoprene unit}}{H_2C = \overset{\overset{\displaystyle CH_3}{|}}{C} - \underset{\underset{\displaystyle H}{|}}{C} = CH_2} \qquad \underset{\text{Isobutylene unit}}{H_2C = \overset{\overset{\displaystyle CH_3}{|}}{\underset{\underset{\displaystyle CH_3}{|}}{C}}} \qquad \underset{\text{Butylene}}{\overset{\overset{\displaystyle H_3C}{|}}{HC} = \overset{\overset{\displaystyle CH_3}{|}}{CH}}$$

Typical Insulating Glass Formulations

Tables 12.16, 12.17, and 12.18 present a few typical sealant formulations used in the insulating glass industry.[23,24] These single component, solvent-free sealants provide immediate handling.

TABLE 12.16 Typical Polybutene Formula

Material	Weight, %
Butyl elastomer	15.6
Carbon black	23.5
Phenolic resin	29.5
Polybutene H-1900 polymer	23.5
Alpha methylstrene polymers	—
Ethylene propylene rubber	7.9

TABLE 12.17 Typical Polybutene Formula with Additives

Material	Weight, %
Butyl elastomer	16.6
Carbon black	25.5
Phenolic resin	25.6
Polybutene H-1900 polymer	17.5
Alpha methylstrene polymers	5.9
Ethylene propylene rubber	8.9

TABLE 12.18 Typical Polyisobutylene Starting Formula for an Insulating Glass Sealant

Material	Weight, %
Vistanex LM-MH	100
Mistron vapor talc	48
Carbon black N990 (MT)	2

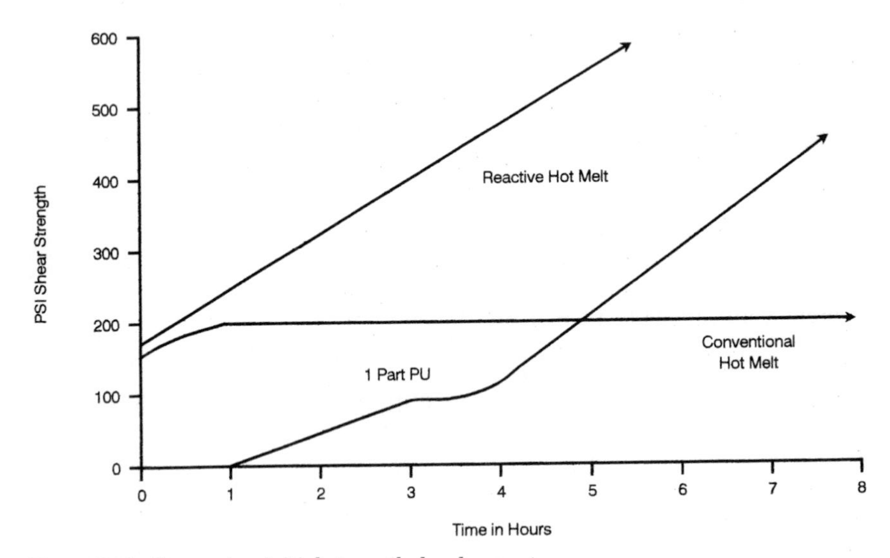

Figure 12.14 Comparing initial strength development.

Conventional hot melts are thermoplastic and do not cure or crosslink. Figure 12.14 compares initial strength development, measured as shear strength.[21]

Therefore, the standard hot melts have limited heat resistance, well below 225°F (107°C), and poor solvent resistance. Also, conventional hot melts are typically applied at 325°F (162.5°C) to 400°F (204°C), and, as a result, they distort heat-sensitive substrates such as thin PVC films. Figure 12.15 is a graph of cured heat resistance of a couple of hot melt sealants.[21]

Typical properties

Application properties	Extrudable at 230–266°F (110–130°C). Pressures of 50 psi
Hardness, Shore A	54–64

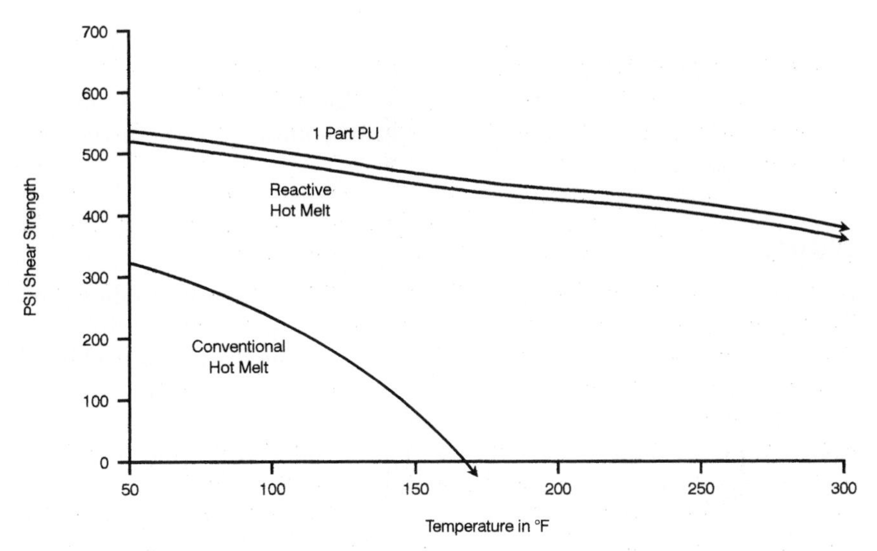

Figure 12.15 Cured heat resistance.

Penetration, mm	12
Dead load, h to failure at 158°F (70°C), 500 g load	4.2
Lap shear, psi (kPa)	407–427 (59–62)
Adhesion	Excellent to glass, aluminum, stainless steel, and galvanized steel
Moisture vapor transmission	0.024 g/24 h/m^2/mm Hg at thickness of 2.4 mm
Specific gravity	1.1
Shear strength	0.2 N/mm^2 at 30 psi at a film thickness of 1–2 mm
Service temperature range	−30 to 200°F (−34 to 93°C)
Colors	Black, gray

Advantages

Hot melts are 100 percent solids, thus reducing transportation costs, with a safer working environment and higher production speeds. Of great importance, there are no solvent emissions and less shrinkage. The hot melt insulating glass sealants require lower energy consumption, therefore requiring no additional curing equipment. Repair is easily accomplished because of their thermoplastic nature and they can be reactivated by heat. They have excellent freeze stability and are easy to handle and store.

Limitations

Hot melt butyl systems have limited shelf life and require specialized equipment to extrude the hot materials, which require strict temperature control. Because of the temperature, many products require extra venting of fumes.

Silicones

With the debut of neutral cross-linking silicones, many applications for insulating sealants, previously considered particularly difficult, have become quite feasible. New solutions now even address problems associated with insulating glass. Prior to understanding the use of silicone sealants, it is important to remember what has been learned earlier in this chapter about other sealant backbones and how they are used in the various edge-seal systems.

Most silicone sealants belong to a family of products called RTVs (room temperature vulcanizing), versions of which have been in use since the early 1960s. The use of silicone sealants in fabricating insulating glass units began in the mid-1970s; these were one-component (no mixing) systems. Because of the requirements for shorter cure times and longer pot life, the two-component product was developed and introduced to the industry in 1978.[25]

Secondary sealants

The secondary sealant in an insulating glass unit is required to perform a two-fold function. It is necessary to have good adhesion to ensure the seal and continuous contact between the two lites of glass by adhering perfectly to the glass and the spacer. The butyl must be able to perform its moisture vapor barrier function by keeping it under continuous pressure, and not allowing it to be subject to movements exceeding its capacity because of the action of rolling or static loads of the insulating glass. To perform these functions, the secondary sealant must possess good mechanical properties and excellent adhesion to glass and to all other components, in addition to resistance to traction and to cutting sufficient to withstand all the rolling and static loads sustained by the insulating glass unit. The secondary sealant must also have superior elasticity and must recover its initial mechanical properties once deformed by rolling loads.

The secondary sealant or silicone, as is the case, should have outstanding resistance to ultraviolet radiation and the capacity to withstand chemical attack from atmospheric agents, such as oxygen, ozone, and other pollutants. The secondary sealant should not contain aggressive volatile products capable of migrating to the interior of the air space of the insulating glass unit and chemically attacking the glass, the butyl, or the other components.

Among the other essential qualities of the silicone sealant are good chemical and physical compatibility with the materials with which it comes into direct contact, the capacity to maintain its chemical, physical, and mechanical properties across a broad temperature range, nonadsorption of water so that the latter does not come in contact with the primary sealant, low permeability of moisture vapor, and resistance to ageing. The intrinsic characteristics of the secondary silicone sealant must be accompanied by those "application characteristics" which ensure that the sealing of an insulating glass unit is as easy as possible, so as to guarantee full utilization of the properties previously mentioned.

Application characteristics

These characteristics are

1. outstanding rheological characteristics of the sealant which render both its extrusion and its application problem free,
2. for two component systems, the capacity to maintain mechanical, physical, and chemical properties intact over a broad base-catalyst mixture ratio range,
3. the ability to maintain homogeneity of the cross-linking characteristics in response to changes in environmental conditions such as temperature, humidity, etc.,
4. the possibility of adjusting and refining the sealing of the insulating glass unit during the manufacturing process, and
5. the ability to select, according to the application requested, fully compatible single- and two-component sealants.

As in the case with secondary silicone sealants, the other components of insulating glass also should be selected for compatibility with the manufacturing and application processes. At this point, it is necessary to divide secondary sealants into two categories according to their chemical and physical nature. The first group, called organic sealants or those based on the chemistry of carbon, includes polysulfides, polyurethanes, and hot melts that possess diverse chemical and physical characteristics and display all the advantages and disadvantages linked to their organic nature. These systems have been covered earlier in this chapter. Therefore, all their attributes will not be repeated.

One-component and two-component silicones

The one-part silicone cross-links with atmospheric moisture (humidity) and releases acetic acid. Optimum conditions for application and cure are 75°F (24°C) at 50 percent relative humidity. Temperatures below 60°F (15.6°C) and 40 percent relative humidity slow down the cure and affect ultimate adhesion and tensile strength.

The new generation of silicone sealants, on the other hand, is characterized by so-called neutral cross-linking. This means that the reaction does not release any aggressive products, but rather small quantities of alcohol and acids with a neutral pH.

The two-component variety is also characterized by cross-linking. The only difference is that the reaction is more rapid and homogeneous because it is controlled by a special catalyst.[26] Two-component silicones are not as dependent on atmospheric moisture as the single-component silicones and will cure in thick cross sections.

Chemical backbones. The basic polymer for the vast majority of silicone sealants is a hydroxy-ended polydimethylsiloxane.[27]

$$HO-\overset{\overset{\displaystyle (CH_3)_2}{|}}{Si}-O-(\overset{\overset{\displaystyle (CH_3)_2}{|}}{Si}-0)_x-\overset{\overset{\displaystyle (CH_3)_2}{|}}{Si}-OH$$

Three silicone sealant formulations suitable for insulating glass and their physical properties are shown in Tables 12.19, 12.20, and 12.21.

Performance. The characteristics of the silicone backbone sealants and the possibilities they offer will be analyzed. To thoroughly explore this issue, it is necessary to review the causes of dew point failure in sealed insulating glass units. To this end, we refer to the results of a study by IFT (Institut fur Fenstertechnik) in Rosenheim, Germany, that addresses the premature causes of dew point failure of insulating glass. The German Federal Government commissioned the IFT to scientifically and independently determine the causes of the excessive instances of dew point failure in insulating glass units in Germany. It was concluded that combinations of environmental factors, such as extreme temperatures, high relative humidity, and extensive exposure to ultraviolet radiation, represent the worst possible conditions for insulating glass and provide a valid testing ground for its quality and that of its components, in particular the secondary sealant. Figure 12.16 illustrates two different silicone edge conditions used in advanced insulating glass technology.

TABLE 12.19 Basic High-Modulus, Acetoxy Formulation

Component	Percent
Silanol polymer	80–85
Fumed silica (treated and/or untreated	6–8
Acetoxy cross-linker	5–7
Tin catalyst	0.05–0.1
Typical Properties	
Skin-over time, min	5–7
Tack-free time, min	10–20
Durometer (Shore A)	25–35
Tensile strength, psi	175–300
Elongation, %	200–400
Modulus, 100%, psi	75–125
Tear strength, lb/in	35–70

TABLE 12.20 Basic Medium-Modulus Oxime Cure

Component	Percent
Silanol polymer	80–85
Fumed silica (treated and/or untreated	5–10
Oxime cross-linker	5–7
Tin catalyst	0.05–0.1
Typical Properties	
Skin-over time, min	50–60
Tack-free time, min	10–15
Durometer (Shore A)	15
Tensile strength, psi	85–120
Elongation, %	>1000
Modulus, 100%, psi	20–25
Tear strength, lb/in	20–25

Another important fact relating to this study did not emphasize or focus on the permeability of the secondary sealant to moisture vapor, but rather the gradual loss of adhesion by the secondary sealant because of the combined action of rolling and static loads and the sealant's ageing induced by physical and chemical attacks by the environmental agents. Considerable loss of adhesion allows infiltra-

TABLE 12.21 Basic Low-Modulus, Neutral Cure

Component	Percent
Silanol polymer	46.0
Calcium carbonate	40.3
N-Methylacetamide chain extender	3.0
Aminoxy siloxane cross-linker	0.7
Typical Properties	
Skin-over time, min	10–20
Tack-free time, min	10–30
Durometer (Shore A)	25–35
Tensile strength, psi	175–300
Elongation, %	200–400
Modulus, 100%, psi	75–125
Tear strength, lb/in	35–70

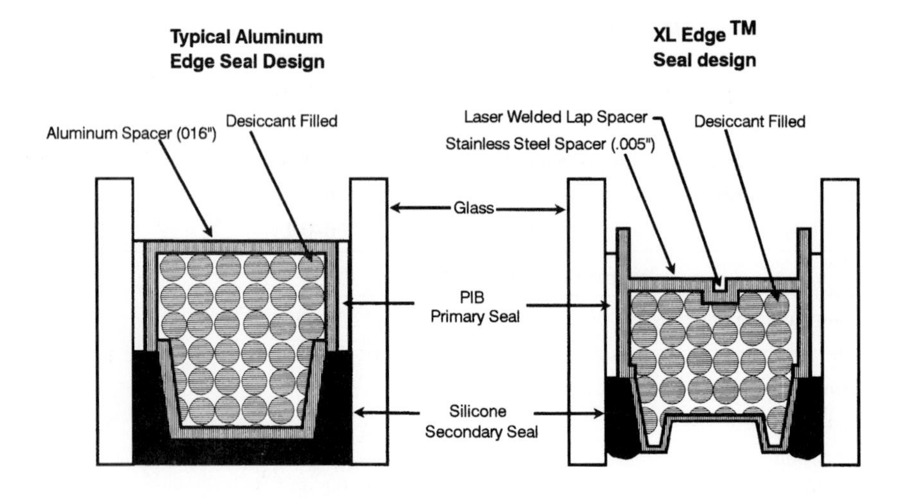

Figure 12.16 Aluminum and XL Edge™ designs.

tion of water in cracks and direct contact of water with butyl which, as noted, does not possess any physical or chemical resistance to this substance.

These phenomena, then, result in direct or indirect loss of adhesion and excessive dew points. Environmental factors that can cause the loss of adhesion by the secondary sealant, individually or in combination, are extreme heat or cold, high humidity or direct contact with water, chemical or physical attack by oxidizing agents (oxygen and ozone), direct exposure to ultraviolet radiation, and chemical and physical incompatibility with materials (glazing systems) which contact the edge sealant.

To examine the effect of temperature and the type of thermal stresses tolerated by the secondary sealant, we make several observations. The maximum temperature observed at the secondary seal was 140°F (60°C) in standard insulating glass units. Thermal differences of 158°F (70°C) were observed between the maximum temperature (summer) and the maximum temperature (winter) reached by the secondary sealant.

After examining the range of temperature extremes recorded, it is important to determine how the secondary barrier reacts to these conditions. The behaviors of organic sealants vary only slightly by type and quality of sealant but tend to be similar. In hot conditions, an increase in temperature causes migration of plasticizer and preservatives. The cross-linking process is prolonged causing increased hardening and modulus and a decrease in elastic qualities. In cold conditions, the extreme drop in temperature causes a decrease in the elastic properties of the sealant. Repeated exposure over time to temperature extremes and sudden thermal shocks adversely affects the adhesive quality of the sealant and causes a breakdown in its chemical, physical, and mechanical characteristics which, by nature, are particularly sensitive to the influence of temperature.

However, because of their chemical and physical nature (which render them less sensitive to temperature changes), neutral cured silicone sealants maintain all their properties essentially unaltered between 140°F (60°C) and 248°F (120°C).

Water is another culprit in the loss of mechanical properties because of the infiltration of moisture vapor into insulating glass units. As already stated, butyl maintains its elastic and adhesive properties with regard to glass only if kept under continuous pressure by the secondary sealant. Swelling results if the secondary sealant absorbs water, decreasing the pressure exerted on the butyl and causing a gradual loss of adhesion to the glass and an alteration of the butyl's primary function as the moisture vapor barrier. Water then condenses on the insulating glass surface. Absorption of water and the resultant swelling are not the same for all sealants. Comparative laboratory tests on sealants immersed in water at 140°F (60°C) for two months have produced the following results expressed as an increase in weight percentage:

1. polysulfides 30%
2. polyurethanes 12%
3. neutral, cured silicones 2–6%

The increase in volume, the root cause of the decease in pressure exerted on the butyl, was even more significant:

1. polysulfides 50%
2. polyurethanes 15%
3. neutral, cured silicones 3–10%

Therefore, neutral cured silicone insulating glass sealants in direct contact with water do not swell appreciably and maintain their elastic properties. Hence, the butyl may effectively perform its barrier function against moisture vapor without losing contact with the glass and without indirect contact with water. Other causes of edge-seal deterioration of some secondary sealant characteristics are attack by oxygen and ozone. By directly attacking the organic sealants, oxygen, and more specifically ozone, continues the cross-linking activity, which increases the modulus and, over time, reduces the elastic properties. Over the long term, with the continuous action of the rolling and static loads to which insulating glass is subjected, loss of adhesion and water penetration may occur.

Conclusion

Polyurethane-based, insulating glass sealants represent another major advance in the technology of the insulating glass industry. Although polysulfides have performed well, their limitations have been overcome by a new family of polyurethane sealants based on Poly BD™ resins. Now, more than ever before, sealants are available that maximize seal life with cost effectiveness. Many insulating glass manufacturers now believe that urethane sealants represent one of best balances of properties available.

Polysulfide sealant-to-glass joints are weakened by air exposure at 140°F (60°C), and some with higher levels of plasticizer fall apart during exposure. The amount of visually assessed interfacial failure for tested joints increases with exposure and is closely related to a discoloration, which develops in the polysulfide sealant and can be seen through the glass at the sight line. This area is permanently damaged and does not recover after drying.

The sealants adsorb large amounts of water (45 to 92 percent) but most of this appears as droplets. Water attacks the interface. Values of diffusion coefficients of water in the various sealants are very similar. The rate of joint weakening is greater than can be accounted for by the rate at which water diffuses into the sealant. The thermodynamic work of adhesion of sealant-to-glass bonds is positive in dry conditions (about 80 ± 20 mJm^{-2}) but probably slightly negative in the presence of water (about -5 ± 25 mJm^{-2}, indicating that water may displace the sealant from the glass.[18] Plasticizer cannot displace polysulfide from glass because the work of adhesion in its presence is positive (26 ± 14 mJm^{-2}).

A sealant formulation is always a compromise where a good balance of properties is sought at reasonable cost. However, none of these factors addresses the unit's durability and, when guarantees are being given beyond 10 years, this can be a serious oversight. Laboratory testing and an extensive field history require a good compromise between performance, durability, and cost. Polysulfide in an insulating glass sealant is best at a 30 percent polymer level.

Hot melt butyl formulations are a versatile group of sealants that have gained their niche in the insulating glass industry. Performance has been well documented over the past 20 years. Both large and small manufacturers have taken advantage of their economy and versatility.

Silicone sealants, not having reactive sites susceptible to chemical attack, maintain their elastic characteristics intact and withstand static and rolling loads in the absence of problems related to loss of adhesion or infiltration of water.

Another environmental factor that can directly affect the adhesive qualities of secondary sealants is ultraviolet radiation. Ultraviolet rays consist of electromagnetic radiation distributed across a wavelength range between 4 and 400 nm (nanometers). Glass is transparent to ultraviolet radiation exceeding 300 nm. In addition to the possibility of direct exposure, it has been calculated that 38 percent of incidental ultraviolet rays reach the adhesive interface of the secondary sealant in an insulating glass unit by reflection, even if it is protected by a coating. This leads to a direct attack on the adhesive interface of the secondary sealant itself.

Organic sealants are particularly sensitive to ultraviolet rays measuring approximately up to 350 nm. In fact, the energy associated with the ultraviolet radiation of this wavelength can be easily absorbed by the organic molecules of these sealants and is sufficient to cause two kinds of attack. One attack causes the homolysis of several links present in the polysulfide molecules (a reaction which, catalyzed by oxygen, also occurs frequently in the polyurethanes). The other is a direct photolytic attack on the highly photosensitive organic polymerization additives or intermediates present in the compositions of polysulfide and polyurethane sealants

References

1. Zachariasiewicz, M. *Poly BD™Resins—High Performance Materials for Insulating Glass Sealants* (Sartomer Company), Elf Atochem North America, Inc.
2. *Technical Information Bulletin, Poly BD™ resin,* Elf Atochem North America, Inc.
3. Duffy, J. W., and J. S. Amstock, The advantages of urethane insulating glass sealants, *U.S. Glass, Metal and Glazing,* 8/1994.
4. ASTM E 774 89, *ASTM 1989 Book of Standards,* Volume 4.07, ASTM, W. Philadelphia, PA, p. 608.
5. ASTM E 773 89, *ASTM 1989 Book of Standards,* Volume 4.07, ASTM, W. Philadelphia, PA, p. 603.
6. Evans, R. M., *Polyurethane Sealants,* Technomic Publications, Inc., Lancaster, PA.
7. Amstock, J. S., *Handbook of Adhesive Bonding,* Chapter 7, McGraw-Hill, Inc., New York, 1982.
8. Patrick, J. D., and Ferguson, H., U.S. Patent.
9. Bertozzi, E. R., Chemistry and technology of elastomeric polysulfide polymers, *Rubber Chemistry Technology,* 2/1959.
10. Singh, H., *Permapol® P-3 Polymers,* Products Research & Chemical Corp., Glendale, CA.
11. *European Insulating Glass Glazing Market,* Morton International Ltd., Coventry, U.K., 1988.
12. U.S. Patent 4,104,189.
13. Coates, R. J., Ph.D. Thesis, University of York, U.K., 1993.
14. Capozzi, G., and G. Modena, *The Chemistry of the Thiol Group,* John Wiley & Sons, New York, 1974.
15. Lowe, G. B., Durability of insulating glass units, *Glazing Today,* 3/1994.
16. *Liquid Polysulfide Polymer Brochure,* Morton International Ltd., Chicago, IL, 1995.
17. Backman, H., *Glass Control AB,* Predicting the Service Life of Insulating Glass Units, Sweden, 1976.
18. Lowe, G. B., Ph.D. Thesis, DeMontfort University, Leicester, U.K., 1993.
19. Hardy, A., and D. Burdett, Failure of sealants, *Adhesive Age,* 1965.
20. Lowe, G. B., *Dynamic Mechanical Thermal Analysis of Sealants for Insulating Glass,* Morton International Publication, 1988.
21. Duncan, K., National Starch and Chemical Company, Bridgewater, NJ, 1992.
22. Panek, J., and J. P. Cook, *Construction Sealants and Adhesives,* 3rd Ed., John Wiley & Sons, Inc., New York, 1991.
23. Amoco Chemical Company, *Amoco Polybutene Caulks and Sealants Formulary,* R0393, 1993.
24. Higgins, J. J., F. C. Jagisch, and N. E. Stucker, Butyl Rubber and Polyisobutylene in Adhesives and Sealants, *Handbook of Adhesives,* 3rd Ed., Irving Skiest, ed.
25. Sanford, A. G., Insulating glass sealants, uses and types, *Glass Magazine,* 1/1989.
26. "Insulating Glass and Sealants" Technology Section, p. 42.
27. Klosowski, J. M. *Sealants in Construction,* Marcel Dekker, Inc., New York, 1989.

Trademarks

Swiggle Seal™ a registered trademark of Tremco, a B.F. Goodrich Specialty Chemicals Company, Beachwood, OH

Super Spacer™ a registered trademark of Glastech, a subsidiary of Lauren Manufacturing Company, New Philadelphia, OH.

Intercept™ —a registered trademark of PPG Industries, Inc., Pittsburgh, PA.

LP™ Polysulfide a registered trademark of Morton International, Inc., Chicago, IL.

Poly BD™ a registered trademark of Elf-Atochem North America, Inc., Philadelphia, PA.

XL Edge™ is a registered trademark of Cardinal IG, Minneapolis, MN.

Chapter

13

Warm Edge-Seal Technology

Kevin Zuege

*Technical Service Manager, Tremco, an RPM-owned Company**

Ramesh Srinivasan

*Chief Engineer, Tremco, an RPM-owned Company**

Joseph S. Amstock

President, Professional Adhesive and Sealant Systems

Introduction

Low-E glass, inert gases (such as argon and krypton), thermal breaks, and warm-edge-sealant technology are what the glass industry has focused on to increase a window's thermal efficiency. Warm-edge-seal technology is one way of improving insulating glass energy performance. The edge-seal area, extending about 2½ in inward, can allow a great amount of heat to pass through the insulating glass unit. The edge seal or spacer/desiccant /sealant system, although important to the unit's long-term integrity, can also reduce the overall window thermal performance.[1] See Figure 13.1 for the three efficiency robbers: radiation, conductance, and convection.

*Kevin Zuege and Ramesh Srinivasan are coauthors of the section on Swiggle Seal™.

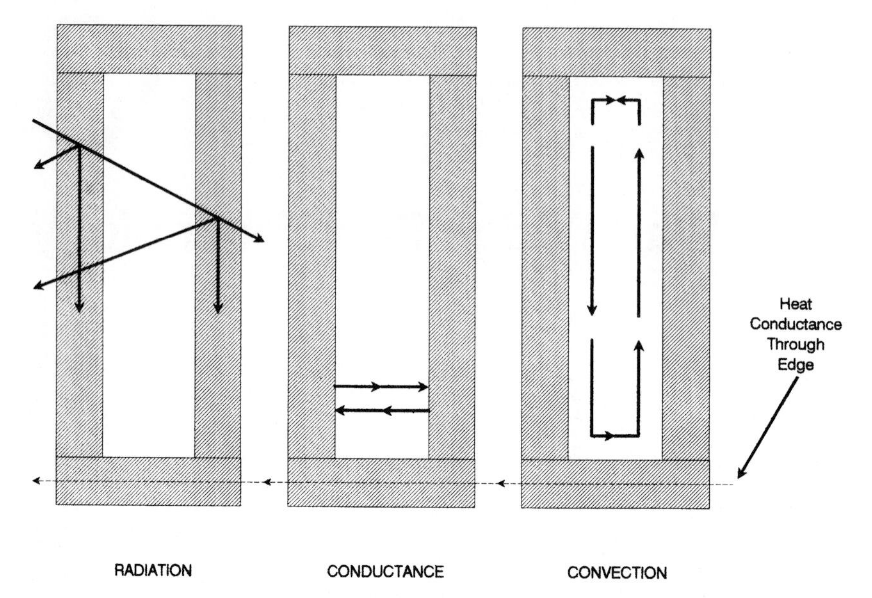

Figure 13.1 Efficiency robbers.

Because the window's edge-seal area is solid (glass/sealant/spacer/sealant/glass), it loses the beneficial effect created by the gap or airspace. Thus, the potential for heat transfer is greater in this area. Just as we have choices among several exotic gases and air, we also have choices of edge-seal systems, such as Super Spacer™, Swiggle Seal™, and the new Intercept™ system.

Choosing an edge-seal system of lower conductivity (or greater thermal resistance) can improve the overall thermal performance of an insulating glass unit. This depends on several factors. Figure 13.2 illustrates the areas affected by center-of-glass frame and edge-of-glass for a typical residential window. The window's overall thermal resistance, therefore, is a weighted average of the thermal resistances of these three factors.

Scientists at the University of Waterloo's Advanced Glazing Systems Laboratory (AGSL) in Ontario, Canada, developed a way of measuring thermal resistance. This entails isolating the edge of the glass from the rest of the insulating glass unit. The AGSL scientists built solid edge-seal sections composed of the various systems to be evaluated. These were mounted in a test fixture that created a temperature differential across each section. Heat flow was measured, and the results were categorized into two groups (see Figure 13.3), edge-seal systems with high thermal conductivity, i.e., low thermal resistance (LOW-TR), and those with low thermal conductivity, i.e., high thermal resistance (HI-TR).

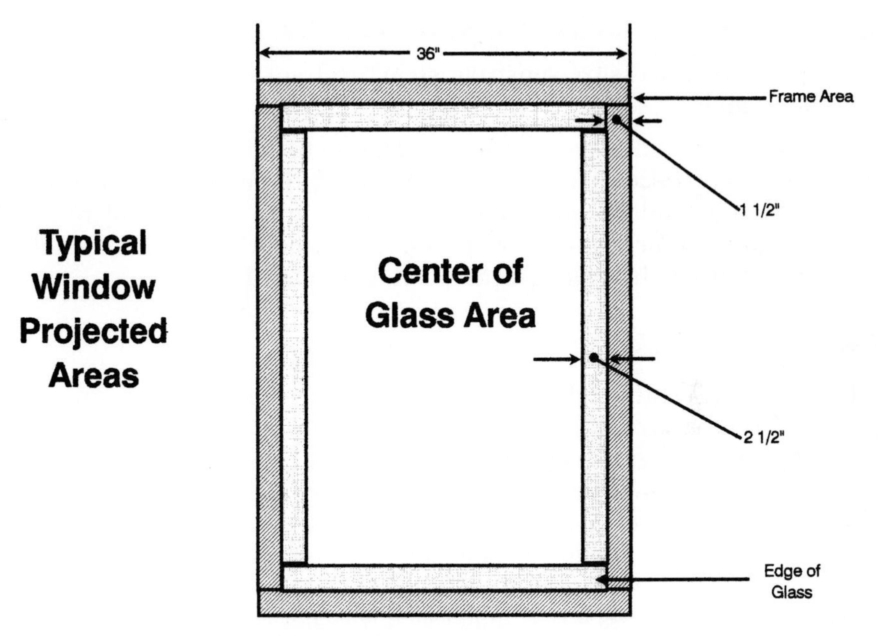

Figure 13.2 Contribution of glass vs. framing system.

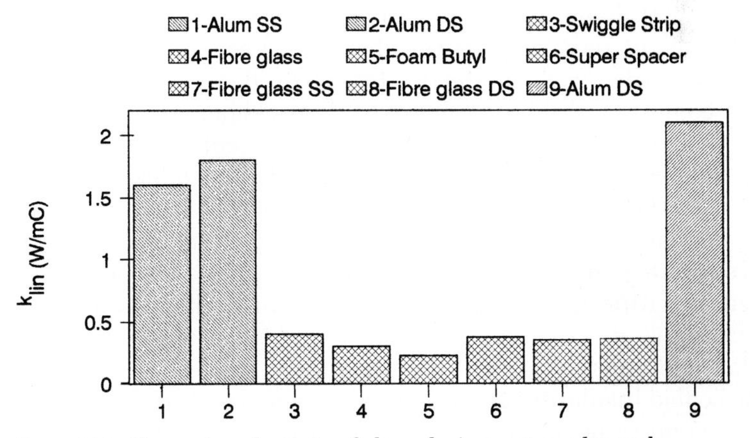

Figure 13.3 Thermal conductivity of glass glazing system edge seals.

Super Spacer™

Product description

This unique product was developed as part of a new edge-seal technology for high-performance thermal glazing.[2] Three special features are incorporated:

1. flexible silicone foam with a low-deflection desiccant fill,
2. preapplied, pressure-sensitive structural adhesive on the spacer sides, and
3. high-performance metallized backing film to help prevent moisture vapor infiltration and gas loss. Designed to be visually unobtrusive, the standard color is a soft neutral gray. Custom colors are available in large quantity orders. This product became available to insulating glass manufacturers in the fall of 1988.

Thermal performance

For a conventional wood-framed residential size window, up to 30 percent of the potential energy savings of high-performance thermal glazing units (Low-E coatings and argon gas fill) are lost because of conductive heat loss through a metal spacer. By substituting Super Spacer™, perimeter heat loss is substantially reduced, and the full potential of super window technology is finally realized. In addition, Super Spacer™ essentially eliminates condensation and substantially reduces thermal glass stress under cold weather conditions.

Edge-seal durability

To address the problem of perimeter heat loss, various efforts have been made over the past 25 years to substitute lower conductive plastic for metal as the spacer material. Primarily because of concerns relating to edge-seal durability, however, these rigid plastic spacers have not been commercially accepted. With the recent introduction of high-performance thermal glazing, there are new durability concerns especially with the use of insulating organic plastic spacers. One concern is an increase in the sealant stress due to increased glass bowing and differential expansion caused by the more extreme temperatures and pressures within these high-performance units. Higher air-space temperatures give rise to a second problem, namely, potential thermal degradation and outgassing from organic plastic materials contained within the sealed insulating glass unit. A third concern arises because of the permeability of plastic spacers and the sensitivity of Low-E sputtered coatings which will quickly be visually blemished if there is any moisture condensation within the unit. A fourth concern is the need to prevent long-term loss of low conductive gas from sealed units.

Rather than design a plastic copy of a hollow profile aluminum spacer, a new style spacer was designed. For superior insulating performance, Super Spacer™ (see Figure 13.4) is made from a low conductive silicone foam which contains a prescribed amount of desiccant

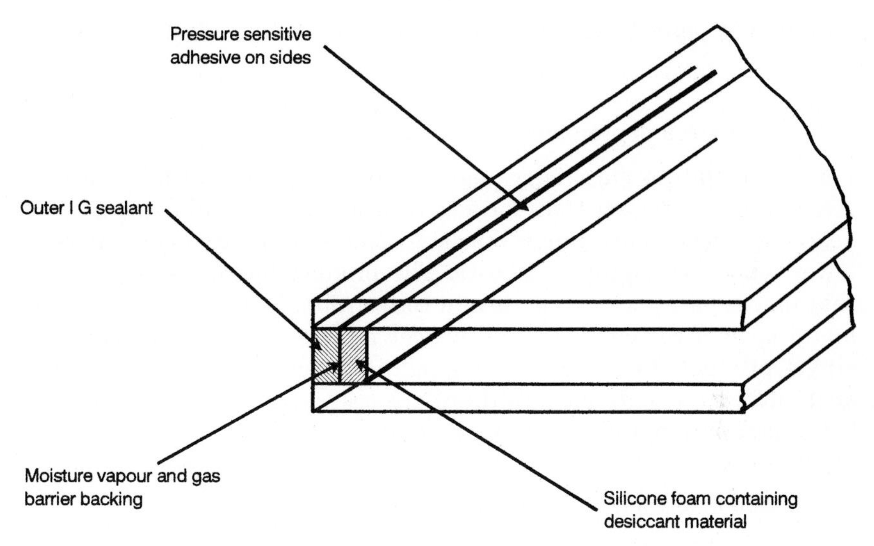

Figure 13.4 Cross section of Super Spacer™.

as a fill. The foam is flexible and can accommodate the increased glass bowing and movement of high-thermal-performance units.

Because silicone is an organic material, the product is very durable and, even after prolonged exposure, the material remains flexible and does not dust, plasticize, or degrade in any way.

Although other outer sealants may be employed, it is recommended that a hot melt butyl sealant be used for an ultralow-permeable edge seal. For conventional units, the uses of thermoplastic hot melt butyl sealants create a number of problems, such as spacer bar migration, cold creep, and loss of sealant adhesion at cold temperatures. Through the use of preapplied pressure-sensitive adhesive, the spacer design eliminates these traditional problems. The Super Spacer™ is firmly held in place by the preapplied structural adhesive on the spacer sides, and there is no problem of spacer migration. When the butyl units are being transported, typically, there is no need for elaborate packaging to hold the sealed glass sheets in position. At cold temperatures, the flexibility of the foam spacer also helps maintain butyl sealant adhesion. Additional protection against edge sealant failure is further provided through a high-performance, multilayer metallized barrier film laminated to the back of the spacer.

The desiccant system for this innovative spacer system has been specially developed for high-performance thermal glazing. To avoid gas adsorption at low temperatures, the low-deflection desiccant system incorporates a combination 3-Å molecular sieve and silica gel desiccant (see Chapter 15). The permeable silicone foam also ensures a rapid dew point temperature drop and very active adsorption by the desiccant.

Insulating glass unit production

For an insulating glass edge seal, the material costs of fabricating a sealed unit are higher than for a conventional edge seal using a metal spacer. However, with Super Spacer™, these higher costs are offset by lower labor and equipment costs. The product has been designed so that four separate components of a conventional edge seal—desiccant, metal spacer, corner keys, and spacer adhesives—are replaced by a single product. Because this desiccated spacer system is a flexible strip, it is easy to apply, and there are cost advantages in using it in both manual and automated production.

Specialized insulating glass unit designs

Because of the unique design features of Super Spacer™, there are advantages in using the product to fabricate specialized insulating glass designs including curved and bent glass units and odd-shaped, sound-insulating, and triple-glazed units with art glass inserts.

Thermal performance

With the introduction of high-performance thermal glazing systems, ultimately, the metal-type spacer can be declared obsolete and must be replaced by a newly designed insulating spacer. There are three factors to be considered:

1. *Accentuated perimeter heat loss.* In cold weather, the combined impact of Low-E coatings and gas fill is to substantially increase the interior glazing temperature by 43°F (8°C) or more. As a result, there is an increased temperature differential between the glazing layers and this higher driving force accentuates perimeter heat loss through the edge seal.
2. *Perimeter glazing condensation.* Despite the higher center glass temperature with Low-E units, there is still a problem of perimeter glazing condensation where a metal spacer is used. An insulating spacer acts as a thermal break and essentially eliminates perimeter condensation even in extremely cold weather.
3. *Increased thermal glass stress.* Low-E coatings intercept part of the solar spectrum causing the coated glass to heat up. On cold, sunny days, the center of the coated glass heats up and expands, but this expansion is constrained by a cold perimeter zone, creating stress in the glass pane. In extremely cold weather, this temperature differential can create sufficient thermal stress to cause glass breakage. An

insulating spacer reduces the temperature differences between the glass center and the perimeter zone, and so thermal glass stress is reduced to a minimum.

Superior insulating performance of silicone foam

For superior insulating performance, Super Spacer™ is made from low conductive foam. Even compared to solid plastic spacers, the spacer has a higher insulating performance. The improvement in thermal performance using this silicone foam product has been determined by computer analysis and direct measurements.

Computer analysis using FRAME program

The overall thermal performance of different glazing assemblies incorporated within small typically sized, wood-framed residential windows is shown in Table 13.1. These programs were generated using the FRAME program, which is a finite difference analysis model developed by Enermodal Engineering Ltd., Waterloo, Ontario. A sample output from the FRAME program is presented in Figures 13.5 and 13.6.

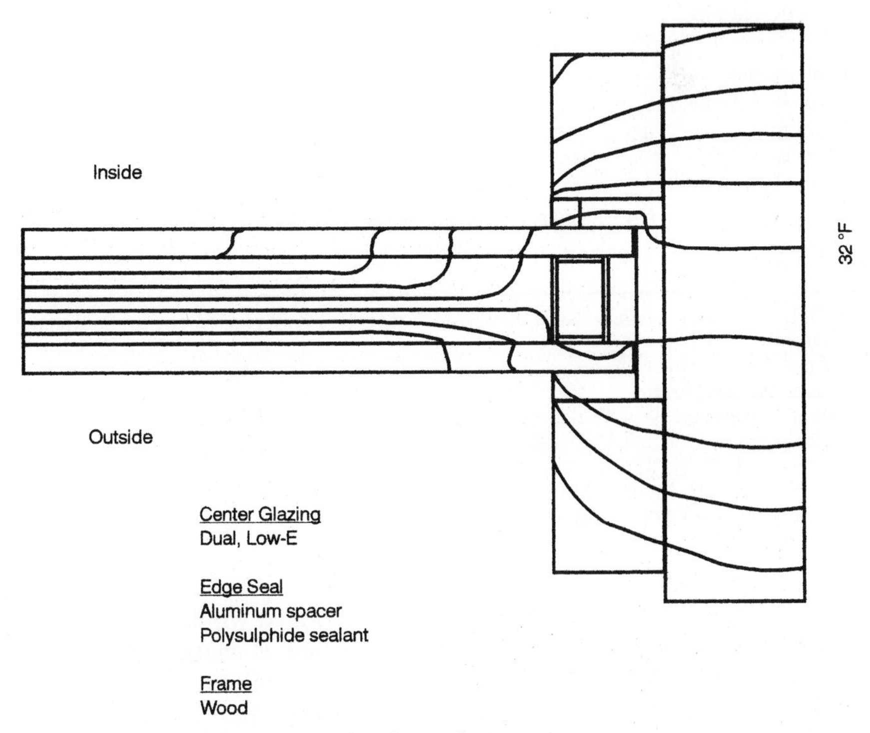

Figure 13.5 Conventional edge seal with metal spacer.

TABLE 13.1. FRAME® Analysis Results*

Center glazing			Overall window performance†			
Unit design	U-value, W/m^2°C	R-Value	Edge seal	U-value, W/m^2°C	R-Value	% Improvement
Air Filled Dual Sealed	2.8	2.0	Metal spacer Polysulfide sealant	2.9	1.9	—
Air filled Dual sealed	2.8	2.0	Super Spacer Butyl sealant	2.7	2.1	6.0
Air filled Dual seal Low-E coatings	1.8	3.1	Metal spacer Polysulfide	2.1	2.6	25
Air filled Dual seal Low-E coatings	1.8	3.1	Super Spacer Butyl sealant	1.8	3.1	36
Air filled Dual seal Low-E coatings	1.6	3.6	Metal spacer Polysulfide sealant	1.9	3.0	33
Air filled Triple sealed Low-E coatings	1.6	3.6	Super Spacer Butyl sealant	1.6	3.5	44
Air filled Triple sealed 2 Low-E coatings	1.1	5.2	Metal spacer Polysulfide sealant	1.5	3.7	46
Air filled Triple sealed 2 Low-E coatings	1.1	5.2	Super Spacer Butyl sealant	1.2	4.7	59
Argon filled Triple sealed 2 Low-E coatings	0.9	6.5	Metal spacer Polysulfide sealant	1.4	4.2	5.2
Argon filled Triple sealed 2 Low-E coatings	0.9	6.5	Super Spacer Butyl sealant	1.0	5.7	6.5

*Data on energy savings based on FRAME analysis study carried out by Enermodal Engineering Ltd., Waterloo, Ontario.

†Input data on window frame:
Type of frame: Wood
Weighting system: 80 percent glazing/20 percent frame
Glazing area: 28.35×28.35 in (720×720 mm)
Frame area: 31.50×31.50 in (800×800 mm)
Center glazing: 3.1 R (0.54 RSI)
Depth of frame: 1.57 in (40 mm)

The isotherm plots are given for a wood-framed window incorporating a sealed insulating glass unit with a Low-E coating and argon gas. Figures 13.5 and 13.6 illustrate conventional edge seals with a metal spacer backed up by a polysulfide sealant and a silicone foam insulating edge sealant backed up by a butyl sealant. The percentage

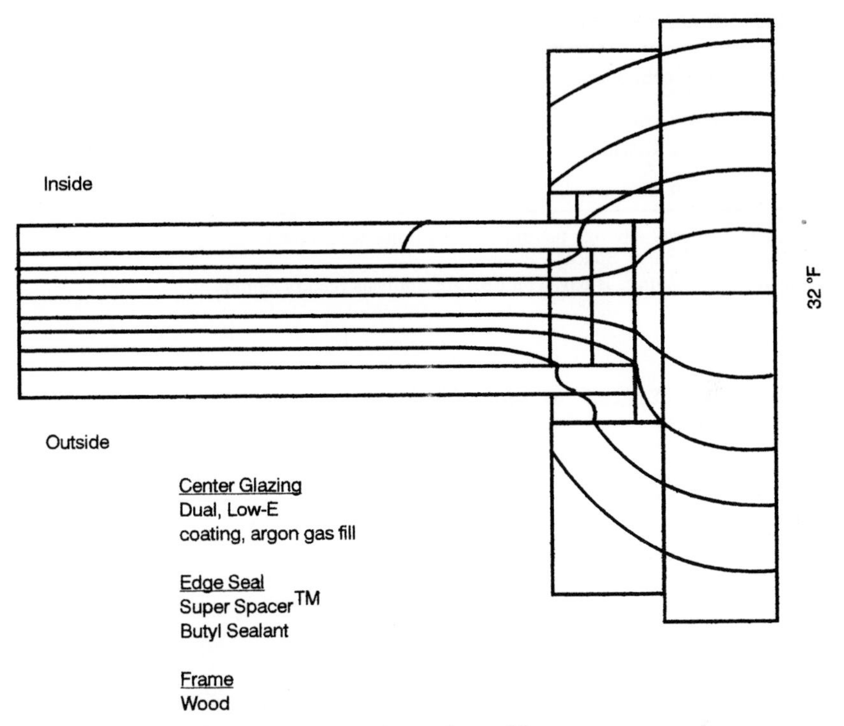

Figure 13.6 Insulating edge seal with Super Spacer™.

contributions of the different energy saving components of super window technology are illustrated in Table 13.2. The figures given are also for the same size residential, wood-framed window with a sealed insulating glass unit incorporating a Low-E coating, argon gas, and an insulating spacer. The FRAME analysis shows that the Low-E coating provides 57 percent of the overall energy savings and is its most important component in reducing heat loss. The insulating spacer is second with 25 percent energy savings, and the low conductive gas is third with 18 percent energy savings. The comparative energy savings provided by an insulating spacer and low conductive gas fill are also illustrated in Table 13.3.

For more energy efficient glazing units, the energy saving benefits of substituting an insulating spacer for a metal spacer are further increased. As shown in Table 13.4, with a sealed insulating glass unit, the overall energy savings for substituting an insulating spacer are only 6 percent whereas for triple-glazed units with double Low-E coatings and argon gas fill, the overall energy savings for substituting an insulating spacer are 28 percent.

TABLE 13.2. Component Energy Savings*

Percentage breakdown of energy savings†	Glazing
Low-E coating	57%
Super Spacer™	25%
Argon gas	18%

*Component energy savings for dual-sealed units with Low-E coating, argon gas, and Super Spacer.

†Data on energy savings is based on FRAME analysis results summarized in Table 13.1.

TABLE 13.3. Super Spacer™ vs. Argon Gas Fill

	% Improvement*	
Glazing	Argon gas	Super Spacer™
Dual-sealed unit, Low-E coating	10	14
Triple-sealed unit, double Low-E coating	10	21

*Data on energy savings is based on FRAME analysis results summarized in Table 13.1.

TABLE 13.4. Energy Savings of Super Spacer™ with Alternative Glazing Designs

Glazing Type	% Improvement with Super Spacer™*
Dual-sealed	6
Dual-sealed with Low-E	14
Triple-sealed with Low-E coating and argon	16
Triple-sealed with double Low-E coating	21
Triple-sealed with double Low-E and argon gas	28

*Data on energy savings is based on FRAME analysis results summarized in Table 13.1.

Direct thermal resistance measurement of glazing edge sealants

The thermal resistance of alternative edge sealants has been directly measured with an experimental test by the Advanced Glazing Laboratory of the Mechanical Engineering Department, University of Waterloo. The experimental procedure and test results are described in an ASHRAE Technical Paper.[3] The key test measurements summarized in Table 13.5 show that the Super Spacer edge seal has the lowest thermal conductivity of the designs evaluated.

In reviewing these tests results, it should be noted that the test procedure measures average heat resistance across the total cross section

TABLE 13.5. Measured Thermal Conductivity of Alternative Edge Sealants

Spacer bar	Secondary sealant	Primary sealant	Glass spacing, mm	R_{eal}, m^{2}°C/W	k,*† W/m°C
1. Aluminum	Silicone	PIB	14.0	0.007	2.00
2. Aluminum	Silicone	—	13.1	0.008	1.64
3. Swiggle Seal‡	Butyl	Butyl	14.1	0.018	0.78
4. Fiberglass	Silicone	PIB	12.8	0.042	0.31
5. Super Spacer	PSA acrylic	Butyl	13.1	0.073	0.18

*Thermal conductivity k = Glass spacing/R_{eal}.
†Foam thermal conductivity = 0.12 W/m^{2}°C.
‡Swiggle Strip is the trademark of Tremco Corporation.

of the edge seal, but the test procedure does not take into account that with the Super Spacer™ edge seal, the low conductive sealant foam (k = 0.12) is located in front of the more conductive material so that overall heat loss through the edge seal is reduced accordingly. Also, the Super Spacer™ test samples incorporated a 0.25-in (6.35-mm)-thickness butyl sealant. A reduced thickness of the sealant material would further lower the thermal conductivity of the edge sealant.

Direct measurement of glazing thermal performance

In connection with a new museum building in Canada, a limited test program has been carried out to measure the improvement in the overall thermal performance resulting from the use of Super Spacer™. Two separate tests were carried out. In one case, the insulating glass edge seal comprised a conventional dual seal with a metal spacer and polysulfide sealant, and, in the other case, the edge seal incorporated Super Spacer™ backed by a butyl outer sealant. Because very high thermal performance was required for the project, the insulating glass units were triple glazed with double Low-E coatings and argon gas fill. One of the Low-E coatings was located on the No.2 face and the other coating was located on the No. 5 face.

Because the objective was to evaluate the thermal performance of a curtain wall glazing system, two separate insulating glass units were located on either side of a wide aluminum glazing frame. This resulted in a low glazing/frame ratio of 73 percent glazing and 27 percent frame which would be typical for a very small residential window. However, because of the comparatively limited perimeter, the weighting system for the edge seal was equivalent to 85 percent glazing and 15 percent frame.

TABLE 13.6. Comparative Thermal Performance: Measured Test Results

	Test Conditions			Measured Temperature	
Edge Seal	Indoor	Outdoor	U-Value	Glass, 15 mm	Frame, 50 mm
Metal spacer	70°F (21°C)	20°F (−7°C)	1.72	52°F (11.0°C)	58°F (14.4°C)
Metal spacer	70°F (21°C)	−15°F (−26°C)	1.69	40°F (4.4°C)	50°F (9.8°C)
Super Spacer™	70°F (21°C)	20°F (−7°C)	1.49	56°F (13.3°C)	59°F (14.9°C)
Super Spacer™	70°F (21°C)	−15°F (−26°C)	1.52	46°F (7.8°C)	52°F (10.9°C)

The thermal performance testing was carried out at the National Research Council of Canada by the Institute for Research in Construction using the general approach outlined in the paper.[4] The test results are shown in Table 13.6.

For the particular design of the metal frame tested, substituting an insulating foam spacer resulted in three main improvements:

1. Overall heat loss was reduced between 18 and 19 percent depending on the exterior temperature.
2. The overall R-value of the window was increased from 3.3 to 3.9.
3. At −35°F (−37°C) exterior temperature, the lowest interior perimeter glass edge temperature at a location ⅝ in (15 mm) from the bottom corner of the window frame was increased to almost 46°F (7.78°C) from 40°F (4.44°C). Typically, the center glass temperature was about 52°F (11.1°C).

Although these improvements are significant, it should be noted that if more energy-efficient seals and framing profiles had been used, the measured improvement in thermal performance would have been higher.

High-thermal-performance glazing

High-thermal-performance insulating glass consists of four key components: Low-E coatings, multiple glazing layers, low-conductive gas fill, and insulating glass edge sealants. Using these components and depending on the number of glazing layers and Low-E coatings, sealed glazing units can be produced with a thermal performance ranging from R-4 to R-12 (center glazing). In contrast, conventional dual glazing is rated at R-2.

Using Super Spacer™, its manufacturers have pioneered the development of new specialized types of high-thermal-performance glazing units, including

1. *Slim line, krypton gas-filled, dual-glazed insulating glass units with a single Low-E coating (R-4).* For conventional air-filled, dual-sealed units, the optimum cavity width for thermal performance is about ⅝ in (15 mm) although the spacing can be reduced to ½ in (12.5 mm) with minimal reduction in thermal performance. Adding krypton gas can further reduce the optimum spacing to ¼ in (6 mm), but krypton is more expensive than argon and so its use can be justified only for specialized applications, such as heritage retrofit projects. For this application, single glazing panes (R-1) in heritage, wood-framed windows are replaced by krypton gas-filled, dual-glazed units with one Low-E coating (R-4) and this retrofit of energy-efficient, slim line units reduces heat loss by 75 percent (center glazing).

To demonstrate the feasibility of the heritage retrofit strategy, a demonstration project was carried out, wherein two large, multipane, wood-framed windows were retrofitted with slim line units. The prototype project was carried out in connection with a major renovation project, where a college building constructed in the 1890s was being transformed into a new administrative office complex. For heritage reasons, existing wood-framed windows on the building had to be retained. The demonstration project showed that by routing out a small amount of the wood frame, slim line dual-glazed units could be retrofitted and the heritage aesthetic of the multipane windows was maintained.

2. *Argon gas-filled, triple-glazed insulating glass units with double Low-E coatings (R-8).* In addition to improving the energy efficiency of conventional, dual-glazed insulating glass units, the new glazing technology of Low-E coatings and low-conductive gas fill can radically improve the energy efficiency of triple- and quad-glazed insulating glass units. For multilayer units, the rule of thumb is that there should be one Low-E coating per air space. By the addition of argon gas fill and two Low-E coatings, the thermal performance of a conventional triple-glazed unit can be improved from R-3 to R-8 (center glazing).

For residential buildings, the use of high-thermal-performance glazing units can be justified for various high-loss applications, such as roof lights and sun space glazing. For commercial buildings, one immediate market for high-thermal-performance glazing is high-humidity buildings, such as museums, art galleries, libraries, and swimming pools. As previously stated, the museum project involved high R-value, tripled-glazed units with argon gas fill and two Low-E coatings. Other long-term, commercial markets for high-thermal-performance glazing include applications where the use of energy-efficient glazing results in eliminating perimeter heating systems.

3. *Slim line, krypton gas-filled, triple-glazed insulating glass units with double Low-E coatings.* As previously explained, the addition of krypton gas can be used to reduce the width of sealed insulating glass units and, with triple glazing, the overall unit width can be reduced to match the conventional 1-in width of dual glazing. The advantage of these slim line, triple-glazed units with double Low-E coatings (R-8 center glazing) is that the energy efficiency of existing windows can be radically upgraded by retrofitting high-thermal-performance units. For triple- or quad-glazed units, a further advantage of using a krypton gas fill is that the extreme pressure fluctuations of thick air-space units are substantially reduced.

Edge-seal durability

The introduction of high-performance thermal glazing accentuates the problem of edge-seal durability. To provide for long-term performance, Super Spacer™'s edge seal incorporates four special features:

1. Silicone material
2. Flexible, resilient foam
3. Ultralow permeable edge seal
4. Low-deflection desiccant system

Silicone material

To address the problem of perimeter-edge loss, various efforts have been made over the past 25 years to use plastic instead of metal for the insulating glass spacer. However, these efforts to use conventional thermoplastic materials have proven unsuccessful because plastics are generally not suitable for rigid spacers. The main problems with conventional plastics include one or more of the following factors: high coefficient of thermal expansion, poor UV stability, high moisture vapor and gas transmission, poor thermal stability with outgassing and stress relaxation, brittleness at cold temperatures, and material degradation in low-moisture environments.

For assured long-term performance, Super Spacer™ is made from a silicone material. Chemically, silicone is unlike organic plastics in that the backbone is made up of silicone/oxygen linkages instead of carbon atoms. As a result, silicone is classified as an inorganic material that has outstanding durability. Listed below are some of key advantages of using silicone material for insulating glass spacers.

1. *Ultraviolet resistance.* Silicone materials resist the deteriorating effects of UV radiation which are destructive to organic polymers.

After prolonged UV exposure, the material does not dust, pulverize, or degrade.

2. *Chemical stability.* Silicone is an inert, extremely stable material and contains no sulphur or other acid-producing chemicals to corrode, stain, or damage even the new sensitive Low-E sputtered coatings.

3. *Long-term flexibility.* Silicone foam retains its flexibility at extremely low temperatures. Its durometer and modulus properties show little change within the operating temperature range of a sealed insulating glass unit.

4. *Stability at high temperatures.* Despite extended exposure to high temperatures, the silicone foam material retains its flexibility and will not soften or harden during the extended life of the sealed unit.

5. *Good structural strength.* In developing this spacer concept, the silicone material has been specially formulated for structural performance and has good tensile strength and tear resistance.

6. *Effect of low humidity.* Silicone is not affected by the low humidity levels within a sealed unit. Very dry air even at extreme temperatures will not leach, harden, or affect the stability of the silicone.

7. *Flame resistance.* Silicone material resists prolonged exposure to fire because of the stability of its silicone/oxygen backbone. Although specific testing is required for each application, the product has the potential to be used for fire-rated glazing systems.

8. *Nonoutgassing.* Silicone-based foam has been specifically formulated and cured to exhibit no outgassing when tested at the rigorous North American specifications.

Flexible, resilient foam

Rigid spacers are generally not appropriate for high-thermal-performance glazing. There are two specific reasons:

1. *Increased glass bowing and sealant stress.* When Low-E coatings are located on the inner glazing layers of multiple-glazed units, the temperature within the air spaces can reach above 140°F (60°C) on sunny days. Conversely, because the Low-E coatings reduce heat loss under cold conditions, temperatures within the air spaces are lower than with conventional units. As a result, there are larger pressure fluctuations within the insulating glass unit, and this creates increased movement and bowing of the glass panes. Also, when the glass panes pivot on a rigid spacer, the problem of increased glass and sealant stress is accentuated.

2. *Increased differential expansion.* With improved thermal performance glazing, there are higher temperature differences and

increased differential expansion between glazing sheets and rigid spacers resulting in increased seal stresses.

To avoid sealant stress, the expansion coefficient of a rigid spacer must match the expansion coefficient of glass. One way of providing the required dimensional stability is to use glass-reinforced, plastic extrusions or pultrusions. However, there are drawbacks with these rigid glass-reinforced plastic spacers including comparatively high permeability, excessive rigidity, comparatively high thermal conductivity, low productivity in conventional insulating glass assembly methods, and high dust levels created when spacers are cut to size.

To avoid these problems with rigid spacers, Super Spacer is made from a flexible, resilient foam. This flexibility can accommodate the increased differential expansion between glazing layers and between the spacer and the glazing layers without applying any additional stress on the outer sealant. Similarly, the increased movement and bowing of the glass panes can be accommodated without applying additional stress on the outer sealant. Because silicone foam is very durable, this friendly accommodation of glass bowing and movement is maintained for the life of the sealed insulating glass unit.

Even though the spacer is flexible, uniform spacing between the glazing layers is maintained because an insulating glass unit acts, in some ways, like an inflated tire. Once the outer sealant is in place, the air within the unit cannot be easily compressed and, even under high wind loads, the foam material is not excessively stressed. In addition, the closed-cell, silicone foam has good resistance to compression set, and this also helps ensure that uniform spacing is maintained.

Ultralow permeable edge seal

For high-performance glazing (Low-E coatings and gas fill), it is critical that the integrity of the edge seal is maintained. There are three factors which affect edge-seal integrity:

1. Low-E sputtered coatings are sensitive to moisture and even with partial edge failure, the coatings can be visually blemished. Low-E hard coatings are not affected.
2. The insulating glass edge seal must be prevented from long-term gas loss and, in contrast to moisture vapor transmission, there is no desiccant material to provide a safety margin. So, any long-term gas loss will lower thermal performance.
3. Insulating with permeable plastic spacers accentuates the potential problems of long-term moisture buildup or gas losses, and this is particularly a problem where comparatively permeable thermosetting sealants are used.

To provide the ultralow permeable edge seal for high-thermal-performance units, it is recommended that Super Spacer™ be backed by a thermoplastic hot melt butyl sealant which has very low moisture vapor and gas permeability. To minimize gas loss from the unit, a $^3/_{16}$-in (4.5 mm) thickness of this sealant should used.

Unlike thermosetting sealants, thermoplastic sealants do not chemically bond to the spacer or glazing layers, and this creates a number of problems, including spacer bar migration and cold creep. However, through the use of preapplied structural adhesive on the spacer sides, Super Spacer™ eliminates these traditional problems because the spacer is held in place by the adhesive. When the units are transported, there is typically no need for special packaging to hold the glazing sheets in position. In addition, in cold temperatures, the flexibility of the foam spacer helps prevent sealant stress and loss of adhesion.

The pressure-sensitive acrylic that is applied to Super Spacer™ has excellent heat, water, and UV resistance. It has good tack which allows it to form a bond with minimum contact time. Also, the adhesive has good shear adhesion properties so that the silicone spacer can hold large sheets of glass together without any structural sealant backing. The peel adhesion of this spacer system is not as high as that of conventional thermosetting sealants. However, because of the flexibility of the foam spacer, peel stresses are reduced ensuring long-term durability of the insulating glass edge seal.

To further reduce moisture vapor and gas transmission through the edge seal, Super Spacer™ is backed by a multilayer, metallized barrier film which also provides an extra safety margin in improving the long-term durability of insulating glass units. The barrier film is laminated to the back of the spacer and, by acting as an intermediary layer, it allows sealants to bond to the back of the Super Spacer™ resulting in improved edge-seal durability.

High-performance desiccant system

The desiccant system for this foam spacer combination has been specifically developed for high-thermal-performance insulating glass units. There are three special features:

1. *Large quantity of desiccant material.* If Super Spacer™ is used in combination with a low-permeability thermoplastic outer sealant, there is more than enough desiccant capacity to ensure a minimum 20-year life for the insulating glass unit.

2. *Low-deflection desiccant system.* To avoid the potential problem of low-temperature adsorption of argon or nitrogen gas, Super Spacer™ incorporates a low-deflection desiccant system consisting of 3-Å molecular sieve and silica gel desiccant material. With 3-Å molecular sieves,

the desiccant pores are so small that the material can adsorb only the very small moisture vapor molecules but not the larger nitrogen or argon gas molecules. The silica gel provides desiccant capacity for solvent adsorption but unlike a large-pore, molecular sieve desiccant, there is a very limited adsorption of the argon and nitrogen gas.

3. *Permeable silicone foam for rapid dew point drop.* Edge-sealant products such as Swiggle Seal™ have a comparatively slow dew point drop and this may cause condensation when units are shipped on cold days. However, with Super Spacer™, this problem is avoided because the high permeability of the silicone foam ensures that the desiccant is very active and there is a rapid dew point drop within the sealed unit.

Durability testing

1. *Edge-seal integrity.* Sealed glazing units incorporating Super Spacer™ and backed up by a thermoplastic hot melt butyl sealant have undergone extensive durability testing and have consistently passed the requirements of both the Canadian CAN-12.8-M76 and U.S. ASTM-C-773/774 specifications for sealed insulating glass units.

To demonstrate the ultralow permeability of the recommended edge-seal design, the same units have been cycled through high-humidity testing three times in a row and have still met requirements for minimum dew point temperatures. Test units have also been cycled through the rigorous ASTM P-1 test (proposed) providing up to 40+ weeks performance without failure. In comparison, a single-seal polysulfide unit provides six to eight weeks durability, or a dual-seal polyisobutylene/polysulfide unit provides 12 to 18 weeks durability. Figure 13.7 shows the results of P-1 accelerated durability testing.

2. *Volatile outgassing.* With organic plastic spacers and high-performance glazing, outgassing is a particular problem because a thin layer of volatile material can condense on the sputtered Low -E coating creating mirror-like distortions. With Super Spacer™, the silicone foam has been specially formulated to prevent outgassing, and these units routinely pass North American specifications.

Insulating glass production

Flexible strip product. For an insulating glass edge seal, material costs are higher because higher-quality materials must be used to ensure improved thermal performance and long-term edge-seal durability. However, with Super Spacer™, these higher material costs are offset by lower labor and equipment costs and improved thermal and acoustic performance. The product has been designed so that four

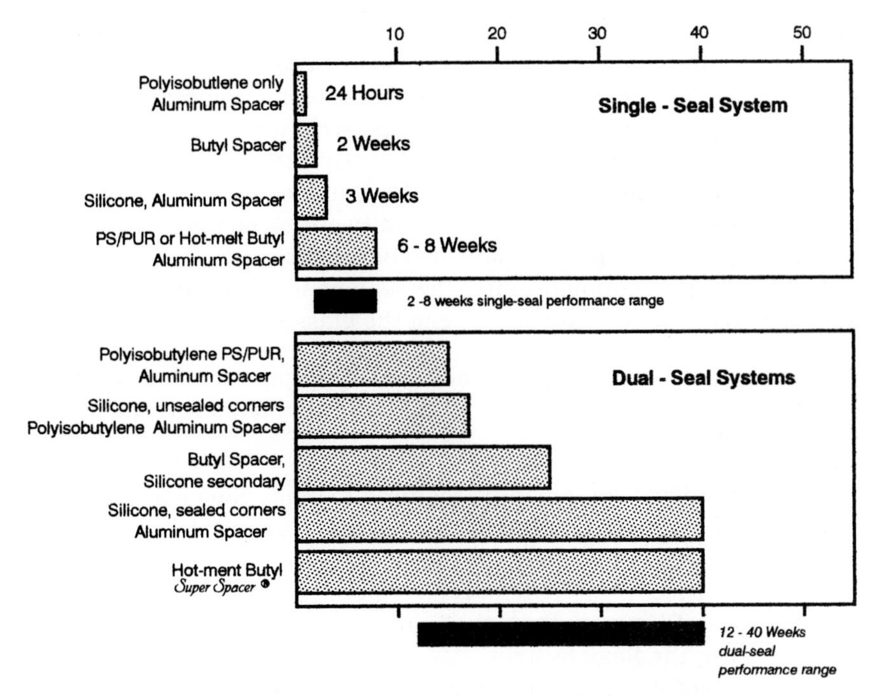

Figure 13.7 P-1 accelerated durability testing.

separate components of a conventional edge seal—desiccant, spacer, corner keys, and adhesive sealant—are replaced by a single product. Because this flexible strip is easy to apply, there are cost advantages in using it for both manual and automated production methods. Gas filling can also be more cost effective.

Manual production methods

Compared to the manual production of conventional units, the substitution of Super Spacer™ eliminates the following tasks: desiccant filling of hollow spacer bars; cutting of rigid spacer bars to length; assembling of spacer bars into frame using corner keys; sealing of the corner assembly with primer or polyisobutylene; and applying the polyisobutylene/adhesive sealant onto the spacer glass assembly.

The manual application of Super Spacer™ can be broken down into three different steps.

1. The spacer is located on the first sheet of glass so that the glass extends beyond the spacer by ¼ in (6 mm) to $^3/_{16}$ in (4.5 mm) and is held firmly in position by the pressure-sensitive adhesive on the spac-

er sides. A circular, notched joint is cut or punched out at the corners so that the flexible spacer can be folded around the corner while maintaining the continuity of the barrier backing. Spacer application starts and ends at the corner, and a tape strip may be applied across the connecting joints to ensure the full continuity of the vapor barrier.

2. After the spacer has been applied, a second sheet of glass is placed on top of the first sheet, and the two lites of glass are pressed together by hand to ensure good adhesive wetout.

3. As with conventional units, the sealant is then gunned into the outward facing open channel between the glass sheets.

Besides reducing labor costs, there are other production advantages in using this type of product. For manual application, these advantages include

1. no specialized equipment is required to apply Super Spacer™,
2. the silicone foam material can be easily notched using simple hand tools.
3. Super Spacer™ is clean and convenient to handle.
4. because of its resilience, there are virtually no problems of permanent set in the flexible strip.
5. for sealant gunning, the adhesive on the spacer sides firmly holds the glass sheets in position and, when using a hot melt butyl sealant, there is no need for additional measures, such as weights or clamps.
6. inventory costs are lower because of the reduced number of separate components required.
7. spacer scrap costs can essentially be eliminated and, by continuing with butt joint connections, small lengths of the material can be used up.

Hand application equipment

Specialized, hand-held equipment has been developed to simplify the installation of Super Spacer™. The applicator ensures that the foam spacer is applied in a straight line and at a uniform distance from the glass edge. For creating neat corner joints, the applicator incorporates a corner notching device. The applicator also features an adjustable roller to accommodate different spacer sizes and an adjustable pressure block to accommodate different glass thicknesses.

Automated production methods

Within the insulating glass industry, there is a general trend toward increased automation which is being fueled by the growing importance of the retrofit market and the resulting need to produce insulating glass units of different sizes. To date, the automated insulated glass production equipment which has been developed has generally copied existing manual methods, and, as a result, the equipment has been unnecessarily complex, sophisticated, and expensive to maintain. Because the Super Spacer™ is a flexible strip, there is the potential for developing new production techniques which can simplify the design of automated equipment.

For fully automated production, La Fond International has developed a fully automated applicator that can produce more than 1000 lites per shift with a single line and in excess of 5000 units per shift with a parallel double line. Combined with a butterfly table and a vertical hot melt gunning system, three operators can potentially produce 2000 units per shift.

Super Spacer™ acts a manufacturing aid for automated and semiautomated production lines. The adhesive on the spacer sides holds the glass in position and allows the glass sandwich to be transported to the sealant gunning station. In addition, because the glass panes are held firmly in position, there is not the same need to control gunning pressure, and thus the task of sealant gunning is also simplified.

A variety of layouts lead to flexibility, speed, and enhanced productivity. The simplicity of using this versatile material is worth noting in the following process overview. There are a few basic steps that must be followed.

1. The glass is cut to size, the Low-E edge deleted as required, and the glass thoroughly washed according to recognized industry standards.

2. The Super Spacer™ is applied by pneumatic hand applicator at a set distance from the glazing edge as illustrated in Figure 13.8. To provide for a continuous vapor-barrier backing, the spacer is notched and bent over the final corner joint.

3. For small glass units, vertical one-person glass matching, as shown in Figure 13.9*a* is feasible with a simple assembly easel or muntin float table. For large-size units, horizontal or vertical glass matching with a two-person team is required. Alternatively, by using a butterfly table, as shown in Figure 13.9*b*, glass matching and adhesive wet out can be combined into a single one-person operation. Where units are manually matched, the spacer side adhesive should be wet out using either simple hand pressure or a roller press.

4. Gas filling is an optional step using this type of edge-sealing material. Refer to Chapter 16, which covers gas filling in minute detail. With slow-fill equipment, insulating glass units must be vertically stacked and gas-filled off-line. With fast-fill equipment, the units can be filled on-line using double assembly easels.

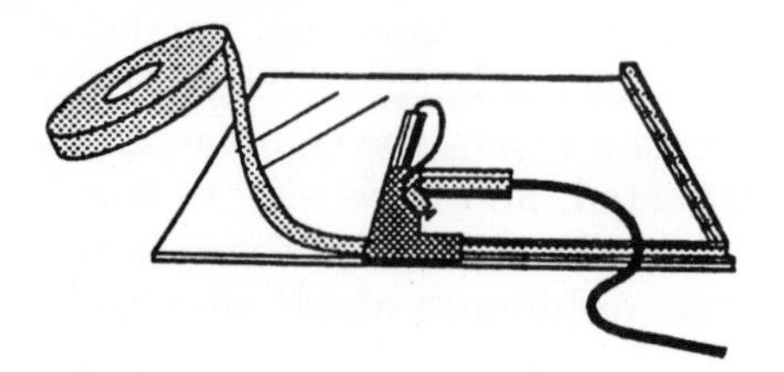

Figure 13.8 Spacer application.

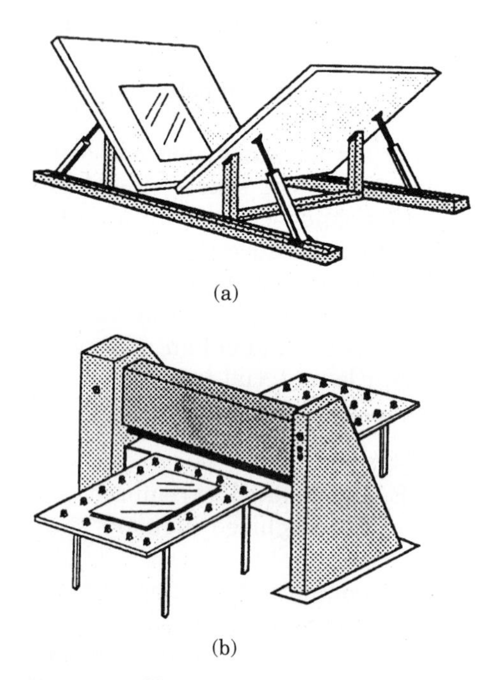

(a)

(b)

Figure 13.9 Glass matching.

5. Super Spacer must be conventionally backed by an industry-approved insulating glass.

6. The specialized application equipment can be configured into a range of different layouts with outputs ranging from 200 to 2000 sealed units per shift. Key equipment options include one, two or four air-float tables; butterfly, roller press, or assembly easels; and the add-on of one or two muntin tables. Figure 13.10 is a typical layout for a 24×30 ft (7.315×9.144 m) production unit which has the versatility to be linked by conventional case-top conveyors for fast and convenient horizontal glass movement between the equipment.

7. Hand-held pneumatic tools, illustrated in Figure 13.11, apply an uninterrupted, sufficient supply of the desiccated silicone foam tape.

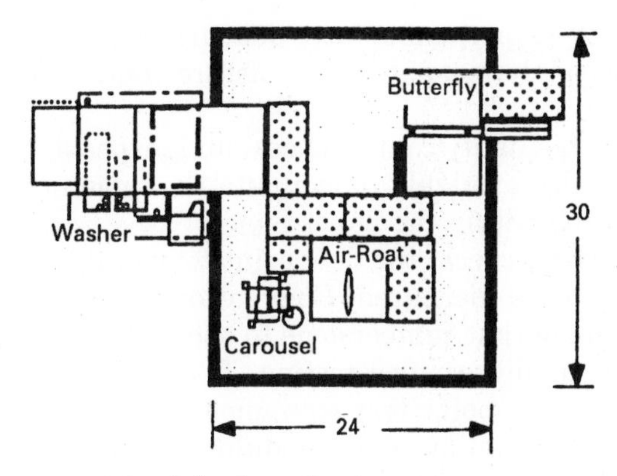

Figure 13.10 Adhesive wet out.

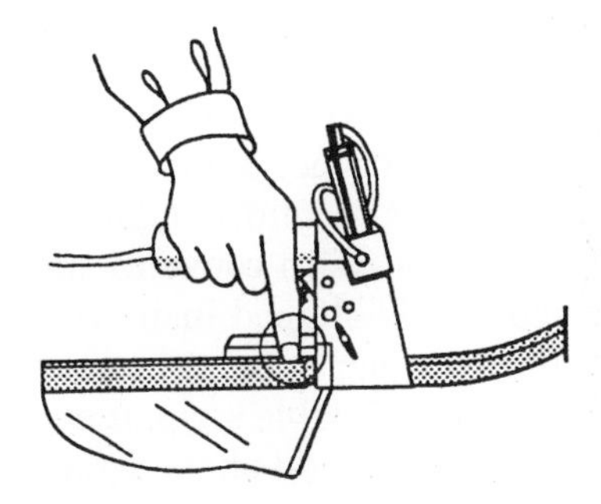

Figure 13.11 Hand-held applicator.

Previously used tools required spacer lay-down, corner notching, punching gas-fill holes, and a final spacer cutoff.[5]

A detailed engineering description is available from the manufacturer, taking into account the various assumptions one would make in planning any particular option.

Quality control testing

For in-plant quality control testing, the moisture content of the desiccant material within Super Spacer™ can be measured either directly or indirectly.

For precise, direct measurement of moisture content, the Arizona Instrument's Computrac TMX moisture analyzer is recommended. This equipment measures moisture content by oven-drying a sample of the material and then collecting and weighing the moisture vapor driven off into a selective moisture trap. With the oven temperature set at 580°F (304°C), laboratory experiments have demonstrated that the equipment can very accurately measure the moisture content of the desiccant material.

For a simple but less accurate quality control check, an inexpensive electronic moisture meter is also available to evaluate the moisture content of the desiccant material within the Super Spacer™.

As an additional quality control check, the dew point temperatures of the sample sealed units can be measured. Conventional dry ice, dew point test equipment is somewhat cumbersome to use and cannot be easily adapted for on-line, quality control testing. To simplify the test procedure, a new type of dew point test equipment was developed, which uses thermoelectric cooling plate modules. Faster and more convenient to use, the thermoelectric equipment can be electronically controlled and does not have to be continually monitored.

Specialized insulating glass designs

Because of the unique features of Super Spacer™, there are advantages in using the product for fabricating specialized types of insulating glass units, including curved and odd-shaped sizes, sound insulating, units and triple-glazed units with art glass inserts. With conventional rigid metal spacers, the production of curved or odd-shaped insulating glass units is comparatively complex because the spacer must be bent into the required shape. Because Super Spacer is a flexible strip, it can be easily placed around the perimeter of the curved or odd-shaped glazing units.

Slot profile for triple-glazed units

For triple-glazed units, Super Spacer™ can be manufactured with a slot profile allowing the spacer to be wrapped around an inner glazing insert. One potential application for the slot profile is the decorative glass market. In addition to simplifying the fabrication of triple-glazed units, other specific advantages include an unobtrusive, neutral gray appearance, neat corners, narrow sight lines, and smaller spacing widths than with conventional metal spacers.

Intercept Insulating Glass

Much of the new warm-edge technology has concentrated on the shape of the spacers, with each supplier introducing its own approach to improved spacer design.[6] One of the industry's major glass suppliers, PPG Industries, Inc., has a unique warm-edge idea called the Intercept™ insulating glass spacer system.[7]

Matrix composition

The unit starts with a thin plated or stainless steel spacer, depending on the manufacturer's preference, that is shaped into a continuous U-channel, with return legs for greater rigidity, surrounded by sealant, and with a desiccated matrix applied to the inside bottom of the U-channel. Figure 13.12*a* and 13.12*b* exhibits this unique design that helps build in superior thermal performance with an aesthetically pleasing low profile.

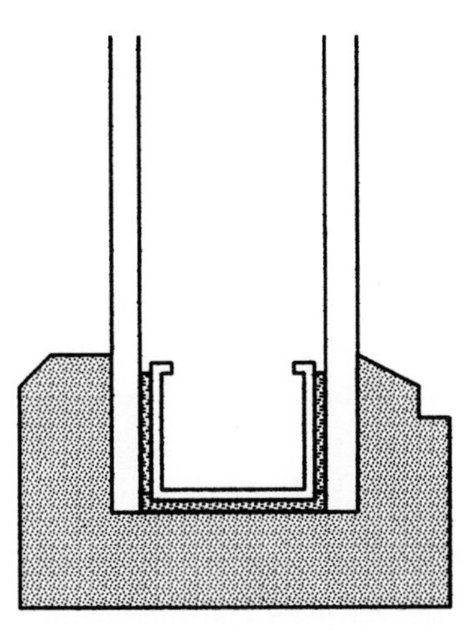

A new type of spacer design features a steel U - shaped channel which is thinner than traditional aluminum spacer and has a lower thermal conductivity. The U - shape has only one pathway for heat transfer, compared to two box - shaped spacers. In this detail the desiccant matrix is omitted for clarity.

Figure 13.12(a) U-shaped spacer design.

Figure 13.12(b) Frame with desiccated matrix.

This type of warm-edge unit achieves its performance by using steel instead of conventional aluminum, which is a significant factor. Steel has lower thermal conductivity than aluminum, and the U-shape has only one path for heat transfer compared with a traditional box-like spacer. The steel spacer is thinner than its aluminum counterpart, again contributing to reduced thermal conductivity.

This type of unit spacer design also offers a structural advantage. When high temperatures occur, the glass in insulating glass units bows in toward the center. This glass movement stresses sealant-to-glass and sealant-to-spacer bonds and can contribute ultimately to seal failure and even glass breakage. The Intercept spacers move with the glass, thus reducing sealant stress that can lead to unit seal failure.

Performance comparison

Not all insulating glass units are created equal. Windows made with new insulating glass technology can save money on energy costs, improving the overall comfort in the structure and providing improved long-term performance for in the future. Table 13.7 outlines this improved thermal performance.

Thermal advantages

1. Excellent U and R insulating values on energy costs.
2. Warm-edge technology reduces condensation on the inside glass surface of the window, as shown in Figure 13.13.
3. Warmer inside glass temperatures allow higher relative humidity in the room with reduced edge condensation, retaining heat better

TABLE 13.7. Window Unit Comparison—Thermal Performance*

	Warm-edge (s.s. spacer)	Warm-edge (tin-plate spacer)	Conventional aluminum spacer
Center U-value	0.26	0.26	0.26
Emissivity (Low-E)	0.10	0.10	0.10
Edge U-value	0.55	0.59	0.63
Warm-edge avg. temp.	38°F (3.3°C)	37°F (3.0°C)	34°F (1.5°C)
Frame U-value	0.23	0.25	0.63
Window U-value	0.33	0.34	0.35
Allow. rh at 70°F (21.1°C)	31%	30%	26%
Min. roomside bottom temp.	30°F (−1.1°C)	30°F (−1.1°C)	30°F (−1.1°C)

*Conditions: 2.5-mm glass, ½-inch spacer, argon gas filled, Low-E, 0°F outside, 70°F inside.

Intercept® Insulating Glass units reduce condensation problems. Compare the *Intercept* I.G. unit (R.) with conventional I.G. unit (L.). (Conditions: Cold side temp.=0°F. Room side temp.=72°F. Room side relative humidity=25%) Both units have low-E coating and argon gas infill.

Figure 13.13 Condensation comparison.

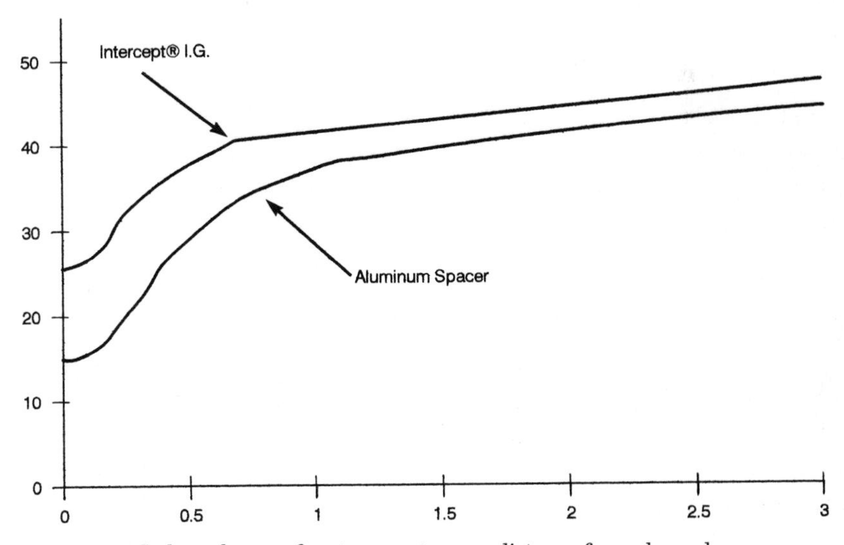

Figure 13.14 Indoor glass surface temperature vs. distance from glass edge.

and increasing comfort. Figure 13.14 illustrates the significance of indoor glass temperatures near the sash.

4. One-piece, sealed, spacer design retains insulating gases better and maintains thermal performance longer.

5. Insulating glass spacers resist seal migration and failures due to thermal cycling.
6. Sight line is enhanced (Figure 13.15), making it virtually disappear compared to the highly reflective, traditional, insulating glass spacer.

Structural advantages

With this construction, the spacer inside the warm-edge unit actually flexes under thermal cycling, with the U-channel absorbing much of the movement instead of stressing the sealants that provide the hermetic seal to the insulating glass unit. The glass edge sealants and spacer tend to remain parallel. The reduced edge stress reduces the likelihood of seal failure, enhancing the life of the insulating unit. The insulating structural advantages are illustrated in Figure 13.16 versus warm-edge construction.

Manufacturing process

Unit manufacturing with this method can be integrated quite readily into an insulating glass manufacturer's production. The system machinery is computer-controlled for automatic integration with

SEE THE DIFFERENCE
The unique spacer design of *Intercept*® insulating glass (L.) makes it virtually disappear compared to the highly reflective traditional insulating glass spacer (R.)

Figure 13.15 See the difference.

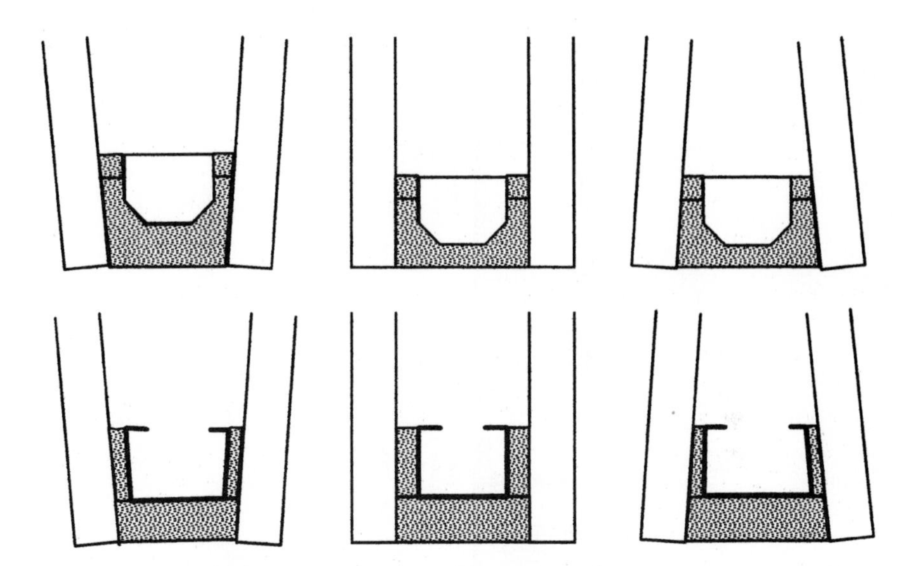

The bond between the glass and sealant and the sealant and the spacer for two different configurations is shown. With standard aluminum spacer construction, top, the seal is stressed when cold temperatures cause glass to bow in, left, or when warm temperatures cause glass to bow out, right. This can result in seal failure and even in glass breakage. In an Intercept® unit, bottom, the U-shaped spacer flexes during thermal cycling. The desiccated matrix was omitted for clarity.

Figure 13.16 Standard spacer construction vs. warm-edge construction.

existing linear insulating glass manufacturing systems. It begins with the steel strip fed from a continuous coil into a die stamping machine. Notches are made in the strip where it will be bent to form continuous corners and ends of the spacer. Holes and/or notches are also punched for gas filling and for internal muntin grids, respectively. A computer programs the master production schedule, setting the sequence and location of the spacer fabrication operations. Fiber optics is used to measure the spacer.

The strip passes into a roll former that shapes it into a U-channel configuration. The rollers are adjustable for channel widths (air-space width). Figure 13.17 shows the industry's first computer-integrated, fully automatic, insulating glass fabrication process.

The desiccated matrix for the air or gas space between the window lites and the edge sealant is applied, by either a single- or dual-seal system. The spacer is folded into shape by hand, the first hands-on step of the operation, ready for placement between the lites of glass to form the insulating glass unit.

The continuous corner construction of the U-channel unit is designed to save production steps and to avoid problems associated with moisture leakage that can occur with noncontinuous, corner key construc-

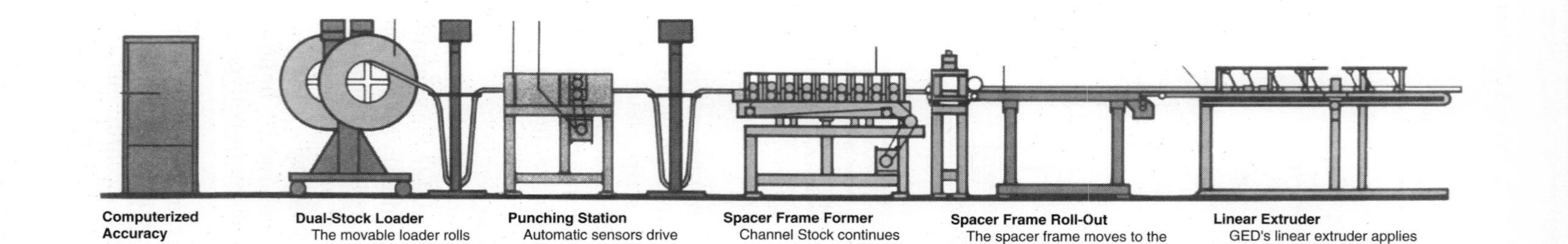

Figure 13.17 Computer-integrated automated insulating glass line. (*Courtesy of Glass Equipment Development Corp., Twinsburg, OH.*)

tion. Production rates for the units can reach 2500 to 3000 units per shift. As for material costs, stainless steel costs about the same as aluminum spacer material, whereas galvanized and plated steels cost less. The insulating glass and manufacturing technology were codeveloped by PPG and Glass Equipment technology, and this technology is available from PPG for license to all insulating glass manufacturers.

Energy codes

With window labeling on the horizon, buyers will be able to compare insulating glass unit energy performance before making purchases. The National Fenestration Rating Council (NFRC) code requirements will become an important influence in window purchase decisions. Warm-edge technologies boost overall window U-value energy performance 6 to 8 percent, which, in many cases, is just what may be required for the overall window to meet rather than miss energy code requirements. An Intercept™ spacer configuration combined with Low-E glass products can substantially improve the window U-value. It is quite possible that energy codes and other performance requirements will take warm-edge technology out of the realm of custom production into the mainstream of insulating glass production.

Performance verification

Comparisons of energy efficiency between fenestration products can be made easily by comparing R-values or measuring edge temperatures, but evaluating warm-edge products must go beyond these basic thermal measures. A manufacturer must be confident that these new spacer constructions will not result in having the insulating glass fail prematurely.

Without the benefit of long-term field experience to verify product integrity, accelerated testing that simulates real life must be done to evaluate seal performance. For conventional insulating glass units, the ASTM-E-773/774 test is performed to evaluate seal durability under controlled conditions. Shown to relate to longevity based on a field correlation study,[8] the test has become the industry standard.

Evaluating the performance of warm-edge insulating glass requires going beyond this step because much of the technology is new. Because this design for insulating glass units incorporates a unique U-shaped spacer and a desiccated matrix not used before, longevity had to be demonstrated in laboratory tests before market introduction.

Testing should always be part of new product development. To assure customers of the performance of a specific warm-edge technology, manufacturers should perform standard procedures to allow product comparison and should submit products for third party testing. Some new procedures nay need to be devised to demonstrate performance when industry requirements are not yet established. In doing so, a manufacturer verifies the integrity of its product and provides customers with greater performance assurances.

Guarded hot box tests

Evaluating warm-edge technology begins with thermal performance. The Guarded Hot Box,[9] as described by ASTM-C-236, is an accepted industry method and should be used for evaluation. Windows up to 6×4 ft in size can be evaluated in the test apparatus.

To measure heat transfer, the fenestration product is mounted in a wall between two chambers, one set at 0°F (−18°C) and the other at 70°F (21.1°C), comparable to the inside of a home during cold winter weather. The warm side is shielded in a box within an anteroom, thus giving its Guarded Hot Box name. Sophisticated thermostatically controlled refrigeration equipment keeps the cold side temperature constant. Because heat moves from warmer to colder areas through fenestration, this test measures the energy needed to keep the temperature inside the warm side box at 70°F (21.1°C), indicating a heat loss.

Surface temperatures are measured by thermocouples placed at prescribed locations on both sides of the fenestration. During evaluation of the warm-edge units, temperatures should also be measured at the glass center, edge, and frame, because of the significance of these measurements to the units' comparative performance. The heat transfer equation used for this calculation procedure is

$$Q = UA(t_i \cdot t_2)$$

where Q = is heat flow per unit time
A = area
t = temperature

The U value is in British thermal units per hour per square foot per °F in the equation.

Seal durability

The thermal performance of a unit with a conventional aluminum spacer is compared with warm-edge units. Although the center R-value for the constructions may be the same, the Guarded Hot Box test demonstrates performance benefits of warm-edge units: improved R-value, warmer room side edge temperatures, higher allowable relative humidity, and improved overall window R-value.

Long-term seal durability is required for insulating glass unit performance. If warm-edge units incorporate new seal technology, as with the U-shaped channel in this new technology for insulating glass units, verifying seal durability is important.

Seal durability is evaluated with the ASTM-E-773/774 test method that determines the three grade ratings, in ascending order, of C, CB, and CBA. These tests are normally performed by independent testing laboratories on insulating glass units of various glass thicknesses, various air-space thicknesses, and industry-recognized sealants. The standard tests are run according to the ASTM test referred to.

Spacer evaluation

As with any new innovation lacking long-term field experience, the physical performance of this spacer concept must be evaluated. Windows are expected to endure usual and high wind loads without mishap. To evaluate stability under wind loads, the warm-edge spacer should be compared with the conventional aluminum box spacer. A test unit should be able to sustain significant wind loading before breaking at the point known as the ultimate wind load. A wind speed of 100 miles per hour equals a pressure of 32 psf.

Center glass deflections in warm-edge units and units with a conventional spacer should almost be identical, and the ultimate wind loads should be very similar, demonstrating that the warm-edge unit spacers will carry equivalent wind loads. The spacers in warm-edge units also should be submitted to a fatigue test that simulates the cyclic stress to which the material will be subjected for more than the suggested life of the unit.

During wind loading, an insulating glass spacer is subjected to a shear force resulting from the differential displacement of the glass lites at the perimeter. Tests have been devised to simulate this condition by subjecting the spacer to repeated shear loading based on displacement of the glass at the unit center. Large displacements corresponding to high wind loads and various smaller displacements are used to simulate a range of wind conditions. The metal stress is plotted over the number of cycles survived during the test procedure, and a curve is developed to show the load frequency profile. The spacer should show no fatigue problem.

Some insulating glass units are subjected to fluctuating wind pressures, high temperatures, and substantial glazing pressures that can push them out of place. A well-designed spacer is flexible enough to maintain its angle to the glass when subjected to such conditions, thereby reducing sealant strain.

To ensure that the U-channel spacer in the Intercept unit would be able to withstand typical weather conditions, it was subjected to a "torture test." The test was used to verify the performance of a standard insulating glass unit. Units incorporating tempered glass were exposed to 80-mph wind loads applied three times a minute and were heated from room temperature to temperatures in excess of 200°F (93.3°C) several times a day.

The units were subjected to two series of tests in which glazing edge pressure was applied to the edge. In the first, the pressure was 9 psi, near the 10-psi maximum allowed, and then an exaggerated edge pressure of 15 psi was applied. Samples were exposed for 140 hours at each of the two edge glazing pressures. To pass the test, the spacer must not move more than ¼ in into the sight line nor beyond the glass edge.

Field exposure

Although accelerated laboratory testing is used to simulate anticipated field conditions, knowing how a new product reacts under actual exposure remains a key component of performance evaluation. Warm-edge production units should undergo weathering at field test sites in various climates to study exposure to high humidity, high temperatures, and high UV conditions.

For maximum exposure, glass should be mounted on racks facing south, inclined at 45° angles, with the units unglazed so that the sealants receive significantly harsher exposure than in a glazed window sash. The units should be periodically removed and air-space dew point measurements taken. The dew point is the temperature at which moisture forms. Dew point measurements show how dry the airspace is in parts per million of water molecules. When the dew point is high, seal failure is indicated, and condensation forms on the inside glass surfaces during cold temperatures. No dew point change should occur in the exposed field units.

Gas retention

For many warm-edge units, argon gas retention is a factor that should be considered in evaluating performance. Although there is no standard performance requirement or test method in place for the U.S. marketplace, a maximum argon loss of about 1.2 percent per year is usually accepted.

Argon gas retention is measured using a modified test method (DIN 1286 Part 2) from the Institute for Window Technology at Rosenheim, Germany. Comparisons should be made against standard aluminum spacers and competitive warm-edge products. Based on the inventor's tests, a loss rate of no more than 1 percent per year, the Intercept units showed argon loss less than 0.2 percent. Not all competitive products did as well.

Desiccated matrix qualification

To remove internal moisture from an insulating glass unit, a desiccant is used, commonly as beads inserted in the box-shaped spacer, where they remain. Warm-edge units incorporate innovative spacer technology and may preclude the use of desiccated beads.

For example, a special desiccated matrix was required for Intercept insulating glass units containing both the desiccant and adhesive for application onto the channel cavity. The adhesive bond of the matrix material had to be evaluated to make sure it would not be loosened and droop into the vision area because of force or heat. Although the performance of the matrix would not be affected, the appearance of the unit would, because the vision area must remain unobscured. Because there are no similar applications in the marketplace, tests had to be developed to evaluate the adhesive strength and stability of the desiccated matrix material.

A dynamic bump test was developed to assure that the matrix bond would not be dislodged over time by the daily bumps and slams of a window being opened and closed. The test simulated a double-hung window opened and closed twice a day over a 20-year period, estimated to receive 15,000 bumps. With specially designed equipment, a closed, full-sized sash, insulating glass unit was raised and then allowed to drop to a closed position in a dramatic slam in an attempt to dislodge the matrix from the spacer cavity. The tests were conducted at temperatures of 120°F (48.9°C), 20°F (−6.7°C), and −20°F (−29°C). The desiccated matrix materials available for use in warm-edge units maintained adhesion in the windows after 15,000 drops at each temperature level, a total of 45,000 cycles. To gauge the thermal stability of the matrix materials, another test was designed to evaluate them under extreme high temperatures, based on a test devised for evaluating glass coating durability.

Both insulating glass units and spacer frames with desiccated matrix materials were placed in an oven at high temperatures to determine if the materials would continue adhering to the inside of the spacer and remain in place. Because residential glazing temperatures generally are not very high, a realistic test time was set for eight days. However, for greater product durability assurance, the spacer frame was subjected to a longer test of several months.

The matrix materials, available for use in the warm-edge units, passed the testing by remaining in place without slumping or displacement from the spacer cavity.

Although developers of warm-edge technology can provide product assurances, careful attention must be paid to the actual practices of the individual window manufacturer. Windows with warm-edge technology, the best available, offer a high level of quality and performance.

Swiggle Seal™

During the summer of 1978, Dr. Thomas Greenlee, a senior chemist, offered the idea of an undulating aluminum spacer to a team working to develop a warm-edge concept for an insulating glass unit. This resulted in the issuance of U.S. patent 4,431,691, Dimensionally Stable Sealant and Spacer Sealant Strip and Composite Structures Comprising the Same.[10] In 1979, during one of the regional trade shows, Swiggle Strip™, as it was then known, was introduced to the insulating glass manufacturing industry in the U.S. A few years later in 1982, it was introduced in Europe. It is a revolutionary system, incorporating desiccant, a specifically formulated sealant, and a corrugated aluminum or stainless steel spacer.[12] Figure 13.18 shows a cross section of the product. It provides reduced heat transfer around the perimeter of the insulating glass unit. The benefit is that high-performance windows in the home are less subject to the ill effects of condensation, resulting in clear view and less potential for mildew to form or rotting to occur in the frame. This type of insulating glass edge seal provides all the functions of an insulating glass unit (except the glass itself) combined into a single product. The all-in-one tape concept also revolutionized the manufacture of insulating glass units.

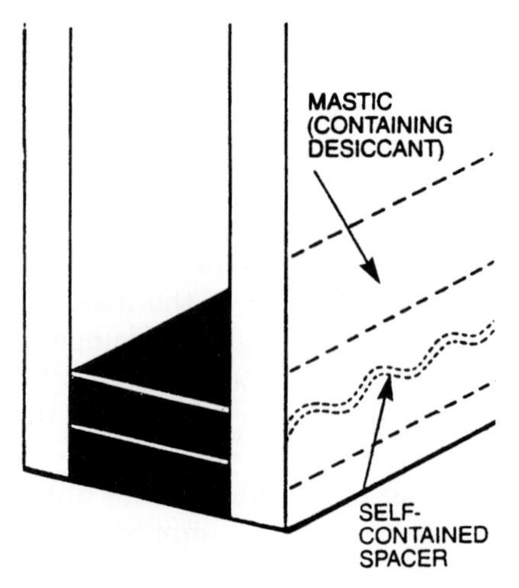

Figure 13.18 Cross section of Swiggle Seal.

Product composition

Swiggle Seal™, as it is now known, is a continuous extruded strip comprised of sealant, desiccant, and integral aluminum spacer. The 100 percent sealant composition is a proprietary formulation of high-performance polymers. A molecular sieve desiccant blend in powder form is uniformly dispersed within the sealant. The integral spacer is a wave form (corrugated) strip incorporated into the sealant/desiccant preparation. Because of its flexibility, the material can be readily formed around corners. Therefore, no corner keys are required. It is supplied in either roll or reel form on a release paper carrier.

Typical product information

Table 13.8 lists the typical properties for use in developing specifications.[12]

Typical performance properties

Table 13.9 presents a table of performance characteristics that meet the performance criteria of ASTM-E-773/774 and CAN/CGSB-12.8-M90.

Argon permeance

The most recognized test method in the world today which measures gas retention is the German specification DIN 1286, part two. Only the warmest sealed insulating glass units can claim proven argon gas

TABLE 13.8. Typical Properties Of Swiggle Seal

Colors	Gray and black
Sizes	Gray—airspace $\frac{1}{4}$–$\frac{13}{16}$ in in increments of $\frac{1}{16}$ in (6–20 mm in increments of 2 mm)
	Black—airspace $\frac{3}{16}$–$\frac{13}{16}$ in in increments of $\frac{1}{16}$ in (5–20 mm in increments of 2 mm)
	Product is available with grooved sight line surface for decorative glass applications
Packaging	Gray—reels available in returnable packaging
	Black—flat rolls (76–110 ft/23–33 m) or reels available in returnable packaging
Storage conditions	Store in original airtight container
	Expose to air only during application
	Procedures for opening and resealing Swiggle Seal are outlined in Form No. F-3331 and should be followed
Shelf life	One year minimum in unopened containers at storage conditions below 140°F (60°C)

TABLE 13.9. **Performance Characteristics**

Properties	Results
Argon permeance	6–8 cc/100 in^2 per atm as measured on a 0.040-in (1.02-mm) film at 77°F (25°C) per ASTM-D-1434.
Moisture vapor transmission (Mocon):	0.09 g H_2O/m^2/24 h for 0.06-in film at 100°F (38°C) per ASTM-F-1249.
Volatiles	Nonfogging in units when tested according to ASTM-E-773 and E-774; meets CanCGCB-12.8-M90.
Peel strength	7–10 pli when stressed at a rate 1 in (50 mm) per min
Yield strength	20–29 psi when stressed at a rate of 2 in (50 mm) per min
Force to compress the mastic	100–200 psi when compressed 2 in (50 mm) per min
Resistance to compression of self-contained aluminum spacer	8 mils thickness, 1300–1700 lb per linear ft, when compressed at a rate of 2 in (50 mm) per min

retention backed by the DIN standard. Neither the U.S. nor Canada, through ASTM or CGSB, has developed a test method for measuring argon gas retention properties in an insulating glass unit. Figure 13.19, in Swiggle Seal, shows how the continuous metal shim and thick cross section of sealant reduce the rate of argon gas loss, whereas conventional units with corner keys and more permeable sealants have a higher potential for argon loss.

Moisture vapor transmission path

The moisture vapor transmission path and resistance to moisture permeability are major contributors to the life expectancy of an insulating glass unit. The tape provides greater consistency to the moisture vapor path because of the solid polymer extrusion in its makeup. Figure 13.20 demonstrates the leakage path of four different systems.

Condensation resistance

Compared to a conventional spacer insulating glass unit, the Swiggle unit provides reduced heat transfer around the perimeter of the glass. One of the most important advantages is reduced condensation. Condensation can shorten the life expectancy of the insulating glass system by allowing entry of damaging water around the edge seal. This water can eventually destroy the seal between the glass and the spacer

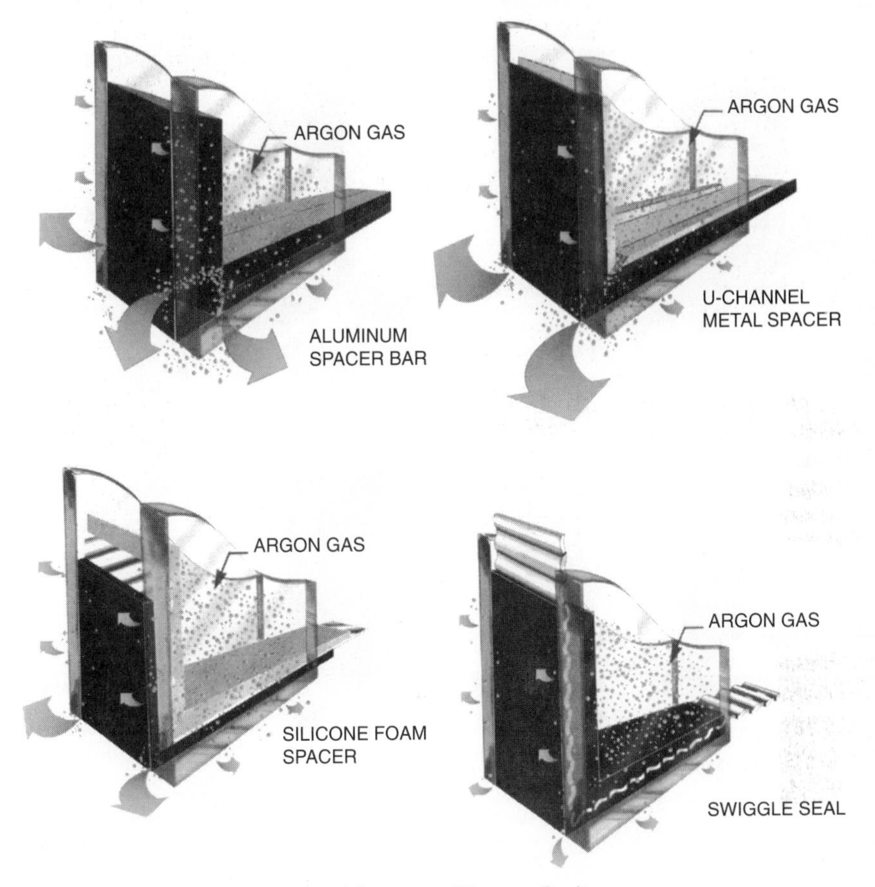

Figure 13.19 Argon gas retention. (*Courtesy of Tremco, Inc.*)

causing the sealed unit to fog permanently. The data in Figure 13.21 is a computer model by an independent laboratory that shows the condensation resistance of the stainless Swiggle Seal™. Figure 13.22 illustrates the condensation factor.

Low-E glass compatibility

Low-E glass combined with this tape system improves the thermal performance of sealed insulating glass units by keeping in valuable heat energy. Low-E coatings can be fragile and moisture sensitive, but the exceptional moisture resistance of this sealant system prevents degradation of these coatings. Daytime and nighttime effects on energy losses are illustrated in Figure 13.23.

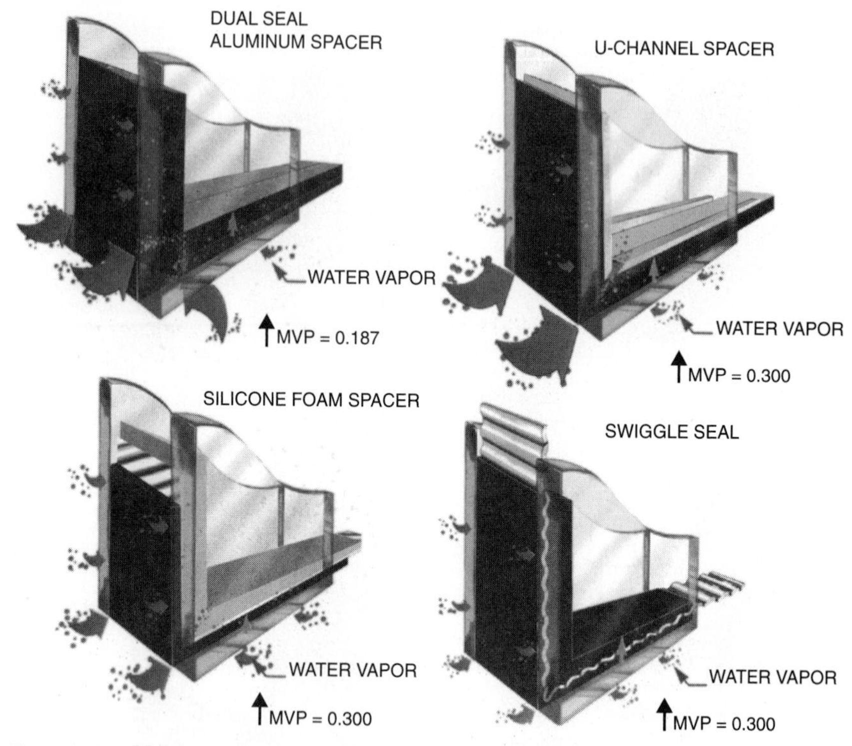

Figure 13.20 Moisture vapor path. (*Courtesy of Tremco, Inc.*)

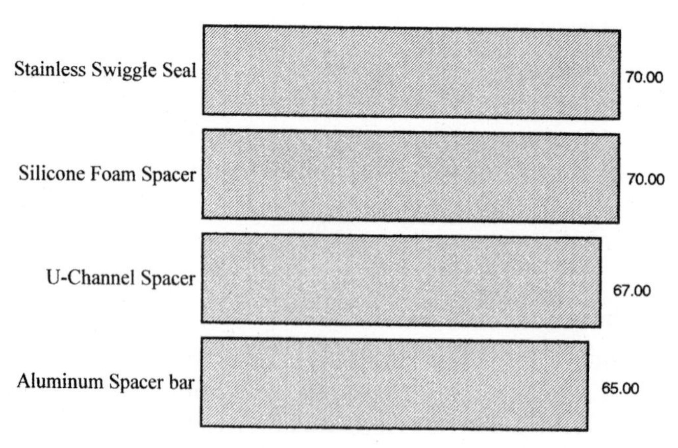

Figure 13.21 Condensation resistance factor.

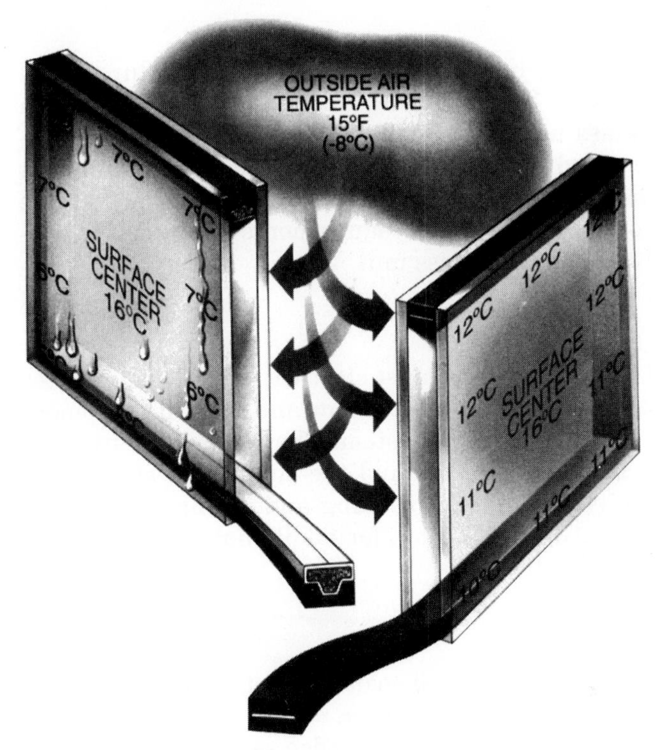

Figure 13.22 Condensation resistance. (*Courtesy of Tremco, Inc.*)

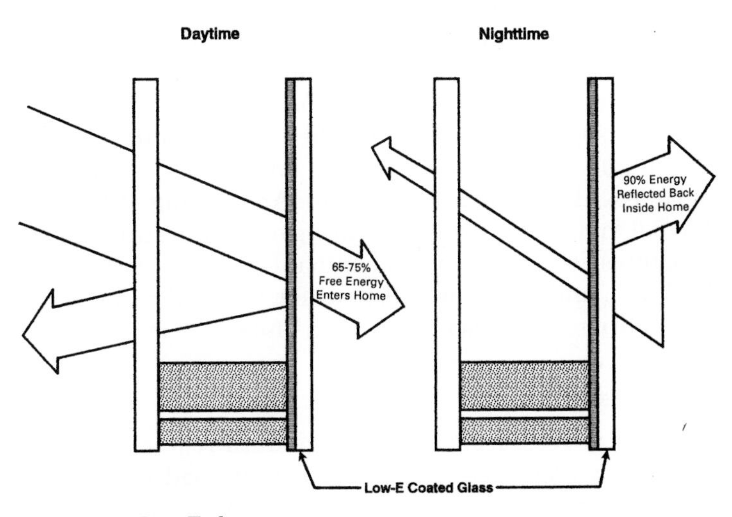

Figure 13.23 Low-E glass.

Manufacturing procedures

This section will outline, in detail, the methods, procedures, and requirements of applying Swiggle Seal™ to a lite of glass to form a sealed insulating glass unit. Also included is information on equipment, maintenance, and adjustments for the associated equipment mentioned. As in any edge-sealant system, you will find that Swiggle Seal has fabrication procedures and requirements necessary to construct a high-quality insulating glass unit.

Glass cleaning

The application of the sealant tape must start with clean and dry lites of glass. The glass washer is the most important piece of equipment in any insulating glass fabrication system. The recommendations of the glass washing machine supplier and the supplier of sheet glass must be followed to ensure proper cleaning and drying of the glass lites. See Chapter 21, Glass Washing, for additional information on the techniques of washing the glass.

Placement of Swiggle Seal

The placement of the composite tape sealant must be applied within the guidelines outlined by its manufacturer.

All lites must be clean and dry, and the edge sealing tape must be applied perpendicular to the glass lite. The side with the release paper must be placed to the outside of the unit, and the beginning and the end of the roll should be cut squarely. It is important that all splices be made at the corners. Place the Swiggle Seal a minimum of 0.062 in (1.5 mm) from the glass edge to ensure that the mastic does not extend beyond the glass edge after unit compression.

Stretching of the sealant material and the internal shim must be kept to a minimum. The applied in-place amplitude of the shim should never be less than 0.062 in (1.5 mm). An opening of 0.032 to 0.156 in (0.8 to 3.9 mm) is required for venting of the unit through the heating and compression stages of production. Use 0.032 in (0.8 mm) for an air-filled unit and 0.156 in (3.9 mm) for a gas-filled unit. See Chapter 16 for additional details on gas-filled units.

Application techniques

Do not remove rolls or reels from the package until required for unit manufacture. The Swiggle Seal should be conditioned so that the material temperature is 65 to 75°F (17 to 25°C). Within this temperature range, it is sufficiently pliable and tacky to make application to the glass easy. This temperature conditioning is achieved most effectively by storing stock material in a clean room. The time required to condition the sealant to the required temperature depends on the temperature at which it has been stored or shipped. In general, one or two days conditioning is required.

When removing flat rolls of Swiggle Seal from the package, ensure that the separating disc is kept underneath the roll. The disc helps support the roll, prevents "telescoping" of the roll, and aids in keeping the tape free of foreign matter when placed on the holder. Place the fingers through the center core of the roll and through the hole in the center of the support disc and lift from the package. Tilting the roll vertically as it is removed also helps prevent telescoping of the roll. Commercially available stands are available.

After removal of each roll or reel from the package, it is important that the desiccant bags and indicator card are placed back in the package and that the package is resealed. This protects the desiccant bags so that they continue to provide protection for the tape. If this procedure is not followed, moisture vapor entering the package is adsorbed by the desiccant in the sealant, thereby reducing its capacity for drying the air space when subsequently used to manufacture an insulating glass unit.

Before handling the Swiggle Seal™, ensure that hands are clean and dry. After meal breaks, it is important that hands are washed to prevent any foreign matter from contaminating the surface. Areas of the application table and dispenser stand, which the tape may contact, should similarly be free from dust, oil, grease, moisture, etc.

Swiggle Seal™ is manufactured with an aluminum or stainless steel shim positioned closer to the outside of the tape than the inside as it relates to the glass unit. This side must be used as the exterior side of the seal of the insulating glass unit to maximize the amount of available desiccant on the air-space side of the seal. The side which has the release paper attached is the side closest to the aluminum spacer, as illustrated in Figure 13.24.

The tape can be applied to the glass working from left to right or from right to left, depending on the preference of the operator. Rolls of the tape should be supported on a stand during application. The stand should consist of a flat surface, inclined at an angle of around 5 to 15° from the horizontal, and should support the whole roll. There should be no sharp edges which could damage the roll. The height of the stand should be adjustable to suit individual operator preference and facilitate application to varying sizes of glass.

When starting a new application, the end of the tape should be cut off squarely. This is particularly important at the beginning of each new roll because the end of the tape is deformed when cut at the time of manufacture.The cut should be made in the plane of the metal

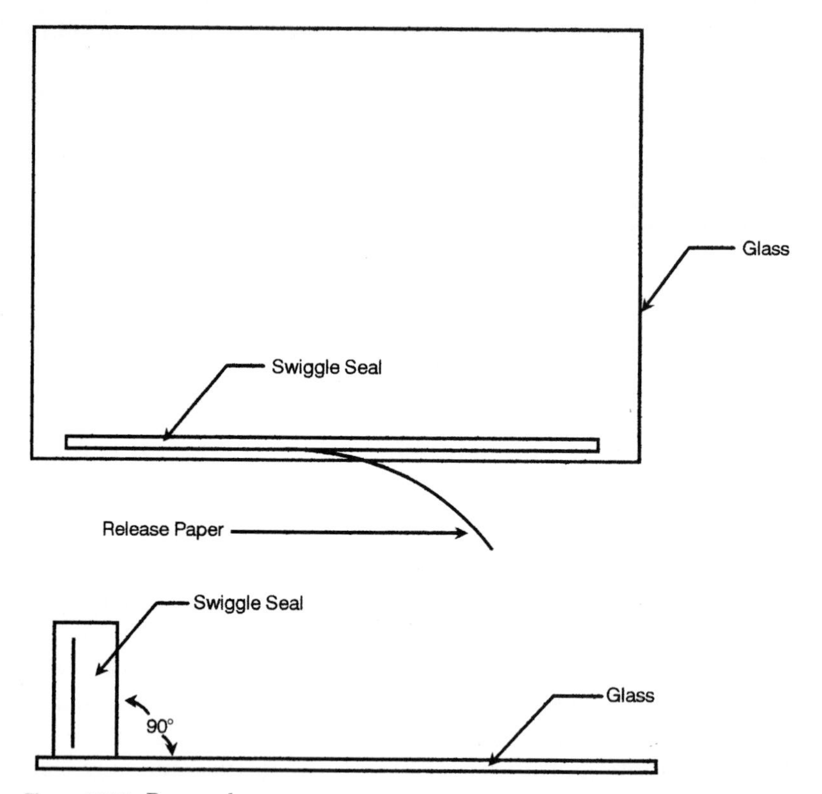

Figure 13.24 Paper release.

shim, not across the plane of the shim. See Figure 13.25. If the end of the tape is not cut clean and square, it is likely that a poor-quality corner joint seal will result. Typically, this material can be cut with a pair of shears. A number of snips are also available, although the most effective tool is a pair of garden shears.

Manual application of Swiggle Seal™ is carried out with the glass horizontal. The glass should be supported on an application table with a vacuum cup, which holds the glass in a fixed position so that the tape can be applied accurately. The table should be capable of rotating the glass so that each edge, in turn, is brought to the operator.

When applying by hand, the tape should be handled to minimize touching or pressing on the sides which are the bonding surfaces. This should be handled on the inside and outside surfaces. Refer to Figure 13.25. Hand contact with the bonding surfaces can result in transfer of oils or contaminants to the surfaces from an operator's hands. In extreme cases, the tape can be deformed by finger pressure to such an extent that proper wet out to the glass is not obtained.

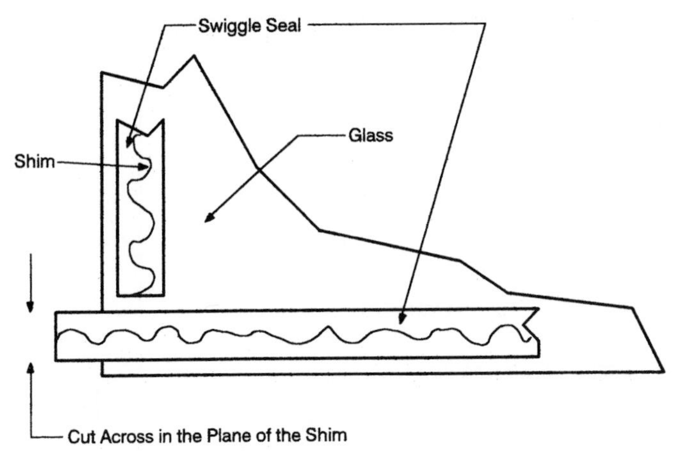

Figure 13.25 Swiggle Seal—top view.

Care should also be taken to avoid other forms of damage to the tape during its use. The loose end of the tape should not be allowed to hang over the edge of the dispenser stand because this might cause damage to the tape.

Peel back the release paper, and take a length of tape in the hands (gripping it by the inside and outside surfaces). Apply tension to the tape slightly to straighten it, and then allow it to relax. Tape application is started at a corner of the lite of glass. The end of the tape is positioned on the glass approximately half the thickness of the tape approximately 0.187 to 0.250 in (5 to 6 mm) from the corner of the lite and with the outside edge of the tape inset by approximately 0.062 in (1.5 mm) from the edge of the glass. See Figure 13.26 for edge positioning.

Continue the application of the tape along the first edge of the glass, pressing it down onto the glass by gripping the inside and outside surfaces.

At the first corner of the lite of glass, ensure that the tape is firmly positioned on the glass. Then, rotate the glass so that the second edge is brought to the operator. Holding the tape so that it is clear of the glass, fold the tape approximately 45° to the inside of the glass lites shown in Figure 13.27, and pinch the outside of the corner with the thumb and finger

Ensure that the tape comes just to the edge of the glass on both sides of the corner so that after compression there will be sufficient material to protect the glass corners. Neat corners with a minimum radius are preferable from an aesthetic viewpoint.

Continue the tape application around the remainder of the perime-

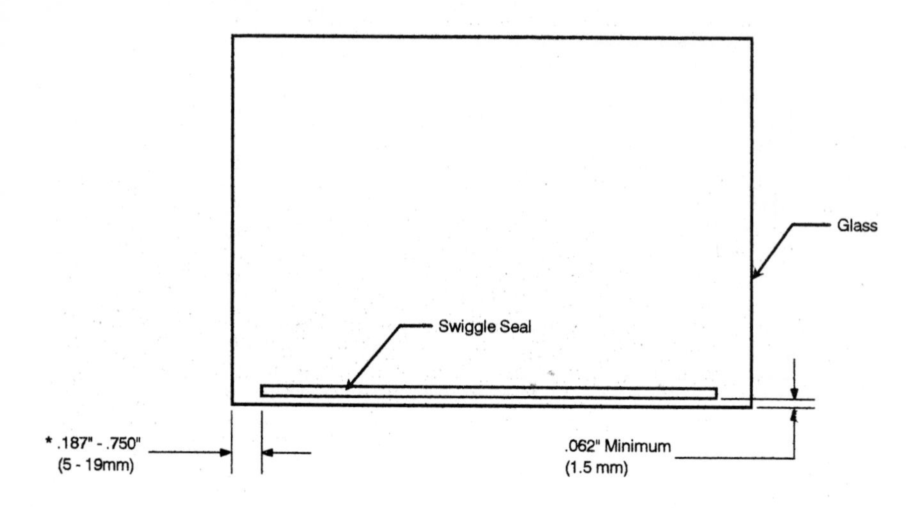

Figure 13.26 Edge positioning.

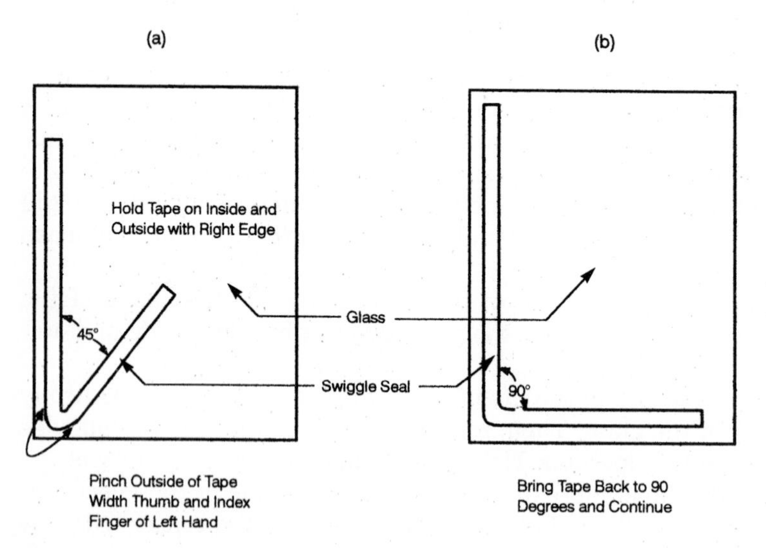

Figure 13.27 First-corner application.

ter of the glass in the same way. When the final corner is reached (the starting point of the tape application), depending on the start point, either bend the end of the tape outside of the glass leaving a small gap which will allow air to escape from inside the unit during compression (see Figure 13.28), or run the tape in a straight line past the start of the tape application leaving a small gap for escape of air. Cut off the tape cleanly and squarely just outside the edge of the glass. If a long "tail" is left on the Swiggle Seal, it may not be possible to tool this neatly after corner joint sealing and the potential for deposit of sealant on the compression rolls of the press exists.

The second lite of glass is now placed onto the Swiggle Seal that has already been applied to the first lite of glass. It is important that the two lites are accurately aligned. Any misalignment (producing an offset unit) will result in insufficient contact area of the Swiggle Seal™ on the glass at some part of the perimeter of the unit. Once the second sheet of glass has contacted the tape, it is not possible to adjust the alignment, because the initial adhesion prevents any movement. For small sheets of glass, accurate placement can, with care, be achieved by one operator. For larger sheets of glass, the assistance of a second operator is required. Squaring stops incorporated in the application table and positioned on adjacent sides of a right angle will assist with alignment. The use of a vertical board for alignment is also effective.

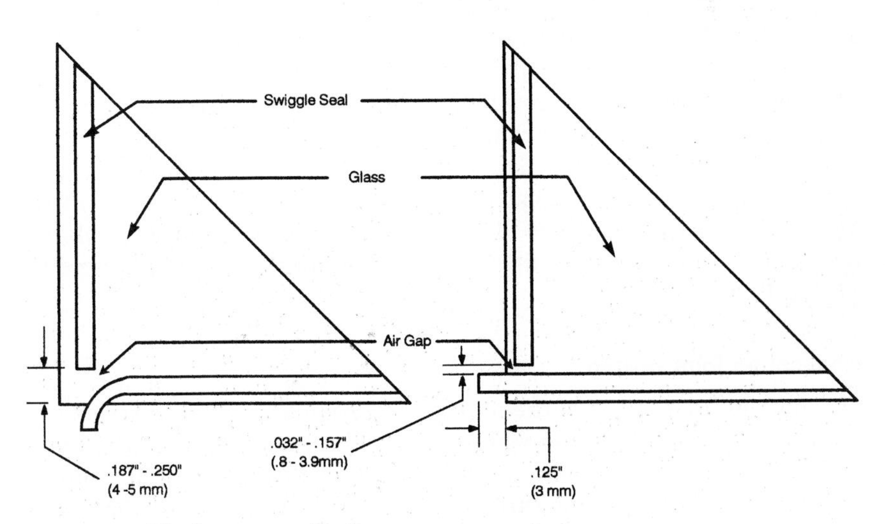

Figure 13.28 Final-corner application.

Assembly of large units is greatly facilitated by carrying out the topping procedure in the vertical plane, so that horizontal and vertical edges are accurately positioned against fixed stops. Vertical assembly can be carried out on a separate vertical frame or on a tilt table, or a tilting option may be incorporated in the application table.

Hand tool application

Begin tape placement the same as with manual application, 0.125 to 0.750 in (3 to 19 mm) from the glass edge, depending on the corner technique and inset to be used. A properly adjusted hand tool will allow the pivot pin to clear the top surface of the glass lite when beginning the tape application. Place the hand tool with the horizontal guide wheels against the glass edge and hold the Swiggle Seal in the left hand approximately 4 to 6 in (100 to 300 mm) above the table surface, 18 to 24 in (450 to 600 mm) to the left of the hand tool and 4 to 8 in (100 to 200 mm) behind the hand tool. Using a slight amount of tension with the left hand off the edge of the lite of glass, move the hand tool from right to left while pressing the hand tool guide wheels against the outer edge of the glass. There is no down pressure required with the right hand. Sufficient pressure is applied with a properly adjusted pressure wheel. Bring the hand tool to the left edge of the glass lite, stopping 1 to 2 in (25 to 50 mm) from the end. Press down on the Swiggle Seal™ at the edge of the glass lite to bond the tape to the glass. Hold the tape on the inside and outside surfaces without touching the bond line. Hold the tape 0.250 in (6 mm) from the glass edge with your left hand, activate the index mechanism of the application table, and rotate the glass lite clockwise with your left hand while the hand tool remains in its last position on the glass. Bring the hand tool to the edge of the glass with the pivot pin approximately 0.437 in (8 mm) from the edge of the glass. In one continuous motion, rotate the hand tool 90° clockwise while bringing the tool and the tape 45° into the glass lite. The hand applicator tool is pictured in Figure 13.29.

Final-corner joint sealing

For most units, sealing of the fourth corner joint should be carried out immediately after pressing while the tape is still warm. In the recommended temperature range, the tape is pliable and will readily adhere. If the tape temperature is below 100°F (38°C), difficulty is experienced in forming a good seal. In this situation, it is recommended that additional heat is applied to the joint surfaces. This is readily achieved by using a hot air gun or a domestic hot air paint stripper. Industrial hot air guns designed for shrink-wrapping or welding of

Figure 13.29 Hand applicator. (*Courtesy of Tremco, Inc.*)

plastics are also suitable. A hot air blast of about five seconds duration, directed at the surfaces to be bonded, is generally sufficient.

For gas-filled units where the entire unit has cooled and the corner must be reheated prior to sealing, use infrared heat lamps (approximately 300 W), placed in close proximity to the corner joint for about 10 seconds.

Completing the seal

The seal should be completed with a simple tool made from nonstick materials by pressing the flared-out end of the Swiggle Seal tape onto the starting end of the strip. Exert pressure in the direction of the starting length of the strip and ensure that a totally homogeneous seal is formed with no visible joint line. Take care not to push the Swiggle Seal™ into the vision area of the unit.

The final seal

Inspect the corner joint, ideally on both sides of the unit, to ensure that a good seal has been achieved with no joining line visible. A mirror will help examination of the underside of the joint.

Special situations

Seal insulating glass units at room temperature. Pressure equalization is required for certain units to relieve stresses in the glass and the tape. For example, where the area is greater than 16 ft^2 (1.5 m^2) or when units should be sealed with the glass in a vertical position to prevent glass deflection, one or both glass edge dimensions are less than 15 in (375 mm). After the insulating glass units have been pressed, place in a vertical plane, and allow to cool to room temperature before sealing with an air gun.

Storage of units

Handle and store units with care to prevent damage and deterioration in quality. The main conditions of storing completed units are to store inside in dry conditions away from direct sunlight. Store units on edges on a suitably resilient soft surface with adequate support to prevent bowing or distortion. Ensure that the bottom edges of both sheets of glass are supported.

Conclusion

Thermal performance is taking on increased significance, and consumers will be value-driven more than ever in the 1990s. Overall thermal improvements of insulating glass units should emphasize areas with the greatest impact, energy saving, and economy.

Benefits to the consumer can be maximized by selecting insulating glass performance factors which match geographical considerations.

Swiggle Seal™ is an essential component in the warm-edge market and offers long-lasting, impressive features: all-in-one simplicity, continuous metal barrier, most desiccating power, and longest moisture vapor transmission path.

Today, there are more than 300 lines operated by window manufacturers across North America making warm-edge units by all three of the major processes. Although developers of warm-edge technology can provide product assurances, careful attention must be paid to the actual practices of the individual window fabricator. Windows with warm-edge technology, the best the industry has to offer, offer a high level of quality and service.

Super Spacer™, Swiggle Seal™, and Intercept™ are components essential to the insulating glass market, offering long lasting impressive features: all-in-one simplicity, continuous barriers, high desiccating power, and longest moisture vapor paths.

References

1. Plavecsky, J., Sealants Div., Tremco, Inc., NWWDA 8/91.
2. Glover M., and G. Reichert, Edgetech, a Div. of Louren Mfg. Co., New Philadelphia, OH, 9/89.

3. Wright, J. L., and H. F. Sullivan, Thermal Resistance Measurement of Glazing System Edge Seals and Seal Materials Using Guarded Heater Plate Apparatus, *ASHRAE Transactions* 95, Pt.2, 1989.
4. Bowen, R. P., DBR'S Approach for Determining the Heat Transmission Characteristics of Windows, BRN 234, IRC, National Research Council, Ottawa, Canada, 1985.
5. "Super Shops," *Edgetech Newsletter,* Winter 94–95, Vol. 5, No. 1.
6. Meyer, P., Sr., Innovation to warm-edge IG fabrication calls for a new approach to desiccants, *DWG Fabricator,* 12/94.
7. U.S. Patents 5,177,916, 5,255,481, and 5,351,451.
8. SIGMA-GRAM SG 2000-90, Sealed Insulating Glass Manufacturers Association. Chicago, IL, 1995.
9. Rupert, M. L., Verifying performance of warm-edge units goes beyond R-values, *Glass Digest,* 12/95.
10. U.S. Patent 4,431,691.
11. Zuege, K., and R. Srinivasan, *Simple Facts Brochure Information,* Beachwood, OH, Tremco, 1996.
12. Swiggle Seal®, Technical Data Sheet, F-8042L, Tremco, 8/95.

Chapter

14

Spacers and Corner Keys

I. Douglas Sherman

Vice-Chairman, Hygrade Metal Moulding Manufacturing Corp.

Joseph S. Amstock

President, Professional Adhesive and Sealant Systems

History and Introduction

Insulating glass spacers create the air space dividing the two lites of glass that form the insulating glass unit. That space provides and increases the insulating value of the vast majority of fenestration products put into place today. Insulating glass has proven to be a cost-effective way to improve the energy efficiency of virtually any structure. This is manifested by the fact that the penetration of insulating glass, or the percentage of new units put into place, has reached an estimated 95 percent of all residential fenestration products and 87 percent of all commercial products, In addition to the improvement in energy efficiency, insulating glass provides an additional level of security and provides an excellent sound barrier.[1]

By simply creating the space between two lites of glass, insulating glass spacers also hold the desiccant that dries the air space between the lites of glass, extracting either water or chemical vapors that could fog the unit. The spacers and their connectors also create a system that supports and enhances the proper sealing of the unit, in addition to forming a solid moisture barrier to improve the integrity and durability of the insulating glass unit.

On August 1, 1865, Thomas D. Stetson[2] was granted a patent for

Improvement of Window-Glass. His claim was to test the effects of uniting sheets of glass A and B by welting or welding them along the edges and at the intermediate points, so as to have the space or spaces enclosed hermetically sealed. Putty and a strip of wood or string were the method of separating the two lites of glass. Thus, the insulating glass industry was born. The aluminum spacer as we know it today was first marketed in the standard T profile in 1954 and was primarily used for insulating glass in refrigeration cases. The oil shock of the early 1970s spurred energy conservation which, in turn, vastly increased the demand for insulating glass units and with it corresponding increases in the number of spacer producers. Around 1977, as insulating glass production continued to grow, new automated lines began to appear, principally from European equipment manufacturers, and the spacer producers began to look for ways to make air spacers that were stronger and more cost efficient. After considerable research dollars were spent, in the early 1980s, the insulating glass industry saw the introduction of the first CO_2 laser welding process,[3] which dramatically improved the strength and fabrication efficiency of spacer frames. The stronger, more rigid laser-welded spacer frames facilitated fabrication, particularly cutting to appropriate length, punching for muntin, and ultimately for bending processes. The last ten years have seen a boom in spacer usage and a proliferation of an ever expanding variety of spacer products and fabrication techniques. Although the primary purpose of insulating glass spacers has not changed since Mr. Stetson's invention, the focus on the air spacer has increased. This increased emphasis results from ongoing improvement in insulating glass and window efficiency, the constant search for increases in manufacturing productivity to improve competitiveness, and the desire by fenestration products companies to differentiate their products in the marketplace.

Types of Spacers

Changing requirements and increased competition have spawned many different spacer systems, each with its own advantages and liabilities. In an attempt to understand these various spacer systems, they have been categorized into four primary areas:

1. Conventional metal spacers
2. Combination sealant type spacers
3. Nonmetal, synthetic spacers
4. Hybrids

Conventional standard metal spacers

Conventional metal spacers are made from a variety of materials, primarily aluminum, but also tin plate, electrogalvanized steel, and stainless steel. The insulating glass spacers have qualities that vary depending on the material used. In general, conventional insulating glass spacers are structurally sound. This has been proven in the field for many years and also in laboratory tests. With appropriate corner keys and connectors, these spacers make rigid frames that support and enhance the ability of insulating glass sealants to effectively seal the units for ongoing durability. Spacers, particularly aluminum and electrogalvanized steel, have coefficients of thermal expansion similar to those of glass. This lessens sealant stress because the spacer expands and contracts in tandem with the glass. Aluminum spacers are relatively easy to fabricate because they are easy to cut and are compatible with virtually all sealants and automated spacer bending machines. They are easy to store, can be handled readily because of their light weight, and are relatively inexpensive. In these days of increased environmental awareness and regulations, aluminum spacers can be recycled.

Conventional or standard air spacers come in a variety of profiles that have been developed or have evolved in response to particular applications and fabrication requirements. These profiles have some similarities in that most are available to produce air spaces from $^{3}/_{16}$ to 1 in, with some profiles available up to 2 in for specific applications. With the exception of spacers used for bending that will be covered later in this chapter, these profiles are produced with serrations in the bottom to allow corner keys to grab and hold in the spacer. In addition, the spacers can be imprinted with the company's identification number, third party certification tests, or other identifying information.

The most popular conventional spacer profile is the standard T profile which is $^{5}/_{16}$ in (8 mm) in height. Figure 14.1 is a standard widely

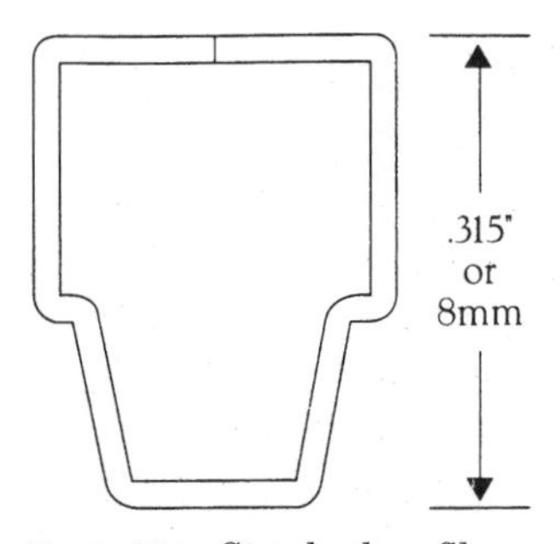

Figure 14.1 Standard profile.

used in single-seal and dual-seal applications and can also be used in bending processes. Any conventional sealing technique, gunning, troweling, and semiautomatic or automatic lines are applicable to this type of spacer.

The standard T profile comes in two permutations. Both have similar shoulder height, but differ in the sharpness of the angle leading to the bottom of the spacer. What has become known as trimline has a 16° angle, whereas the standard is 9°. The principal difference is that the 16° spacer takes a greater amount of the sealant. Trimline, illustrated in Figure 14.2, is widely used in butyl hot melt applications, in addition to all other designs.

As the industry developed, variations in spacer design were created, namely, the PIB (polyisobutylene) Groove Spacer, which is also $\frac{5}{16}$ in (8 mm) high. The butyl bead profile is similar to the standard profile except that it has an added deep grove on the sides running the length of the spacer on the shoulder. Figure 14.3 shows this configuration that facilitates the application of the PIB by hand and reduces a potential sight line problem caused by sealant overrun. The advent and increased usage of PIB extruders has reduced the demand for this type of spacer groove.

The rectangular spacer, designed in the 1970s, has the standard height of $\frac{5}{16}$ in (8 mm) and was developed for use with a hot melt linear extruder. The spacers are connected with folding locking corners

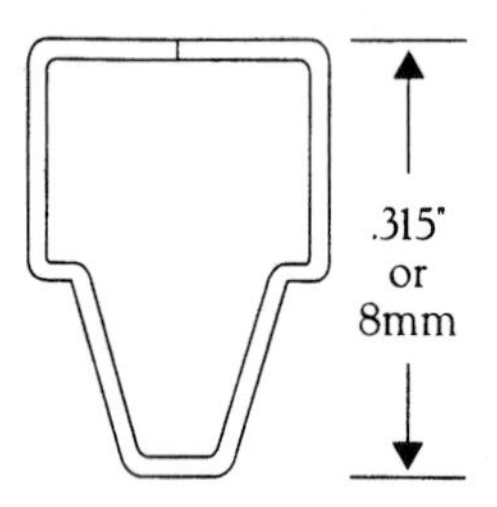

Figure 14.2 Trimline profile.

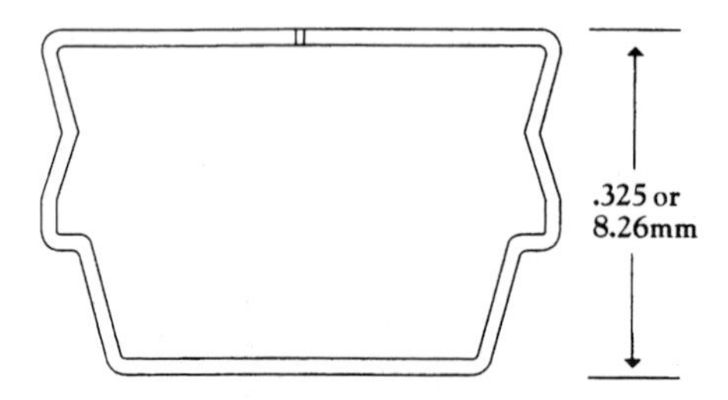

Figure 14.3 Butyl bend profile.

keys, lie flat on the extruder belt, and are bent and locked into a frame following sealant application. This profile is not recommended for bending. A variation of the rectangle (Figure 14.4) has been designed with a slight bump on the top of the spacer to discourage sealant blowby. Also, with improvements in equipment and manufacturing techniques, the popularity of this profile has declined.

One of the most popular spacer profiles after the standard T is a low-profile spacer, which is a modification of the rectangular one, with a height of ¼ in (6 mm). This shoulder creates a dam that prevents the U-shaped extruded hot melt butyl from migrating into the insulating glass air space, and the indentation for the PIB facilitates dual sealing. Its popularity has risen because it provides a low sight line, thus interfering to a lesser degree with the view through the insulating glass unit. In addition, this design allows a reduction in the bite necessary for the sash to cover the sealed unit. Figure 14.5 is a modified top and illustrates the above characteristics.

Because of its popularity with automated systems, the 250P was designed along with a trimline version. Named for its 0.250-in shoulder specification, it has a height specification of 0.315-in, provides a larger surface for the PIB, and also creates a larger surface to directly contact the glass (standard, trimline, etc.). (See Figure 14.6.)

There are other popular conventional spacer bars and certain proprietary variations of the profiles mentioned above. One, the decora-

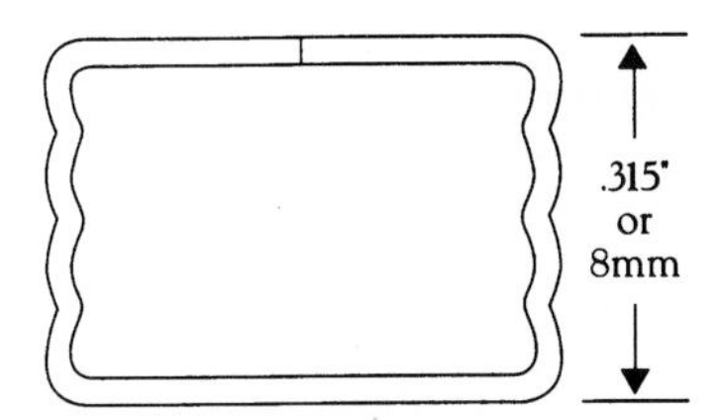

Figure 14.4 Rectangular profile.

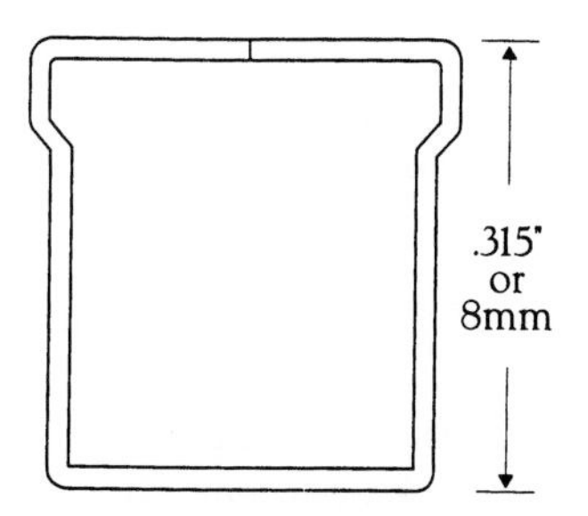

Figure 14.5 MT (modified top) profile.

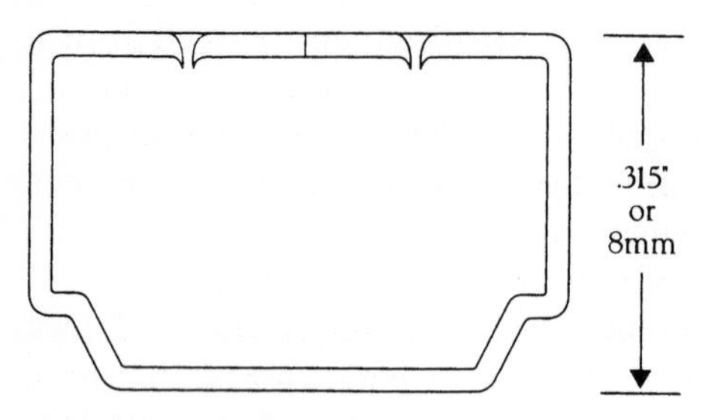

Figure 14.6 250P profile.

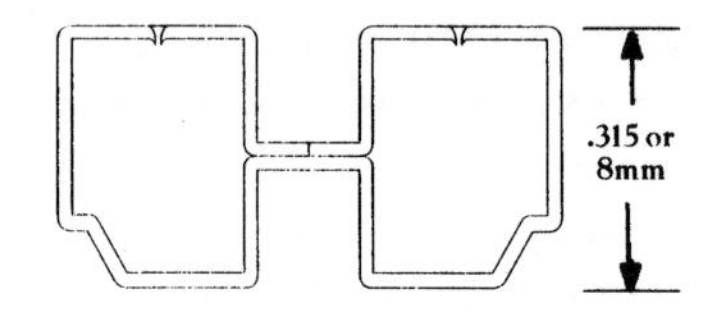

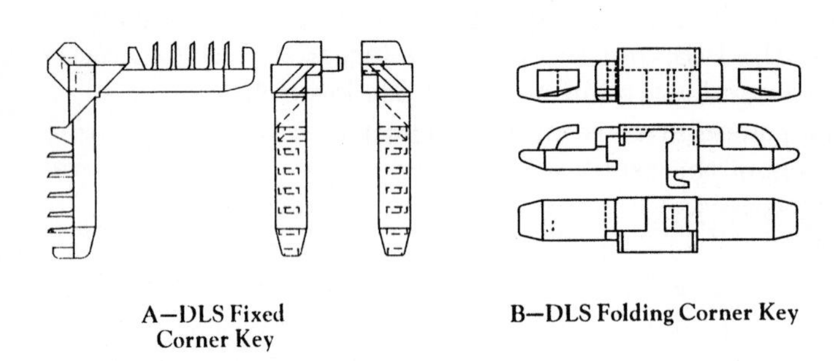

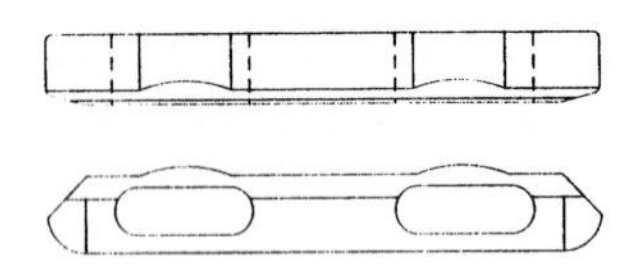

Figure 14.7 DLS profile.

tive lite spacer (DLS) with a standard height of 5⁄16 in (8 mm), is a specialty spacer that permits the placement of a center lite of glass in the center between the inside and outside lites (see Figure 14.7). A common practice is to insert a decorative lite of glass, such as stained or art glass. Some manufacturer's debridge the center portions of this profile, creating a thermal break unit by filling the void with a nonconductive material.

As discussed in the previous chapter (Chapter 13), the characteristics of the warm-edge technology have been detailed.

Materials and finishes. The predominant materials and finishes are an important factor that must be discussed when manufacturing conventional spacers. One of the more popular materials for a spacer is anodized aluminum. Anodized spacers are available in clear, dark bronze, black, and gold colors. The qualities of an anodized spacer include a more attractive satin or less shiny finish than nonanodized or mill finish aluminum. The anodized coating makes handling easier and less susceptible to scratching and will not show fingerprints that may result from manufacturing. Anodized spacers will not oxidize as does mill finish, nor do they carry mill contaminants, such as oil, that may be incompletely removed. Furthermore, anodized aluminum is noted for its excellent sealant compatibility. Mill finish aluminum (3000 series alloy) spacer has some other qualities. It tends to be less expensive than anodized, and it is useful if the insulating glass fabrication process includes dip soldering of the corners or butt welding the ends of the spacer in certain automated processes. In general, mill finish spacer bends well and is also recommended by some sealant manufacturers, particularly for silicones.

Spacers are also available with tops painted a given color to match the window frame or muntin bar or other architectural specification. These are primarily mill finish spacers with the top painted. The painted spacers have qualities similar to a plain spacer, but care should be taken to avoid spacers with paint overrunning down on the side's flanges. To avoid potential failure, sealants should adhere to the aluminum, not to painted surfaces. From the mid-1960s to the middle of the 1970s, aluminum was a plentiful and low-cost material. Today, aluminum is a commodity traded worldwide, subject to large price fluctuations, which encouraged the industry to explore lower-cost raw materials and systems (i.e., tin plate steel, the Intercept System, Swiggle Seal, and Super Spacer). However, aluminum spacers are still the primary choice.

Cold rolled carbon steel was used early on but was replaced by electrogalvanized steel because of its corrosion resistance and continues to be specified in Heat Mirror® applications. This will be covered in greater detail in Chapter 17 on Films.

Production methods. Taking flat coil stock (aluminum, steel, stainless steel, etc.) and gradually forming the material through a series of dies to the desired shape or form is known as roll forming. With the exception of extruded aluminum, which is mostly produced in Europe, or other spacer alternatives, roll forming continues to be the primary method of manufacturing spacers. Figure 14.8*a* through 14.8*g* gives a brief pictorial description of the spacer manufacturing process.[4]

Prior to 1980, conventional roll-formed air spacers were produced with a zippered butt seam or lock seam design in raw material gauges from 0.0175 to 0.020 in. The zippered butt seam was produced by knurling the edges of the raw aluminum stock and having the edges meet up on the final forming pass. A lock seam is achieved by interlocking one edge of the raw material to the other, thus creating a lock. This method of spacer manufacture gave way to laser welding technology, which allowed the use of lighter gauge materials from 0.012 to 0.020 in. Figure 14.9 shows five different configurations of welds used in the production of air spacers.

Corner keys. Corner keys and related connectors are required in spacers systems to assemble spacer frames, except in certain manufacturing processes and alternative spacer systems. A corner key with the appropriate fit will prevent the desiccant from spilling out. The majority of corner keys have fins that permit easy insertion, but will prevent the key from being pulled out of the spacer, particularly if the spacer has serrations on the bottom to hold the key.

Early keys were stamped aluminum or die cast zinc. They had good adhesion qualities but only provided a moderately good lock fit. These types of keys are used today only in specialized applications where corner soldering is required. With improvements in plastics, modern keys are made primarily of type 6 nylon or polyester. Type 6 nylon is available in a variety of colors. Over the years, design changes occurred to accommodate the various sealants, particularly butyl inject and wraparound designs.

As with air spacers, many types of corner keys and connectors have been developed to meet specific requirements. The most popular corner key is the fixed 90° key that can be used with virtually any gunning or spatula method of sealant application. Figures 14.10 and 14.11 illustrate two types of fixed corner keys.

Flex corner keys are similar to the fixed key except that the center of the key is flexible allowing fabrication of spacer frames other than those with 90° angles. Folding/locking keys are available for use with hot melt butyl linear extruder applications. For those applications requiring additional sealants in the corners, butyl inject keys are available that allow more sealants to be injected into the corner. See Figure 14.12. Wraparound keys are used to produce a continuous shape from spacer to corner. These keys are used primarily in Heat Mirror applications and sometimes in butyl inject systems. The key in

(a)

(b)

(c)

Figure 14.8 Description of spacer manufacturing.

(*d*)

(*e*)

(*f*)

Figure 14.8 (*Continued*)

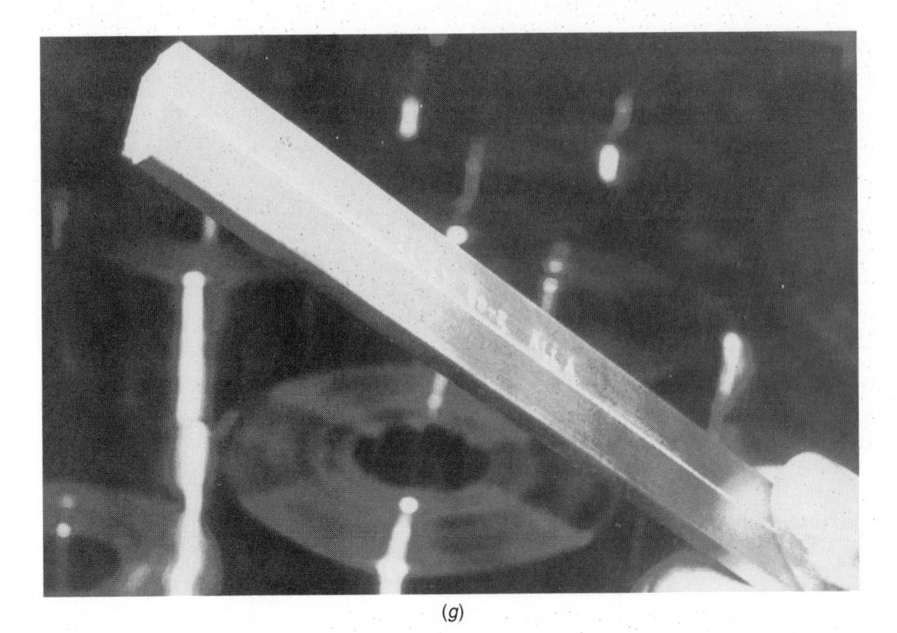

(g)

Figure 14.8 (*Continued*)

Figure 14.13 illustrates two styles, one with a fin and the other with a grip lock. Beside the myriad variations of corner keys, straight connectors are available that are used for 180° connections, primarily in conjunction with the spacer bending process (Figure 14.14). Certain types of straight spacer connectors allow desiccant to pass through the spacer, as shown in Figure 14.15.

Muntin bars. Internal muntin bars, hereafter referred to as muntin bars or muntin, are purely decorative. They are an aesthetically pleasing option sought by consumers and date back to the 1970s when manufacturers developed a method to mimic the popular mullion-style, true divided, lite windows, giving consumers the advantages of insulating glass while improving production efficiency. The concept of providing colonial-looking units with the energy efficiency of insulating glass has made the insertion of internal muntin grids a popular and lucrative upgrade for insulating glass and window manufacturers. Internal muntin bars have two additional benefits. They permit easy cleaning of both the interior and the exterior of the insu-

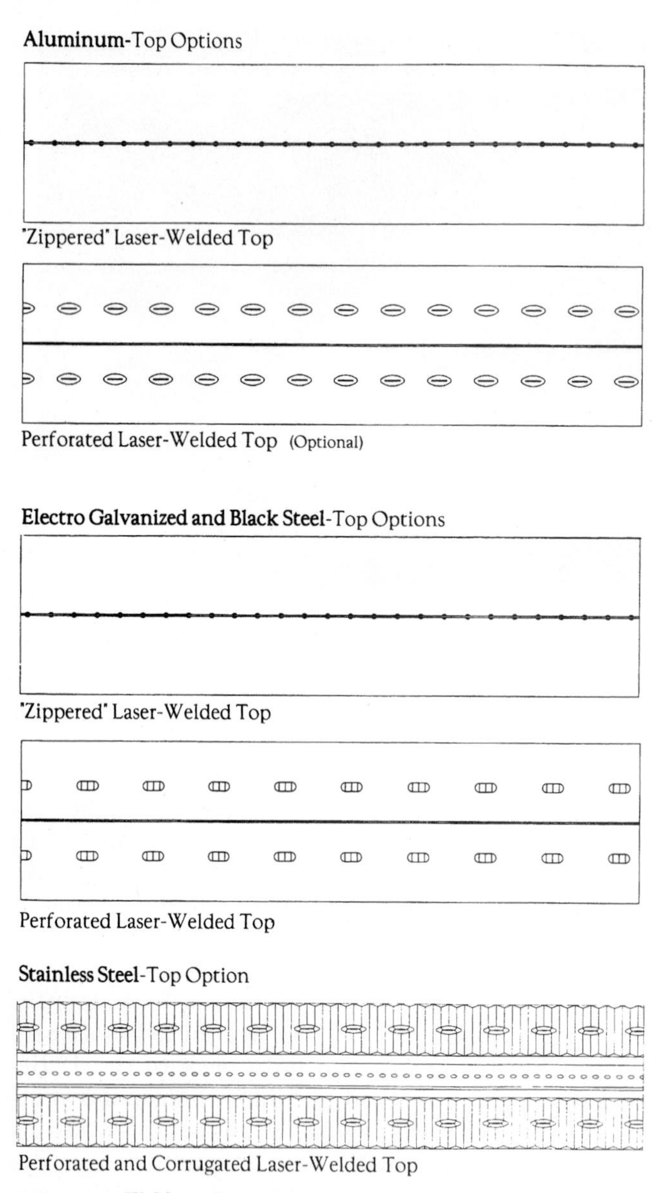

Figure 14.9 Weld configurations.

lating glass unit, and the grid lines produced by the muntin bar provide an additional security barrier, notifying people of the existence of the insulating glass unit in the opening.

The most popular internal muntin is flat, rectangular muntin, known as a colonial muntin or grid, is produced from painted alu-

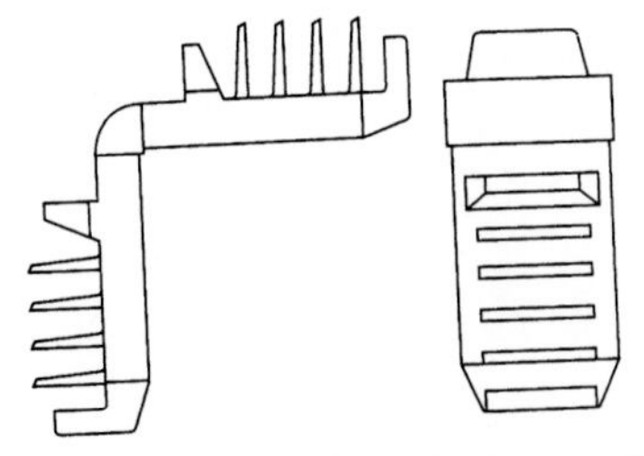

Figure 14.10 Standard fixed corner key with fin.

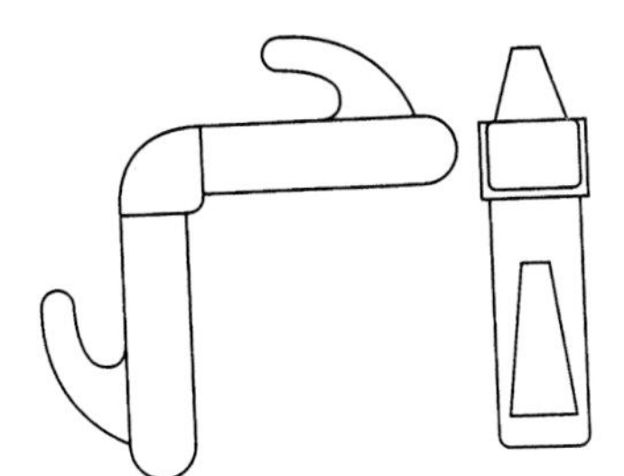

Figure 14.11 Standard fixed corner key with hook.

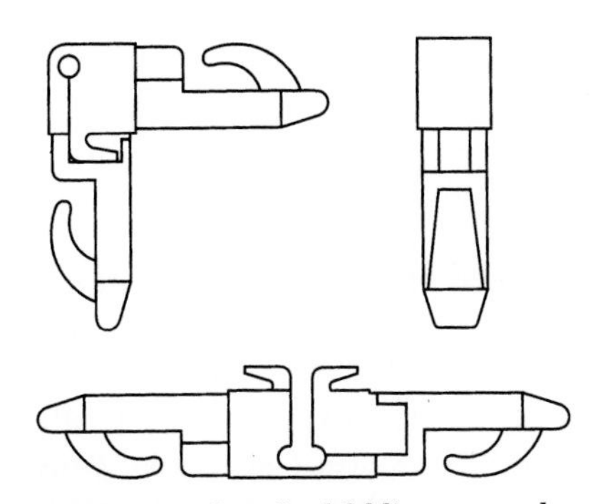

Figure 14.12 Standard folding corner key.

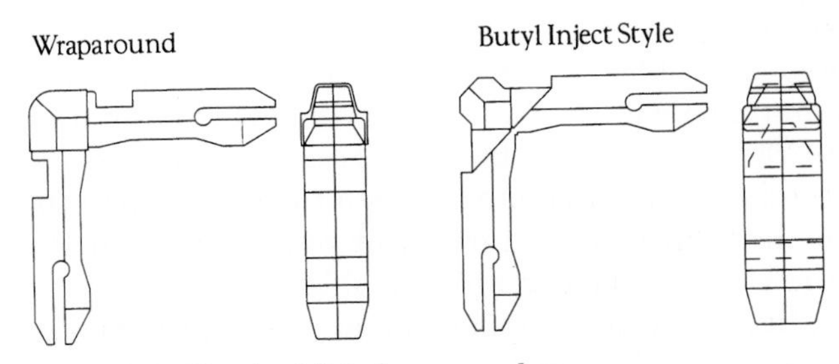

Figure 14.13 Standard Grip Loc corner key.

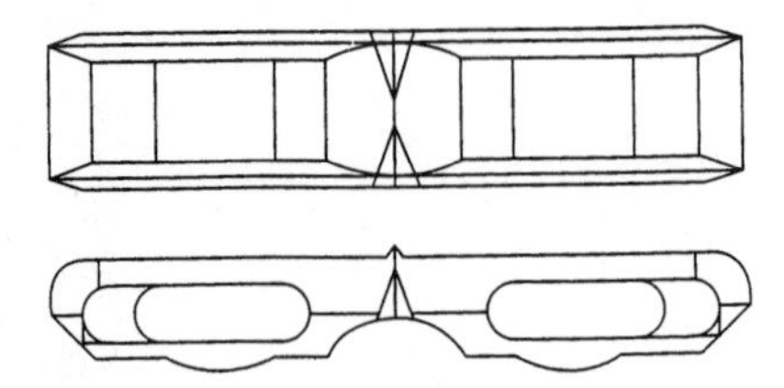

Figure 14.14 MP straight connector.

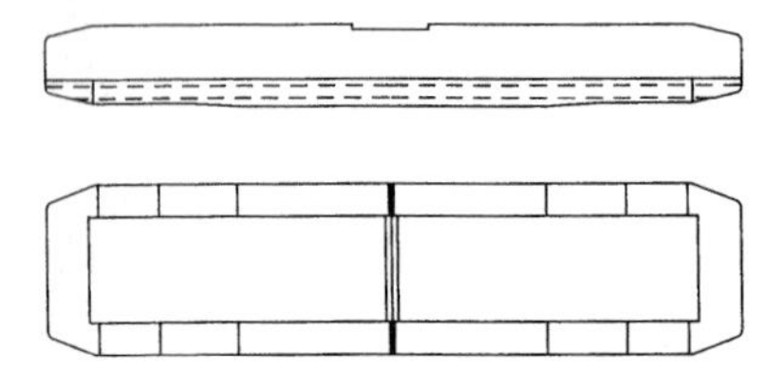

Figure 14.15 Trimline LP 250 straight connector—pass desiccant.

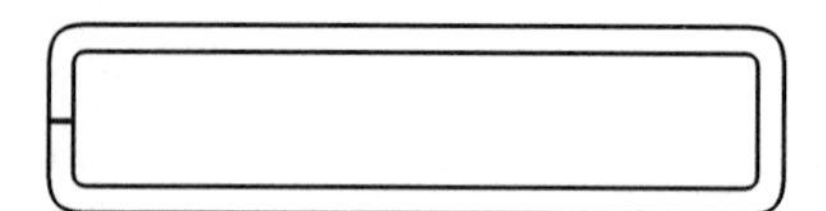

Figure 14.16 Muntin bar.

minum coil stock. Care must be exercised in selecting the paint, which can outgas and cause fogging. With advances in polymers and resins, modified polyesters were introduced in the mid-1980s. These new polyesters are higher-solids systems that have improved roll-forming qualities and provide more scratch-resistant surfaces, compared to the earlier acrylics. See Figure 14.16. Rectangular muntins are typically arranged in grid patterns but are also fabricated in diamond patterns and bent to various radii for use in conjunction with grid patterns to form sunburst and other intricate styles. Flat muntin comes in a wide range of profiles with thicknesses varying from $^{3}/_{16}$ to $^{9}/_{16}$ in (8.0 to 14.28 mm) and widths (face of the window) varying from $^{3}/_{16}$ up to 1 in (8.0 to 25.01 mm). A thickness of $^{3}/_{16}$ in (8.0 mm) is preferred because it permits a sufficient gap between the muntin and the glass to minimize thermal conductivity and problems with the muntin touching the glass on larger units. A fabricator's choice in determining the width is based on aesthetics and preference, usually driven by marketing.

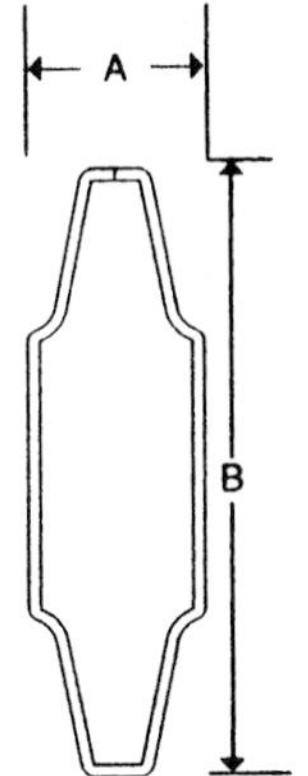

MILLIMETER
DECIMAL IN INCHES

Size	A	B
5.5 x 18	.217	.709
8 x 18	.315	.709
8 x 25	.315	.984

Figure 14.17 Contoured muntin bar.

A recent trend in internal muntins is the use of the contoured or Georgian style. See Figure 14.17 for this configuration, which saw its beginning in the North American market in the early 1990s. State of the art laser welding made the North American version of the extruded contour bar lighter, yet extremely strong and rigid, producing the effect of classic beauty and quality. The most popular size is $\frac{3}{16} \times \frac{23}{32}$ in (8×18 mm), followed by $\frac{3}{16} \times 1$ in (8×25 mm), and $\frac{7}{32} \times \frac{23}{32}$ in (5.5×18 mm). Contoured muntin is fabricated to create designs similar to those of flat muntin, as shown in Figure 14.17.

Flat muntin is typically provided in thicknesses up to 0.018 to 0.020 in including the paint and, for certain applications, in thicker gauges up to 0.024 in. The thickness of the material from which the muntin is made has nothing to do with the aesthetics of the assembled grid but does play a role in certain assembly processes, particularly bending.

Muntin fabrication. Muntin connectors or intersects, joiners, and keepers are available in a wide range of designs to satisfy the preferences of muntin fabricators. Figures 14.18, 14.19, 14.20, and 14.21 are typical of industry use today. The manufacturer of insulating glass should consult with the spacer supplier for the correct specifications.

One of the most difficult muntin issues pertains to cleaning and handling. No magic cleaning fluid yet exists. Methyl ethyl ketone (MEK) and alcohols are used. However, the Environmental Protection Agency (EPA) rates MEK as a hazardous material, and it also destroys the integrity of the paint. Disturbing the painted surface prevents the paint from holding up and can subject the unit to the

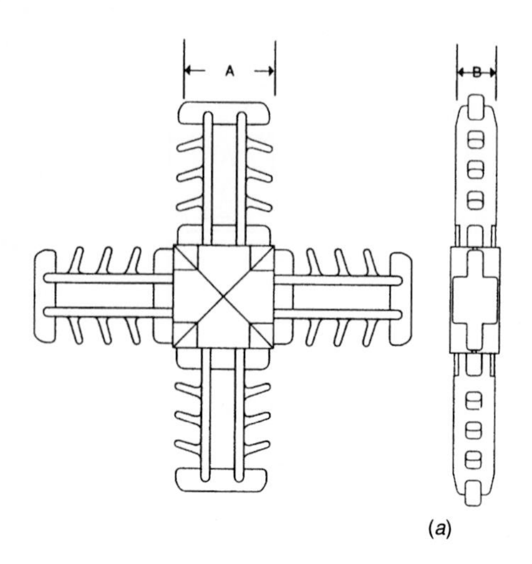

MILLIMETER
DECIMAL IN INCHES

Size	A	B
5.5 x 18	.686	.183
8 x 18	.686	.283
8 x 25	.970	.283

(*a*)

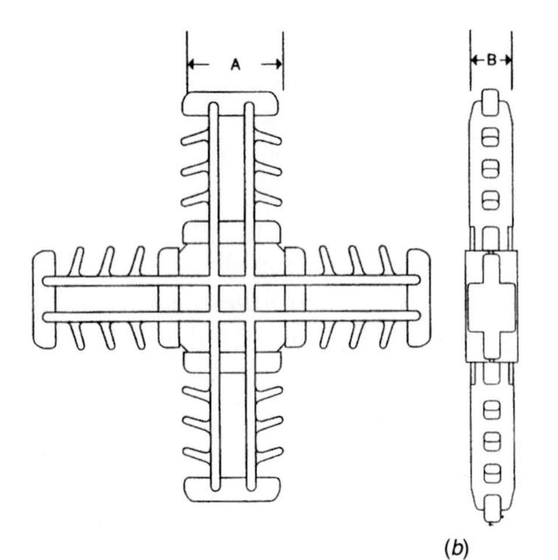

DECIMAL IN INCHES

Size	A	B
5.5 x 18	.686	.183
8 x 18	.686	.283
8 x 25	.970	.283

(*b*)

Figure 14.18 Contoured muntin bar. (*a*) External intersect. (*b*) Internal intersect.

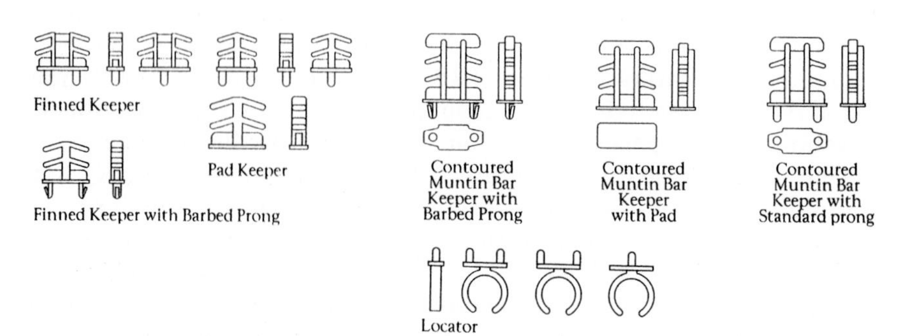

Figure 14.19 Muntin bar locator.

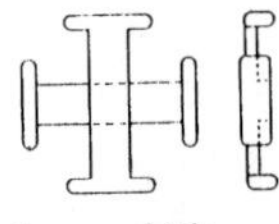

A. Fits Sizes 3/16 x 9/16, 3/16 x 5/8, 1/4 x 9/16, 1/4 x 5/8
B. Fits Sizes 3/16 x 3/4, 3/16 x 13/16, 1/4 x 3/4
C. Fits Sizes 3/8 x 5/8, 3/8 x 3/4
D. Fits Sizes 3/16 x 1", 1/4 x 1", 3/16 x 7/8, 1/4 x 7/8
E. Fits Sizes 3/8 x 7/8, 3/8 x 1"

Figure 14.20 Internal muntin bar clip.

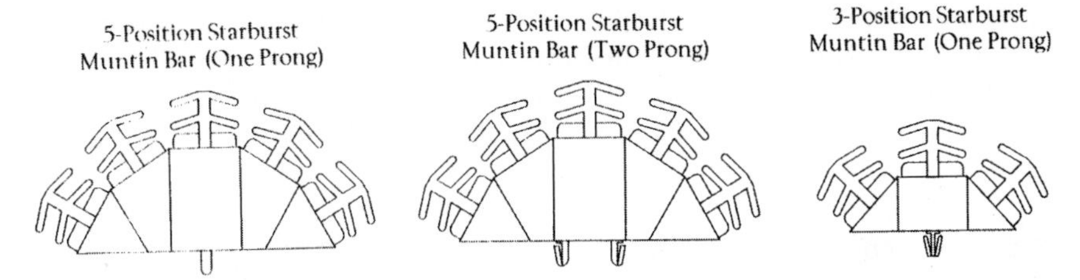

Figure 14.21 Starburst muntin bar keeper.

risks of outgassing. One possible suggestion is wiping the muntin lightly with common, mild, household glass cleaner.

Muntins must be handled carefully to prevent marring of their finish. Saws, cutting, and assembly surfaces should be kept clean and free of metal shavings. Touch-up paint is normally available to manufacturers should mishaps occur. When cutting, the blades of the saw should be sharp to avoid splitting the seams. Also, the muntin should be notched with the seam up. Having the seam up allows the blade to gather some momentum and make a cleaner cut.

Length specification and cutting. Spacers and muntins can be purchased in lineals or cut to size. Lineals of spacers and muntins, provided in either 12 ft 8 in or 13 ft, are most appropriate for insulating glass manufacturers that manufacture few stock sizes of windows. This can help to improve inventory turns and hold down inventory obsolescence. This potentially reduces the waste of spacer cutting, but results in a large potential inventory investment, as even the best designed programs can result in slower inventory turns and the greater likelihood that a particular size or sizes will become obsolete. Cutting to length can be done from 6 in to 24 ft, although spacer suppliers may require certain minimum orders for a particular size.

Cutting of the spacer bar and muntin is absolutely critical to the productivity of any insulating glass manufacturer. Burrs on the ends of the spacers can score the glass and result in a high spot that can cause the glass to fracture and could interfere with corner key insertion, thus lowering production. The equipment suggested for cutting is a chop or radial saw that operates at a minimum speed of 3600 r/min. Additional specifications include a 10 in blade with 80 carbide-tipped teeth. Cutting lubricants should not be used because they may contaminate the muntin and spacer, ultimately causing loss of adhesion of the sealant or causing the sealed unit to outgas.

Spacer bending. Spacer benders in use today can be classified by the production output of various size companies.[4] A bent corner can provide a better corner seal than a keyed corner and provides a continuous and uninterrupted wall of impermeable material around the corner. A good seal is necessary to prevent moisture from entering the unit. The bent corner can also retain gas much better than a keyed corner, if sealed properly.

Types of bends

Roll bending. Roll bending is the method used to form aluminum or steel air spacer material into large-radius circles or semicircles. This method is necessary to form spacer material economically along several different bend radii which are characteristic of custom-shaped windows, such as arch tops. Roll bending is usually a manual operation with very low volumes. Figure 14.22 illustrates a spacer being roll-bent.

Compression bending. In compression bending, the spacer material is clamped against a static or stationary die and then compressed or wiped around the bend die with a wiper die. This method is used to produce a tight-radius bend in the spacer. Most windows require

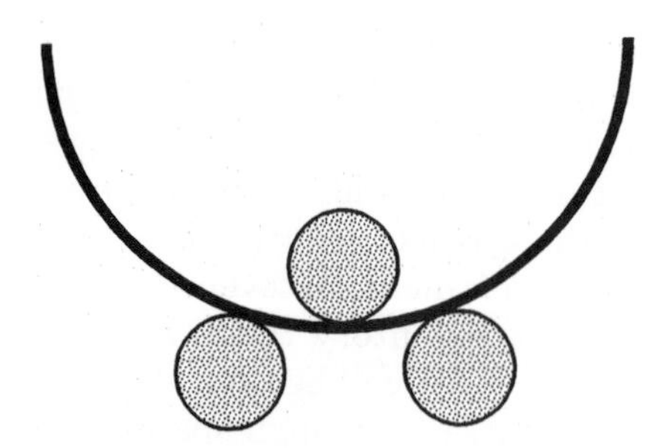

Figure 14.22 Roll bending.

tight-radius bends to avoid sight line problems, and, this being the case, we will focus on the compression method of spacer bending.

The outer wall of the spacer will actually collapse until it comes in contact with the inner wall of the spacer. In Figure 14.23, we illustrate compression bending.

Another important advantage exists to crushing the spacer profile in the bend area. The gas filling process is often performed by drilling two holes along one of the longest legs normally opposite the open corner, shown in Figure 14.24.

As Figure 14.25 reveals, the open corner is adjacent to the two legs of the frame that are filled with desiccant. The crushed spacer in the

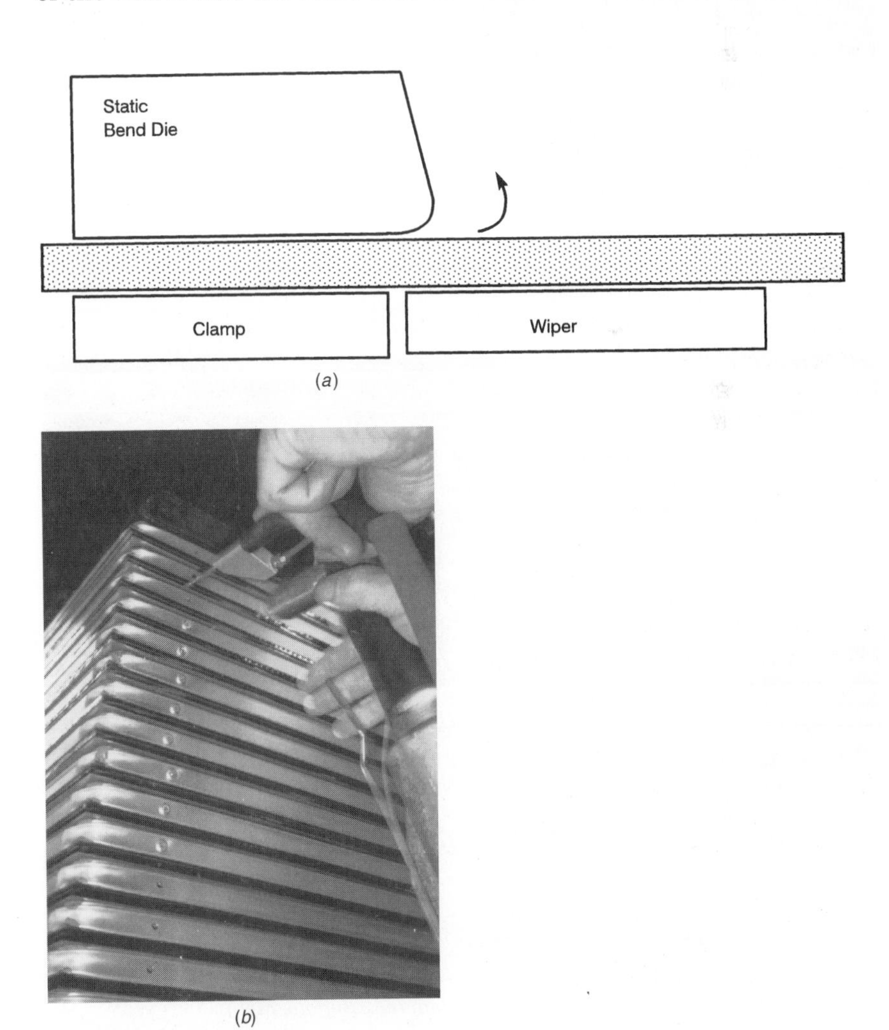

Figure 14.23 (*a*) Compression bending. (*b*) Compression bent spacer bar.

Figure 14.24 Gas filling through spacer.

Figure 14.25 Gas filling. (*Courtesy of FDR Design, Inc., Buffalo, MN.*)

bend area will not allow the desiccant to flow into the spacer leg containing the gas filling holes. If desiccant were allowed to enter the leg containing the holes, it could enter the insulating glass unit through the holes.

Inside crush. Figure 14.26 reveals an inside crush bend.[6] There is an obvious controlled crush along the inside bend radius. The primary sealant shoulder must remain flat and maintain its width consistently around the bend area.

Bend dies. Each spacer width will require a set of bend dies machined to the proper dimensions. Bend dies require heat treating and should be chrome plated to avoid frequent replacements and to achieve maximum performance. Bend die removal is instantaneous and requires no hand tools. An illustration of die removal is shown in Figure 14.27 from a semiautomatic spacer bender.[6] The operator is responsible for manually loading each piece of spacer stock onto the machine. The machine's functions are to automatically drive the spacer and perform each cut, drill, and bend operation.

In some machines, after the finished frame is removed from the machine, the computer will determine whether or not the existing tail stock is long enough to produce the next required frame. If the tail stock is long enough to produce the next frame, the machine will automatically advance the tail stock and produce another frame.

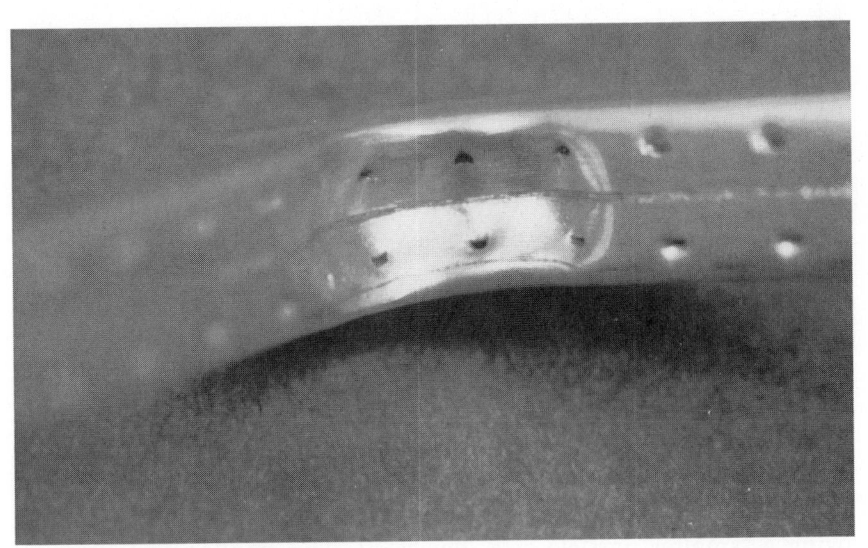

Figure 14.26 Inside crush bend. (*Courtesy of Tools for Bending, Inc., Denver, CO.*)

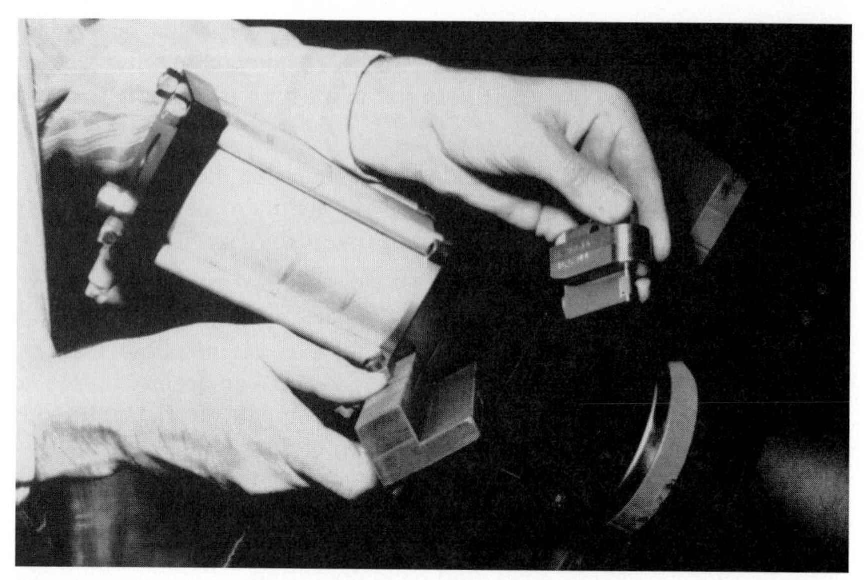

Figure 14.27 Bender die removal. (*Courtesy of Tools for Bending, Inc., Denver, CO.*)

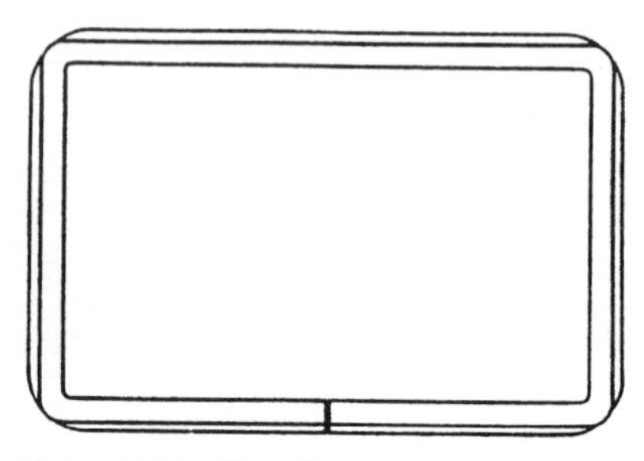

Figure 14.28 4 bends.

4 bends. 4 bends are produced in the frame and the seam or opening is left along one of the sides. This configuration (Figure 14.28) can produce high frame quality because all four corners are bent.

Filling this configuration with desiccant presents a production problem. There is no open corner to insert the desiccant, and the desiccant will not flow around a bent corner.

It is possible to drill holes in the leg of the spacer and blow desiccant into the spacer. This process is usually very slow, and it requires using a "smaller particle size" desiccant. The smaller particle sized desiccants are almost always more expensive to purchase than those with standard-sized particles.

Figure 14.29 3 bends.

3 bends. 3 bends (Figure 14.29) is the configuration most commonly produced in North America. 3 bends are performed in the spacer to produce a frame with one open corner. This method is faster than producing frames than 4 bends. Leaving one corner open allows you to fill two legs with desiccant very quickly and very efficiently. The insulating glass manufacturer is assured of a good fill ratio because the desiccant is blown into the unrestricted ends of the spacer legs.

Bending machines capable of producing the 3-bend configuration are also capable of producing specialty shapes, such as trapezoids and triangles.

L-bends. The primary reasons for producing a frame from two "L" bends are as follows:

1. Optimization. Whenever a piece of leftover spacer is not long enough to make a complete unit, an L-configuration can be used to produce one-half of a unit. This is one of many ways to optimize spacer material on a spacer bender.
2. Some of the very large frames cannot be produced in one piece because of limited stock lengths. An L-bend will remedy this problem.
3. Sometimes, very large sizes are awkward to handle, but two L-shaped units are easy to handle. See Figure 14.30.

J-bends. Two "J" bends can also be pieced together to produce a frame, as shown in Figure 14.31. The pieces are connected with a straight key connector instead of a corner key connector. This configuration has the same advantages of the L-bend and it also produces a frame with 4 bends.

Trapezoids and triangles. Figure 14.32 shows trapezoids, triangles, octagons, and pentagons normally produced from 3-bend or other multibend configurations. The corners normally use flexible corner

Figure 14.30 L-bends.

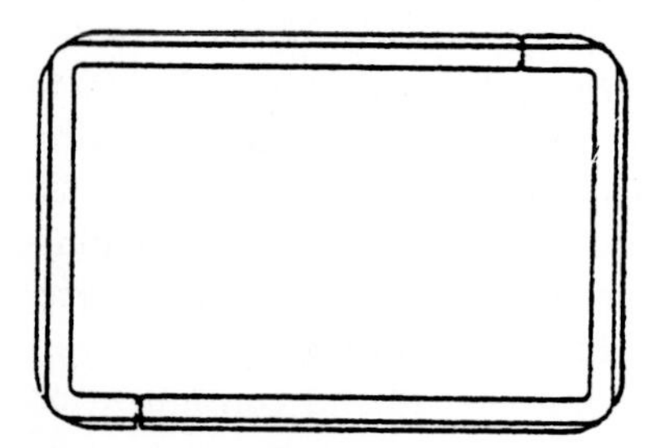

Figure 14.31 J-bends.

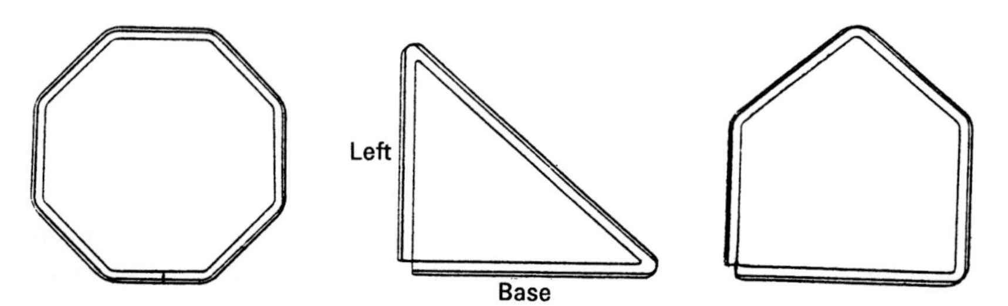

Figure 14.32 Triangle, octagon, and pentagon.

keys. Triangles are generally produced with two bends while leaving the 90° corner open. Like the trapezoid, a triangle can have two legs filled with desiccant after bending and then be keyed with a conventional corner key. Other shapes are possible on spacer benders. It should be noted that a majority of companies using spacer benders produce special shapes.

Desiccant filling. The open corner of a bent frame is inserted into the manifold of a desiccant filling machine. Under low air pressure, the

Figure 14.33 Automatic spacer bender and desiccant filler.

desiccant is quickly blown into two legs of the frame. This process of desiccant filling is fast and fairly clean. Figure 14.33 is a picture of an automatic spacer bender and desiccant filler.[6] The fill is reliable and consistent because the desiccant is blown directly into the end of the spacer profile with no restrictions and the spacer is bent automatically. The open corner, then, can be "closed" with a conventional corner key.

Some of the insulating glass manufacturers who produce dual-sealed units with a spacer bender ultrasonically solder the open corner using a metal corner key. Other manufacturers of dual-seal systems usually butyl inject the open corner with a space key designed for the butyl injection.

Bendable spacer. A bendable spacer requires specific characteristics for performance: proper alloy and hardness, sufficient elongation, proper profile, and a solid weld with perforated top and no serrations.

The increasing popularity of spacer bending has created some problems for companies which manufacture spacers. Not all materials are bendable. The spacer material should be specified as ¼ hard to prevent its outer wall from rupturing during the bending process. The metal's elongation should be between 8 and 15 percent.

If the material is to be anodized, then the initial elongation should be even higher than 15 percent. The anodizing process actually can lower the elongation by 2 to 3 percent. Most spacer profiles bend without any problems. Spacers with a discrete PIB groove usually end up with some minor distortion along the PIB groove of the bend area. Spacers to be used for bending should have a solid weld with perforations. The solid weld will not break during bending, and the perforations will allow the desiccant to perform its function. To order the appropriate material, the manufacturer of insulating glass should always know beforehand if he intends to bend the spacer.

Earlier in this chapter we discussed profiles and their intended use. The standard and low profiles are excellent candidates for spacer bending.

Seam designs. Although most spacer materials used on benders have a solid weld and perforations located on the "top" or inside of the spacer, two alternative designs are also being used.

Mechanical lock and bottom lock. The spacer is roll-formed into a lock-seam condition to eliminate welding. This seam design requires no perforations. In the bottom weld design, the spacer material is welded along the "bottom" or outside of the spacer. Such material must be perforated on the top to allow the desiccant to work.

Conclusion

Conventional spacers and muntin bars and their associated connectors come in a wide variety of configurations to meet the fabrication and application needs of the insulating glass manufacturer. An insulating glass manufacturer should communicate with its suppliers and let them know the fabrication techniques being followed, whether dual sealing, spacer bending, or gas filling. Keeping equipment in top shape assists production, prevent mishaps, and contributes to quality insulating glass units. Constantly look for means to improve the quality.

References

1. Sherman, I. D., I.G. Spacers and Muntin Bars—Issues and Answers, SIGMA-1996-Educational Seminar, Hygrade Metal Moulding Manufacturing.
2. U.S. Patent 49,167.
3. Vinopal, T. M., Allmetal, Inc., verbal discussion, 7/30/96.
4. Photographs provided courtesy of Hygrade Metal Moulding Manufacturing Corp., Farmingdale, NY.
5. Ekren, D., A brief review of spacer bending, *U.S. Glass, Metal and Glazing,* 5/1990.
6. Photographs provided courtesy of Tools for Bending, Inc., Denver, CO.

Chapter

15

Desiccants

Thomas J. Dangieri

Development Specialist, UOP Molecular Sieves

Joseph S. Amstock

President, Professional Adhesive and Sealant Systems

Introduction

This chapter will be limited to the four main desiccant systems used in the insulating glass industry. They are silica gel, molecular sieves, silica gel/molecular sieve blend, and the new binder system for desiccants (discussed in Chapter 13). For each product, discussions will ensue as to what they are, how they work, and their advantages and limitations in an insulating glass unit. A desiccant is a substance that has a high affinity for water, such that it will extract moisture from substances in the immediate area. Solid desiccants are included in a larger class of materials called adsorbents.[1]

The surfaces of some adsorbents used in insulating glass units are polar in nature. This means they have positively and/or negatively charged sites which result in the adsorptive forces described above. The more polar the substance being adsorbed, the greater the attraction between that substance and the adsorbent. This can be represented more simply by the series shown below, where the substance to the left is adsorbed more strongly than the substance to the right.

WATER > ORGANIC SOLVENTS > NITROGEN > OXYGEN > ARGON

TABLE 15.1 Dew Point vs. Water Content At 1 atm, 14.7 psi

Dew point	Water content
−20°F (−29°C)	400 ppm
−60°F (−51°C)	32 ppm
−90°F (−70°C)	6.5 ppm

Desiccants are used in insulating glass units to prevent the inside surfaces of the two lites of glass from fogging due to condensation of moisture vapor or organic vapors that may be in the air space. Table 15.1 illustrates several dew point temperatures and associated water contents of interest to the insulating glass producer.[2]

Silica Gel

With due respect to chronology, the first adsorbent to be discussed is silica gel although today it is rarely used alone. Its uses in insulating glass units date back to 1957, when a couple of independent insulating glass manufacturers began making sealed units.

Silica gel is manufactured by reacting sodium silicate and sulfuric acid producing silicon dioxide with residual sodium sulfates that are removed by stringent washing. Hence, the final product, silica gel, is a chemically inert silicon dioxide of 95.5+ percent purity.[3]

The type of washing performed in the manufacturing process (see Figure 15.1) gives rise to two classical forms of silica gel. The first type, produced by acidic washing conditions, is known as *regular density* silica gel. This is the classical adsorbent grade such as that used in insulating glass units. The second type of silica gel is known as *intermediate density* silica or buffer grade and is shown for academic purposes only. Some comparisons of the surface characteristics of these two silica types are given in Table 15.2. Although the applications for the intermediate density silica are quite diverse, it is not capable of producing extremely low dew points and does not have application in the insulating glass industry.[4]

The basic theory of silica gel adsorption is that it is a capillary condensation action enhanced by maximizing the surface area of the silica gel. The actual mechanism and the degree of adsorption that occur in the molecular sieves are substantially different and warrant a separate discussion.

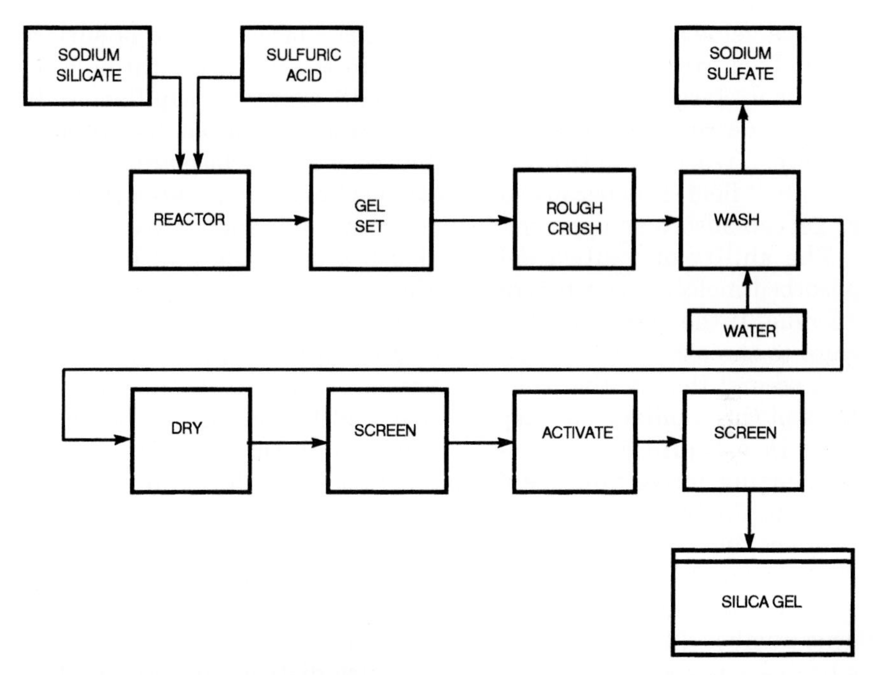

Figure 15.1 Silica gel manufacturing process.

TABLE 15.2 Silica Gel Pore Structure

Adsorbent grade	
Small pores	20–300 Å
Large pores	750–800 m^2/g
Pore volume	45cc/g
Density	45 lb/ft^3
Buffer grade	
Large pores	120–140 Å
Small surface size	300–350 m^2/g
Large pore volume	1.15 cc/g
Density	25 lb/ft^3

Silica gel is an adsorbent which provides a large amount of surface area. This extended surface is built into the silica gel through an extremely large porous construction with countless capillaries. In fact, if you could open these capillaries and spread out the internal surface, a teaspoon of silica gel contains enough surface area to cover a football field. It is because of this tremendous surface area that silica gel can adsorb an appreciable amount of water and hydrocarbons.

The ability of a silica gel to adsorb a molecule and hold that adsorbed molecule is a function of the surface area and the pore size opening. If the pore size (the diameter of the capillary) is reasonably close to the size of the molecule being adsorbed, capillary condensation occurs; that is, the molecule actually condenses within the confines of the capillary and deposits itself on the surface of the adsorbent in a liquid state. Hence, the pore diameter controls the condensing forces which subsequently control the minimum dew point attainable.

In keeping with our discussion on pore size, now let us turn to the pore size analysis of a silica gel adsorbent. The pore size of the silica gel used in insulating glass units is nominally referred to as 23 Å. This number 23 is actually a mean pore diameter and, in reality, the silica gels possess a rather wide pore size distribution. Figure 15.2 shows a typical pore size distribution curve for silica gel. You will note

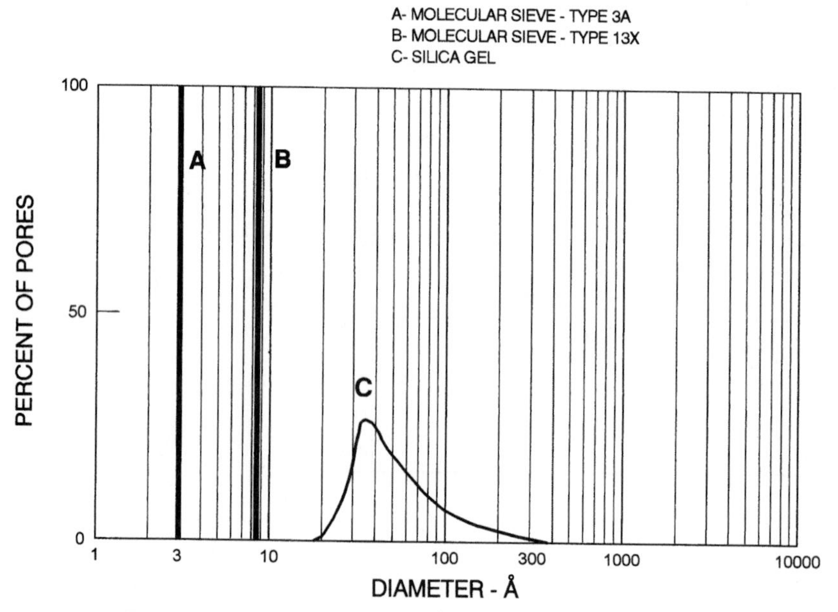

Figure 15.2 Desiccant pore size.

that a very small percentage of the pores actually exceed 300 Å. The molecular size of the various adsorbates which exist in the enclosed air space of an insulating glass unit may range from water at 2.65 Å to some larger hydrocarbons, such as toluene at 6.7 Å or xylene at 7.1 Å. Because silica gel has a wide pore size distribution, the concept of molecular exclusion is not applicable, but silica gel prefers to adsorb some materials more readily than others as do all of the adsorbents used in insulating glass applications.

This preferential adsorption is based on the boiling point of the molecule to be adsorbed. Because water has the highest boiling point and is the most polar molecule in an insulating glass unit, it will be adsorbed preferentially over the hydrocarbons in the sealed unit.

Advantages

1. *Coadsorption.* Size selection works very well if the pores are precisely controlled as in a molecular sieve. If a molecule is too large to fit into the crystal structure, nothing will make this molecule, even in large concentration, adsorb preferentially over a small molecule. Silica gel has large pores that are shaped like a "V." Of course, the pore size is smaller as you descend into the V. It is for this reason that 3A and 13X or silica gel are blended. 3A removes only water and 13X or silica gel gets all the larger molecules. Unfortunately, the industry does not have the luxury of removing only one adsorbate. In practice, both water and various hydrocarbons cohabit in the air space. Depending on the temperature, ultraviolet light, quality of the seal, and the type of sealant used, the proportion of water to hydrocarbon molecules can vary considerably at any given time. For this reason, there are some benefits in having a mixed adsorbent system with small and large pore sizes, which will exhibit an affinity to adsorb any given adsorbate.

2. *Regeneration at moderate temperatures.* As the facilities of the insulating glass fabricator have continually become more sophisticated over the past 25 years, many fabricators have become interested in regenerating adsorbents that may have been overexposed to humid conditions prior to use in insulating glass units. This can be carried out with an air temperature of 350 to 400°F (177 to 204°C), if the air is scrupulously dry. However, at no time should the silica gel be exposed to prolonged contact with air above 500°F (260°C). This may result in damage to the internal pore structure. The weaker water to silica gel bond that makes moderate regeneration possible may also become a detriment to window performance under certain climatic conditions. A warm day with direct sunlight on the window may increase the temperature of the air space between the two lites of glass to a point where moisture begins to desorb off the surface and back into the air space. If the temperature then drops quickly, the

moisture may condense before it can be readsorbed. This phenomena is referred to as the *reflux factor.*

3. *Economics.* Relative to the other adsorbents, silica gel has definitely the lowest cost on a per pound basis. However, economics alone is not the sole issue. With each adsorbent, the economics on a per pound basis must be carefully balanced with both the advantages and limitations of performance.

Limitations

1. *Moderate dew point depression.* The functional requirements of a sealed insulating glass unit have become very stringent. The insulating glass manufacturer now requires that the adsorbent remove water to a level of concentration that will provide a much lower dew point at low temperatures. The dew point requirements of −80 to −100°F (−62 to −73°C) have made the use of silica gel impractical because of the amount of silica gel required. Because it is not possible to make a silica gel with a uniform pore size, the very existence of some pore distribution places limitations on maximum dew point depression. It is agreed that dew points below −70°F (−56°C) cannot be reached with silica gel because this adsorbent has very low adsorption capacity for both hydrocarbons and water at less than −40°F (−40°C) dew point. The Sealed Insulating Glass Manufacturers Association (SIGMA) has specified this dew point level in their pass/fail criteria for their highest performance rated windows. This dew point must be achieved when the internal window temperature is above 80°F (26.7°C). ASTM-E-774-88 also specifies that a class "C" performance window exhibit no frost or chemical dew point at a measured temperature of −30°F (−34°C) after the windows have been exposed to accelerated weather conditions, as outlined in ASTM-E-773-88.

2. *Coadsorption limitations.* Having an adsorbent whose pore openings are large enough to permit coadsorption of different size adsorbates is a definite advantage. However, when coadsorption occurs on a single adsorbent, there are definite limitations. This is particularly true in insulating glass applications where the boiling points of the adsorbates differ substantially. Concomitant with the preference of the silica gel to adsorb the higher boiling molecule first is its willingness to release a less polar hydrocarbon molecule to adsorb additional water.

This is also partially true in the case of a large-pore molecular sieve, which will be discussed later in this chapter. This limitation does not become detrimental to the insulating glass' performance unless a steady rate of moisture infiltration exists or until most of the adsorbent capacity has been used for water adsorption. At this point, the hydrocarbon vapors are released into the air space seeking an

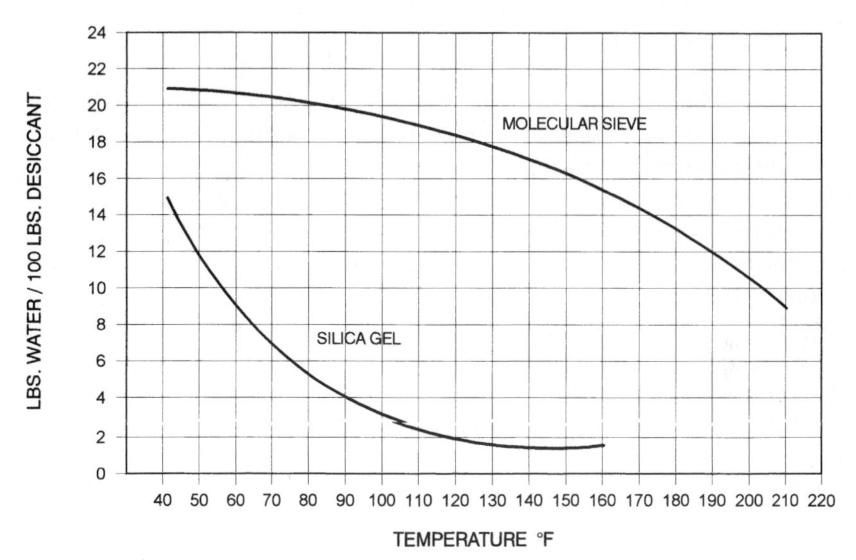

Figure 15.3 Effect of temperature on silica gel and molecular sieve.

equilibrium state until the hydrocarbon partial pressure becomes sufficiently high enough to result in the condensation of hydrocarbons on the insulating glass unit.

3. *Temperature versus capacity.* The last limitation of silica gel is that its capacity to adsorb molecules is substantially reduced at high temperatures. Under strong sunlight, insulating glass units commonly achieve air-space temperatures above 150°F (65.6°C). At that temperature, silica gel has lost most of its capacity to adsorb water or hydrocarbons. See Figure 15.3.

However, the capacity of the air to hold the unabsorbed vapors is greatly increased at those temperatures and fogging may not occur unless there is an extreme temperature swing and the rate of temperature decline is greater than the rate of readsorption by the silica gel.

Molecular Sieves

As early as the 1700s, it had been observed that naturally occurring zeolite minerals release water upon heating and reabsorb water on cooling. In nature, zeolites are not found in a very pure state. In 1948, Union Carbide Corporation research scientists began attempts to manufacture synthetic zeolites for use in industrial applications, notably air separation.

The synthetic zeolites became known as molecular sieves because the pore dimensions could be closely controlled so that molecules could be separated by size, similar to a screen separating different sizes of bulk solids. The years 1953 and 1954 brought about field testing of some 30 distinct, pure zeolite species. In 1959, R. M. Milton was granted composition of matter patents on the two most widely used molecular sieve zeolites in the industry today, the type A and the type X molecular sieves.[5] Relative to silica gel, molecular sieves are newcomers to the insulating glass industry.[2] Figure 15.4 is a picture of a typical spacer bar filled only with molecular sieves on the left and a molecular sieve/silica gel blend on the right.

Molecular sieves are highly refined, synthetic versions of naturally occurring zeolite. The manufacture of these molecular sieves is carefully controlled in a complicated series of chemical operations shown in Figure 15.5. A solution providing a source of silica, one providing a source of aluminum, and a source of sodium hydroxide are premixed at known concentrations and metered into a vessel called the gel makeup tank. The gel solution is mixed thoroughly and then pumped to another tank, where the temperature is elevated with slight agitation for a prescribed period of time. In this crystallization tank, molecular sieve crystals form and grow in a controlled, precise manner. The exact concentration of the three solutions and the temperature and time in the crystallizer determine which zeolite will be synthesized, type A or X. When the crystallization is complete, it is washed with water, filtered, and stored as a wet cake.

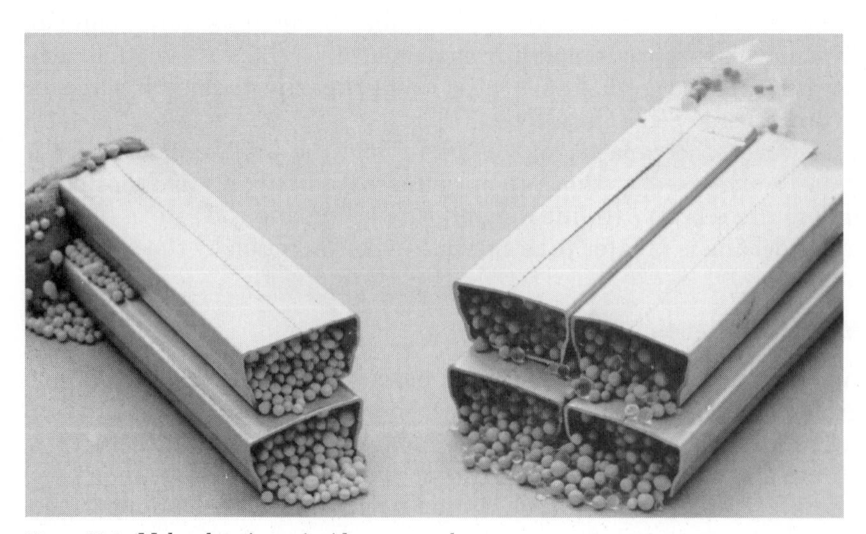

Figure 15.4 Molecular sieves inside a spacer bar.

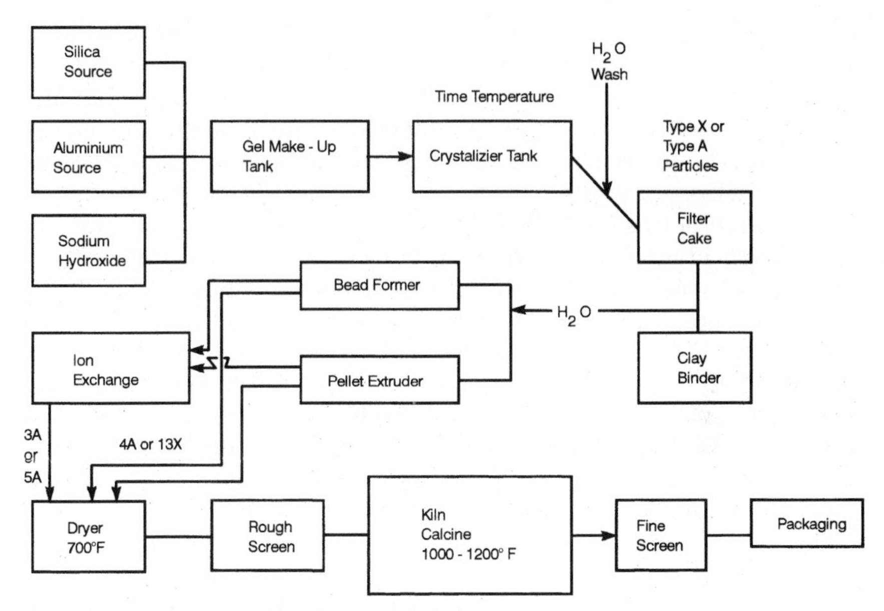

Figure 15.5 Manufacture of molecular sieves.

At some point, the filter cake is conveyed to another tank, where a clay, used for bonding the sieve crystallites together, and water are added in exact proportions. The clay and zeolite are thoroughly mixed and then conveyed to an extruder which pushes the mixture through a die to form a pellet (cylinder shape). Alternately, the zeolite mixture could be conveyed to a bead forming apparatus to form uniform spherical particles. If the desired product is a type 4A or 13X, the synthesis is complete; the beads or pellets are dried to remove bulk water, screened for a rough size cut, and then calcined in a kiln at 1000 to 1200°F (538 to 649°C). The calcination allows the binder and the zeolite crystal structure to combine into a strong particle that resists breakup and attrition (which produces dust) and provides a product that can be handled and transported.

If the desired product is a type 3A or 5A molecular sieve, which is either the potassium or calcium form, respectively, of the sodium type A, another step must be added at the point when the zeolite is a filter cake or after the foaming step. This is called an ion exchange. The 4A molecular sieve is mixed with a potassium or calcium solution in a tank or column with some elevation in temperature. The exchange of some of the sodium in the crystal structure for calcium ions produces a type 5A, or a potassium exchange for sodium in the crystal forms type 3A. The capacity of a molecular sieve to undergo this exchange process is key in precisely controlling the pore size opening. These exchange products must also be dried, screened, and fired in the kiln.

After final screening, the pellets, beads, or the molecular sieve powder, free of any binder, is quickly packaged warm into drums, lined sacks, or bags, which are air tight and water resistant, to prevent any readsorption of moisture into the molecular sieve. Molecular sieves are usually delivered to the end user with less than 2 percent by weight of adsorbed water.

As a result of this controlled manufacturing process, all the pores in zeolite are identical in size and can be tailored through cation exchange or modification to the crystal structure to have either 3-, 4-, 5-, or 8.5-Å pore openings. The typical internal surface area of a molecular sieve desiccant is about 750 m^2/g. This is equivalent to more than 40 football fields in a cupful of desiccant. Type 3A has a uniform 3 Å opening; type 4A a uniform 4 Å, and type 5A has a uniform 5 Å opening. 13X has an 8.5 Å opening. There are 100,000,000 angstroms in one centimeter. Historically molecular sieves used in insulating glass units were Na, A, K, and Na X.

Molecular sieves adsorb various adsorbents differently. First is the *molecular sieve* effect, which applies to the exterior surface of the crystal. Depending on the sieve under consideration, the external surface will present *windows* of precise molecular size. If an adsorbate molecule exceeds this size, it will be excluded from the interior of the crystal and will be able to adsorb only on the external surface. Surface adsorption, which also occurs on amorphous solids, such as silica gel, is a weaker, less selective phenomena.

Once inside the crystal, millions of intracrystalline voids and channels are available to the molecules for further adsorption. Molecular sieves are composed of atoms of silicon, aluminum, and oxygen. The collective ionic charges of these molecules should add up to zero in a stable crystal. To balance the positive and negative charges, positively charged ions of potassium, sodium, or calcium are present in the crystal. The vast collection of positively and negatively charged atoms create "areas" of charge density commonly known as adsorptive sites. The attractive forces between adsorbent and absorbate are very strong and large amounts of energy (typically heat) are needed to break down these attractions and regenerate the sieves. The charge density sites are so strong that few layers of adsorbate molecules may exist as the adsorbate concentration increases. Finally, near the full capacity of a particular adsorbate, pore filling may occur that is akin to capillary condensation.

If a molecule of water and benzene (for example) both enter a 13X crystal, they have an equal chance of finding a charge site and adsorbing into the crystal, because both molecules have positive and negative densities. However, because molecules, such as nitrogen, argon and the like, do not have distinct charge densities within the mole-

TABLE 15.3 Adsorption of H_2O, Air, Argon, SF_6, Krypton, Xenon, and Solvents on Desiccants

Desiccant type	Pore size, Å	Adsorbs	Excludes
3A molecular sieve	3	H_2O	All others
4A molecular sieve	4	H_2O, air, argon, krypton	SF_2, xenon, solvent
13X molecular sieve	8.5	All	None
Silica gel	20–300	All	None

cule, they may adsorb weakly on the crystal and are displaced easily when a more charged molecule approaches. In any adsorptive process using sieves, energy in the form of heat is given up to the surroundings. All the molecular sieves mentioned so far can adsorb at least 20 percent of their weight of water at room temperature. Insulating glass units constructed with molecular sieves may reach dewpoint depressions of −100°F (−73°C) with very few grams of desiccant.

The substances that may be present in insulating glass units that could be adsorbed by the desiccant are water, air, argon, krypton, xenon, SF_6, and organic solvents. Argon, krypton, xenon, and SF_6 (sulfur hexafluoride) are gases used in gas-filled units to improve the U-value and sound transmission properties of the unit. Organic solvents may come from the sealant material, cutting oils, and paints from internal muntin bars. Table 15.3 outlines which of these substances are adsorbed and which are excluded from the four desiccant materials. Type 3A molecular sieves adsorb only water and exclude all others. Type 4A adsorbs water, air, argon, krypton, and excludes xenon, SF_6, and solvents. Type 13X and silica gel adsorb all these and exclude none of them.

The choice desiccant material depends on the properties of each desiccant. The properties important to desiccant selection are water capacity, solvent capacity, and air adsorption. Water capacity and solvent capacity depend on the temperature of the desiccant and the dew point required. Figure 15.6 shows the solvent capacity of the four desiccant materials at 70°F (21°C) and a dew point of −40°F (−40°C). Only type 13X molecular sieve and silica gel exhibit any capacity for solvents at these conditions.[6]

Molecular Sieve/Silica Gel Blends

Blends are available from most manufacturers of 3A or 4A sieves combined with silica gel. Blends take advantage of the hydrocarbon capacity of the silica gel. By combining a 3A sieve with silica gel, only

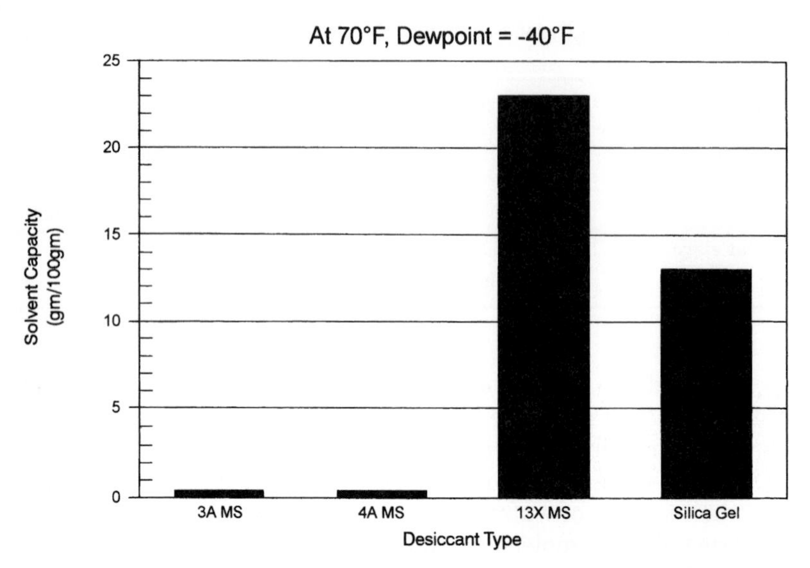

Figure 15.6 Solvent capacity.

water is adsorbed on the sieve leaving the gel active for adsorbing hydrocarbons.

Molecular sieves and the silica gel sometimes separate on shipment. This will not occur if the density and the particle sizes of the sieve and silica gel are well matched. By comparison, 3A and 13X desiccants are blended together as powders in the correct proportions and supplied as one bead, so there is no chance of separation.

With desiccants at 70°F (21°C) and a dewpoint of −34°F (−37°C), type 13X has the highest capacity at about 23 g of solvent per 100 g of desiccant. Silica gel is second at 12.5 g. Adsorbents have characteristic pores into which the molecules of the substance are adsorbed. This size can be selected so that the molecules are either adsorbed or excluded as a function of the size of the molecules of the substance. Approximate diameters of some of the molecules important to insulating glass manufacturers are shown in Table 15.4.[5]

The water capacity of insulating glass desiccants is shown in Figure 15.7.

The water capacity of type 13X is the highest at 16 g of water per 100 g of desiccant at these conditions. Type 4A is the second highest and 3A the third at 13.5 g per 100 g of desiccant. Silica gel has less than 1 g of water capacity per 100 g of desiccant.

The water or solvent capacity changes with temperature. The higher the temperature of the desiccant, the lower is its water or solvent capacity. The lower the required dewpoint, the lower is the water

TABLE 15.4 Diameter Reference in Angstroms

Material	Pore size, Å
Water	2.65
Nitrogen	3.64
Argon	3.40
Oxygen	3.46
SF_2	5.50
Organic solvents	
Benzene	5.85
Isobutane	4.50
Toluene	6.70
Xylene	7.1

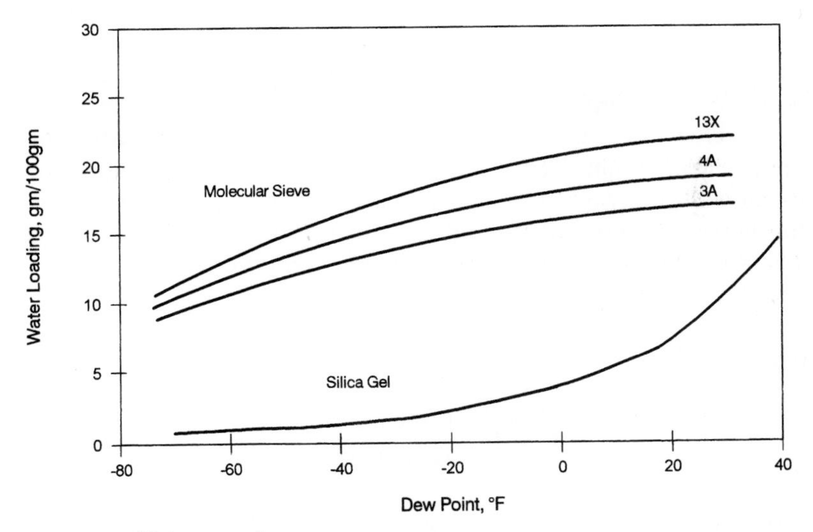

Figure 15.7 Water capacity.

capacity of the desiccant. The four desiccants, 3A, 4A, 13X, and silica gel, maintain the same capacity ranking at lower and lower dew points, with 13X the highest. Silica gel has practically no water capacity at the low dew points. Only at very high dew points does silica gel have appreciable water capacity. At these high dew points, very little protection from fogging is available.[6]

Air Adsorption

The air adsorption and desorption characteristics of insulating glass desiccants are important because they affect glass deflection. Deflection is the bending of the glass inward and outward from the desired parallel position. When the unit is assembled and sealed, it contains air at the same temperature as the surrounding air. The pressure inside is the same as the unit. If the pressure changes so that the inside and the outside pressures are unequal, deflection occurs. If the pressure inside is lower than outside, the unit is said to have negative pressure, and the glass deflects inward. If the pressure inside is higher than outside, the unit has a positive pressure, and the glass will deflect outward. The pressure difference can be caused by a change in the barometric pressure outside the insulating glass unit, relative to the barometric pressure when the window was sealed. Barometric pressure changes can occur due to changes in weather or elevation of the unit when installed. Barometric pressure goes down as elevation increases, so a unit manufactured at low elevation and installed at high elevation will have a positive pressure between the glass lites.

Temperature changes also create deflection. Temperature change causes a pressure change inside the airspace which results in glass deflection. This effect is known as "Charles' Law," which states the pressure is proportional to the absolute temperature. Because air sealed between the two lites of glass behaves as an ideal gas, the relationship between temperature and pressure in an undeflected sealed insulating glass unit is given by the following equation:

Charles' Law

$$P'' = P' \cdot \frac{(460 + T'')}{(460 + T')}$$

where P'' = pressure at assembly, psi
P' = pressure in the field, psi
T'' = temperature at assembly, °F
T' = temperature in the field, °F

Let us look at an example for a sealed insulating glass unit assembled at 77 F° (25°C) and atmospheric pressure and then put into service at 0°F (−18°C):

$$P'' = 14.7 \text{ psi} \qquad T'' = 77°\text{F} \ (25°\text{C})$$

$$P' = \text{unknown} \qquad T' = 0°\text{F} \ (-18°\text{C})$$

$$P = 14.7 \cdot \frac{(460 + 0)}{(460 + 77)} = 12.6 \text{ psi}$$

The result is a decrease in pressure of 2.1 psi, which will result in negative deflection. An increase in temperature of the sealed unit, such as encountered on a warm day, results in positive deflection.

Figure 15.8 illustrates the inward and outward changes in the sealed insulating glass unit because of the Charles' Law effect. An increase in temperature between the glass lites creates a positive pressure difference and the glass deflects outward. As the temperature between the lites decreases from the original sealing temperature, the internal pressure decreases, creating negative pressure and inward deflection.

Desiccants that adsorb or desorb air can add to the deflection. If a desiccant adsorbs air, it will adsorb more as the temperature decreases. Adsorbing these molecules of nitrogen and oxygen from the air space of the unit further reduces the pressure between the lites of glass causing more deflection. Because air adsorption is completely reversible, the desiccant gives up or desorbs air as the temperature rises and returns to the original sealing temperature. The desiccant could also add air molecules adsorbed on the desiccants during manufacture and shipment to the interior air space, when the temperature rises, resulting in an outward deflection.

Molecular sieves 4A and 13X and silica gel all have pore sizes large enough to adsorb air. Figure 15.9 shows a measure of the air adsorption and desorption on selected window desiccants.

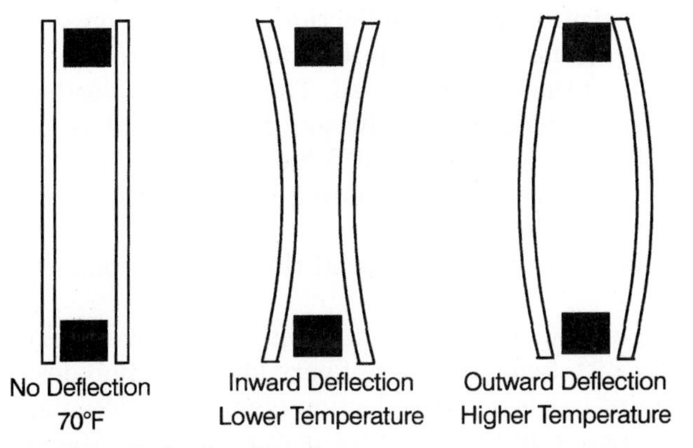

Figure 15.8 Deflection, free edge.

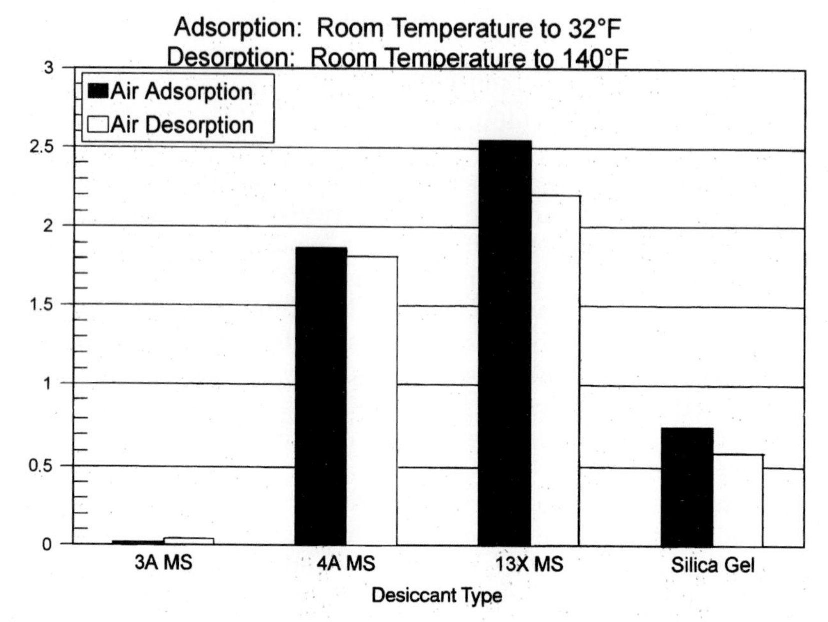

Figure 15.9 Adsorption/desorption of IG desiccants.

The adsorption data is obtained by cooling a sample of the desiccant in a test tube at room temperature to 32°F (0°C) in an ice-water bath. The volume change in cubic centimeters is divided by the weight of the desiccant in the tube to obtain the air adsorption. The desorption data comes from similar experiments where the desiccant is heated to 140°F (60°C) in a hot water bath. This data shows that 13X has the highest air adsorption, followed by 4A. Type 3A has almost no air adsorption, and silica gel is in between.

The effect of air adsorption by low-deflection desiccants on glass deflection in insulating glass units is quite small compared to the effect of the contraction of the air space itself. The deflection has been calculated using the equations of K. R. Solvason[7] from a standard test unit measuring 14×20×¼-in air space. See Table 15.5.

Another example of proper desiccant loading based on the typical test unit size is shown in Table 15.6.

The percentage reduction in the air space at the center is compared in Figure 15.10.

The first bar in Figure 15.10 is for a unit with no desiccant. The second bar is for units with two long sides of very low deflection desiccant, Type 3A. The third bar is a low-deflection blend of desiccants, a 3A/13X blend. The reduction in the air space is due mostly to the air-space contraction, because of a change in temperature, whereas the

TABLE 15.5 Insulating Glass Deflection (Free-Edge Conditions) of ASTM Test Unit—14 × 20 × ¼ in

Width = 14 in
Length = 20 in
Air space = 0.250 in (¼ in)
Glass thickness = 0.1875 in (3⁄16 in)

	Air-space conditions at 0°F (−18°C)			
	Control = no desiccant	4 Sides 3A	4 Sides 3A/13X	4 Sides 13X
Original pressure, psia	14.70	14.70	14.70	14.70
Undeflected pressure, psi	1.89	1.98	2.47	4.89
Center deflection, in	0.026	0.028	0.035	0.067
Center glass stress, psi	1465	1536	1931	3720
Breakage probability, lites/1000	0	0	1	21
Reduction in air-space center, %	10	11	14	27

TABLE 15.6 Desiccant Selection—A Balance of Properties

	Water capacity	Solvent capacity	Air adsorption
3A molecular sieve	3	XXX	333
4A molecular sieve	33	XXX	XX
13X molecular sieve	333	33	XXX
Silica gel	XXX	3	X

desiccant makes only a small contribution. This calculation is done at the free-edge condition. This limiting condition assumes that the glass edge is unrestrained by the frame or sealant. The opposite limiting condition is the clamped edge, which assumes that the glass edge is held perfectly tight to a rigid spacer frame. The actual or realistic condition is somewhere in between the free-edge and clamped-edge conditions. The Solvason equations are designed to calculate only the limiting conditions.[7]

In this example, breakage probability has increased for the case with four sides of type 13X molecular sieves. Excessive filling with 13X should be avoided in small insulating glass units. Table 15.6 evaluates the available desiccant materials for the three properties that have been discussed.

A “3” means that the desiccant is good with respect to that property, “33” equals better, and “333” is best. An X means that the desiccant is

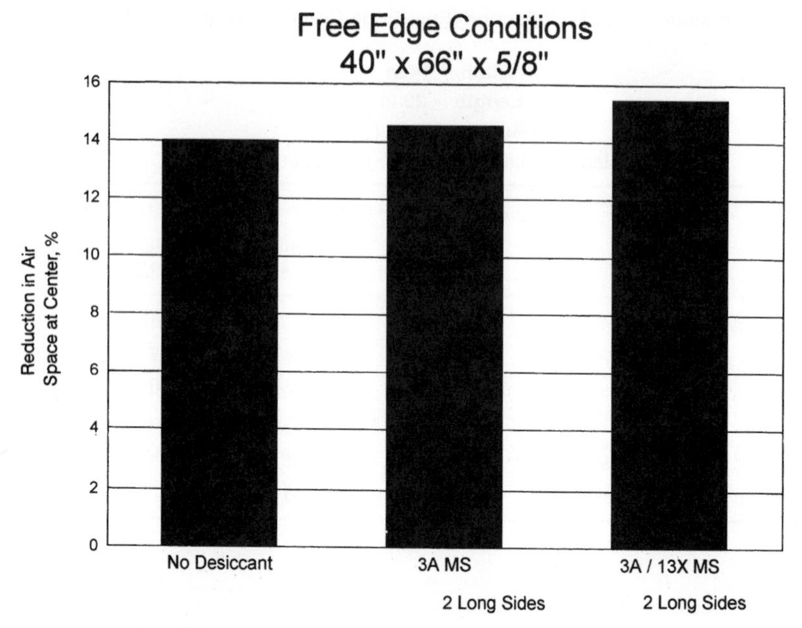

Figure 15.10 Insulating glass deflection.

poor with respect to the property, two Xs are poorer, and three Xs are the poorest. No desiccant has a "3" in all three properties. If the solvent capacity is not needed, type 3A molecular sieve has 3s in the remaining properties and is the optimum choice. Type 13X has the highest water capacity and solvent capacity, which is fine, but it also has the highest air adsorption, which is poor.

Type 4A has the second highest water capacity, but it has no solvent capacity and adsorbs air. Silica gel has good solvent capacity, but its water capacity is very low at low dew points, and it adsorbs some air. Thus, 3A has all the desired properties when solvent capacity is not needed. If solvent capacity is required, no single desiccant material has all the desired properties. The optimum desiccant is a mixture of materials that will give high water capacity, sufficient solvent capacity, and low air adsorption. There is a blend of types 3A and 13X (or 3A and silica gel) that gives these properties.

Air Adsorption

Air adsorption can be further eliminated by using a 3A sieve to replace 4A and 13X sieves for water adsorption. 3A excludes nitrogen by virtue of its smaller pore size. If organic vapors are present, they are adsorbed by incorporating some amount of large-pore adsorbent, such as silica

TABLE 15.7 Effects of Molecular Sieve Commercial Window—40×66×⅝ in
Width = 14 in
Length = 20 in
Air space = 0.250 in (¼ in)
Glass thickness = 0.1875 in (3⁄16 in)

Air-space conditions at 0°F (−18°C)	Control = no desiccant	2 Sides 3A/silica gel blend	2 Sides 13X molecular sieve
Original pressure, psia	14.70	14.70	14.70
Undeflected pressure, at 0°F (−18°C), psi	12.81	12.78	12.23
Negative pressure, psia	−1.89	−1.92	−2.47
Original air space, in	0.06250	0.6250	0.6250
Center of glass deflection at 0°F (−18°C), in	−0.089	−0.091	−0.117
Actual air-space thickness, in	0.528	0.534	0.508
Center of glass, in	606	616	794
Air-space thickness reduction, %	−14	−14.5	18.7

gel or 13X molecular sieve. Both adsorbents contribute to some air adsorption, as seen in Figure 15.9. In Table 15.7, we compare 13X and a 3A/silica gel blend in a typical 40 × 66-in unit with a ⅝-in air space. In both cases, the unit is filled with desiccants on two long sides.

In the control with no desiccant, again we see the effects of temperature on an ideal gas, as the pressure inside the unit has been reduced by −1.89 psi or 12.9 percent. The insulating glass unit deflects inward resulting in a 14 percent loss of air space. The pressure in the unit manufactured with 13X is reduced by an additional 4.1 percent resulting in an air-space reduction of 18.7 percent. Consequently, the change of pressure and deflection in the insulating glass manufactured with 3A molecular sieve/silica gel blends is essentially unchanged from the control.

Determination of Adsorbent Use Levels

One of the most asked questions is how much desiccant should be used in an insulating glass unit. There are several ways to answer this question, and the answers can teach us about desiccants. The answers fall into three categories:

the *experience approach,*

the *water balance approach,* and

equivalency requirements.

The *experience approach* provides the short answer to the question of how much desiccant to use. Experience says that most manufacturers use two long sides or one long side and one short. This seems to provide the needed desiccant for satisfactory service.

The *water balance approach* might be thought of as the more scientific approach. You should use enough desiccant to remove the water and the solvents sealed in the unit during manufacture plus enough desiccant to remove the water that permeates in during the life of the unit. Only a small amount of desiccant is needed for the initial dry down. For example, a 30 × 30 × x½-in unit, sealed at 80 percent relative humidity at 70°F (21°C), contains 0.10 g of water. If two sides are filled, the unit would contain about 90 g of desiccant. Of this 90 g, less than 1 g of molecular sieve desiccant would be required to dry the air/gas to −40°F (−40°C) dew point. More than 10 g of silica gel would be required to achieve that dew point.

Additional water permeates into the sealed unit under a *relative humidity driving force.* This driving force is the result of the higher relative humidity on the outside than on the inside of the unit. The amount of desiccant needed to remove the water that permeates in during the life of the unit can be calculated by a water balance. To calculate the amount of desiccant needed, you must know the rate of water ingression or moisture vapor transmission rate (MVTR), the design life of the unit, and the desired maximum dew point during the life of the sealed unit. The MVTR varies with

A. the type of sealant,

B. insulating glass construction techniques,

C. moisture vapor transmission path area,

D. moisture vapor transmission path length, and

E. workmanship

The MVTR of various sealants can be determined by standard tests to provide a quantitative measure. The MVTR path area and path length refer to the path that moisture vapor must take through the sealant to move from the outside to the inside of the unit. In principle, this can be measured. However, the actual area and path will depend greatly on the last item, workmanship. Construction technique refers mainly to the design of the corners of the unit. This is typically the area where moisture vapor permeation is most rapid. Workmanship can have a great impact on the corner permeability as well.

To satisfy *equivalency requirements,* you must have enough desiccant. Equivalency is required of certified insulating glass manufacturers in the North America. Production units in the United States

TABLE 15.8 Desiccant selection

Sealant	Recommended desiccant
Hot melt butyl	3A molecular sieve
Polysulfide, single seal	3A/13X molecular sieve/blend or silica gel blend
Polysulfide/PIB	3A molecular sieve, 3A/13X molecular sieve blend, or silica gel blend
Polyurethane	3A molecular sieve
Silicone/PIB	3A molecular sieve

must be equivalent to units tested under ASTM-E-773/774. The equivalent production units must have an equal or greater weight of desiccant per inch of edge (or the perimeter) than the certified test units. If you pass certification testing with four sides filled, for example, then you may have to use 0.62 g of desiccant per inch of perimeter. Table 15.8 outlines the various desiccants recommended when used with a specific sealant edge seal.

The equivalent production units, then, must have at least 0.62 g/in. A 30 × 40-in unit would have 140 in of perimeter and require at least 87 g (140 in × 0.62 g/in) of desiccant. Up to this point, the discussion has focused on desiccants that are formed into beads or granular particles and then are added into a channel system by various mechanical methods. In the early 1980s, another method of introducing desiccant into the insulating glass unit appeared on the scene. This involved blending molecular sieve or silica gel as powder into an organic compound and extruding this mixture around a metal band for structural support. This idea was the forerunner of the Intercept system.

Gas-Filled Insulating Glass Units

The principal gases being considered for gas filling are argon for improving insulating values and sulfur hexafluoride for sound control. These gases have several key physical properties with a bearing on the selection of the desiccant. Argon is an inert gas whose molecules have an approximate diameter of 3.5 Å, roughly the same diameter as nitrogen. Because it is inert, it has no polarity and, therefore, should be less readily adsorbed than nitrogen. Table 15.9 also compares the argon adsorption of a 3A/13X sieve blend and a 3A silica gel blend. Although both will fulfill the requirements for hydrocarbon adsorption, the molecular sieve/silica gel blend will adsorb one-fourth as much argon. The same results will hold true for sulfur hexafluoride gas-filled units. Gas-filled units should be filled with the following desiccant types, accordingly:

Sealant System	Desiccant Type
Dual-Seal	
PIB/silicone	100% 3A molecular sieve
PIB/polyurethane	100% 3A molecular sieve
PIB hot melt butyl	100% 3A molecular sieve
PIB polysulfide	100% 3A molecular sieve
Single-Seal	
Polysulfide	Blend of 3A/13X molecular sieve/silica gel
Polyurethane	Blend of 3A/13X molecular sieve/silica gel
Butyl hot melt	Blend of 3A/31X molecular sieve/silica gel

Table 15.9 compares the adsorption of air (78 percent nitrogen) and argon on various desiccants at 32°F (0°C) and 5°F (−15°C). By comparing adsorption at both temperatures, you can see that lower temperatures result in higher adsorption. As expected, you readily see that the 100 percent 3A sieve adsorbs no argon and almost no air at either temperature because of its small pore size. Contrast this with the adsorption results on a 13X molecular sieve and you will see that both gases are strongly adsorbed.

TABLE 15.9 Pressure Reduction for Insulating Glass Units

Percentage pressure reduction due to adsorption			
Adsorbent	Gas	32°F (0°C)	5°F (−15°C)
3A molecular sieve	Argon	0	0
	Air	0.2	0.39
4A molecular sieve	Argon	2.5	4.6
	Air	5.7	11.0
13X molecular sieve	Argon	3.3	5.5
	Air	6.3	12.0
3A/3X blend 75/25	Argon	0.9	1.5
	Air	1.8	3.5
3A/silica gel 75/25	Argon	0.2	0.6
	Air	0.32	0.62

Desiccant Matrix

The PPG patented Intercept process for insulating glass unit fabrication presents another unique set of demands.[8] This system utilizes a polymeric desiccant/adhesive matrix with the following properties and has been discussed in detail in Chapter 13.

1. desiccant capacity to remove water and other condensables from the air space, sufficient to meet the ASTM-E-774 specification to level CBA,
2. low-deflection desiccant to maintain insulating properties and avoid glass deflection and seal stress,
3. adhesion to "open channel profiles" over a wide range of window conditions,
4. low volatiles to avoid hydrocarbon fogging, and
5. ultraviolet stability.

In addition to the properties given above, the following are typical physical properties:

Typical Properties

Desiccant content %	~ 50.0
H_2O ads %	> 10.0
Slump @ 190°F (87.8°C)	None detected
Density, lbs/gal	> 9.0
Color	Grey or black

The drying rate is the amount of time during which the desiccant removes moisture from the air space resulting in a reduced frost point within the insulating glass unit. The earlier a frost point of −90°F (−67.8°C) is reached, the earlier the finished unit can be shipped to a customer without risk of condensation during transportation or installation. The matrix ensures that, in less than 20 hours, the frost point will reach −94°F (−70°C) in air-filled units. Due to their polymer characteristics, all desiccant matrix systems contain some volatiles. Unfavorable conditions, such as high temperature, cause volatiles to be released and result in chemical fogging within the insulating glass unit, particularly in coated glass.

As with any component used in the manufacture of insulating glass units, ease of fabrication is of utmost importance from an economic viewpoint. The matrix systems are readily pumpable and lines can run at a maximum of 80 ft per minute. For wider spacers ($^{25}/_{32}$ in), this requires that a minimum of 2.5 lb (1.12 kg) of a matrix be pumped through the system per minute.[9]

Quality Control and Handling of Desiccants

It takes pure common sense to handle desiccants. Desiccants rapidly pick up water from the atmosphere. Air has an infinite supply of water relative to a small amount of desiccant. Therefore, care must be exercised to keep the two apart when possible.

Incoming shipments should be examined for damage to the containers which might breach the container's moisture barrier. Damage to a drum near the rim is especially important because it may result in damaging the seal between the top of the drum and the lid. If there is any question about the product, check the temperature rise of the material using a test kit generally supplied by the desiccant manufacturer. When desiccants adsorb water, they give up energy in the form of heat, referred to as the heat of adsorption. The amount of heat given off is related to the residual moisture content and physical properties of that particular desiccant. It is not an absolute number to be used in comparing the performance of one desiccant to another. It is important that you use only the test kit supplied by the manufacturer of the desiccant you are using, because every desiccant is different and every test kit is different. As the test is somewhat technique dependent, you should closely follow the manufacturer's instructions and also make sure that everyone in your laboratory runs the test the same way. The temperature rise of the desiccant should be checked on:

1. new drums upon opening,
2. partially used drums stored overnight, and
3. any material left exposed for extended periods of time, including open drums and filled spacer bars.

It should be remembered that the desiccant's primary function in the insulating glass unit is to adsorb water and that its performance will decrease with preadsorbed moisture. Keeping accurate records of temperature rise test results will provide meaningful data in case of any field problems, and keeping records also helps to ensure that the tests are actually being run.

Conclusion

Adsorbents will always play an important role in the manufacture of insulating glass units. An informed understanding of some of the properties and functions of desiccants will aid you in choosing the right product to meet your needs.

References

1. Gallion, W. R., W. R. Grace & Co., Baltimore, MD, 1990.
2. Breck, D. W., *Zeolite molecular sieves: structure, chemistry, and use,* John Wiley & Sons, Inc., New York, 1974.
3. Thomas, J. F., W. R. Grace & Co., Baltimore, MD.
4. Cohen, A. P., and T. J. Dangieri, Desiccant Use in Insulating Glass Units, UOP Molecular Sieves, Des Plaines, IL, 1989.
5. U.S. Patents 2,882,243; 2,882 244.
6. Cohen, A. P., Molecular Sieve Desiccants and Gas Filling for Insulating Glass Windows, UOP Molecular Sieves, Des Plaines, IL. Presented at SIGMA, 8/1989.
7. Technical Paper 423, Division of Building Research, National Research Council, Ottawa, Ontario, 1974.
8. U.S. Patents 5,177,416; 5,255,481; 5,351,451.
9. Desiccant Matrix System, W. R. Grace & Co., Baltimore, MD, 1995.

Chapter

16

Inert Gases

Randi L. Ernst

President, FDR Designs, Inc.

Introduction

Gas is defined as a state of matter with no definite volume or shape, whose molecules move freely to fill any available space.

When the first two pieces of glass were sealed together, it was not for energy savings. Figure 16.1 illustrates a single lite of glass with a thermal value of R-1. Figure 16.2 shows two lites of glass hermetically sealed forming an insulating glass unit with a thermal value of R-2.

The two pieces of glass were sealed together to eliminate having to clean the two inner surfaces of glass. The thermal performance of two clear pieces of glass sealed together is the same as traditional two pieces of glass, one the window, the other a storm sash.

Sealing the layers together created a new problem, condensation of water vapor between the layers of glass. Desiccant was introduced into the sealed cavity to super dry the air and sealants were improved to prevent water vapor from entering the truly dried cavity.

This presented an opportunity because we have a closed cavity with very good seals. If the extremely small water vapor molecules could be excluded, it would be easy to trap a gas other than air.

U.S. Patent 2,756,467[1] was for the introduction of dried nitrogen into a glass cavity. Nitrogen comprises 80 percent of air so that the thermal performance remained the same. U.S. Patent 3,683,974[2] was for the insertion of a fluorocarbon gas into a glass cavity. The fluorocarbons are better insulators than air but are not sunlight stable, and scattered reports surfaced of a colored fog trapped between the glass layers. Carbon dioxide was also tried, but carbon dioxide, being acidic, reacted with the spacer and glass coatings.

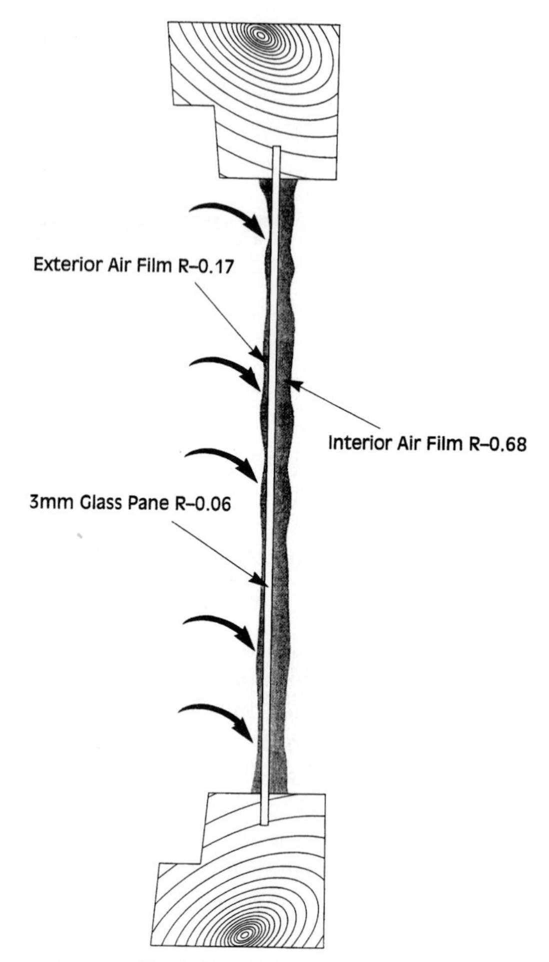

Figure 16.1 Single lite of glass.

Types of Gases

The noble gases, argon, krypton, and xenon, offered the most practical solution—they are stable, inert, and less thermally conductive than air. Of the three, argon is the most plentiful—almost 1 percent of the air we breathe and, consequently, the most economical to use. In Figure 16.3 is a listing of the properties of the noble gases and their positions in the periodic chart of elements.

Noble metals are metals that resist corrosion by water and acid, i.e., gold, silver, and platinum. Noble gases are in group VIII of the table of elements. None of the noble gases chemically combines with

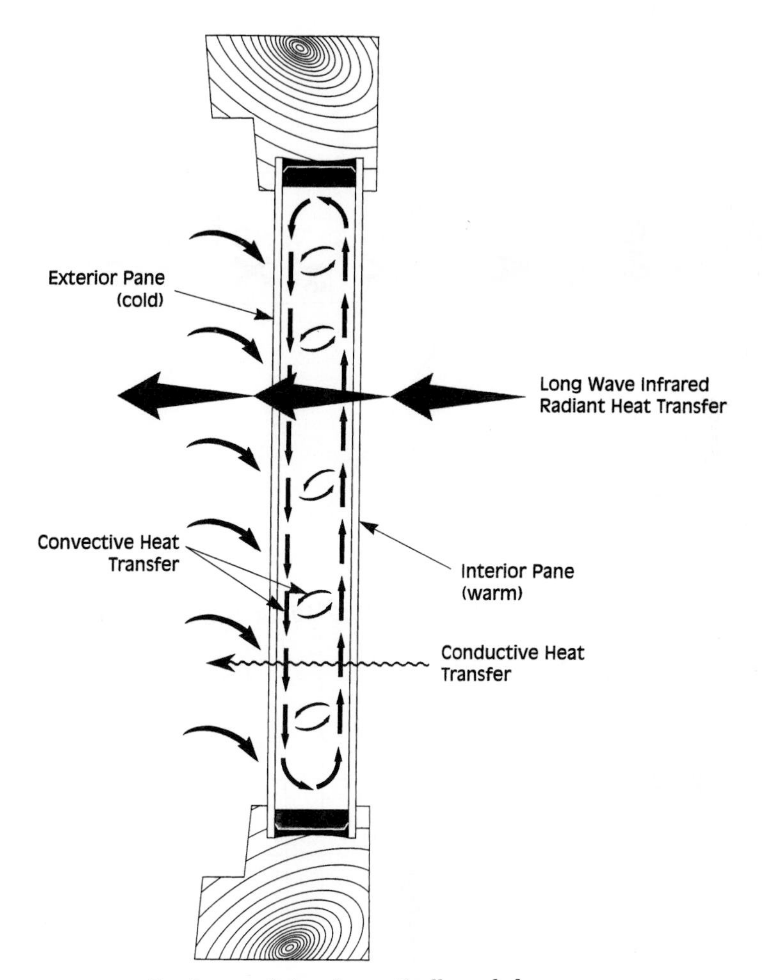

Figure 16.2 Two layers of glass hermetically sealed.

any element. All except helium (with two) have eight electrons in each atom's outer shell. No atom can hold more. Figure 16.4 depicts an argon atom.

Because the "insulation" in multiple-glazed windows is the entrapped gas, increasing the gas space thickness increases the overall R-value. For example, multiple-glazed units with half-inch "air" spaces have higher R-values than those with quarter-inch "air" spaces. However, the effect of increasing the thickness of the gas space is limited.

Beyond a certain thickness, interpane convection increases, carrying heat from the inner to the outer pane through air circulation.

Name	Formula	Molecular Wieght	Specific Gravity	Boiling Point °F	Conductivity
Air	O_2N_2	28.96	1.00	–317.8° to –312.4°	.0150
Argon	Ar	39.95	1.38	–302.6°	.0100
Krypton	Kr	83.8	2.89	–244.0°	.0053
Xenon	Xe	131.3	4.61	–162.6°	.0032
Carbon Dioxide	CO_2	44.01	1.52	–109.4°	.0092
Sulfur–Hexafluoride	SF_6	146.054	5.32	–82.7°	—

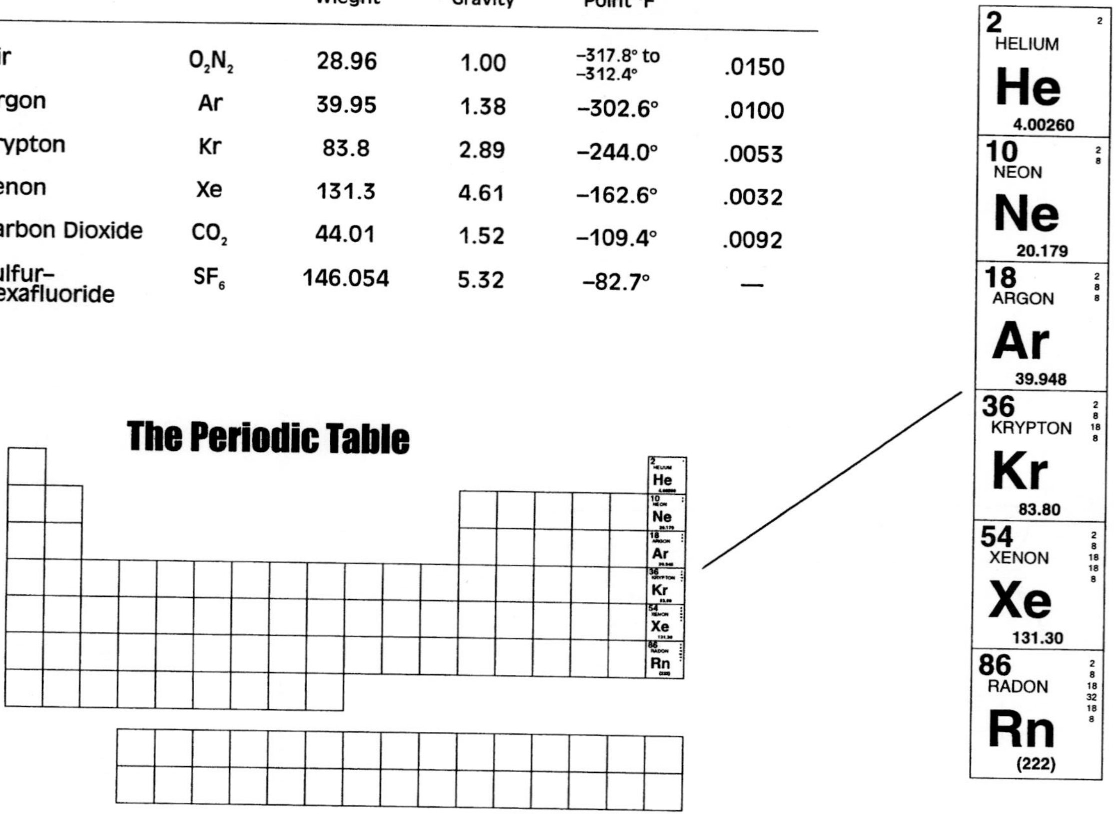

Figure 16.3 The noble gases.

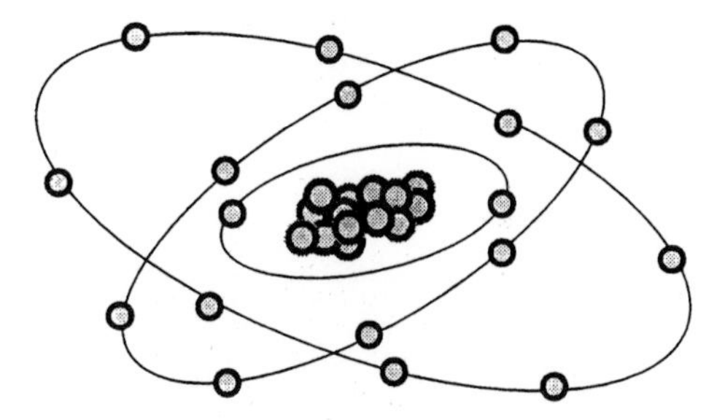

Figure 16.4 Argon atom.

Further increases in gas space thickness do not result in an increased R-value.

The exact thickness at which convection becomes a dominant heat transfer mechanism depends on the height of the glass unit, the gas used, and the temperature differential between indoors and outdoors.

Clear glass lets long-wave radiation pass through readily. Fifty percent of the energy transferred out a clear glass window is due to radiation. Twenty-five percent of the energy is lost due to conduction. Until the door could be shut on the large radiant transfer, gas filling would not really be viable.

Sulfur hexafluoride (SF_6), a manufactured gas, showed some ability to absorb or capture some of the long-wave radiation, and studies were done to see if that could slow the radiant heat transfer. The results showed minimal improvement with SF_6.[3]

What finally made gas filling viable was the introduction of low-cost, Low-Emittance coated glass and plastic film. This revolutionary coating effectively blocked the radiant heat transfer. Now, with radiation transfer under control, argon was used to reduce the conductive loss. Figure 16.5 depicts a hermetically sealed unit with one Low-Emissivity coated glass pane.

Table 16.1 lists a series of R-values using different combinations of glass and gas.

Glass manufacturers could produce units made with Low-E glass and an argon fill, a product with twice the thermal performance of clear sealed units, using the same basic manufacturing process they were already using.

Gas filling works for the following reasons:

1. Argon, krypton, and xenon are less conductive than air.
2. Krypton and xenon will yield roughly the same performance as argon but in two-thirds to one-half the cavity width.

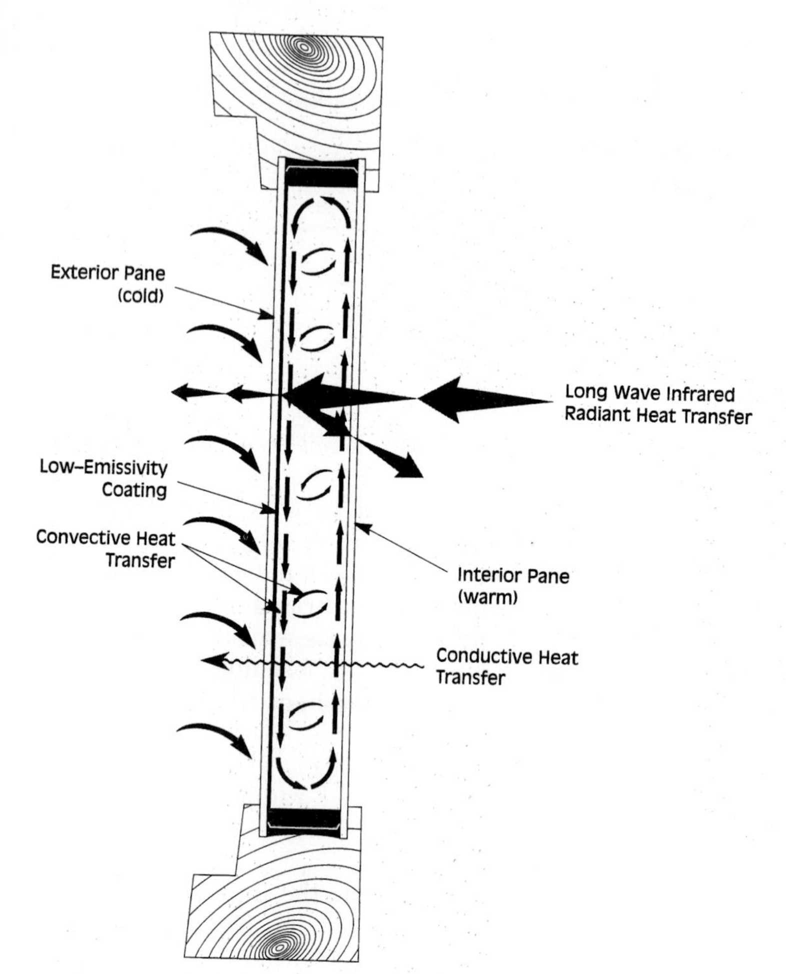

Figure 16.5 Sealed unit with Low-E glass.

TABLE 16.1 R-Values of a Sealed Unit

Air space	R-Value
Clear air	2.0
Clear argon	2.2
Low-E air	3.3
Low-E argon	4.0

3. Krypton gas allows a narrow, high-performance unit to be manufactured. Krypton is used most often with three layers, the center layer being a plastic film as shown in Figure 16.6.
4. Xenon is seldom used because of its rarity and subsequent high cost.
5. SF_6, $C0_2$, and the CFCs have disappeared from the market, CO_2 because of its acidity and SF_6 and CFCs because they are not sunlight stable and deplete the ozone layer.

The initial selling point of these high-performance windows was energy savings, but people quickly discovered that it was more com-

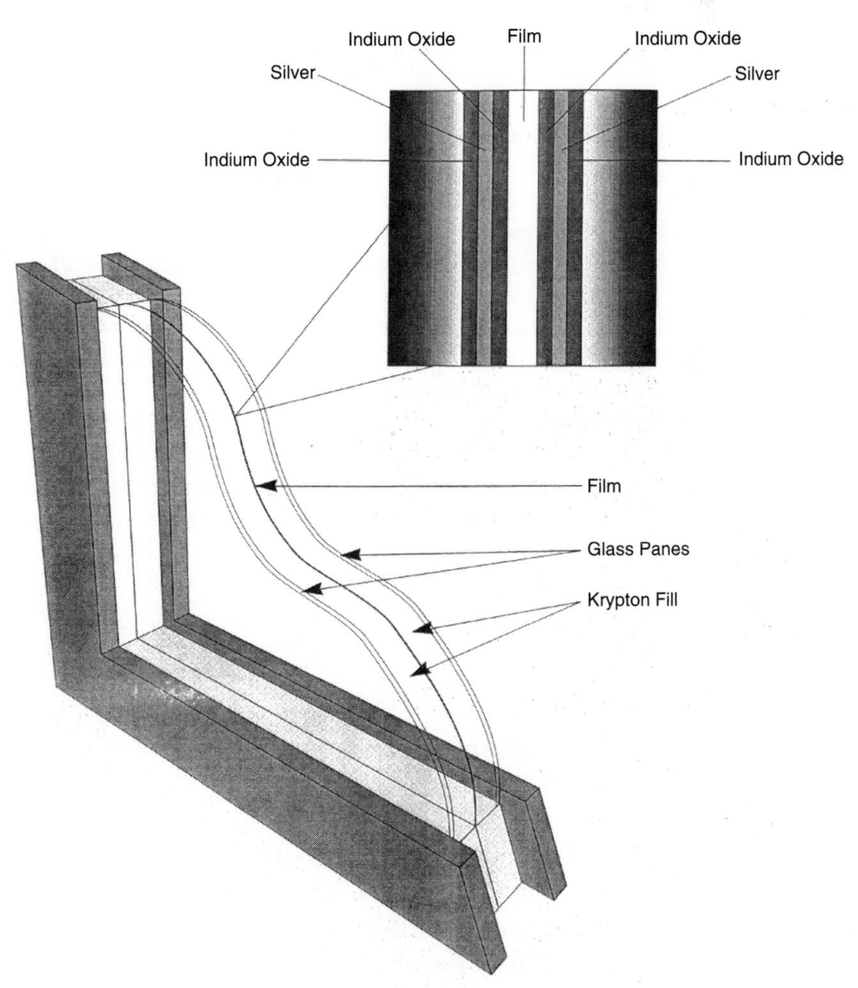

Figure 16.6 Insulating glass unit with plastic film interlayer.

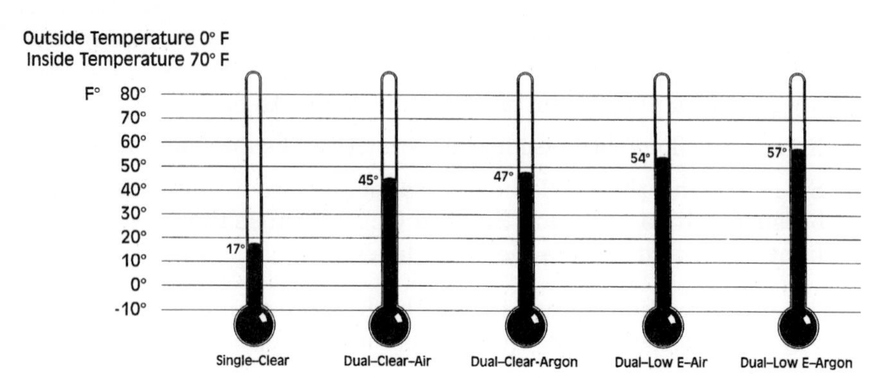

Figure 16.7 Surface temperature of glass.

fortable to be near them because of the warmer surface temperature of the glass (Figure 16.7).

Even though energy issues have taken a back page, sales of high-performance windows have continued to grow. In 1970, insulating glass represented 14 percent of the market. Today insulating glass has almost 90 percent of the market and is still growing. In less than 15 years, Low-E coatings have captured two-thirds of the market and it has been estimated that, by the year 2000, 88 percent of the windows produced will contain a Low-E product.

Argon filling has ridden on Low-E's coattails. Because gas filling is effective, inexpensive, and relatively easy to do, it has been assumed that a Low-E product will include a gas fill. Most manufacturers offer two products, a standard clear glass unit and a high-performance, Low-E, gas-filled unit.

Mechanics of Filling

There are three general methods for gas filling insulating glass:

1. Assemble the unit in a gas-filled environment, requiring a sealed, gas-filled, assembly press (Figure 16.8).
2. Place the assembled unit in a chamber, pump all the air out of the chamber (and the units), and then replace the air with gas (Figure 16.9).
3. Insert one or more lances into the insulating glass cavity via access hole(s), and exchange the air trapped in the cavity with gas (Figure 16.10).

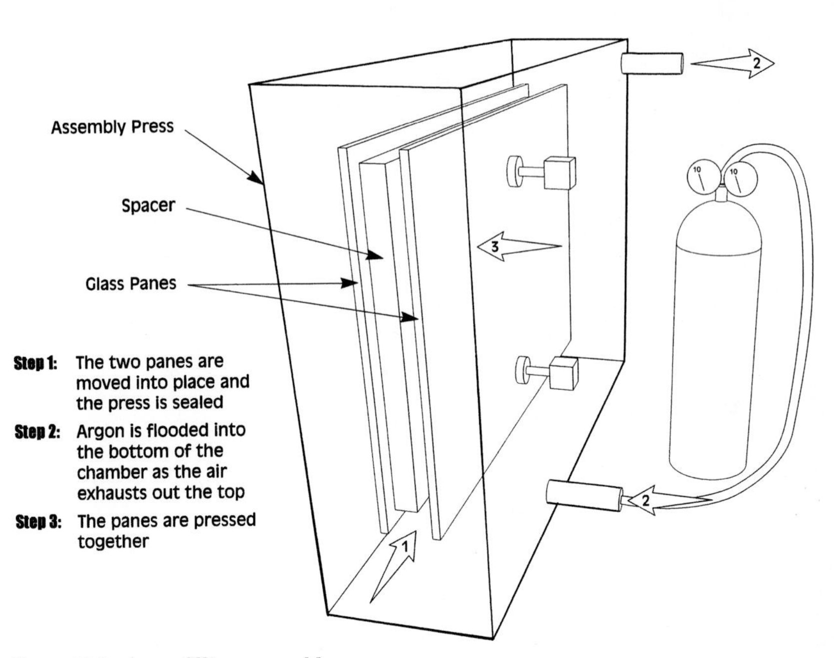

Figure 16.8 A gas filling assembly press.

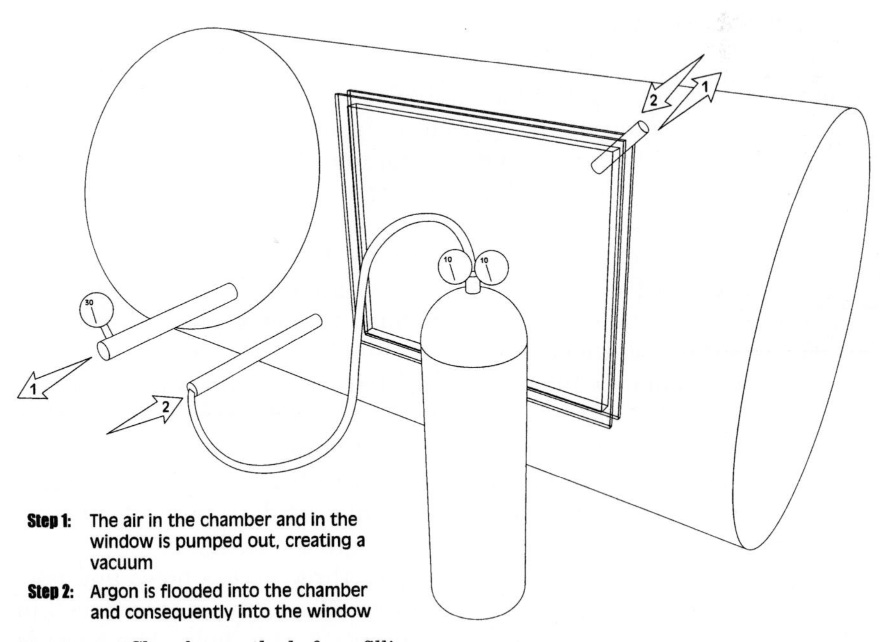

Figure 16.9 Chamber method of gas filling.

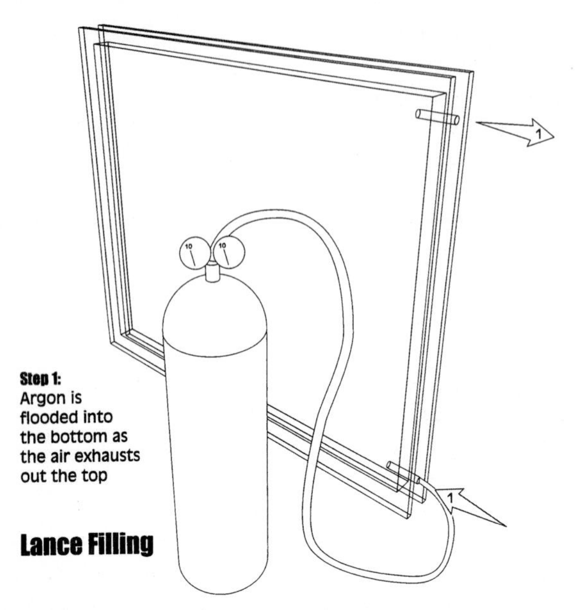

Figure 16.10 Lance method of gas filling.

Because we cannot see air and argon, it can be difficult to comprehend how gas filling of insulating glass is done. The actual concept is really quite simple. Argon is about 40 percent heavier than air. For a short period of time, air will actually float on top of argon. By carefully introducing argon at the bottom of an insulating glass unit, it is possible to float the air up and out the top, the key ingredient being careful introduction of the argon. If argon is turbulently introduced, lamination will not occur. Then, the window must be filled by stirring in large amounts of argon and eventually diluting the mixture of air and argon to mostly argon. By having an accurate sensor monitoring the concentration of argon in the exhaust, we know when the window is 95 percent full of argon.

Once the window is filled and sealed, a minute amount of gas will gradually seep out of the unit from the top and from all sides. As it dilutes over time, the gas will not have different concentrations from the top to the bottom of the window, but, rather, will mix completely together. If the window is 90 percent argon and 10 percent air, this same concentration is present everywhere in the cavity.

Lance filling is done with one, two, or three access holes and one, two, or three lances. In the case of timer fillers, gas is introduced in the bottom and air is floated up and out the top hole. The cycle is complete after a predetermined amount of time has elapsed.

Sensor fillers place a lance in the top hole that monitors the exhaust air. When a high concentration of argon is sensed, the filling cycle is terminated. To obtain faster fill times, a suction pump is used to extract the exhaust air at the same time the gas is being injected into the insulating glass unit at the bottom. Even faster speeds are obtained by putting two suction lances in the top of the unit.

Standard lances were developed to handle the standard rectangle spacers. Alternative spacer materials necessitated new lance designs.

Foam Super Spacer™, a ribbon spacer/desiccant system, requires punching holes in the foam or piercing the foam with reusable bushings or pointed lances.

Swiggle Seal™, a ribbon sealant/spacer/desiccant system, uses soft silicone tubes that are threaded into the cavity. A laminar insert can be placed in the tip of the tube to yield smooth nonturbulent introduction of the heavy filling gas. Alternatively, Swiggle Seal™ can be filled with the one-hole filling method described below.

In some cases, PPG Intercept™, a patented roll-formed spacer/desiccant system, shown in Figure 16.11, requires smaller-diameter lances.

The Intercept™ process has a gunned-in desiccant matrix so gas filling works best if it can be done through just the one final assembly hole. Special lances were developed that could fill Intercept™ through one hole. With one-hole filling, no lamination or floating of the air occurs, so that the sensor must be accurate to determine appropriate fill levels or timers must be set to longer times for a greater volume of argon to purge the cavity.

Other U.S. patents pertaining to insulating glass are 4,773,453; 5,351,451; 5,177,916; and 5,255,481, available through the U.S. Patent Office.

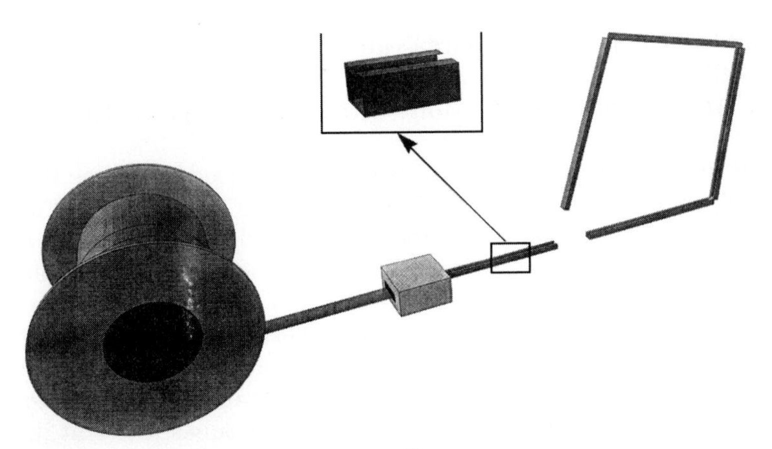

Figure 16.11 Intercept™ method of gas filling.

Quality assurance can be accomplished by drawing samples from the insulating glass cavity and testing them for residual oxygen with an oxygen analyzer or testing for the specific gases with a gas chromatograph.

Presently there is no practical method of checking a sealed insulating glass cavity for gas content. U.S. Patent 4,848,138[4] is for a small strip that would change color in the presence of air, but it has never been commercially developed.

A U.S. Department of Energy research grant DE-FG0293ER81522 funded research for a process to determine the oxygen in a sealed insulating glass cavity but never went beyond the initial funding. Cardinal IG developed a device using the principle of Raman spectroscopy for their use but its sophistication and cost exclude it as a candidate for field use.[5]

The method of choice will probably be video thermography, currently under U.S. Department of Energy evaluation at Lawrence Berkeley National Laboratory.[6] Thermography offers a method of analyzing thermal performance of the whole package—gas, glass, and spacer.

The industry is presently looking at setting production standards and test procedures to assure properly fabricated products. The conclusion reached by those researching standards is that gas is lost only if and when the edge seal around the glass perimeter fails. When the seal fails, the glass will show visible fogging because of the introduction of water vapor.

Consequently, unless the unit has visibly failed, the consumer should be assured that most of the argon has been retained. To maintain their "value," units should have over an 80 percent fill.

Typical Standards

Typical internal industry standards call for a minimum 80 percent initial gas fill when the window is shipped. This initial fill rate is not regulated. Most fabricators ship units with a 90 to 95 percent initial fill rate.

A Low-E unit with just an "air" fill is still a very good unit. Gas filling typically adds 15 percent to the performance of the glass. For example, if the center of glass R-value of an air-filled unit was 3.5, a gas-filled unit might have a center of glass R-value of 4.0. As the gas escapes, the R-value will decrease linearly from 4.0 to 3.5.

A breather tube is a large-diameter tube inserted in the insulating glass unit to equalize the pressure in the insulating glass with the pressure surrounding the insulating glass unit at installation. Refer to Chapter 20 for detailed information on breather tubes. Once at its installation elevation, the tube is pinched and sealed (by soldering or

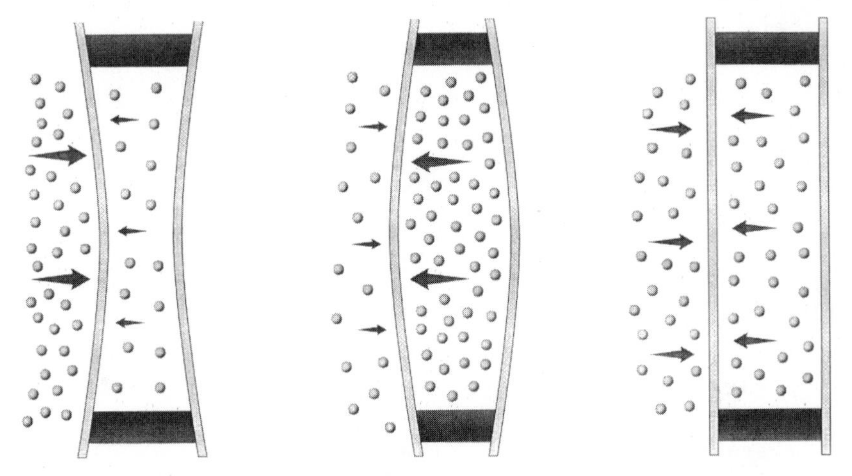

Figure 16.12 Effects of pressure changes.

some other form of barrier). In some cases, breather tubes are left open during shipping to allow rapid equalization of the insulating glass unit as it is transported. Figure 16.12 shows the effect of changing pressure on a sealed unit.

If used with gas filling, the breather tube must be sealed after the insulating glass unit is filled with gas, opened at the installation elevation, and then, resealed after the pressure has equalized. If the unit is going to a lower elevation, breaching of the insulating glass cavity will dilute the gas concentration. At a higher elevation the concentration is unaffected because the excess pressure is expelled from the cavity of the insulating glass.

Capillary tubes normally cannot be used with gas filling. The purpose of a capillary tube is a controlled, small leak in the sealant system. As a result, the gas fill will migrate out and exchange itself with air. The myth that the capillary tube is small enough in diameter to keep gas in is not true. It is small enough to slow down the exchange from seconds to perhaps hours but not small enough to trap the gas molecules from escaping the interior of the insulating glass unit. Nor is it small enough to significantly reduce the flow rate through the tube because of a pressure differential. With a difference of 1 in Hg, the flow rate will be 20 cc per minute.

A manufacturer in Colombia builds and gas-fills units with open capillary tubes at sea level and then transports them to Bogota at 10,000 ft elevation. It takes several days to truck them up the mountain, and once they are at the higher elevation, the tubes are sealed. Figure 16.13 shows the change in air pressure per square foot caused by moving the unit from sea level to 10,000 ft. The manufacturer is

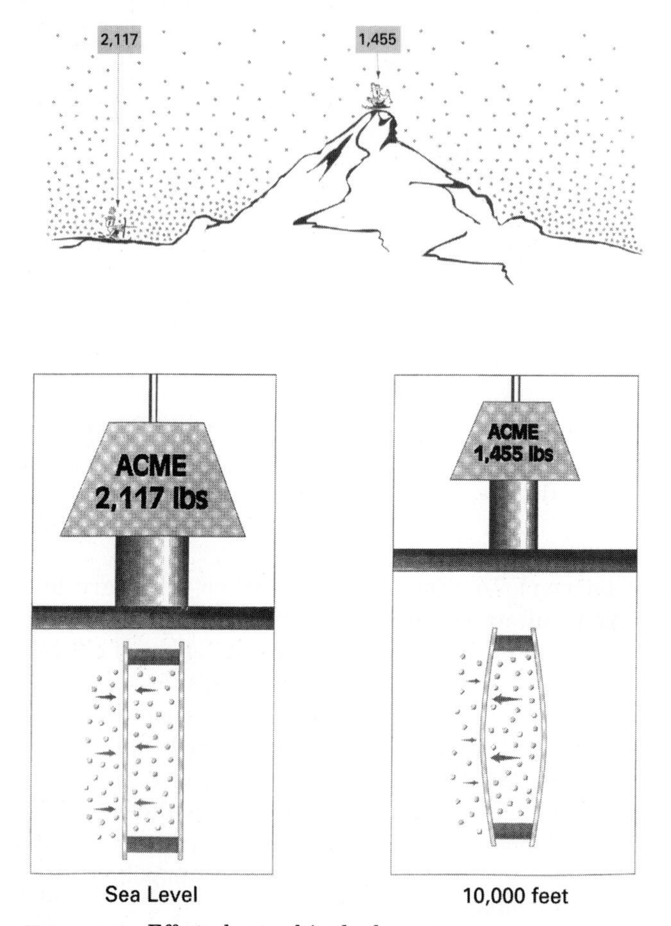

Figure 16.13 Effects due to altitude change.

not installing a capillary tube for use in the traditional sense, where it is permanently left open. It is actually a small-diameter breather tube. The purpose of the tube is to vent excess pressure. Once at the installation elevation, the capillary tubes are permanently sealed. See Table 16.2 for the molecular dimensions of gases measured in angstroms.

Figures 16.14 and 16.15 are photographs of gas filling machines for multiple insulating glass units. The units are filled laterally with hoses inserted through the spacers, and the gas flow is regulated by pressure and flow valves.

TABLE 16.2 Molecular Dimensions of Gases

Gas	Size, Å*
SF_6	5.8
Nitrogen (80% in air)	3.6
Argon	3.5
Water	2.8
Organic solvents	5.5–7.5

*1 angstrom = 0.000000000393 in (3.93×10^{-10}). A human hair measures 0.002 in or 5,089,058 Å. If an argon molecule were the size of a golf ball, a human hair would be 23 miles thick.

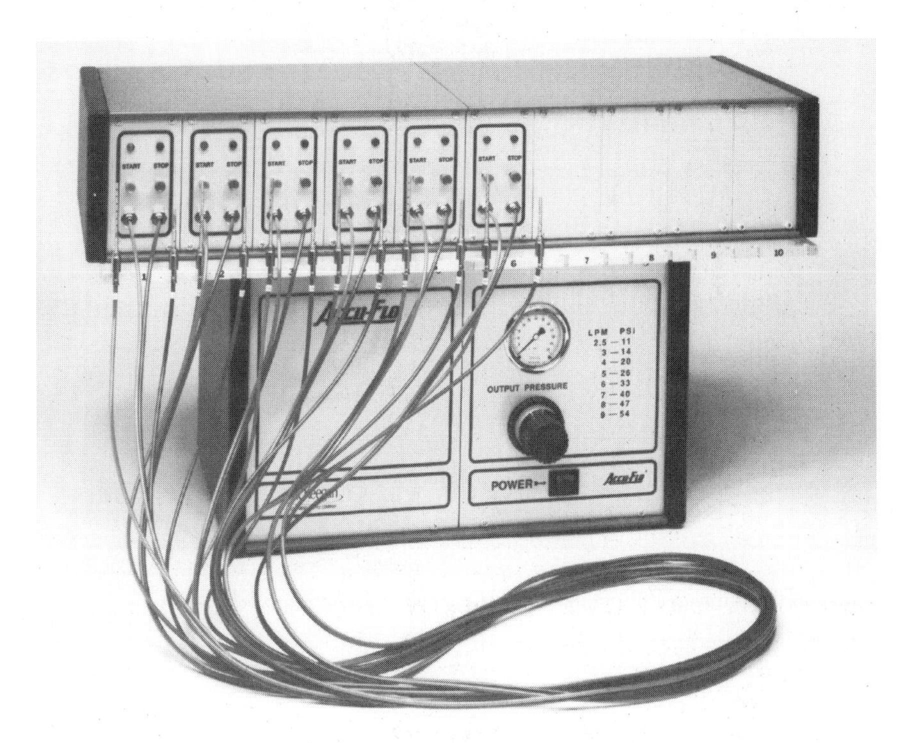

Figure 16.14 Multiunit gas filling machine. (*Courtesy of McKeegam Sales, Inc.*)

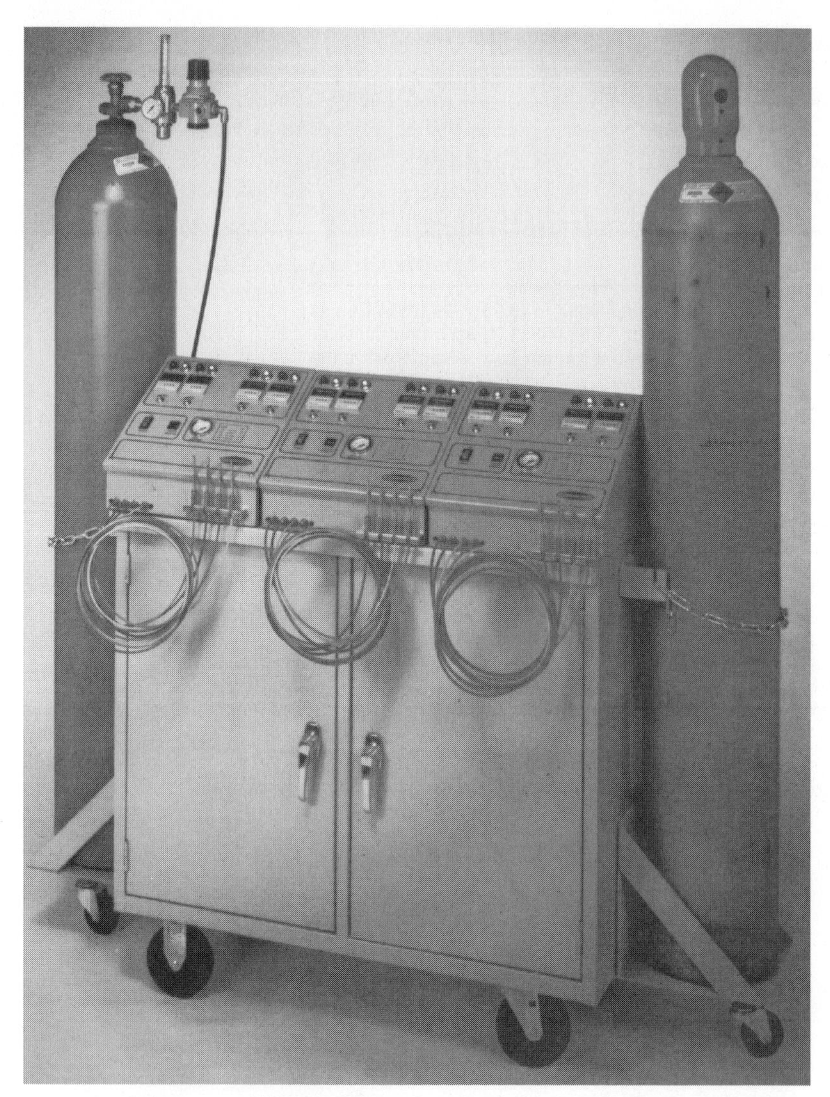

Figure 16.15 Large gas filling machine. (*Courtesy of McKeegam Sales, Inc.*)

Conclusion

Argon gas increases a window's ability to insulate. Because argon gas does not conduct heat as readily as air, it helps keep heat in during the winter and out in the summer. Argon gas will not hurt the building occupants if it leaks or if the window is broken. Argon gas exists naturally in the air you breathe. It is both nontoxic and inert. Argon and krypton gas are clear, colorless, and odorless. Even if installed side by side, just by looking at them, you cannot see the difference between a window filled with gas and one not filled with gas.

There are no real guarantees that the gas is still in the window. Retention of the gas depends on the quality of the sealant system. If the seal fails, the gas will escape. Numerous studies show that gas typically leaks out at a rate of less than 1 percent per year. In more than 20 years, the unit will still have 80 percent of the gas fill trapped inside it.

Argon and krypton block no portion of the sun's light. House plants are not affected.

Although energy savings are important, the real benefit of gas filling is the increased comfort from the inner glazing temperature. Raising the inner glass temperature eliminates or substantially reduces condensation on the glass. A warmer surface is also more comfortable to be near.

The answer is simple as to why we gas-fill. Gas filling is a low-cost procedure which provides a major improvement in the product's thermal performance. Argon is the gas most frequently used for gas filling because it is inexpensive and readily available. Krypton is a good candidate though more expensive. Krypton gas is particularly impressive in glazing units where small thickness is of prime importance. Argon and krypton are both refined from air. You are breathing about 1 percent argon right now.

References

1. U.S. Patent 2,756,467.
2. U.S. Patent 3,683,974.
3. Really, D. A., and M. Rubin, The Effects of Infrared Absorbing Gases on Window Heat Transfer: A Comparison of Theory and Experiments, 1989.
4. U.S. Patent 4, 848,138.
5. Pipino, A. C. R., An Optimized Single-Channel System for Quantitative Raman Spectroscopy of N_2 and 0^2, *Cardinal IG Applied Spectroscopy,* 44, 7, 1990.
6. Beck, F. A., B. T. Griffith, D. Turler, and D. Arasteh, Using Infrared Thermography for the Creation of a Window Surface Temperature Database to Validate Computer Heat Transfer Models, March 15, 1995, Window Innovations 1995, Conference LBL-36975.

Chapter

17

High-Performance Film

Donald E. Holte

*P. Engr., Sr. Vice President Marketing, Visionwall Technologies, Inc.**

Joseph S. Amstock

President, Professional Adhesive and Sealant Systems†

Introduction

In the past ten years, a whole new family of building windows has emerged. These windows tend to be classified under the broad heading "High-Performance Windows." The common characteristic of high-performance windows is their improved thermal resistance (reduced conductance) and visible light-to-solar heat gain coefficient ratio greater than 1.0. Figure 17.1 is a photograph of a cross section illustrating two different designs.

This system provides a wide range of excellent shading coefficients, solar heat gain coefficients, and visible light transmissions through the use of Low-E coatings and glass types. The primary difference between the two systems is that the four-element system offers the user about 25 percent more insulation value. As one guideline for the

*Donald E. Holte, P. Engr. is coauthor of High-Performance Windows based on the article titled "A Parametric Study of the Impact of High Performance Windows (and Curtain Walls) on Building Heating and Cooling Loads, Energy Use & HVAC System Design."

†Joseph S. Amstock authored the section on Heat Mirror.

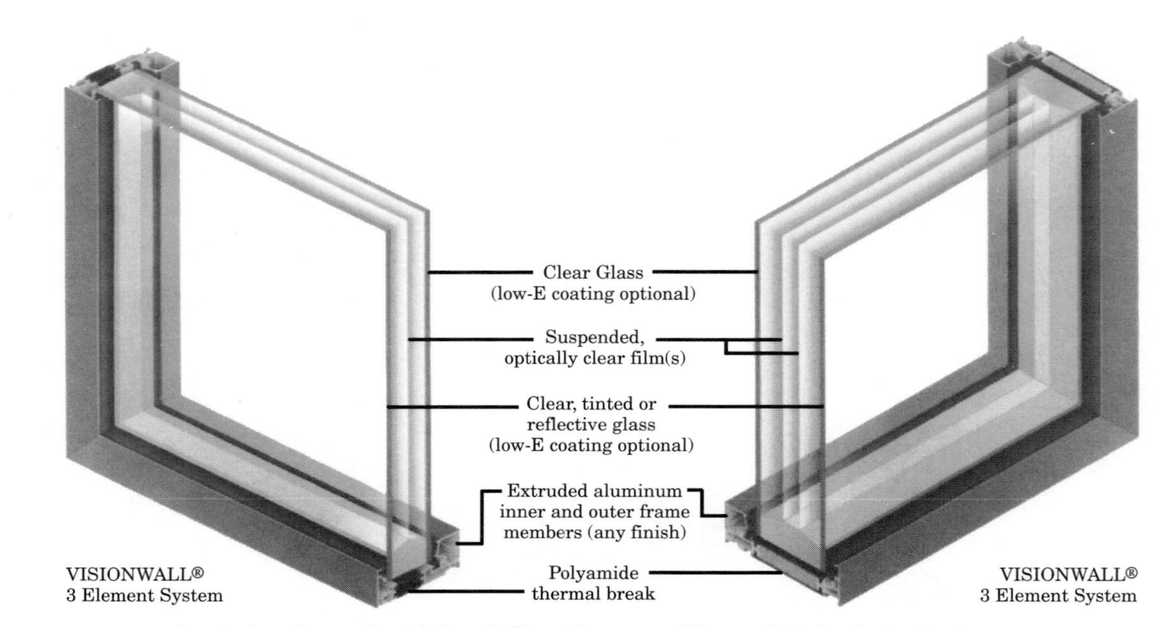

Figure 17.1 Insulating glass unit with Low-E film. (*Courtesy of Visonwall Technologies, Inc.*)

Figure 17.2 Winter design temperature.

selection of the best system for your project, refer to the winter design temperature map in Figure 17.2. If your building is located in an area where winter design temperature is below 0°F (−18°C), consider the use of the four-element system; otherwise, the three-element system will provide adequate thermal performance.[1]

The original technology was developed in Switzerland, in 1989, by Geilenger AG in response to the energy crisis to meet today's and the future building energy code. This innovative window technology enables architects and building owners to create buildings with large expanses of glass without sacrificing energy efficiency. For too long, insulating glass units have been the weak link in a building's thermal envelope. Size and placement of heating and cooling systems are often dictated in large part by the lack of performance of sealed insulating glass windows. Heat loss through inefficient windows often requires the use of perimeter heating systems to compensate for this loss. High-performance windows eliminate cold drafts and provide comfortable interior spaces. In the summer months, uncontrollable solar heat gain through windows requires more energy to keep the building cool.

The high R-values eliminate perimeter baseboard heating systems and reduce heating plant requirements 40 to 50 percent. Teamed with low shading coefficient values to lower cooling load requirements by 20 to 30 percent, the HVAC capital cost savings from using Visionwall can be significant.

Design Characteristics

The high-performance window and curtain wall systems are available in two basic systems as illustrated in Figure 17.1. By combining two glazing units employing the designer's choice of clear, highly reflective glass with Low-E coating technology and a highly insulated aluminum frame, the building owner reaps the benefits of a truly high-performance envelope.

This high-performance product offers an ideal combination of high thermal insulation, low shading coefficient, high visible light transmission, and excellent sound attenuation. The data listed below shows the features and benefits of this system.

Features	Benefits
High insulation values ▪ Complete window R-values from 3 to 7 (Rsi 0.5 to 1.2)	▪ Reduced electrical and mechanical capital costs ▪ Lowered energy consumption ▪ Greater occupant comfort ▪ Elimination of perimeter heating systems
Low shading coefficients and high visible light transmission ▪ A wide selection of glass and film combinations to fit every design requirement	▪ Reduced lighting and energy costs ▪ Increased occupant productivity ▪ Better use of natural light ▪ Reduced cooling loads
Excellent sound attention ▪ STC values from 36 to 45	▪ Quiet interior space ▪ Noisy sites usable ▪ Improved occupant comfort
Factory glazed and tested	▪ High quality control ▪ Reduced installation time and costs
Superior condensation resistance ▪ High-performance frame and glazing	▪ Permits use of windows in high-humidity environments ▪ Better hygiene in health care facilities ▪ Elimination of moisture damage to adjoining walls

Description of Work

A typical 10-story building, described in Figure 17.3 and Table 17.1, was analyzed to determine the impact of high-performance windows (curtain walls) on the heating and cooling loads and energy demand and consumption. The analysis was done with the Trane Trace 600, Version 11.09, computer program. Information about the impact of high-performance windows on HVAC system design and indoor comfort was mainly from practical experience with more than 100 build-

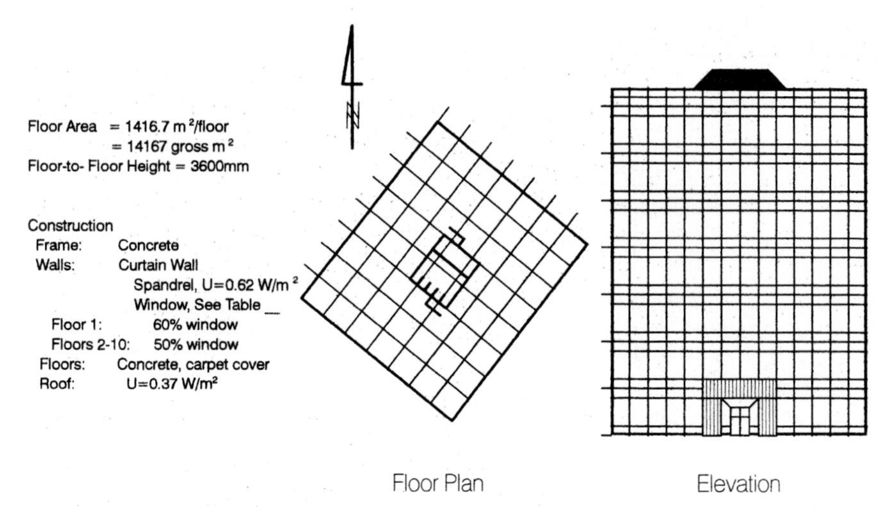

Figure 17.3 Building physical data.

TABLE 17.1 Building Use, Internal Loads, and HVAC System Data

	Main floor	Floors 2 to 10
Use	Retail	Office
People	27.87 m^2 per person	23 m^2 per person
Lights	30.46 W/m^2	16.9 W/m^2
Misc. electrical	2.69 W/m^2	8.07 w/m^2
HVAC system	Variable air volume (VAC) Hot water reheat 4 zones 100% outside air economizer Air-cooled condensing unit	VAV hot water reheat perimeter VAV interior 9 zones per floor 100% outside air economizer Centrifugal water chiller with cooling tower
	Hot water heating with natural gas fueled hot water boilers	
Lighting control	Continuous dimming in perimeter (office) spaces; on-off control in interior	
Schedules	As given in ASHRAE Standard 90.1-1989,* Table 13.3.	

*American Society of Heating, Refrigerating, and Air-Conditioning Engineers, Inc. (ASHRAE) Standard 90.1-1989, Energy Efficient Design of New Buildings Except New Low-Rise Residential Buildings.

ings which have been built with high-performance windows and curtain walls during the last six years.

In this chapter the word "window" refers to the curtain wall frame and transparent glazing unit (i.e., the window in the curtain wall).

The building was analyzed for the four window types shown in

TABLE 17.2 Window Data

Type	Description	Overall heat transfer coefficient (U) W/Cm²–°C	Btu/h²–°F	Solar heat gain coefficient (SHGC)	Visible light transmission (Tvis)
1	Double glazed —clear glass —clear glass	3.28	0.58	0.64	67%
2	Double glazed —tinted Low-E glass —clear glass	2.51	0.44	0.29	43%
3	Medium performance (3 elements) —tinted Low-E glass —uncoated glass —clear glass	1.63	0.29	0.22	40%
4	High performance (4 elements) —tinted Low-E —uncoated film —uncoated film —clear Low-E glass	0.90	0.16	0.12	17%

TABLE 17.3 Geographic Location Data

City		Design temperature*		Heating degree days (HDD)		Cooling degree days (CDD)	
Name	Latitude	Winter (°C db)	Summer (°C db/wb)	66°F (18.3°C)	50°F (10°C)	66°F (18.3°C)	50°F (10°C)
Edmonton, AB	53° 34″N	−34	29/19	59.8	3472	56	688
Toronto, ON	43° 41″N	−21	32/23	4218	2163	224	1201
Denver, CO	39° 35″N	−21	34/15	3379	1473	315	1451
St. Louis, MO	38° 39″N	−16	37/24	2700	1173	815	2329

*From ASHRAE 1993 Fundamentals Handbook.

Table 17.2. The data shown in the table is for a complete (frame plus glazing) 1200-mm-wide × 1800-mm-high window at NFRC/ASHRAE design conditions.[2]

To examine the effect of outdoor climate, the building was analyzed for each of the four different window types in four geographic locations (cities), as shown in Table 17.3.

Cooling plant size

High-performance windows, with their improved (lower) solar heat gain coefficient (SHGC), significantly reduce the building cooling plant size in all four geographic locations.

Cooling plant size reduction, as shown in Figure 17.4, is approximately the same for all geographic locations. The cooling plant size is based on the building's peak cooling load, which is the sum of the internal heat plus ventilation air cooling plus heat flow through the envelope of the building because of the temperature difference plus solar heat gain through the windows.

The amount of solar heat gain through a window is a function of the window SHGC (which is incident angle dependent) and the amount of solar energy striking the window. The peak cooling load in all of the 16 building variations (four window types×four locations) occurred at hour 16 in either month 7 or 8, except for the Denver clear sealed insulating glass window building which peaked at hour 17. At the same time of the day and year, the solar energy striking a window and flowing through a window is approximately the same in all of the four locations, and thus the peak cooling load reduction is nearly the same for all geographic locations.

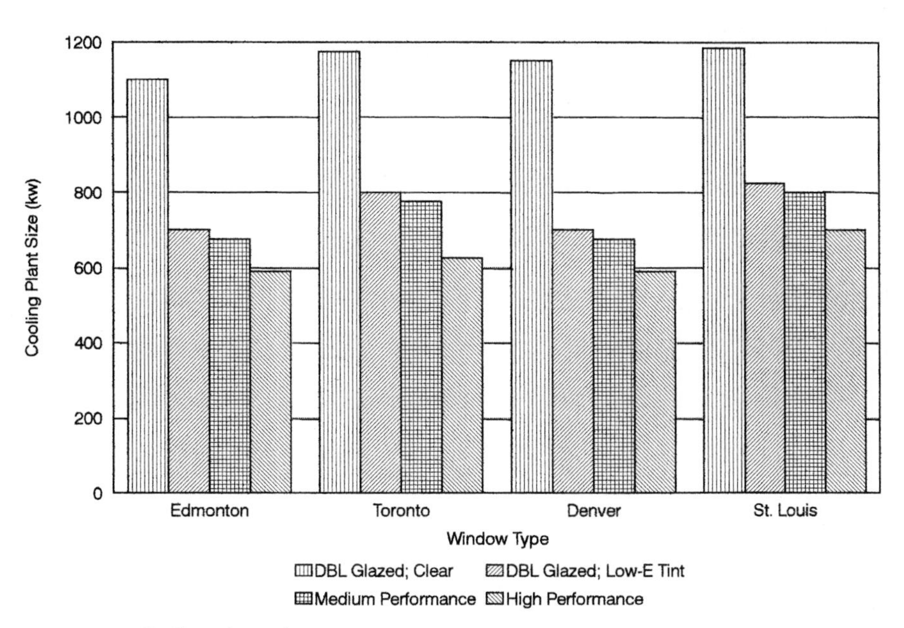

Figure 17.4 Cooling plant size.

Heating plant size

As shown in Figure 17.5, the high-performance window (curtain wall) system, with its lower U-value, reduces the heating plant size by 40 to 50 percent over conventional insulating glass units in all four locations.

Electric power consumption

Electric power consumption, as shown in Figure 17.6, is roughly the same for all window types. The slight increase in electric power consumption of the high-performance window (curtain wall) building over the building with medium-performance windows or Low-E insulating glass is caused by increased electric power consumption of the cooling plant (fan, pump, chiller) motors and perimeter office lighting systems. The high-performance window building loses heat by conduction, and, consequently, the cooling plant motors have to run longer. Because the building also lets in natural (sun) light, the perimeter office systems use more electric power. The high electric power consumption of the clear sealed insulating glass window building is a result of increased cooling plant operation caused by the window's high SHGC.

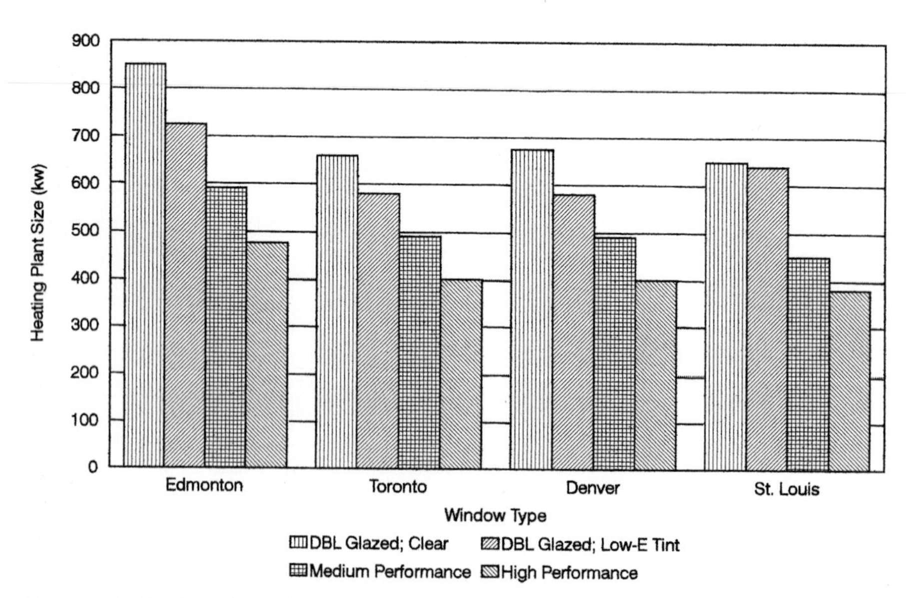

Figure 17.5 Heating plant size.

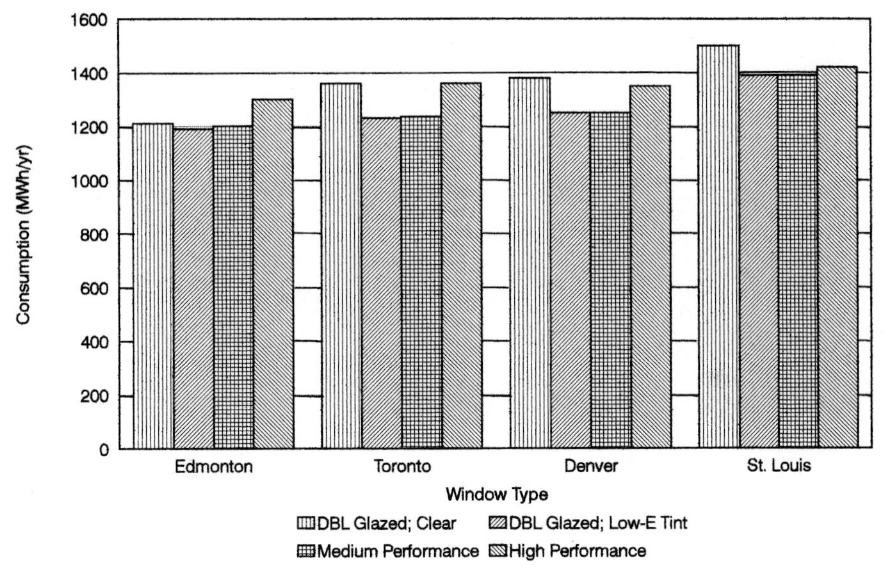

Figure 17.6 Electric power consumption.

Electric power demand

The reduction in electric power demand, shown in Figure 17.7, from the clear glass window building to the high-performance window building is attributable to the window's SHGC and the consequent reduction in the cooling plant size.

The anomaly occurring in the Edmonton building between the Low-E, tinted, sealed insulating glass units and the medium-performance windows is caused by the simultaneous occurrence of the lighting and cooling plant peaks. The building analysis was done simulating a daylight-controlled, continuous type dimmer on the lights in the perimeter offices and, because the medium-performance window has visible light transmission lower than the Low-E, tinted, sealed insulating window, there was a demand attributed to the medium-performance window building.

Natural gas consumption

Annual natural gas consumption, as shown in Figure 17.8, exhibits a similar pattern in every city except Edmonton.

The annual natural gas consumption of the clear sealed insulating glass window building in Edmonton is greater than the consumption

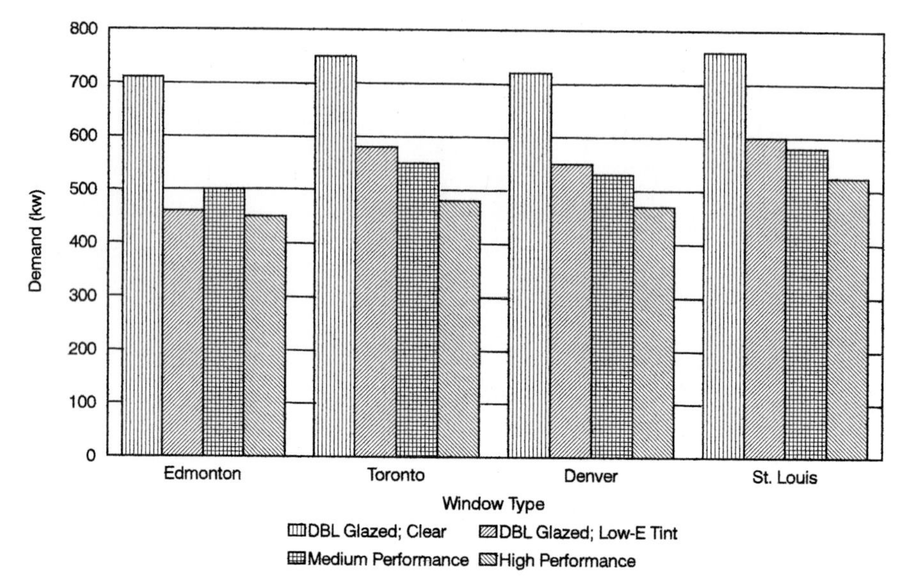

Figure 17.7 Electric power demand.

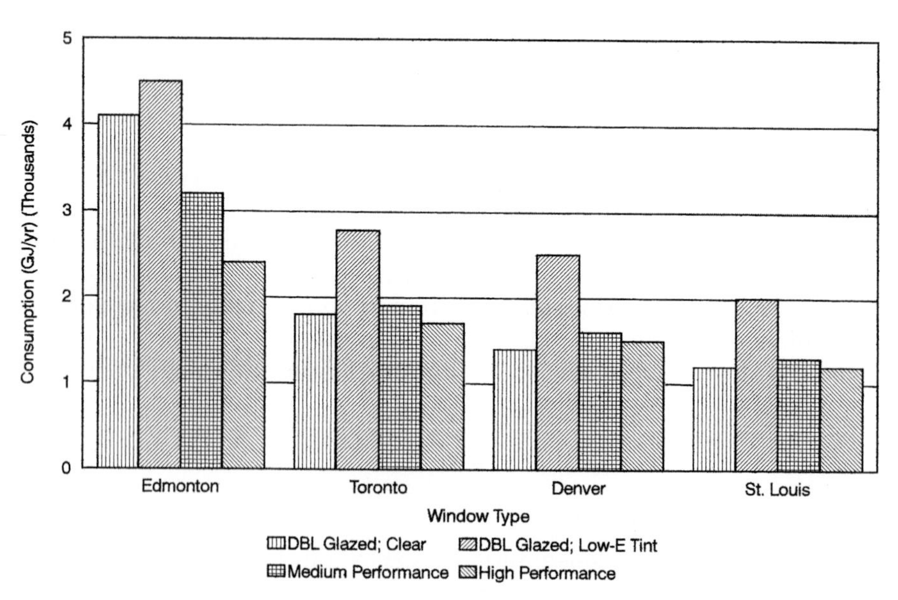

Figure 17.8 Natural gas consumption.

of the medium-performance window building, whereas in the other three cities it is less. Edmonton's climate is much colder than the other cities, and the heating energy savings resulted from the better (lower) U-value of the medium-performance windows with their higher SHGCs.

The increased natural gas consumption of the Low-E, tinted, sealed insulating window building is caused by the fact that the reduced heat loss resulting from the window's lower U-value (lower than clear insulating glass unit) does not offset the loss in "free" solar heating resulting from this window's lower SHGC.

Total energy consumption

In all four locations, the Low-E, tinted, insulating glass window building is the least energy efficient (see Figure 17.9). The small improvement in U-value and SHGC and the resultant reduction in cold weather heat loss and warm weather heat gain over the clear insulating window does not compensate for the reduction in "free" solar heating during the heating season. Also, the reduction in visible light transmission results in less daylight in perimeter spaces and a consequent increase in lighting system energy consumption.

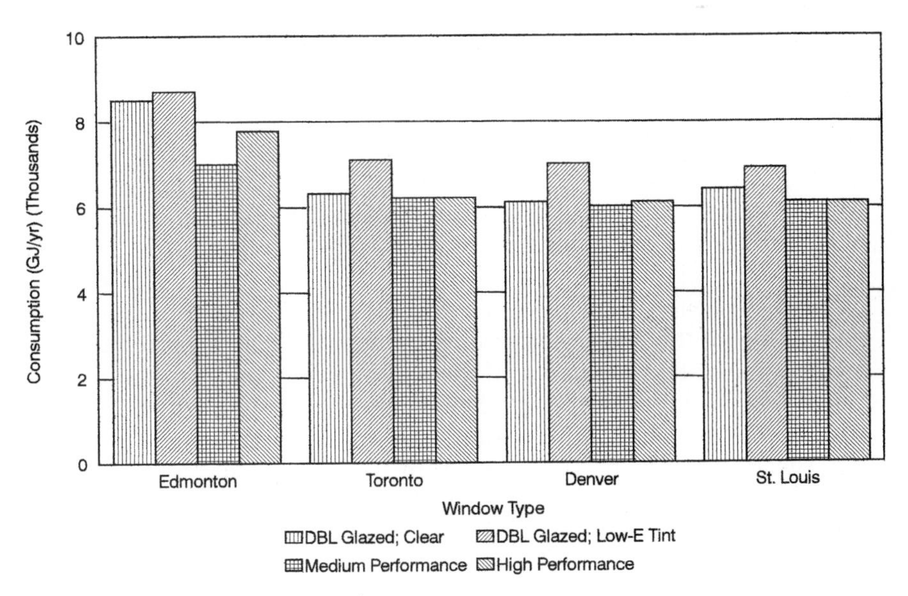

Figure 17.9 Total energy consumption.

HVAC System Design

If the high-performance window (curtain wall) system performance is such that indoor glass surface temperatures do not go below approximately 60.8°F (16°C), then the most significant HVAC system design changes are the elimination of perimeter heating systems and the simplification of supply air distribution systems. There are many buildings in which system designers have taken advantage of high-performance window and curtain wall characteristics to greatly simplify HVAC systems.

Comfort and use

The impact of surface temperature, wall heat loss rate, drafts, room radiative exchange, and other pertinent space design criteria have been well established and reported on by organizations such as ASHRAE.[3]

Windows and curtain walls with improved thermal performance greatly improve space comfort and usability. With indoor glass and frame temperatures near occupied space temperature at all times, occupant comfort is enhanced and condensation on interior glass and frame surfaces and the resultant hygiene and surface damage problems occurring in high-humidity spaces, such as hospitals, swimming pools, water treatment plants, museums, and archives, are basically eliminated.

The elimination of perimeter heating equipment reduces cleaning and maintenance problems and allows the space to be used right up to the window/wall line.

Cost

Most building owners and developers consider the first, or capital, cost of the building of prime importance. Although everyone favors reduced energy consumption and cost, it is very difficult to finance a projection on predicted energy savings. For this reason it is important that the overall cost of the high-performance window and curtain wall building remains basically the same as if the building were built with a conventional insulating glass system. The windows and the curtain wall can cost more, but the overall building cost must not increase. For example, considering the building shown in Figure 17.2. Table 17.4 shows how a high-performance curtain wall building can be, and is, first cost effective. In addition to physical equipment changes in the HVAC system, high-performance windows and curtain wall systems with their much warmer indoor surfaces allow higher relative humidity in the occupied space. This improves the space comfort and hygiene and the space can be kept at a lower dry bulb temperature.[4]

TABLE 17.4 Building Cost Comparison

Cost Item	Curtain Wall Type: Double glazed Low-E tinted glass Conventional thermally broken aluminum frame (U = 3.28 W/m^2 – °C) (SHGC = 0.29)	Curtain Wall Type: High-performance Low-E tinted glass High-performance aluminum frame (U = 0.90 W/m^2 – °C) (SHGC = 0.12)
Curtain wall	5,676 m^2×\$275/m^2 = \$1,560,900	5,676 m^2×\$375/m^2 = \$2,128,500
Heating system		
Perimeter heating	1,504 m×\$160 m = \$240,600	0
Boilers, pumps, piping, etc.	14,167 m^2×\$30/m^2 = \$425,000	60%×\$425,000 = \$225,000
Cooling and ventilation system	14,167 m^2×\$60/m^2 (203 tons) = \$850,000	81%×\$850,000 (165 tons) = \$668,500
Electrical system		(–\$10,000)*
Total comparative cost	**\$3,076,500**	**\$3,062,000**

*Electrical system costs are less due to smaller motors on heating and cooling system equipment.

SHGC = solar heat gain coefficient.

TABLE 17.5 NFRC/ASHRAE Design Conditions

	Outside temp., °C	Inside temp., °C	Wind speed, m/s	Wind direction	Direct solar, W/m^2	T sky, °C	E sky
U-value	–18	21	6.7	0 windward	0	–18	1.00
Solar	32	24	3.4	0 windward	782	32	1.00

NFRC (National Fenestration Rating Council)/ASHRAE design conditions are shown in Table 17.5.

Heat Mirror

Additional technologies are available for insulating glass units that improve thermal properties of the unit even further than sputtered, coated Low-E glass. Heat Mirror produced by Southwall Technologies,

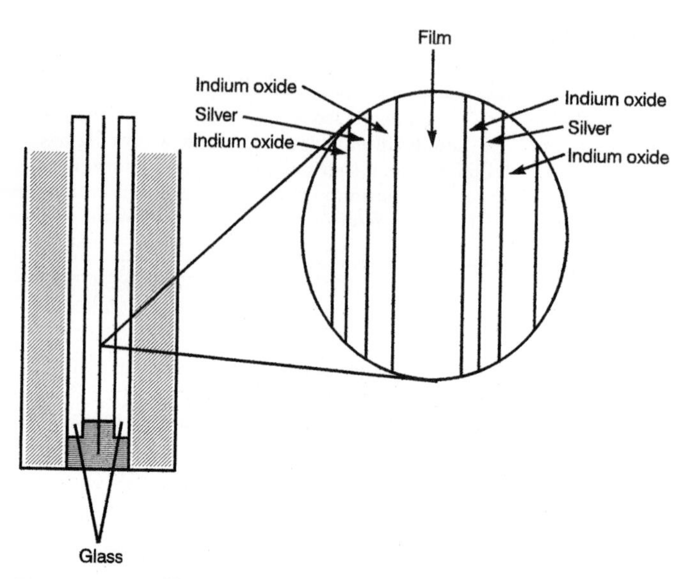

Figure 17.10 Heat Mirror™ unit. *(Courtesy of Southwall Technologies.)*

Inc.,[5] is a concept in low-emissivity glazing slightly different from the sputtered or pyrolitic coated Low-E glasses. Rather than coat the glass, a coated polyester film is suspended between two lites of glass and heat shrunk into place by the fabricator of the insulating glass unit. Heat Mirror units, shown in Figure 17.10, have two air spaces rather than one, which gives improved insulation performance. By coating the polyester film, you can have two air spaces with the same weight.

This product was developed in part at Massachusetts Institute of Technology in the early 1970s and introduced to the glass industry in 1981. The process calls for coating the polyester film using vacuum deposition of a metallic coating that is similar to the process used for sputter coating Low-E.

Heat Mirror has several different coatings to meet the various needs of the industry. The emissivity of the film varies from 0.05 to 0.15 depending on the type of unit with two ½-in air spaces. Heat Mirror 44 on clear glass achieves a U-value of 0.21. It provides a high level of solar control but has a slightly reflective appearance. There are other grades of film that offer higher light transmission in applications where winter solar heat gain is desirable. The coatings are available in clear, bronze, gray, green, blue, or black.

Fabricating Heat Mirror units

Approximately 20 firms are manufacturing insulating glass units incorporating Heat Mirror. A specially designed system for fabricating insulating glass units and the system along with the equipment and the film are provided to these select manufacturers under a license agreement.

Fabricating a unit with Heat Mirror is similar to fabricating a sealed unit with sputter-coated Low-E, except for the additional heat shrinking step, which is done to remove any visual impurities from the film. Also, two spacers are required when mounting it inside the unit, but the edge stripping required for sputter-coated Low-E is not required. Necessary precautions are the same as with sputter-coated Low-E, except that the shelf life is longer because the film is stored in a tightly wound roll. The coated material is more sophisticated. You have to use more care and have a clean, dry environment.

High-performance market

The film is higher in cost, but it addresses a market segment that demands the highest performance. It is used in residential applications but may be used in commercial applications with high cooling loads. The film will provide reduced air conditioning costs and will bring in much more visible light than high-performance reflective glass.

Different types of film can be used in the same building to tune the building's glazing for the highest performance possible on each side of the building. Many of the buildings are characterized by two lites if not three different types of film. Figure 17.11 shows the energy balance of the film.

Meaning of numbers

Just as we live in varied climates, a family of products is produced to provide the optimum thermal performance for virtually any application in a wide range of climates. The higher the number (i.e., Heat Mirror 88), the higher the solar transmission. The film with the highest numbers provides the higher levels of visible light transmission and solar heat gain. The lower-numbered products (HM 66,55,44) provide the best control of solar heat gain for optimum performance in the southern climates or in glass-intensive structures, such as sun rooms or commercial buildings. HM 88 and the recently released Twin-Coated 88 (TC88) are designed for northern climates, where warmth from the sun is desirable for supplemental heating. HM 66 and 77, on the other hand, are designated for building elevations

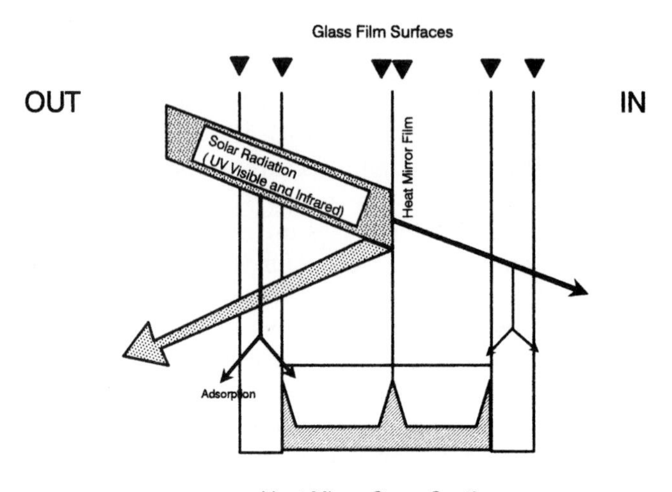

Figure 17.11 Energy balance.

where overheating is the primary concern. They transmit a high percentage of visible light but filter out the sun's invisible solar heat to control excessive heat gain. Both products appear similar to HM 88 but provide solar shading normally associated with the best varieties of dark, tinted glass often used on large commercial buildings.

HM 66 and the recently introduced Solar Control 75 reduce excess heat gain by 40 percent, compared with ordinary clear insulating glass, and by 38 percent compared with the most popular types of clear, Low-E coated glass. HM 33, 44, and 55 provide even greater levels of solar heat control for large glass-intensive structures, such as commercial building skylights and glass-enclosed sun rooms.

Sloped glazing

Heat Mirror insulating glass offers exceptional performance in sloped glazing. In sloped glazing applications, where convective currents within the sealed air space become the dominant form of heat transfer, Low-E coatings and argon gas filling, commonly used in many types of insulating glass units, have little effect. In fact, the insulation performance of these types of products can deteriorate up to 40 percent in sloped applications, whereas the performance of Heat Mirror remains virtually the same. The suspended film splits the unit air space into two thinner cavities, which helps to create a more effective barrier to convective heat.

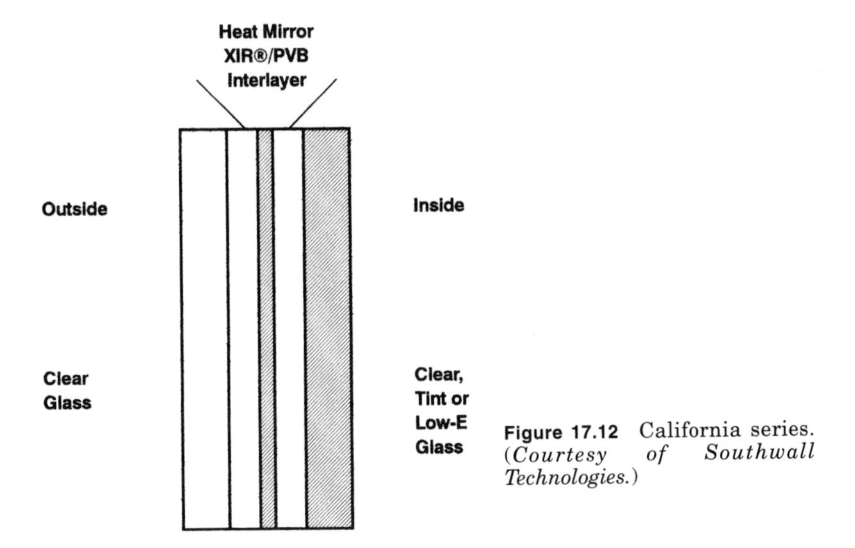

Figure 17.12 California series. (*Courtesy of Southwall Technologies.*)

The Heat Mirror film is a center-mounted layer. Reflectivity in the lower-numbered products is typically not an important consideration when glass units are constructed with tinted glass in the outboard position. In such cases, the tinted glass generally determines the exterior appearance of the glass unit. This significantly reduces any exterior reflectivity which might exist in the case of clear glass and, with the addition of tint, further reduces the shading coefficient.

Figure 17.12 illustrates a new series of products which consist of a thin, wavelength-selective XIR coated film sandwiched between two layers of PVB and glass, basically creating a laminated product which can be used either monolithically or as an insulating glass unit. The glass allows more than 70 percent visible light transmission while reflecting more than 50 percent invisible heat, permitting high-quality, energy-efficient design and low visible reflectance with maximum light transmission and minimum solar heat gain. It may be bent, used in structural glazing applications, or used in sloped glazing.

Insulation properties

Insulation performance is measured to determine how effectively a window keeps heat in a home. Better insulation in windows means lower energy costs and, just as important, improved comfort on cold winter days. The R-value is a measure of the resistance to heat flow. Therefore, a higher R-value means better insulation performance. Figure 17.13 charts the insulation performance.

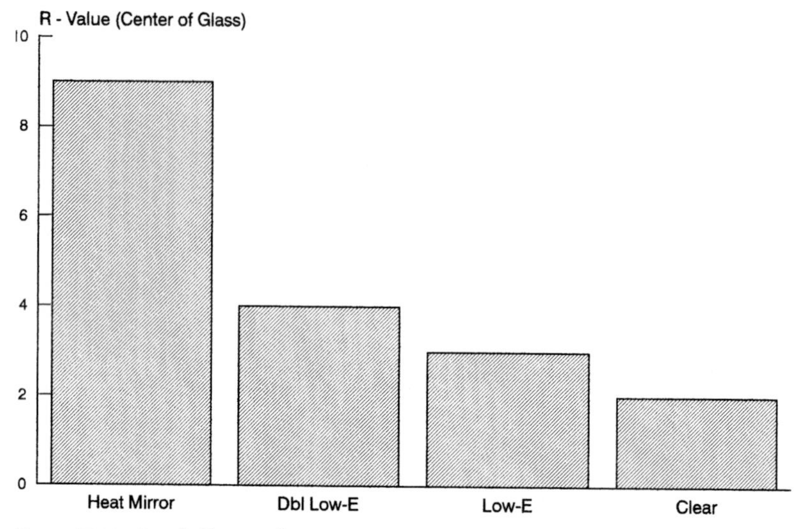

Figure 17.13 Insulating performance.

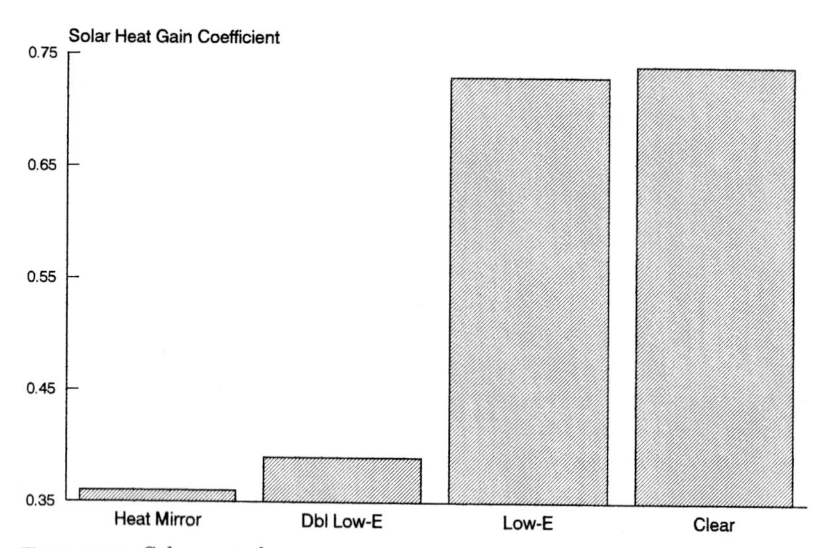

Figure 17.14 Solar control.

Solar control

Solar control is measured by a value called the shading coefficient. The shading coefficient is a relative comparison of solar heat gain. Lower values indicate lower levels of heat gain and, therefore, improved solar control. Figure 17.14 graphically demonstrates this feature.

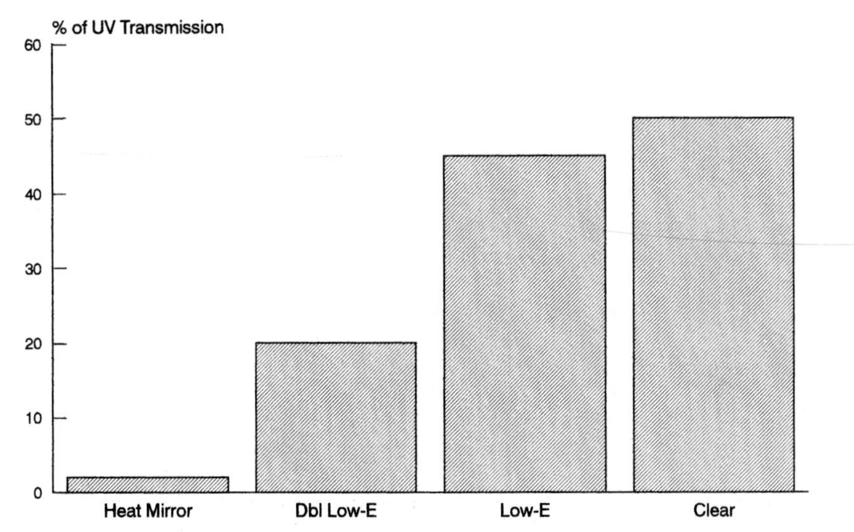

Figure 17.15 Ultraviolet protection.

Ultraviolet protection

We know how dangerous ultraviolet radiation can be and how depletion of the ozone layer increases the incidence of skin cancer because of ultraviolet radiation. Ultraviolet is the main source of color fading. Exposure to UV radiation can damage carpets, furniture, and draperies. Glass products are available which will almost eliminate the transmission of ultraviolet radiation. See Figure 17.15.

Noise control

As covered more extensively in Chapter 8, we know that noise in our urban environment is almost constant. Just as heat and light can be transmitted through windows, noise often finds its way into a home through the windows. Glass products are available today that reduce noise intrusion through the windows. The noise control properties of these product types are measured in Sound Transmission Classification (STC) ratings shown in Figure 17.16.

Gas filling improves insulation

Another technology that improves the thermal properties of Low-E glass even further involves filling coated insulating glass units with argon gas. This technology has been used in Europe for several years but has only recently been introduced in the U.S. market. In every case, the argon gas makes a unit insulate better because the argon is

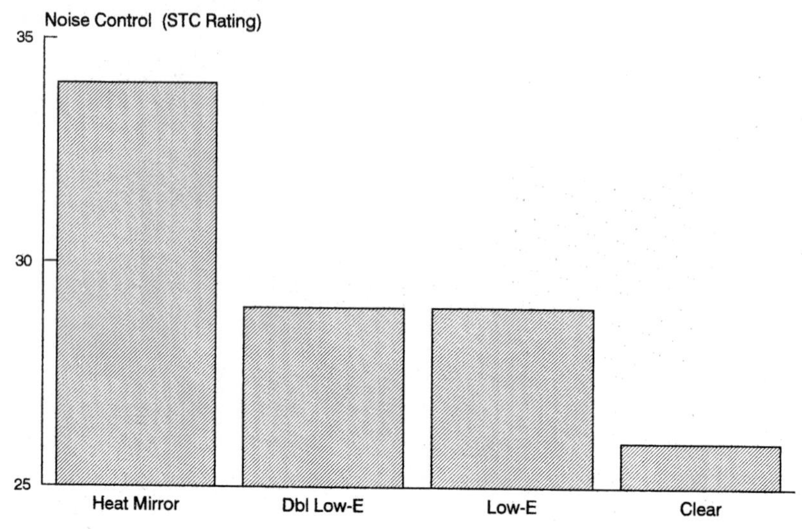

Figure 17.16 Noise control.

less conductive than air. Chapter 16 covers the inert gases. It is possible to achieve a U-value of 0.19 with gas-filled units that have two layers of coated film with a 0.11 U-value.

Gas retention

Manufacturers are still skeptical about the ability of an insulating glass unit to retain gas over time. From a theoretical standpoint, argon gas can provide added improvement in conjunction with Low-E glass, but some manufacturers do not have the confidence level in the sealing systems to maintain the gas in the unit over a period of years.

Conclusion

High performance

The term "high-performance window" needs to be defined relative to a specific building in a specific geographic location and the building owner's views of the relationship between capital cost and operating (energy) cost.

In climates where heating loads dominate and outside air temperatures are low enough that perimeter heating is required for comfort in buildings with conventional insulating glass windows, it will probably be the window's overall U-value that is paramount in defining high performance. In more moderate climates, where both heating and cooling loads are of equal importance and perimeter heating is not required for comfort in buildings with conventional insulating glass windows, the window's SHGC will probably be more important than its U-value in defining high performance.

High-performance window and curtain wall systems, whose thermal resistance is three to four times greater than standard insulating glass windows, are energy efficient and cost effective in cool climates where perimeter heating is required.

If building capital cost is the main concern, then windows or curtain walls with the lowest U-value and SHGC are the best choice. If energy consumption and its related environmental impact and long-term operating cost are the main concern, then a detailed analysis will probably be required to select the right combination of U-value and SHGC for a specific building project.

Heat Mirror

As with new innovations for tomorrow, we see research developing improved products. The illustration in Figure 17.17 shows a cross section of a warm-edge hybrid. Heat Mirror products have expanded the boundaries of performance available to the architect and engineer.

With the center of the glass ranging in U-value from 3.8 to 9.0 and solar shading performance ranging from a shading coefficient of 0.66 in clear HM 88 to 0.23 in gray HM 44, the right balance of insulation, shading, and aesthetics can easily be attained.

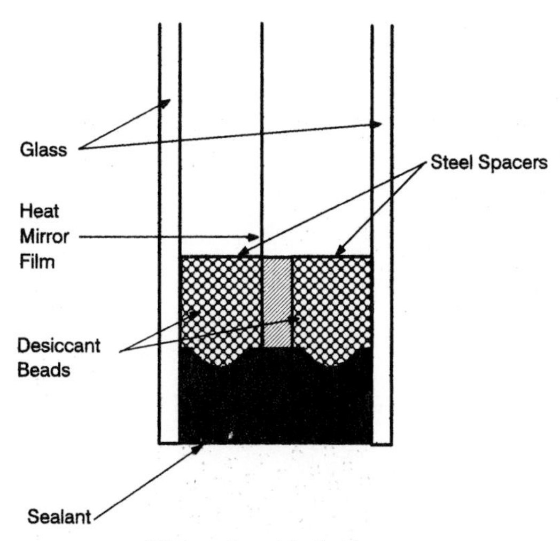

Figure 17.17 Warm edge: A hybrid.

References

1. American Society of Heating, Refrigerating and Air Conditioning Engineers, Inc., Energy Efficient Design of New Buildings Except New Low-Rise Residential Buildings (ASHRAE) Standard 90.1, 1989.
2. NFRC (National Fenestration Rating Council)/ASHRAE Design Conditions.
3. Jackman, P. J., Perimeter Heating System Performance, *ASHRAE Transactions,* 1991, 97, Part 2.
4. Awbi, H. B., and G. Gan, Predicting Air Flow and Thermal Comfort in Offices, *ASHRAE J.,* 1994.
5. Technical Literature published by Southwall Technologies and Glazing Selection Guide for Replacement Windows 100-0119, Southwall Technologies, Inc., Palo Alto, CA, 5/1996.

Chapter

18

Low-Emissivity Coatings

Steven J. Nadel

*Chief Technologist, BOC Coating Technology**

Joseph S. Amstock

President, Professional Adhesive and Sealant Systems

Introduction

This chapter discusses current trends in advanced low-emissivity glazing, comparing product requirements and trends in Europe and North America driven by residential and commercial architectural requirements. A brief description of the sputtering process and associated vacuum deposition equipment is given. Low-emissivity glazings which combine solar control with thermal insulation and high visible transmission are now available. The relative performance of these different products will be discussed. Advanced low-emissivity glazings under development utilizing high-index materials are also presented.

The introduction of inexpensive, neutral-colored Low-E coatings for windows, in the early 1980s, marked the fastest acceptance of a new technology by the building industry since England's Pilkington Glass Limited invented float glass in the 1950s. But Low-E technology actually predates float glass by almost 10 years.

The coating was invented by PPG (then the Pittsburgh Plate Glass Company) and the Mellon Institute during World War II for draining efficiency-robbing static electricity off the glass faceplates of radar

*Contributed extensively to the data included in this chapter.

screens. Several years would pass before anyone became aware that this same coating could increase the thermal resistance of a window.[1]

One of the first glasses coated with this transparent material worked well as the intended electrical conductor. It was quickly dubbed NESA glass, *nonelectrostatic solution A,* by its inventors to differentiate the coating from their experiments. The product is still marketed under the same name. Additional applications for the durable coating were quickly introduced. The coating was put to limited use, creating broad fields for transparently energizing panel lighting fixtures and wireless neon lights.

Unfortunately, the introduction of metallized glass became a painful experience for building owners and users alike because the product had not been thoroughly tested, and many installations soon discolored.

The use of large-scale sputter deposition of thin films for architectural glazing was pioneered nearly 20 years ago. With the development of the planar magnetron sputtering cathode in the early 1970s, the introduction of large-scale deposition created the capability to deposit a wide variety of thin-film materials, enabling the creation of coatings with flexible variations in appearance and thermal performance. This led to the issuance of a U.S. patent.

Figure 18.1 shows a schematic representation of the sputter coating process using a planar magnetron cathode. The key innovation which made sputter deposition an economical process was the use of mag-

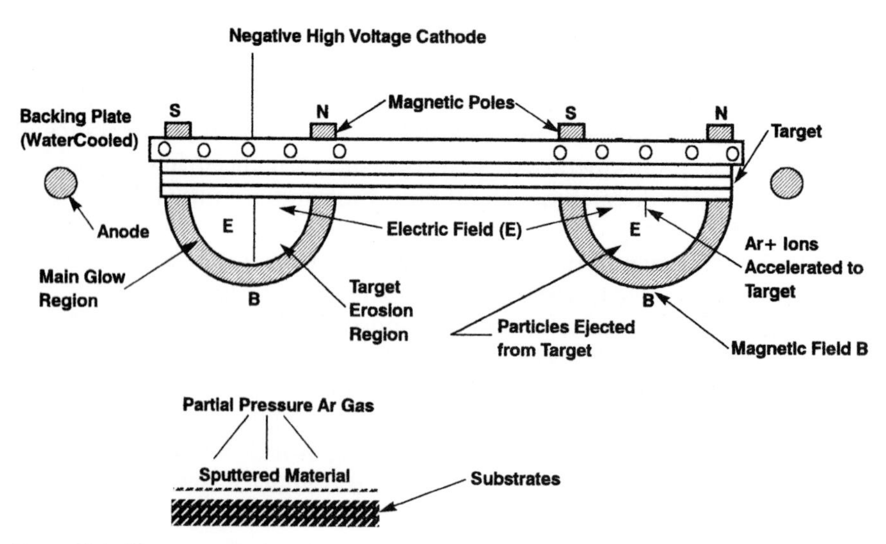

Figure 18.1 Planar magnetron.

netic fields to confine the plasma of backfill gas ions and electrons near the surface of the target.

Low-emissivity (Low-E) glass has a special surface coating to reduce heat transfer through the window. These coatings reflect 40 to 70 percent of the heat normally transmitted through clear glass, while allowing the full amount of light to pass.

Like all other metals, the aluminum commonly used for reflective glazing also had a very low emissivity, which translated into an insulating value higher than the insulating level offered by existing pyrolytic coatings. Unfortunately, aluminum was rendered useless as a heat reflecting mirror by the overcoat used to protect the metal from corrosion. A search was initiated for a reasonably priced, noncorroding metallic coating that had some transparency.

Gold was being deposited in thin enough layers to become reasonably transparent and affordable. It was marketed first throughout Europe as the first Low-E product that reflected light in a uniform color. However, because of this color's cost, it restricted the product's use to commercial projects because of its obvious color and relatively high expense.

The planar magnetron was extremely successful in enabling the deposition of metal films and many useful reactive compounds. However, the planar geometry of the magnetron cathode resulted in some significant processing limitations, especially for the reactive deposition of heavily insulating materials. Because the erosion zone is located in a ring pattern, areas outside of the ring are not sputtered and can build up insulating reactive compounds, when target materials are sputtered in nitrogen or oxygen. Most of these oxide compounds are nonconductive. Therefore, they can build up an electric charge from the plasma until the point at which the materials break down with a destructive arc. The process has kept sputtering from being applied to highly insulating materials, such as silicon oxide or nitride, which have desirable optical or durability properties.

In 1989, BOC Coating Technology introduced a new magnetron cathode geometry, the rotating cylindrical magnetron cathode, or C-MAG cathode. The rotating geometry ensures that all surfaces of the target material are sputtered, eliminating the ability of reactive layers to build up on the cathode. Figure 18.2 shows a schematic of the C-MAG cathode configuration. Currently, the common configuration for this source is two rotating tubes in the place previously occupied by a single planar magnetron.

A batch process was used to coat the glass, because the high vacuums required for sputtering could be pulled only in perfectly sealable, noncollapsible chambers. Costs were reasonable because the metal sources in the chambers did not require changing with each load of

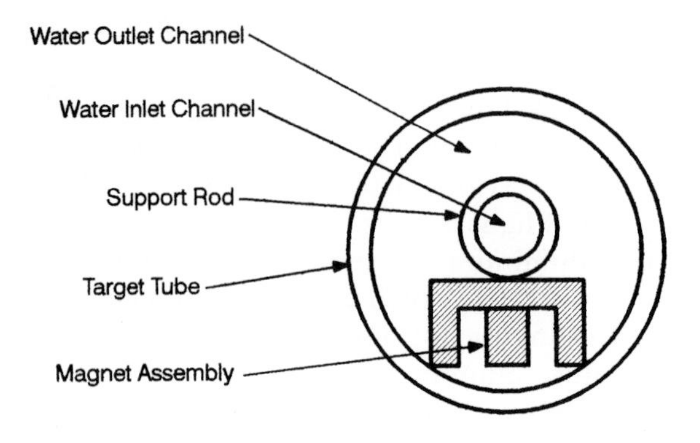

Figure 18.2 C-MAG rotatable cathode cross section.

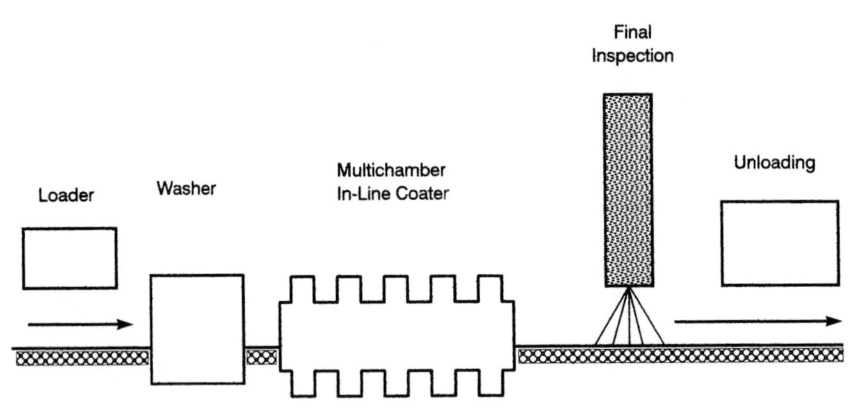

Figure 18.3 Plant schematic-flow chamber. Schematic of an in-line plant for manufacturing Low-E coatings.

glass. It was known that a multilayer stack composed of an antireflection coating, an ultrathin silver layer, and an antireflection coating deposited on a glass would produce a very low emissivity, transparent coating. Figure 18.3 demonstrates the flow through the multichamber unit.

Low-Emissivity Coatings

Conventional Low-E coatings are high visible transmission products designed for residential application to increase the insulation value of double-glazed units in the cold climates. These coatings utilize a thin layer of silver, dereflected by transparent oxide layers, to reflect infrared energy while maintaining its high visible transparency. In addition, a thin sacrificial metal is deposited between the silver and top oxide layer to protect the silver. The region of the infrared front, 5 to 50 μ in wavelength, is in the region in which a heated room will radiate thermal energy. In this region, a typical low-emissivity coating will reflect from 85 to 95 percent of the thermal energy back into the room.

For reducing heat loss, the Low-E coating is typically glazed on the number 3 surface of the insulating glass unit, the number 1 surface being the outside surface of the lite outside of the building. However, the ability to reflect infrared energy can be adapted to the commercial solar control market to provide products that combine higher visible transmission and lower reflectance. In solar control applications, such as commercial architectural use, the Low-E can be glazed on the number 2 surface. This position takes maximum advantage of Low-E reflectance in the infrared portion of the solar spectrum.

In cold climates, such as Canada and Northern Europe, energy conservation regulations are typically written to include a solar gain component based on the glazing direction of insulating glass window units. It took several more years for the companies to shake the bugs out of their proprietary mounting systems. Gerlinger went on to manufacture a triple-glazed unit with an interior plastic film to give the highest insulation value of all. See Figure 18.4.

Meanwhile BOC Coating Technology in the United States, Pilkington in England, and Leybold-Heraeus A.G. in Germany were thinking about how to continuously lay down multilayer Low-E coatings on glass. The major problem lay in getting the glass sheet through the seals into the vacuum chamber and pumping the air out of the chamber fast enough to maintain continuous production. The answer was moving the glass through a series of vacuum chambers. Each in-line vacuum chamber illustrated in Figure 18.5 reduced the pressure around the advancing glass more until it reached a high vacuum in the magnetron coating chamber. Then, an equal number of chambers were used to ease the glass out to the atmosphere at the other end.

In effect, a sheet of glass was quickly jerked through a series of air locks on its way to the coating chamber because any one chamber did not have to remove much air. Thus, the pressure difference over each

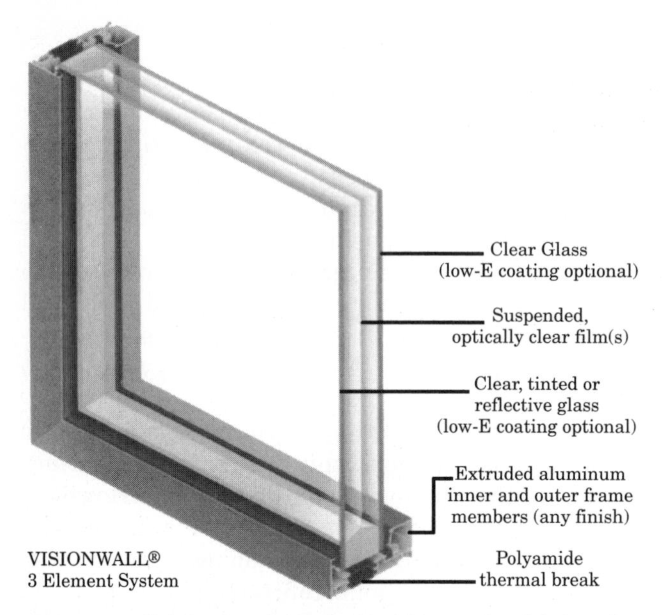

Figure 18.4 Triple-glazed insulating glass unit with interior plastic film. (*Courtesy of Visionwall Technology.*)

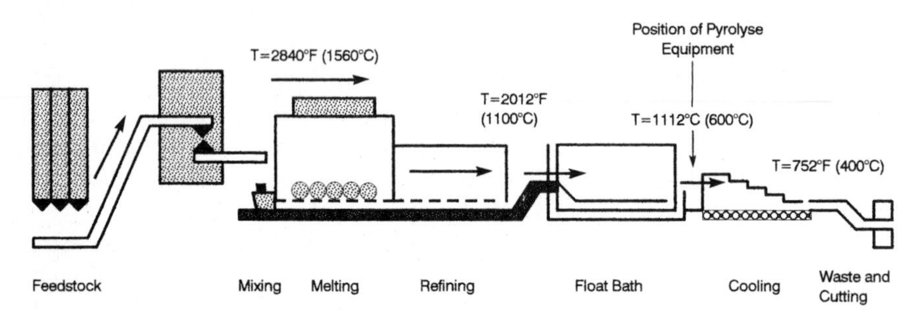

Figure 18.5 Pyrolyse equipment for Low-E.

of the in-line seals was minimal. Any number of coating sections were placed in a row to produce the multilayered stacks. As with the Low-E coatings on polymer films, these coatings would also corrode the coated glass sealed in double-glazed units.

3M quickly patented a means of preventing corrosion in a Low-E silver stack on polyester film. Ultraviolet-stabilized polypropylene was used to overcoat the Low-E material because the polymer was transparent to the impinging infrared radiation. The silver in the Low-E stack could continue to work as a reflector to the infrared, which reached it through the polypropylene. The thermal performance was not as good as the performance associated with the uncoated stacks used in insulating glass units, but the Low-E coated film worked wonders when adhered to single lites of glass.

The timing of the introduction of these products was perfect. Official publications began to educate the architects, designers, and the buying public about the mysteries of Low-E insulating glass units.

The Next Generation of Coatings

Adding a Low-E surface to a sealed insulating glass unit reduces radiant heat transfer across the air gap by more than 90 percent. However, heat still moves in the air sealed between the two plates of glass. Replacing the trapped air with a more viscous, insulating gas lowers this traffic.

Germany began producing argon-filled insulating glass units, which increased the window's insulation level by 25 percent over an air-filled Low-E unit, shortly after they introduced Low-E windows. Argon is an inert gas that is more viscous and less heat conducting than air. More important, it is an inexpensive gas used in the metal welding industry. Refer to Chapter 16 for a detailed study of inert gases. Sulfur hexafluoride was also used as an insulating gas because of its sound-control properties. The gas addition raised the insulation value of an insulating glass, Low-E unit by as much as 30 percent. There was concern that the window's insulation value would decrease as the argon fill permeated through the organic seals over the years, but low insulation values were noticed even after the seals were improved by changing the edge configuration to a dual-sealed unit. The problem was improving filling because too much air remained in the sealed unit after the argon injection. Several proprietary filling techniques have since been patented to overcome this difficulty. There is still some concern over the permeability of sealants, but the higher-quality double seal used today does an adequate job of containing the series of gases during the warranty period of the units.[2]

Krypton is an even better insulating gas, but it has remained expensive because it is difficult to wrest it from the atmosphere. Nevertheless, Lasituku in Finland was the first to fill insulating glass units with the high-performance gas. The gas fell out of favor because of its cost. Now interest is turning to gas fills again, as the demand for improved insulating glass windows increases.

Three spectacular advances signaled the next generation of Low-E coating and glazing systems. These technical advances heated up the competition and made Low-E coatings more popular in the building trades.

First, Pilkington and Libbey–Owens–Ford in the United States announced a pyrolytic coating that rivaled the thermal and visual performance of the sputtered silver-based coating. The emissivity was lowered from 0.40 to nearly match the silver coating's low emissivity of 0.10. This meant that both coatings produced the same high insulating values when applied to glass. During this time, UK building regulations were changed to define Low-E coatings as treatments with an emissivity below 0.20. At this time, tin oxide was formed by chemical vapor deposition. Economy was improved as well, because the new process allowed faster production line speeds. Of course, the new pyrolytic treatment would not corrode, so single-glazing applications were still practical.

Then, BOC Coating Technology announced a sputtered silver coating, which significantly lowered both the emissivity (by a factor of 4 over other commercially available sputtered coatings) and the solar heat gain (for air-conditioning applications), while maintaining high daylight transmission. The emissivity drop raised the bulk insulation value of an insulating glass by 30 percent, compared with the new pyrolytic coating, and by 10 percent compared with the usual sputtered silver coatings. The drop in solar heat gain drew the window even with reflective glazing in heat rejection but without suffering the high daylight losses that come with reflective glass. Durability was also improved to the point where no special handling was required during production.

The coating performance was improved by adding a second heat-reflecting layer of silver separated by an improved antireflection layer. The extra layers added a measure of complexity, which increased cost. Ironically, the additional thickness caused a color change from neutral to a uniform, light green with increased viewing angles. However, vacuum-based coating technology is controllable enough to expect that this drawback will quickly fade.

At the end of 1989, Southwall Technologies became the first North American roll coater to set up volume production of krypton-filled units. They suspended two of their 0.10-emissivity polyester films in a new insulated glass unit forming three krypton gas spaces, which yielded the lowest bulk U-value yet for glazing at 0.12 Btuh/°F ft^2 (0.68 W/K m^2). Normally, uncoated double glazing has a U-value of 0.48 Btuh/°F ft^2 (0.72 W/K m^2). Daylight transmission was kept at a respectable 62 percent with a shading coefficient of 0.52. Krypton was used because its insulation value peaks at a narrow gap dimension near ¼ in (6 mm). This meant that the overall insulating glass thickness could remain nearly the same as ordinary insulating glass units. Southwall uses an uneven gap spacing of 0.312–0.125–0.312 in (8–3–8 mm). Like Gerlinger AG in Switzerland, they also designed the unit with a substantial thermal break at the edge seals to prevent any chance of condensation forming at the glass perimeter near the metal edge spacers.

A wide variety of Low-E coatings have been developed to meet the demands to optimize performance for visible transmission, low emissivity, high or low shading coefficient, coating durability, etc. The various coating stacks currently in production are shown in Table 18.1.

The cost of a framed unit was about 50 to 75 percent more than the cost of an ordinary, installed, insulating glass, Low-E unit and frame (or, for the insulating glass unit without the frame, four times more than an air-filled, Low-E unit), mostly because of the high cost of krypton, which is currently about 10 times the price of argon. Both the American and Swiss window units can generate a net seasonal heat gain from just northern light in severe winter climates.[3]

The BOC advance means that emissivities are now near their theoretical limit, and solar heat rejection is within sight of its theoretical limit. The only path left for heat loss through the bulk of a window is via conduction and convection, but now this loss has now been minimized by the switch to krypton.

TABLE 18.1 Structure of Standard Coatings.

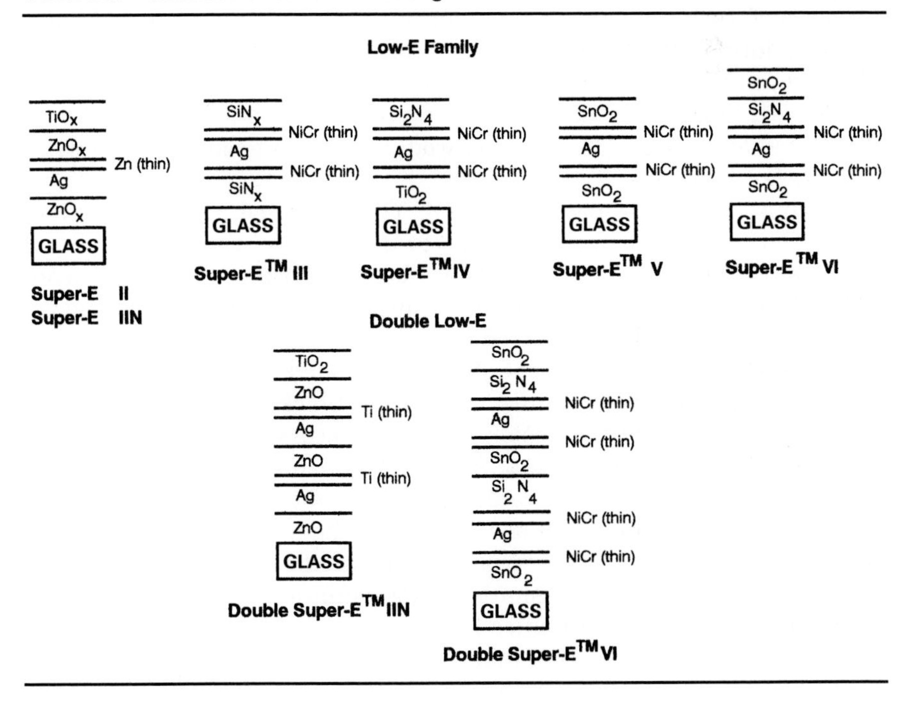

How Low-E Works

There are really four types of Low-E treatments for insulating glass units and windows, but only one of them is based on the ultrathin, antireflected silver mirrors. The see-through silver mirror works as a radiation trap by reflecting thermal radiation while remaining transparent to daylight and, in some cases, to the incoming solar heat. The other three types of coatings work on different but related principles explained later.

The greenhouse effect myth

Why do we need this high technology radiation trap? Ordinary glass is supposed to do that. Greenhouses become warm because of the greenhouse effect, in which the shortwave solar energy can enter, but the absorbed solar heat that reradiates in the form of longwave infrared radiation is trapped. We need the Low-E coatings because ordinary glass does not act as a radiation trap. It acts only as a convection trap.

Although the greenhouse effect, indeed, is responsible for keeping the planet warm, it is not at work in greenhouses. To understand how the greenhouse effect works in buildings means seeing how infrared heat propagates. Infrared heat is really like light—it travels in straight lines, reflects from shiny surfaces, and is absorbed by most other surfaces. Any object that is warmer than its neighbor will emit this radiation, much like a candle broadcasts light in all directions, but the infrared radiation that emits from any object that does not glow with light has a much longer wavelength than light. So, if you could see far infrared "light," you would see an unfocused scene with very fuzzy shadows.

Ordinary glass has been incorrectly characterized as a far-infrared radiation trap because it is opaque to the infrared radiation emitted from warm building materials. Glass loses its transparency to wavelengths longer than those contained in the solar spectrum, as shown in Figure 18.6, but here infrared opaqueness means black.

In fact, all materials (except shiny, unfinished, polished metals and Low-E coatings) are a highly absorptive black in the far-infrared regime, regardless of their color. The glass necessarily acts as a radiation sponge in the presence of reradiated far-infrared, only to remit and convect the soaked up radiation at the outside surface after the energy readily warms the glass so that the heat can conduct through the dense glass. This long-wavelength infrared radiation is formed in greenhouses when the transmitted solar radiation strikes any opaque surface, as illustrated in Figure 18.7.

The surface absorbs the short-wavelength energy and heats up until the incoming solar energy equals the outgoing energy. Slightly

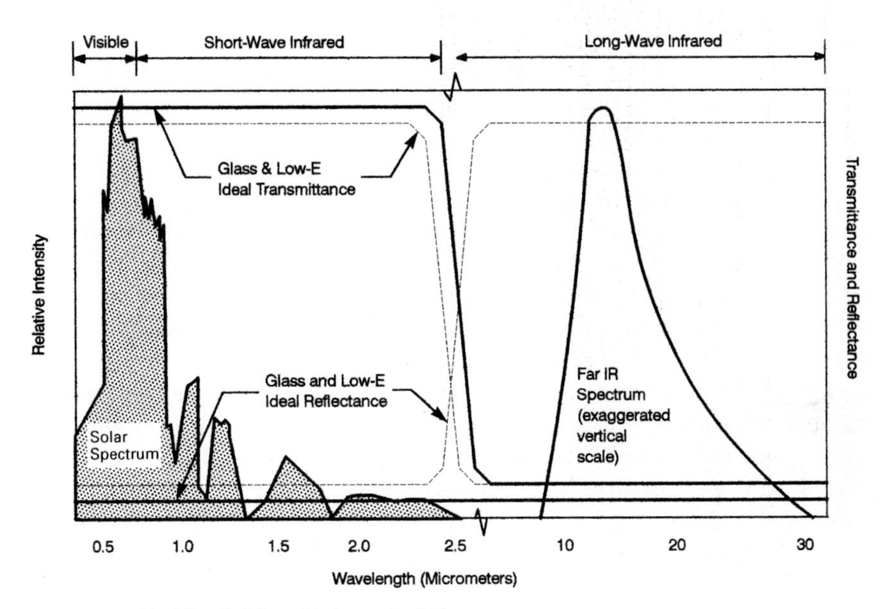

Figure 18.6 Residential Low-E characteristics.

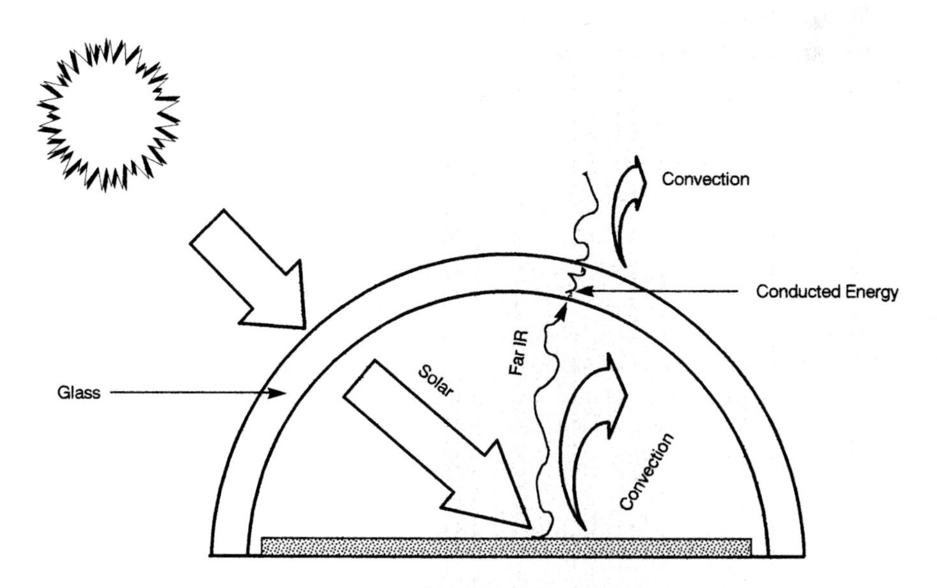

Figure 18.7 Long wavelength in IR.

more than half of this outgoing heat is far-infrared radiation, and the remainder is energy that is wafted into the air via convection. The warm greenhouse air also convects some energy to the glass, which also conducts through the glass to the outside surface.

High solar transmission, Low-E glazing is the first true radiation trap. It behaves like ordinary glass was thought to respond to its high longwave infrared reflectivity which bounces the thermal radiation, originating in the space, back into the space. Since less radiation gets out, the net radiation goes up, even after accounting for the reduced solar transmission of the Low-E coated glass which results in an even warmer greenhouse.

Figure 18.6 shows the entire process. The white spike is the incoming solar energy, of which half the energy lies in the visible region; the gray bump is the reradiated far-infrared. The top dark solid curve shows how well ordinary glass transmits solar energy at various wavelengths, and the bottom dark solid curve shows how the glass reflectance remains low at all wavelengths. The gray-colored, dashed curves show the behavior for a perfect Low-E coated product. Notice how the material acts as a selective reflector—solar transmission is still quite high but solar reflectance is low. Then, the reflectance increases for the far-infrared wavelengths while the transmission goes to zero. Low emissivity in the presence of low transmission means high reflection. Low emissivity also means that any energy held in the coating has a difficult time leaving as radiation—the energy cannot emit from the surface, hence the name of the product.

Thick-film principles

Two of the other three types of Low-E glazings are made from semiconductors. Whereas the above films require multiple layers of ultrathin material, transparent semiconductor films can be made in a single, relatively thick layer (several millionths of a meter thick), which avoids the quality control issues associated with thin films. Some of these transparent semiconductors are tin oxide, indium oxide, cadmium tin oxide, and indium tin oxide. The first type of thick film is deposited on warm glass in a vacuum. The second type is sprayed or chemically deposited on hot glass in atmospheric conditions.

Microgrid principles

The last type of selective reflector is mechanically produced. Metal sheets are etched to create openings of approximately a micron, or a millionth of a meter (2.5 μm), to allow solar radiation to pass through, but not the longer-wave infrared, assuming that the widths of the lines are not too narrow to prevent IR reflection. The technique is sometimes used to improve the performance of thick films, such as indium tin oxide, but the technique remains a laboratory curiosity because of its cost.

Thin-film pros and cons

Each of the four Low-E treatments has strengths and weaknesses. The sputtering process used to lay down thin films on glass demands a high degree of control over the various metal deposition rates to maintain the film's thickness at 0.02 mm. On the other hand, this amount of control makes it easy for the operator to change the film's composition with the twist of a dial. A full spectrum of shading coefficients for different climates and applications becomes possible because of this flexibility. The vacuum magnetron sputtering machinery, which deposits both the metal and the two antireflection coatings, is precise enough to guarantee uniformly thick layers, in turn assuring a uniformly colored coating.

Silver is used as a base because it has the best solar transmission. The antireflection coating has historically been titanium or indium oxide, but now zinc tin oxide is emerging as the material of choice because it more readily resists ultraviolet attack and has much harder and improved abrasion resistance during assembly.[4]

Product performance

The major advantages of Low-E coatings are their climatic adaptability and increased thermal resistance. The coating's shading coefficient and daylight transmission are easily tuned for different climates during the deposition process. This flexibility has inevitably led to so many offerings and variations that determining the right combination of coating and glazing configuration has become confusing.

All Low-E products save energy when compared with similar uncoated products because of the higher insulation value imparted by the coating. Low-E coated insulating glass was initially attractive because it minimized condensation problems at the center of the glass and outperformed triple glazing. The initial coatings lowered the winter nighttime U-value of insulating glass with a ½-in (12-mm) air space to 0.31 Btuh/°F ft^2 (1.76 W/K m^2), which is about 26 percent better than triple glazing fabricated with the same air-gap dimensions. Today, Low-E glazing thermal performance has moved far beyond this, to the point where U-values are low enough to noticeably increase the occupant's thermal comfort in rooms with large windows.

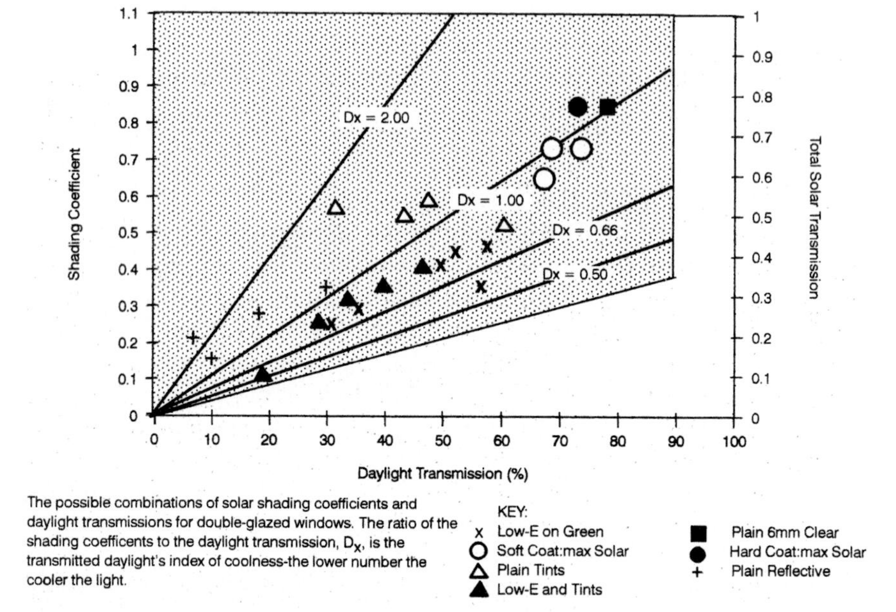

Figure 18.8 Combinations of shading coefficients and daylight transmission.

Light and Heat Performance Criteria

Performance is best evaluated against a set of absolute standards. Because Low-E products are beginning to reach the limiting laws of physics, it is interesting to measure various products in terms of their ultimate performance bounds.

Figure 18.8 shows the combinations of solar shading coefficients and daylight transmissions possible for sealed, insulating glass units. The corresponding total solar transmission, the fraction of solar radiant heat at a normal incidence that is transferred through the glazing by direct transmittance and inward-flowing absorbed energy, is given on the opposite axis. The shading coefficient is also given as the total solar transmission divided by 0.87. All insulating glass units must lie inside the shaded areas. The upper right-hand corner of the shaded area represents two lites of high-transmission, 3-mm water-white glass. Water-white glass does not appear green even at its edges because of its extremely low iron content. The shaded area edges that slope upward from zero show the minimum amount of inward-flowing solar heat that can possibly accompany a given amount of daylight, assuming all of the near-infrared component is rejected. For instance, the graph shows that the lowest possible shading coefficient for a product with 50 percent daylight transmission is 0.18.

Products used for solar heating should lie near the shaded area's top right corner of Figure 18.8. Windows used for light and heat control should lie near the bottom sloped line. The single square near the top right corner shows the performance of ordinary 6-mm double glazing. The circles near the top represent some of the Low-E products designed for maximizing solar gain and day lighting in residences. The four triangles in the middle occupy the range covered by ordinary tinted glazings, which is limited and far from the shaded area edges.

The ratio of the shading coefficient to the daylight transition Dx is the transmitted daylight's index of coolness—the lower the number, the cooler the light. Four reference lines for Dx = 2.00, 1.00, 0.66, and 0.50 are shown as dashed lines radiating from the origin of Figure 18.8. The lowest possible Dx value is 0.36, which forms the bottom edge of the shaded area. Figure 18.8 shows that reflective insulating glass generally has the highest Dx values, followed by lower values for tinted insulating glass units, clear units, and green insulating glass units. A Low-E coating added to the inside of the outside lite or a Low-E film suspended in the air space lowers Dx dramatically. The lowest Dx value of 0.61 corresponds to the new, double-silver, Low-E coating on a light-green insulating glass unit.

Coolness index performance comparisons

The energy use dependencies on Dx and glazing U-value for offices in both hot and cold climates were assessed by the Windows and Day Lighting Group at Lawrence Livermore Laboratory.[5] In cold climates, where cooling loads are low, usually a lower window U-value interferes with free cooling via conduction out the windows, particularly on the south and west sides, where solar impact is high. However, the Low-E products showed cooling load decreases because their lower Dx values compared with their uncoated counterparts. Northern exposures depended little on Dx or U-values. The study also found that incremental energy savings diminish as Dx decreases; that is, changing from a bronze tint to a Low-E coated bronze tint is much more effective than going from a green tint with an already low Dx to a green tint with a Low-E coating. For heating applications, a higher Dx indicates warmer daylight, which means more solar heating. Surprisingly, the study found that a lower U-value saves more energy than a higher Dx.

The same general trends for cooling loads in a cold climate were found in a hot climate. The largest savings in cooling energy are attained by changing from a glazing with a high Dx (2.0) to fenestration with a moderate Dx (1.0). However, greater savings are possible at low Dx values when using large apertures.

U-value measures

The last germane measure is the window's ultimate U-value. A U-value of 0.0 is theoretically obtained by evacuating an insulating glass, Low-E unit, assuming that the coating's emissivity is zero. Several realities increase the U-value. Real emissivities do not fall below 0.03, real insulating glass edges must be sealed, and practical double glazing must resist the atmosphere's formidable pressure with internal compression supports. The sealed edges and supports will always cause thermal bridging. The supports must become slimmer as they become more numerous or thicker as they become less populous. The resulting thermal short circuits suggest a limiting U-value of approximately 0.033 Btuh/°F ft^2 (0.19 W/K m^2) over a normal insulating glass thickness, assuming that the supports slim down to granular microcells. The ultrahigh-performance Low-E windows with four glazings, two Low-E coatings, and krypton gas fills had reached U-values as low as 0.125 Btuh/°F ft^2 (0.68 W/°K m^2). The highest U-value of 0.32 Btuh/°F ft^2 (1.82 W/K m^2) for a Low-E unit is for an insulating glass, air-filled window.

U-value variations with IG unit design

A window's U-value is altered by more than the Low-E treatment. The following graphs show how various treatments and geometries determine the window's resistance to heat flow.

Emissivity effects

Figure 18.9 shows how emissivity alters the U-value for single-, double-, and triple-glazed insulating glass windows, when the $\frac{5}{32}$-in (4-mm) lites are spaced $\frac{15}{32}$ in (12 mm) apart.

The curves do not extend all the way to the right because uncoated glass has an emissivity of 0.84. Low-E coatings change single glazing's U-values the most, but only in climates where condensation is not a problem (water clinging to a Low-E surface cancels out its low emissivity due to the water's high emissivity). Note that the U-value for single glazing nearly matches the U-value for uncoated double glazing when the single glazing's emissivity is 0.02. Reducing the already low emissivity from 0.15 to 0.02 lowers the U-value for double glazing by about 23 percent.

Gas fill effects

Filling insulating glass units with exotic gases significantly lowers the U-value. Figure 18.10 shows the effect of gas conductivity on heat flow through double and triple glazing.

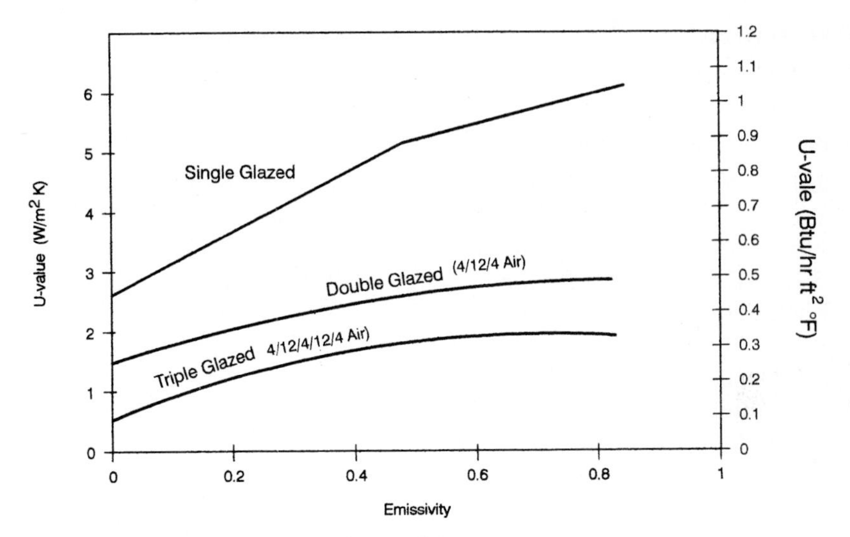

Figure 18.9 U-value variation with emissivity.

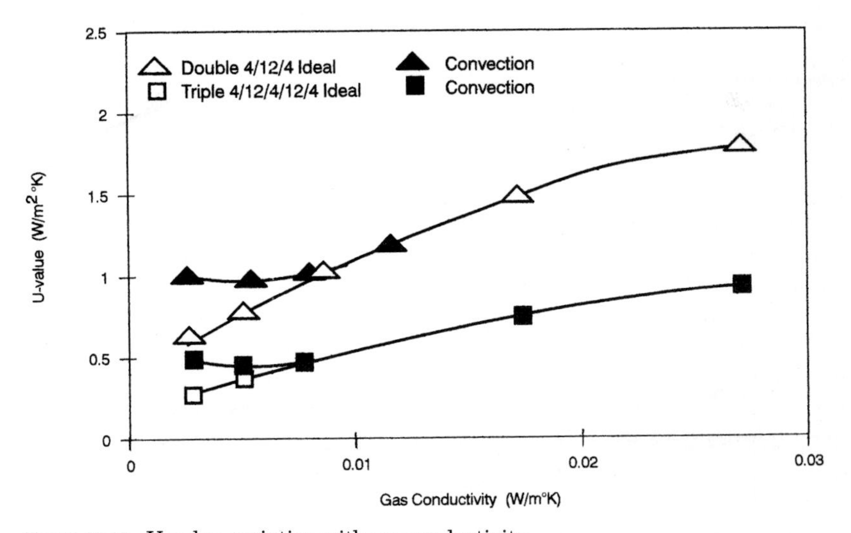

Figure 18.10 U-value variation with gas conductivity.

The curves not filled with data points show idealized behavior when there is no convection. In reality, convection is always present. The extension lines with solid data points show how convection losses overshadow the inconsequential heat flow of a low-conduction gas and raise the idealized U-value in $^{15}/_{32}$-in (12-mm) gapped units. For reference, the conductivity of air is 0.024 W/m^{2}K, argon is 0.016, and krypton is 0.0085.

Multiple air-space effects

Adding more lites creates more dead air spaces and added insulation but at the ultimate expense of light transmission. Figure 18.11 shows how the wintertime U-value decreases with the number of ordinary and Low-E coated lites for air-, argon-, and krypton-filled units.

Each additional pane is spaced 0.5 in (12 mm) away from the next lite. The thick gray curve shows how the Low-E daylight transmission falls at a faster rate than the U-value after the third lite is added. The thinner gray curve represents the uncoated lites. The shading coefficients are also shown as dashed curves for the same materials. Extra lites usually increase the volume of trapped gas, which can cause ruptured edge seals when the gas expands because of ordinary summer temperatures or reduced atmospheric pressure at a site significantly higher in elevation than the manufacturing site. Using krypton minimizes this problem because the gas gives the best insulation when the gap dimensions are the narrowest.

Edge geometry effects

The last variable that significantly affects the performance of an insulating glass unit is the edge geometry. The metal edge spacers used to seal most insulating glass units can seriously raise a window's effective overall U-value, depending on the window's size and nominal U-value.

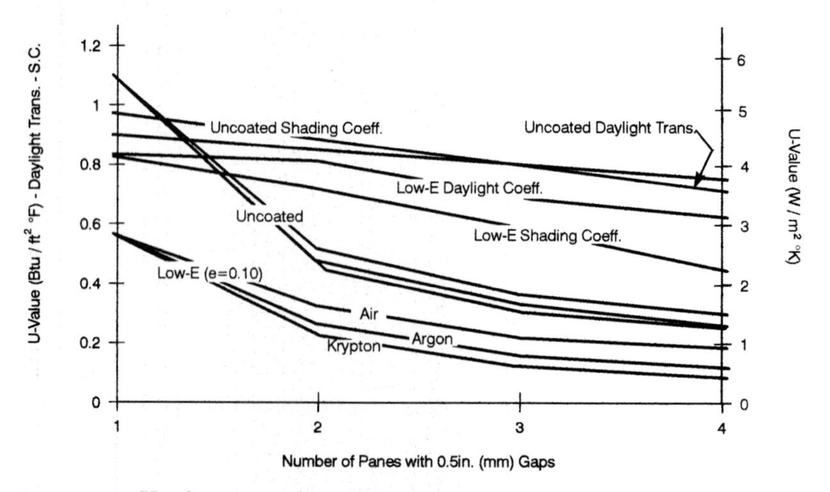

Figure 18.11 U-value vs. number of lites—winter.

The amount of heat lost through the highly conductive metal edge spacer is not significant compared to the loss through the central portions of insulating glass. The situation is reversed for an R-value near 5°F ft^2/Btuh (0.88 K m^2/W) to the point where 25 percent is lost through the edges rather than through the bulk of the insulating glass unit.

Pyrolitic Treatment

A pyrolytic treatment is a virtually colorless, thick coating, usually produced by enveloping hot glass with inorganic vapors. The latest process produces a stable, hard surface without iridescence. The other advantages of pyrolytic Low-E coatings are as follows:

1. ultrahigh operating temperatures over 572°F (300°C) (for solar collectors or industrial uses, such as windows that overlook high-temperature manufacturing areas),
2. highest insulating glass shading coefficient at 0.79 (for solar heating applications) with the same daylight transmission as thin coatings,
3. theoretically, the most economical Low-E product to manufacture,
4. corrosion proof (this is necessary for single glazing and vented, insulating glass units),
5. abrasion proof (for easy cleaning and long life),
6. simple installation (the coated glass handles and glazes like ordinary glass because edge stripping is not necessary),
7. posttempering possible, and
8. good color rendition.

Limitations

The so-called hard coat product offers inherently less variety because the deposition process gives less control over the coating's thermal and optical characteristics than the vacuum sputtering process used for depositing thin coatings.

Other disadvantages include the following:

1. intermediate emissivities of 0.17 to 0.20 produce slightly higher U-values than the thin coatings and
2. slight surface haze (this effect becomes visible in bright sunlight).

A thick coating is a transparent semiconductor, usually indium oxide, cadmium tin oxide, or indium tin oxide, which is deposited at high temperature on glass as a single, relatively thick layer (several millionths of a meter thick). The thick layer avoids the quality control issues associated with thin coatings and pyrolytics, but actually ends up costing more than even the multilayered thin coatings because the semiconductor material is expensive and relatively large amounts of material are required. Thick coatings are not in common use yet because they are expensive. Some coatings add a slight color to the glass. Nevertheless, the material is worth waiting for because of the following:

Advantages

1. corrosion proof (for single glazing and vented, insulating glass units),
2. abrasion proof (for easy cleaning and long life),
3. ultrahigh operating temperatures over 572°F (300°C),
4. handles and glazes like noncoated glass, and
5. lower emissivities than the pyrolytics.

Low-Emissivity Coatings for Winter Climates

For the residential market, very high visible transmissions, at least 75 percent for $^{5}\!/_{32}$-in (4-mm) glass, an insulating glass unit, and a neutral color are required. The basic product for heating-dominated markets is a single silver layer coating, using either tin oxide or zinc oxide to dereflect the silver. These products typically achieve a visible transmission of 84 to 87 percent, with solar transmission above 60 percent. Emissivities of 0.10 are typical for tin oxide-based coatings. Zinc oxide-based coatings achieve lower emissivities of 0.08 or better. Corresponding U-values for insulating glass units with ½-in (12.7-mm) argon-filled gaps are 1.5 W/m^2-C° or less. Due to the high solar transmissions, shading coefficients are high at 0.75.

The typical visible and thermal performance factors for these coatings are shown in Table 18.2. The net nighttime heat loss, when there is no solar gain available, is reduced considerably from 108 W/m^2 to 70 W/m^2. (All thermal performance values presented are center of glass calculations performed using Windows 4.1.)[6]

Highly durable products have also been developed which replace these oxide layers with silicon nitride or combinations of silicon nitride and titanium dioxide. These are extremely corrosion- and abrasion-resistant materials. These products achieve similar thermal performance with greatly enhanced resistance to abrasion, humidity, salts, acid, and alkaline attack. The deposition of materials, such as silicon nitride, is made possible by the use of the rotatable magnetron source. Typical coating stacks utilizing Si_3N_4 are shown in Table 18.3, and the corresponding thermal performance values are shown in Table 18.4.

TABLE 18.2 Base Single Low-E

Parameter	S-E II IN	S-E V	S-E VI
T Vis	87	84	84
R film Y	5.1	6.0	4.0
a*	−4.0	−2.6	−0.5
b*	−7.7	−4.1	−3.4
Rg Y	6.0	5.0	5.0
a*	−4.0	−2.7	−1.1
b*	−10.1	−7.6	−7.6
T sol	66.3	61.8	61.2
Rf sol	20.0	20.0	20.8
Rg sol	16.9	15.2	15.6
E normal	0.073	0.098	0.096
U-value (Ar)	1.45	1.54	1.53
SC	0.78	0.75	0.75

TABLE 18.3 Silicon-Based Coatings

Super-E III
Glass/Si_3N_4/NiCr Ag NiCr/Si_3N_4
Super-E I
Glass/TiO_2/NiCr Ag NiCr/Si_3N_4

TABLE 18.4 Thermal Performance of Silicon-Based Coatings

	Super-E III	Super-E IV
%T	76	82
%Rg	4	6
%Rf	8	4
%T sol	62	63
%Rg sol	12.4	15.5
%Rf sol	14	19
Em	0.16	0.10
K(Ar)	1.92	1.51
SC	0.78	0.78

Low-Emissivity/Low-Shading Coefficient Glazings

To provide thermal insulation in winter and solar shading in summer requires products with a low emissivity and low shading coefficient. The commercial glazing market can accept products with more moderate visible transmission levels and more color options than the residential market. For commercial applications, coatings based upon modifications of the single silver layer coatings can be used.

The Sunbelt Low-E products are based upon a single layer of silver where either the silver or the protective metal layers are thickened to lower the visible transmission of the coating. As measured on $\frac{5}{32}$-in (4-mm) clear glass, visible transmission can be reduced from a value of greater than 84 percent, for residential applications, to approximately 50 to 60 percent for commercial solar control applications. The Sunbelt series film stacks are based on the use of silver and tin oxide (SnO_2).

The Sunbelt series consists of three products. Table 18.5 shows the visible and thermal performance values for these products. The Sunbelt Neutral coating has a very low visible reflectance and little color and is finding increased used in commercial Low-E applications. This product can be used in markets which restrict the external visible reflectance of glazing systems. There are also blue and silver colored coatings. The silver coating offers the lowest emissivity and U-value and the most solar shading. However, the external visible

TABLE 18.5 Typical Product Optical Specifications

Optical parameter	Neutral	Blue	Silver
Visible transmission	57	60	49
Film reflectance			
Y	4	7	34
a	5	−1	1
b	7	−25	−10
Glass-side reflectance			
Y	10	21	38
a	1	−3	−2
b	−4	−13	−3
Solar transmission	40	45	34
Film-side R	22	20	56
Glass-side R	15	20	50
Emissivity	<0.10	<0.10	<0.05
Shading coefficient (6-12-6 IGU)	0.51	0.47	0.38

reflectance level is more comparable to the brighter, stainless-steel-based, solar control coatings.

Although a clear insulating glass unit will transmit 75 percent of the available solar radiation, an insulating glass unit with the Sunbelt Neutral product glazed on the number 2 surface will transmit only 40 percent of the solar radiation. Hence, a low shading coefficient can be achieved with a high visible transmission product, with minimal visible reflectance to the outside of the building (less than 15 percent for an insulating glass unit).

Even better performance can be achieved with a more sophisticated coating design utilizing two silver layers. "Double Low-E" is basically two Low-E film stacks superimposed. This coating provides the highest performance combination of high visible transmission, low U-value, and low shading coefficient (see Table 18.6). As a result, it is now a major Low-E product in the U.S. market and is growing in the European market.

Emissivities of less than 0.05 (i.e., infrared energy reflection levels greater than 95 percent) are typically achieved, resulting in U-values below 1.4 for argon-filled units. Shading coefficients of less than 0.5 can be achieved. As a result, an insulating glass unit glazed with double Low-E on the number 2 surface reduces the transmission of solar energy from 75 percent for a clear glass insulating unit to 41 percent (see Figure 18.12). At the same time, visible transmission for the sealed insulating glass unit is more than 70 percent. This makes the double Low-E coating suitable for both commercial and residential markets. For winter applications, a double Low-E insulating glass unit filled with argon gas can achieve a U-value of less than 1.4W/m^2 for a $\frac{5}{32}$-in glass (4-mm glass, $\frac{1}{2}$-in [12.7-mm] air space).

The use of two silver layers results in a coating system in which visible appearance (reflectivity and color) are much more sensitive to physical thickness variation of individual coating layers. Computer optical modeling has been used to determine the physical thickness uniformity

TABLE 18.6 Zinc Oxide-Based Double Low-E

Glass/Zn/Ag Ti/ZnO/x/ZnO (where x can be $TiSO_2$, SnO_2, or Si_2N_4)				
T vis		75%	T sol	41%
Rf	Y	4	Rf sol	34%
	a	−2.3	Rg sol	27%
	b	−2.7		
Rg	Y	4.8	En	0.045
	a	−0.3	U(air) = 1.77	U(air) = 1.39
	b	−0.5	SC	0.47

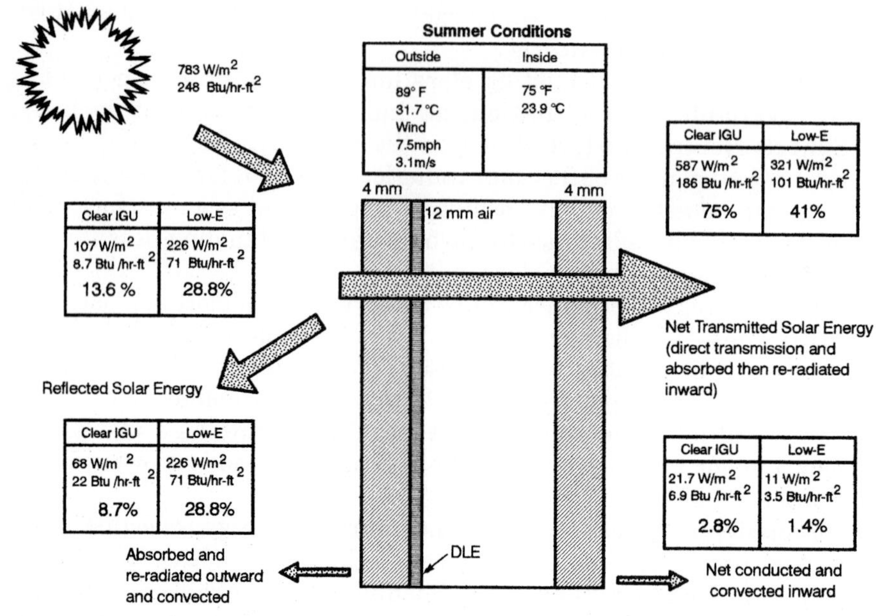

Figure 18.12 Double Low-E summer solar shading.

TABLE 18.7 Comparative Sensitivity Analysis of Zinc Oxide-Based Single vs. Double Low-E*

Parameter	%T	%RfY	a	b	%RgY	a	b
Tolerance							
Single Low-E							
+0.5	7%	9%	<3%	1%	6%	3%	1%
+0.75	10%	10%	4%	<2%	<7%	5%	2%
+1.0	10%	10%	6%	<3%	7%	6%	3%
Double Low-E							
+0.5	2%	3%	<1%	1%	4%	1%	<1%
+0.75	3%	5%	<2%	<3%	4%	2%	1%
+1.0	4%	6%	<2%	2%	<5%	<3%	<2%

*Percentages are thickness tolerances to achieve desired optical tolerance.

required to achieve high uniformity of appearance. By using ellipsometric measurement of optical constants of individual film layers deposited on production systems, we can predict the variation in appearance associated with total layer thickness variation in a coating stack.

Table 18.7 shows the comparative thickness and associated appearance uniformity for a typical ZnO-based single- and double-silver layer coating. This table shows the required physical thickness tolerance needed to control visible transmission, film- and glass-side reflectance to within ± 0.5 to 1.0 and to control reflected color (Hunter a, b units) to within ± 0.5 to 1.0 units. As can be seen, the physical thickness uniformity for a double Low-E production system must be at least twice as exact as a conventional single Low-E system to achieve comparable uniformity. To meet these requirements, modern production systems are designed to achieve better than ± 2 percent physical thickness uniformity.

Advanced Low-Emissivity Coatings

Further advances in low-emissivity performance require the capability to reduce the emissivity, while controlling the shading coefficient for winter heat gain or summer solar shading. This requires the use of thicker silver layers to further reduce the emissivity. However, thicker silver layers alone will typically result in lower visible transmissions and higher reflectance. This can be overcome by utilizing transparent oxides, such as TiO_2, which have a higher index of refraction. This allows the silver layer to be more efficiently dereflected in the visible region. In Figure 18.13, the results of optical modeling of such stacks indicate the increasing index of refraction required to maintain coating appearance (visible transmission and color) as silver thickness is increased to reduce emissivity.

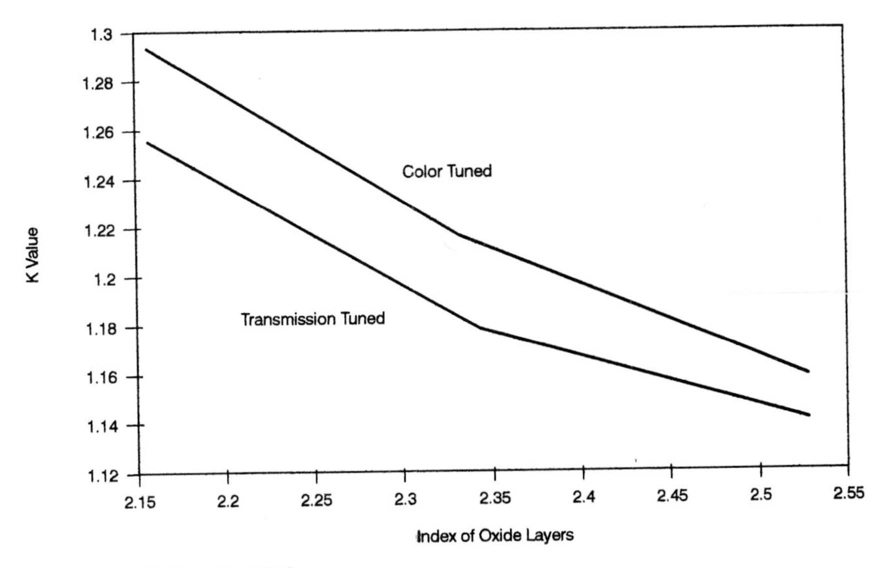

Figure 18.13 Index of oxide layers.

Single silver layers utilizing either an all-TiO_2 stack or combining TiO_2 with ZnO for sputtering efficiency have achieved emissivities of 0.05 to 0.07 with visible transmissions above 82 percent and high solar transmissions. U-values of 1.4 to 1.5 approach the level of the double-silver layer product, and shading coefficients remain high at 0.7. These products are optimized for climates with very cold winters and cool summers, where the lowest U-value and maximum solar gain in the winter are preferable to solar shading in the summer. These products will provide an optimized value of K-effective according to German energy legislation. Table 18.8 shows the coating stacks currently under development and typical performance values achieved with this coating. Figure 18.14 contrasts the solar spectrum of a TiO_2-enhanced, single Low-E to the spectrum of a standard zinc oxide-based single and double Low-E.

Double-silver layer coatings have been computer-modeled based on similar designs. Emissivities of less than 0.02 can be achieved with visible transmissions of 65 to 80 percent. U-values of 1.3 W/m²-°C are achievable. Solar transmissions of 30 to 40 percent result in shading coefficients between 0.35 and 0.50. These products can provide optimum winter U-values and summer shading coefficients. Computer modeling results for these stacks are shown in Table 18.9.

Unfortunately, high-index materials, such as TiO_2, have typically

TABLE 18.8 TiO_2-Based Single Low-E

	Mixed Ti/Zn*	All Ti†
%T vis	82.8	84.4
%Rf Y	4.9	6.5
a*	−0.6	−7.7
b*	−3.3	−10.6
%Rg Y	5.2	7.5
a*	−1.1	−7.4
b*	−9.3	−12.4
T sol	58.9	58.1
Rf sol	22.7	27.7
Rg sol	16.6	21.8
En	0.068	0.051
U-value (Ar)	1.47	1.42
SC	0.74	0.37

*Mixed ZnO/TiO_2 Low-E: glass/TiO_2/ZnO/Ag Ti/ZnO/TiO_2.
†All-TiO_2 Low-E: glass/TiO_2/Ag Ti/TiO_2.

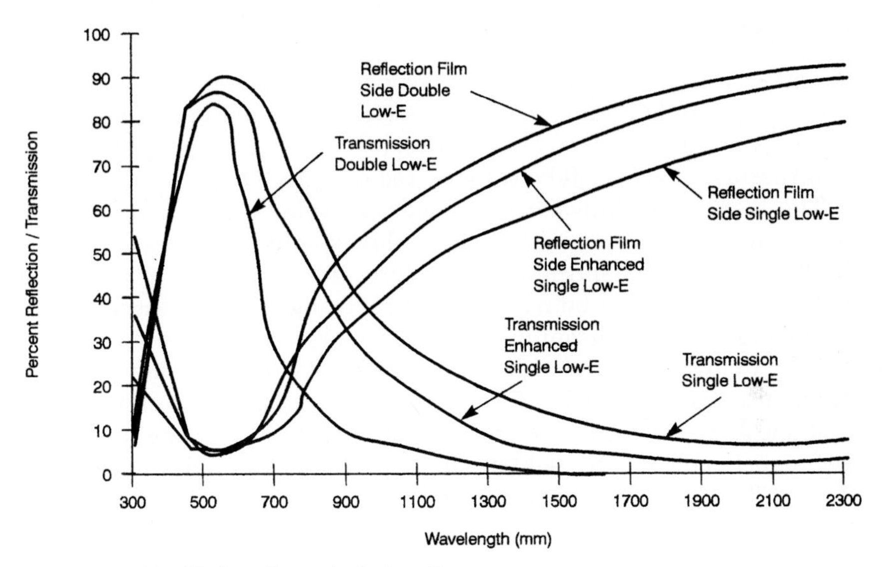

Figure 18.14 Double Low-E vs. single Low-E.

TABLE 18.9 Advanced Double Low-E

Parameter	All TiO_2	All TiO_2	SnO_2/TiO_2
T vis	78	69	65
T sol	42	31	29
Rf sol	40	51	50
Rg sol	30	40	40
En	0.02	0.01	0.01
U-value (Ag)	1.34	1.32	1.32
SC	0.48	0.36	0.34

slow sputtering rates. The use of the rotatable magnetron can enhance the productivity of these systems. However, future work in enhancing the deposition rate of high-index materials will be required to bring the production cost down to that of current coating systems. Closed-loop operation in mixed argon/oxygen atmospheres will be required to achieve high reactive sputtering deposition rates by operating in the transition region of the hysteresis curve between metallic and oxide depositions. Such processes are currently under development for production systems.

Conclusion

Architectural glazings control the flow of solar energy into and out of a building, while adding a broad selection of visual appearance. The glazings act as selective filters, choosing which wavelengths of light and thermal energy will be transmitted or reflected.

The use of low-emissivity coatings for solar control applications opens up the opportunity to achieve low shading coefficients with a higher visible transmission. Therefore, maximum use of available natural daylight can be made, while reducing summer thermal heat loads and winter thermal losses. The Low-E coatings are more selective, providing high visible transmissions with very low infrared transmissions. Figure 18.15 compares the transmission of solar control and Low-E products over the solar spectrum from the ultraviolet region between 300 and 400 nm in wavelength, through the visible at 400 to 700 nm, to the near-infrared between 800 and 2500 nm wavelength. The increasingly spectrally selective operations of these coatings are clearly shown.

Future demands for energy conservation and design flexibility will continue to require more advanced products. The capability to control both solar heat gains and initial losses will continue to be a critical design feature for building construction. Future products will be called upon for higher levels performance by additionally maximizing use of available natural daylight for user comfort and to reduce energy demand for lighting. Table 18.10 charts a variety of manufacturers producing Low-E glass.

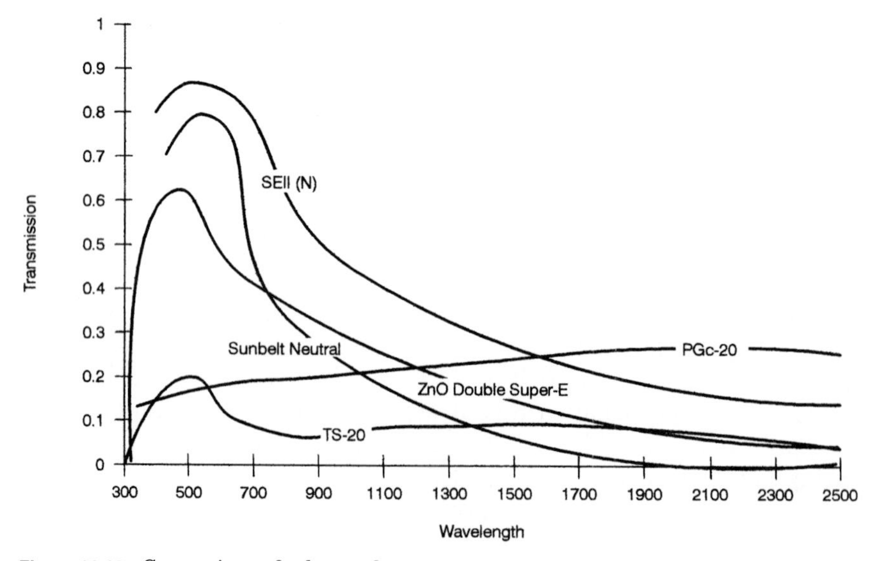

Figure 18.15 Comparison of solar performance.

TABLE 18.10 Low-E Performance Chart*

Manufacturer	Product	Type	Emissivity	U-value	Shading coefficient	Visible light transmission	Solar energy transmission	R-Value
Airco	Super-E™	Sputter	0.10–0.15	0.32	0.71	74%	50%	3.1
AFG	Comfort-E™	Pyrolite		0.39	0.79	74%	55%	2.6
Cardinal IG	LoE	Sputter		0.32	0.64	73%	47%	3.1
Cardinal IG	LoE Sun** (green glass)	Sputter		0.31	0.34	52%	22%	3.2
Ford	Sunglas® HR-P™	Pyrolite	0.35	0.40	0.86	80%	75.5%	2.5
Ford	Sunglas® HR-S™ (1/8 in glass)	Sputter	0.13	0.32	0.81	76%	58%	3.1
Glaverbel	Comfort®	Pyrolite	0.13–0.15	0.33	0.68#2 0.78#3	69%	59%#2 68%#3	2.9
Guardian	Low-E	Sputter	0.10 (less than)	0.32	0.71	74%	52%	3.1
Interpane	I PLUS Neutral R (5/32 in glass)	Sputter	0.10	0.29	0.71	75%	62%	3.4
Interpane	I PLUS Neutral R (5/32 in glass) gas-filled	Sputter	0.10	0.22	0.71	75%	62%	4.5
PPG	Sungate®100	Sputter	0.12 (less than)	0.31	0.74	72%	52%	3.2
PPG	Sungate®200	Pyrolite	0.31–0.35	0.40	0.79	69%	51%	2.5
PPG	Sungate®100 (bronze/glass)	Sputter	0.12 (less than)	0.31	0.49	42%	32%	3.1

*Numbers based on IG unit with a 1/2-in air space and 1/4-in glass unless otherwise noted. Numbers may vary depending on construction of the I-unit, and are based on manufacturers' figures. The chart contains only a portion of Low-E products manufactured. It serves as an informational guideline.

**Preliminary numbers.

References

1. Johnson, T. E., *Low-E Glazing Design Guide,* Butterworth-Heineman, Stoneham, MA, 1991.
2. Glaser, H. J., Coated Heat-Insulating Glasses, *Int. J. Glass Technology,* 2, 1193–99, 1989.
3. Arasteh, D., S. Selkowitz, and J. R. Wolfe, The Design and Testing of a Highly Insulating Glazing System for Use with Conventional Window Systems, *J. Solar Eng., Transactions of the ASME,* 111, 1989.
4. Gillery, F. H., The Significant Advantage of Using Zinc-Tin Oxide to Anti-Reflect Silver in Low Emissivity Coatings, PPG Industries Glass Research and Development Center, Pittsburgh, PA, 1989.
5. Sweitzer, G., D. Aresteh, and S. Selkowitz, Effects of Low Emissivity Glazing on Energy Use Patterns in Nonresidential Daylighted Buildings, Lawrence Berkeley Laboratory Report LBL-21577, Berkeley, CA, 12/1986.
6. Windows 4.1, Lawrence Berkeley Laboratories Report, Berkeley, CA, 1995.

Trademarks

Sunbelt is a registered trademark of PPG Industries, Inc.

Chapter

19

Breather Tubes

Joseph S. Amstock

President, Professional Adhesive and Sealant Systems

Introduction

Breather tubes are made of metal, usually stainless steel or aluminum. There are basically two types:[1]

1. wide-bore (0.063-in inside diameter) tubes used for shipping purposes to reduce pressure effects. These are sealed or removed prior to installation.
2. narrow-bore (0.021-in inside diameter) tubes which are left open to reduce stresses during the service life of the units.

Capillary breather tubes, typically manufactured of stainless steel, are about 12 in long with an inside diameter as listed above. The selection of this length and diameter is based on two different scientific theories. These chemical theories are very involved. The theories build upon the concepts of free mean path and kinetic diameter of moisture vapor pressure. The diameter of the tube is small enough to allow the pressure to equalize rapidly between the outside and the inside of the insulating glass unit and yet small enough to inhibit the passage of water vapor into the insulating glass unit. It would be foolish to imply that no water vapor can travel through the capillary breather tube. It is reasonable to suppose that most of the water will travel the length of the tube because of the surface tension of the liquid/vapor. It is the length of the tube which gives the breather tube its name "capillary." The breather tube just described has not always

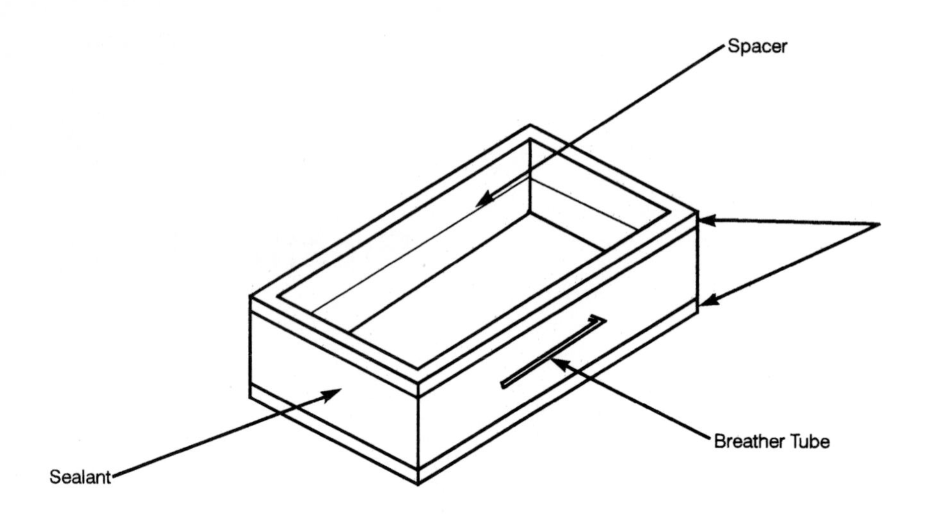

Figure 19.1 Typical breather in insulating glass unit.

been the type used. Figure 19.1 shows a typical breather tube in an insulating glass unit. The tube works in two ways during the service life of the sealed unit:

1. exhaling air during the day as the unit heats up, and
2. inhaling air during the evening as the sealed unit cools down.

It has been shown that any moisture present in the air will also be inhaled through the breather tube. The majority of this moisture enters the air space and is irreversibly adsorbed by the desiccant. When the adsorbent finally becomes saturated with moisture, the sealed unit must be replaced. For a unit with no breather tube, the saturation rate depends only on moisture intrusion through the glass/sealant interface. However, for units installed with breather tubes, both moisture intrusion mechanisms are present, and the adsorbent becomes saturated at a faster rate.

Installation of the Capillary Breather Tubes

There is still room for the design/application improvement of capillary breather tubes and careful consideration must be paid to the installation techniques.

The breather tubes must achieve a high-quality seal around the tube itself and within the sealing compound around the perimeter of the insulating glass unit. Whether the sealant used is a single-seal, a dual-seal, or a hot melt, the integrity of the seal around the outside diameter of the breather tube is of the utmost importance. The seal cannot have any voids, dips, or gaps if this quality is desired.

The diameter of the breather tube is so small that extreme care must be taken to keep the ends free of dirt which will plug the tube. Once it is plugged, there is no appreciable value in having a breather tube. The necessity of assuring that the tubes are stored in a clean environment and in a uniform manner cannot be overemphasized. Even a small grain of sand, skin cells, etc., may render the breather tube ineffective.

The installation of the actual breather tube is very simple. Approximately 8 in of the tube is slid inside the air spacer. Then the corner key should be gently pressed into the spacer. The corner key will act as a wedge to hold the breather tube in position. In the case of units manufactured with corners simply bent, a small hole can be drilled or punched to accommodate the narrow-diameter tube. The tube should be left pointing straight out while the sealant is applied. The tube must not be bent or moved in any fashion until the sealant cures. If the breather tube is wiggled while the sealant is curing, then the seal will most likely be broken around the tube itself. About 4 in of the capillary breather tube should be left outside of the sealed insulating glass unit until it has fully cured. Then the tube should be bent 90° to lay flat against the spacer from which the tube is projecting. It is essential that the capillary tube be pointing downward and that the portion inside the sealed insulating glass unit be inside the horizontal spacer member. It is recommended that the insulating glass manufacturer place a "breather tube sticker" on the glass surface indicating correct orientation, so that craftsmen installing the units cannot unknowingly install it with the breather tube pointing in any direction but downward. It is also advisable that the free end is taped, with masking tape, against the air spacer to ensure that the seal around the tube is maintained. Properly sized breather tubes are essential for good performance.[4]

Deflection

Just how detrimental is deflection to an insulating glass unit?[2] There are three major factors to be considered:

1. Glass under excessive pressure differentials between the interior and exterior of the insulating glass units is obvious. What must be determined is how much pressure the glass can withstand? This is where the situation can become complex. It was mentioned earlier that the ideal gas law must be considered. If we assume a constant-

density gas, then as the temperature increases, so does the pressure. One problem is predicting (for each piece of glass) what the expected temperature rise and fall is, which of course depends on geography and how and where the sealed unit is transported and stored. The greater difficulty may be in our assumption of constant density. The ideal gas law states

$$P = pRT$$

where P = Pressure (lb/ft^2)
p = Density (lb/ft^3)
R = Gas constant (ft lb/lbm°R)
T = Temperature (°R = °F + 460)

Also note that the density is defined as

$$p = \frac{M}{V}$$

where p = Density (lb/ft^3)
M = Mass (lb)
V = Volume (ft^3)

The reason that we cannot necessarily assume constant density is because, as the glass bows in or out due to pressure changes, the volume of the air inside the insulating glass changes. Because density equals mass divided by volume, as the volume changes, the density will change. Whether you do or do not assume constant density, there are still two other reasons why predicting glass deflections is difficult. The first reason is that the density of the air we breathe varies. It depends on both temperature and elevation. The second reason has to do with accurately determining what stress levels glass can withstand. Data is available to estimate these stress limits, but it is statistical in nature. Because of this statistical nature, there are several variables to consider which make specific predictions for glass failure very difficult.

2. *Edge-Seal failure.* The integrity of the edge seal in an insulating glass unit is fundamental. If the seal is broken or was not originally present, then failure is assured.

Now the question is how detrimental is bowing in and out of the two pieces of glass to the edge seal of the insulating glass. Statistically, testing can be done to determine if seal failure will occur in an extremely short time after production, but this type of fatigue stress is difficult, at best, to analyze over a long term. If one chooses to ignore edge-seal stresses, there are indications that the longevity of the insulating glass unit may be sacrificed.

3. *Aesthetics.* In a purely technical sense, the aesthetics of bowing

glass are unimportant, but in terms of reality and customer satisfaction, aesthetics must be considered. Little needs to be said regarding this topic, except that glass deflection is a very subjective matter and, from an architectural/customer standpoint, should certainly be considered.

Morton International Inc.[3] conducted a series of extensive tests to help answer the question regarding deflection and compression testing at high and low temperatures. Various window units were made using different glass thickness with and without desiccants and with and without breather tubes. The results of these tests are shown in Figures 19.2, 19.3, 19.4, and 19.5.

Definition of Terms

The field performance of an insulating glass unit depends on several variables. This section describes each of the terms used in predicting service life.

1. E = estimated lifetime. This would be the service life of a unit without a breather tube. This depends upon the rate of moisture intrusion through the glass/sealant interface.
2. D = desiccant capacity expressed as the maximum amount of moisture that can be adsorbed per length of a desiccant fill. This depends upon the adsorption type and the spacer size (see Chapter 15).
3. P = perimeter length of the spacer filled with desiccant. The product of the perimeter length and the desiccant capacity gives the total moisture content allowed in the unit.
4. C = climatic factor, which depends upon the geographical location and orientation of the unit. This factor depends on the humidity of the air and the daily temperature differences. Table 19.1 lists climatic factors for several U.S. cities.
5. V = volume of air in the internal air space of the sealed unit. This depends on the glass and spacer dimensions. The product of the internal air volume and the climatic factor gives the moisture intrusion rate entering through the breather tube into the unit.
6. L = lifetime of a unit with a breather tube. This depends upon moisture intrusion rates through the breather tube and through the sealant/glass interface.

Predictive Model

The expected lifetime of an insulating glass unit with a breather tube can be predicted by the following model:

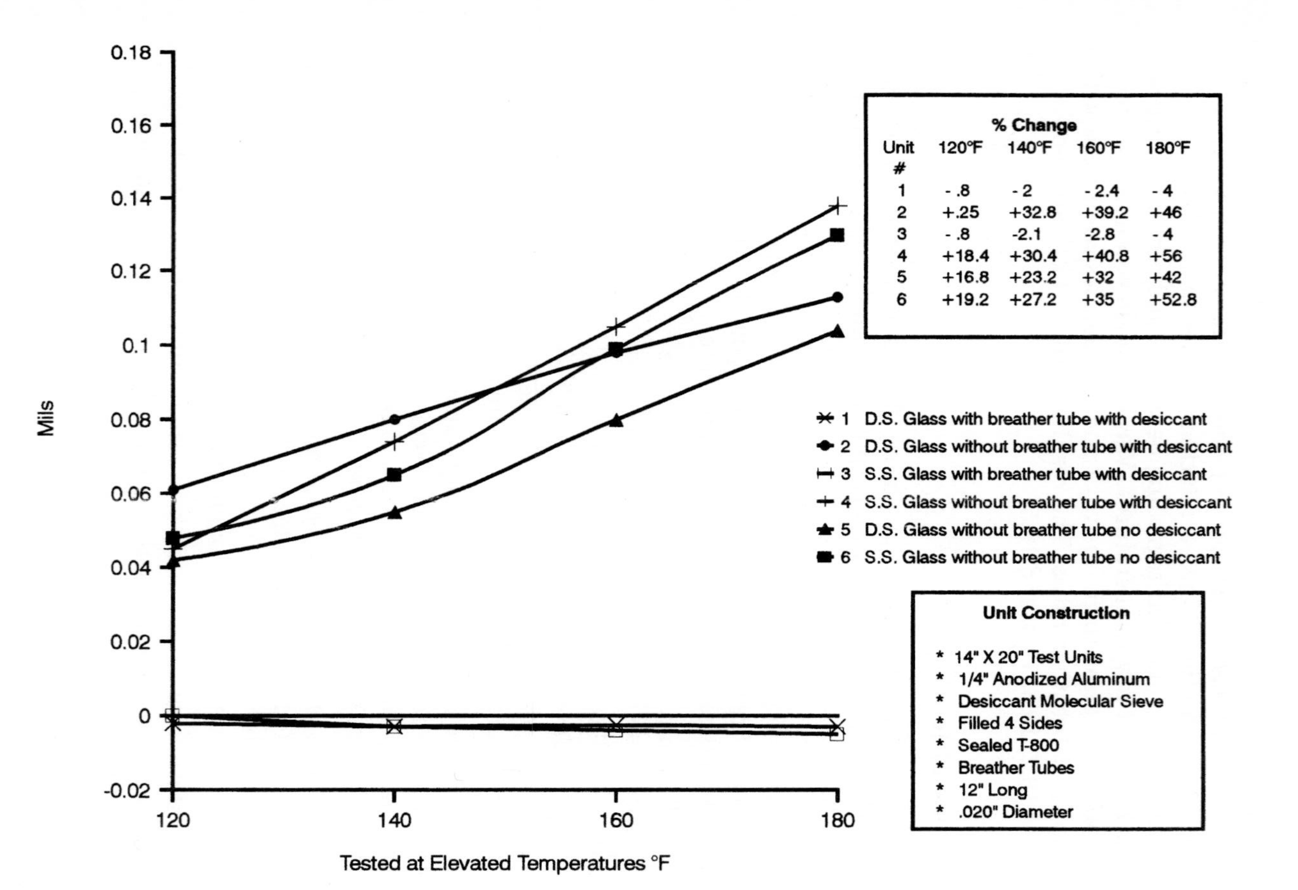

% Change

Unit #	120°F	140°F	160°F	180°F
1	- .8	- 2	- 2.4	- 4
2	+.25	+32.8	+39.2	+46
3	- .8	-2.1	-2.8	- 4
4	+18.4	+30.4	+40.8	+56
5	+16.8	+23.2	+32	+42
6	+19.2	+27.2	+35	+52.8

Figure 19.2 Test results—deflection of $\frac{1}{4}$ inch unit.

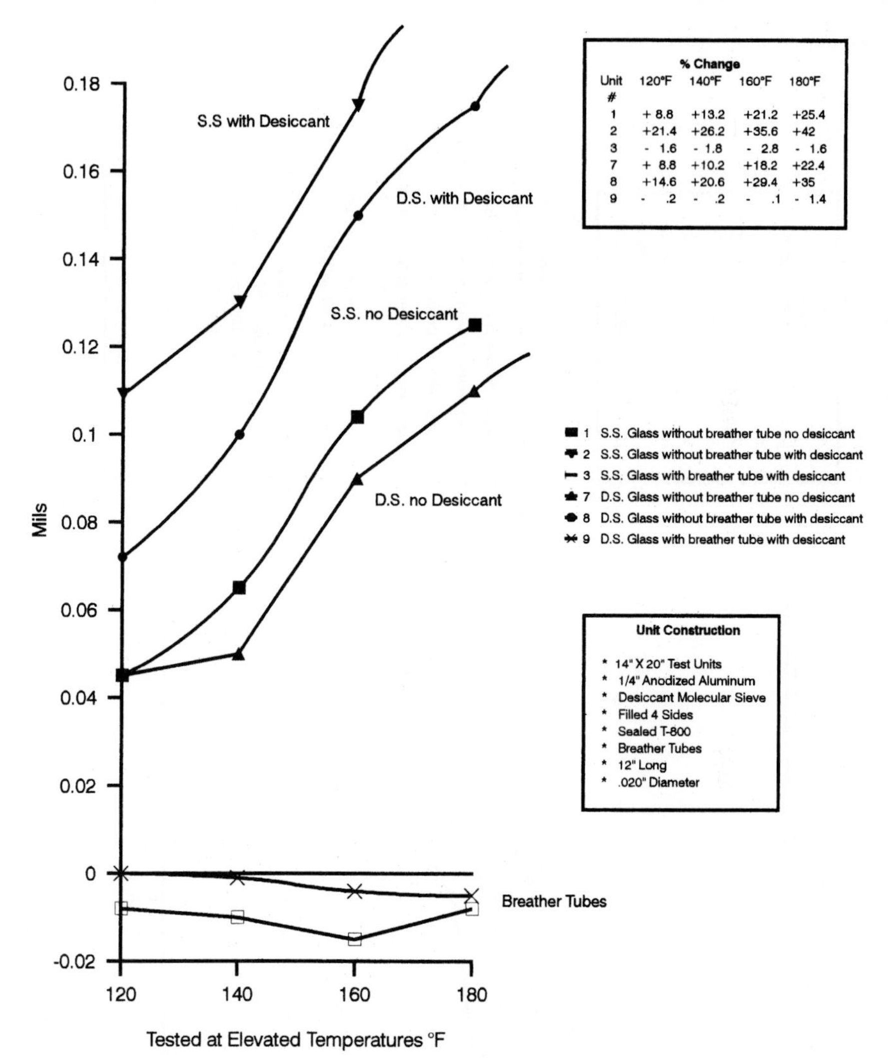

Figure 19.3 Test results—deflection of $\frac{1}{2}$ inch unit.

$$L = \frac{E \times D \times P}{E \times C \times V + D \times P}$$

where L = lifetime with a breather tube (years)
E = estimated lifetime without a breather tube (years)
D = desiccant capacity (milligrams of water/inch)
P = perimeter length of desiccant fill (inch)
C = Climatic factor (milligrams/year/cubic inch)
V = volume of internal air (cubic inch)

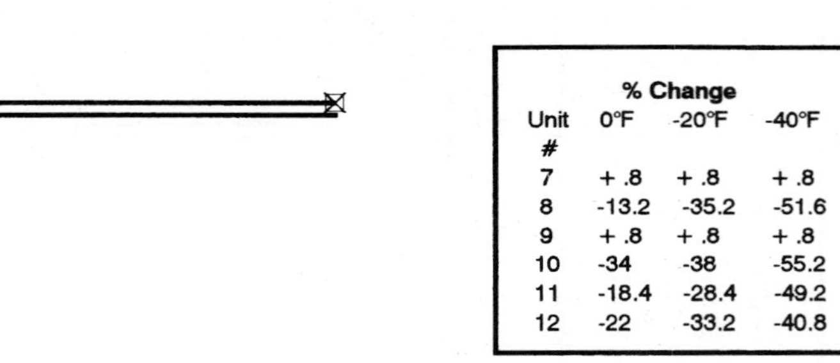
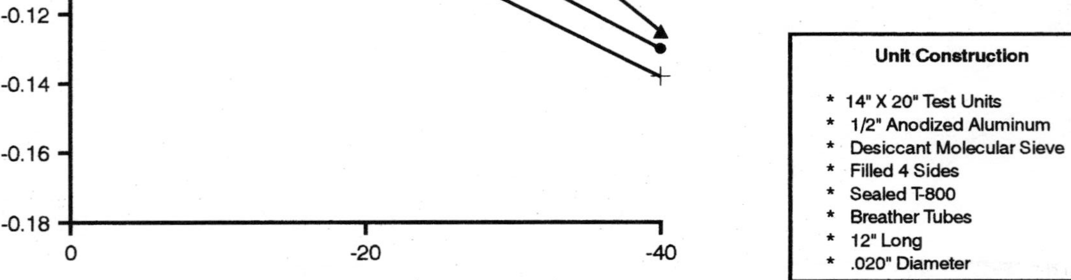

% Change			
Unit #	0°F	-20°F	-40°F
7	+ .8	+ .8	+ .8
8	-13.2	-35.2	-51.6
9	+ .8	+ .8	+ .8
10	-34	-38	-55.2
11	-18.4	-28.4	-49.2
12	-22	-33.2	-40.8

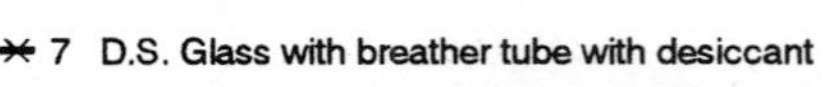

Unit Construction

* 14" X 20" Test Units
* 1/2" Anodized Aluminum
* Desiccant Molecular Sieve
* Filled 4 Sides
* Sealed T-800
* Breather Tubes
* 12" Long
* .020" Diameter

Figure 19.4 Test results—contraction of $\frac{1}{4}$ inch unit.

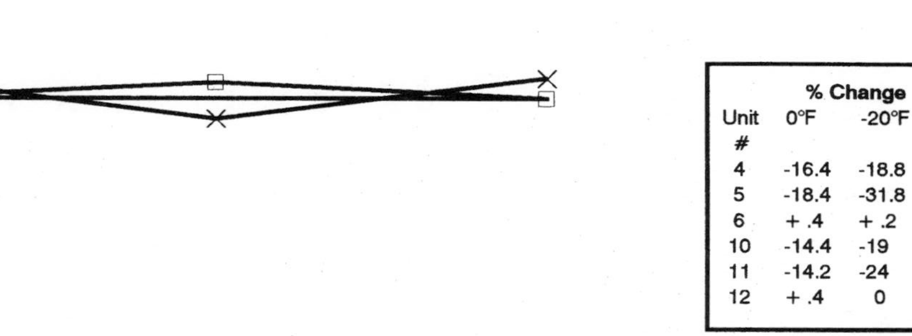

% Change

Unit #	0°F	-20°F	-40°F
4	-16.4	-18.8	-20.8
5	-18.4	-31.8	-35.4
6	+ .4	+ .2	0
10	-14.4	-19	-19.2
11	-14.2	-24	-26.2
12	+ .4	0	+ .4

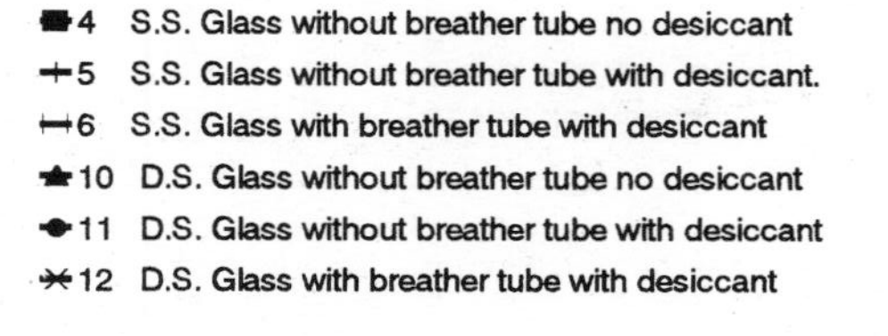

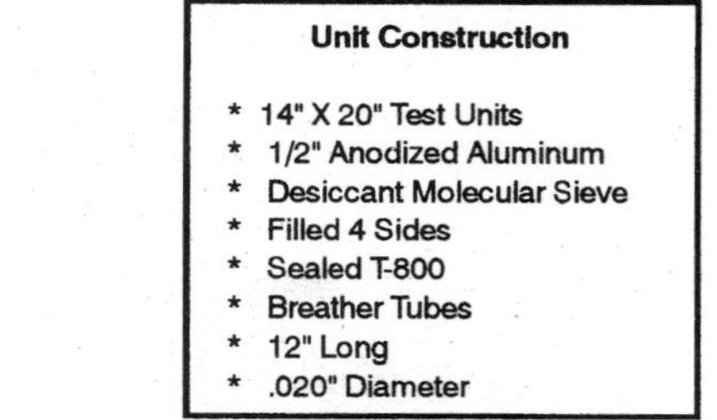

Figure 19.5 Test results—contraction of $\frac{1}{2}$ inch unit.

TABLE 19.1 Climatic Factors (C), in Milligrams of Water per Year per Cubic Inch for Various Cities

City	Shaded/Sunny	City	Shaded/Sunny
1. Anchorage, AK	0.42/0.57	11. El Paso, TX	0.90/1.13
2. Las Vegas, NV	0.62/0.83	12. San Francisco, CA	0.99/1.38
3. Billings, MT	0.68/0.89	13. Wichita, KS	0.17/1.55
4. Seattle, WA	0.72/0.91	14. Honolulu, HI	1.17/1.61
5. Chicago, IL	0.78/1.09	15. Fort Worth, TX	1.19/1.80
6. Caribou, ME	0.80/1.03	16. Nashville, TN	1.36/1.64
7. New York, NY	0.81/1.16	17. Raleigh, NC	1.47/1.89
8. Denver, CO	0.86/1.06	18. New Orleans, LA	1.65/2.21
9. Fargo, ND	0.88/1.14	19. Miami, FL	1.67/2.27
10. Los Angeles, CA	0.88/1.34	20. Brownsville, TX	1.76/2.45

This predictive model was derived using the following assumptions:

1. The predominant factors influencing the amount of moisture entering a breather tube are
a. the amount of air inhaled which depends on the glass dimensions, spacer size, and daily temperature extremes of the internal air volume and
b. the moisture content of the inhaled air (as determined by its temperature and relative humidity).
2. One hundred percent of the moisture inhaled is irreversibly adsorbed by the desiccant.
3. The service time of the unit without a breather tube for a particular location.

True example

This example illustrates the effects of the principal parameters. A series of sealed insulating glass units is to be installed west of the Rocky Mountains, using a 24 in × 48 in × ¾ in unit. It is expected to last 20 years with breather tubes and if one short and one long side is filled with molecular sieve adsorbent. The owner wants to predict the lifetime of these units if installed with breather tubes. The installed units will be shaded from the sun.

Using the predictive model above, one can estimate the lifetime of the units if they are installed with breather tubes.

$$E = 20 \text{ years}$$

$$D = 187 \text{ mg/in}$$

$$P = 70 \text{ in } (23 \times 47 \text{ in})$$

$$C = 0.86 \text{ mg/year/in}^3 \text{ (from Table 19.1)}$$

$$V = 380 \text{ in}^3 \; (23 \times 47 \; \times {}^3\!/_8 \text{ in})$$

$$L = \frac{29 \times 187 \times 70}{20 \times 0.86 \times 380 + 187 \times 70} = 19.34 \text{ years}$$

To obtain the same lifetime without breather tubes, it would be necessary to add about 35 percent more desiccant to compensate for the additional moisture intrusion through the breather tube.

Climatic factors

Any predictive model is sensitive to changes in geographical location and orientation. Figures 19.6 and 19.7 show the variation of climatic factors throughout the U.S. for shaded and sunny installations,

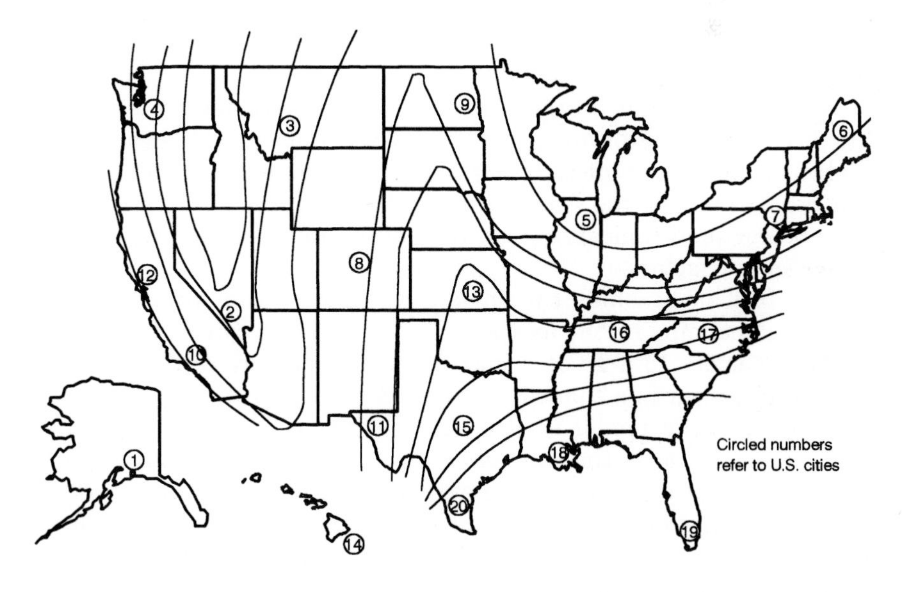

Climate factors for various U.S. locations - shaded installation (Lines connected locations of equal climate factors - MG./YR./CU.IN.)

Figure 19.6 Map of climatic factors—shaded installations.

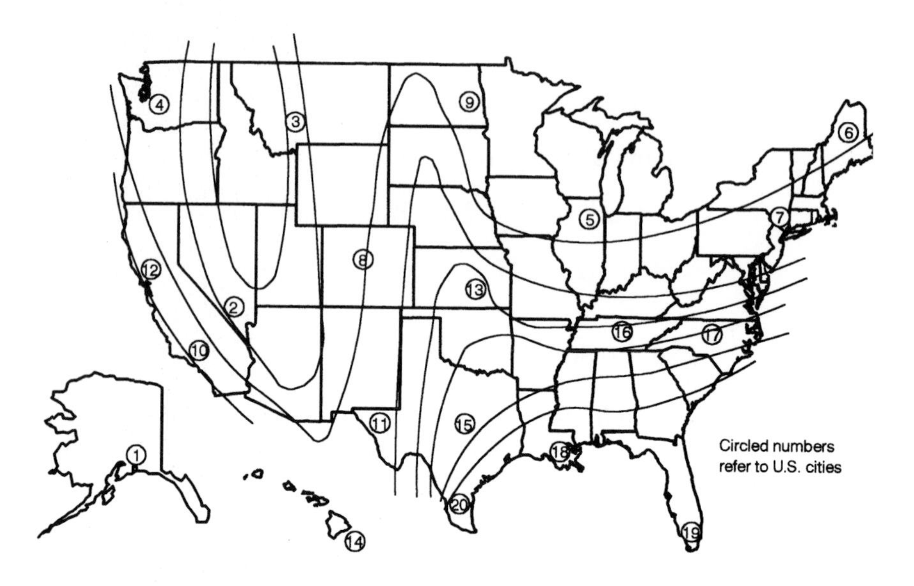

Climate factors for various U.S. locations - sunny installation (Lines connected locations of equal climate factors - MG./YR./CU.IN.)

Figure 19.7 Map of climatic factors—sunny installations.

respectively. The circled numbers refer to the particular cities listed in Table 19.1.

As an example, for northern regions, the climatic factors are quite low because of the dryness of the cold air. For coastal regions, the high humidities are offset by the moderate temperature differences. For desert regions, the extreme daily temperatures are offset by the dry air. For semitropical regions, the hot, humid weather gives rise to high climatic factors.

The effect of sunshine can raise the climatic factor from 25 to 40 percent depending on the location. This increase is caused by heating of the air space to a greater maximum temperature during the day than would occur in a shaded installation.

The predictive model is also very sensitive to changes to glass dimensions. Because the ratio of volume to perimeter length increases with the size of the glass, the overall effect is to decrease, markedly, the unit lifetime with a breather tube. However, offsetting effects would be achieved by increasing the spacer size because the desiccant capacity increases proportionally.

The predictive model illustrates the effects of climate and glass dimensions on the lifetime of sealed insulating glass units with breather tubes.

Conclusion

The advantages of breather tubes can be seen, and we should now examine how they affect accelerated testing and field usage. Results from ASTM-E-773 test show that breather tubes test well compared to the same units without breather tubes. The only disadvantage noted with the breather tubes is the result of poor sealant application around the breather tubes. Voids were evident. Because this is an application problem, it behooves the manufacturer to exercise better care or a better technique. In addition to laboratory testing, outdoor exposure is also very important. From the data presented, it can be seen how capillary breather tubes offer many advantages for alleviating positive and negative buildups in sealed insulating units. If used correctly, it is felt that breather tubes offer an extra dimension to the manufacturing of long-lasting units.

References

1. Booth, R., A Guide to Predict the Lifetime of Insulating Glass Units with Breather Tubes, *U.S. Glass, Metal and Glazing,* 7–8/1985.
2. Fore, G. D., Testing Engineer, *Engineering Bulletin, Capillary Breather Tubes,* Redman Building Products, Inc., Bryan, TX, 9/87.
3. Scherrer, M. J., and J. Sheldon, How to Alleviate Pressure Build-Up by Using Capillary Breather Tubes for Insulating Glass, *U.S. Glass, Metal and Glazing,* 2–3/1985.
4. Schoofs, G. R., Will Leave-Open Breather Tubes Impact the Insulating Glass Market?, *Glass Digest,* 10/1984.

Chapter

20

Sloped Glazing and Suspended Glazing

Joseph S. Amstock

President, Professional Adhesive and Sealant Systems

Introduction

A growing number of architects are expanding the design of curtain walls by constructing buildings with sloped and overhead glazing.[1] With proper planning, these "unconventional curtain walls" can provide the energy savings, performance, safety, and aesthetic appeal that architects require.

Sloped glazing can be defined as glazing systems tilted more than 15° from vertical. Merely tilting a conventional vertical system will not result in a successful, satisfactory, sloped glazing system. Only after special consideration is given to such factors as loads, deflection, thermal stresses, and safety can the potential for a successful sloped glazing system be realized. The use of sloped glazing in overhead skylights, atriums, barrel vaults, and solarium applications presents a distinct challenge for today's architects. Overhead glazing systems must be designed to reduce the risk of injury from glass fallout should the glass break for any reason. Because sloped and overhead glazing is subject to wind, snow, and live and dead loads, building codes typically require the use of laminated architectural glass or costly screens to protect the occupants from any breakage.

Glass size limitations recommended by the American Architectural Manufacturers Association (AAMA) for types and thicknesses of glass in sloped glazing are shown in Table 20.1.

Sloped glazing is a glass roof and, like any other roof, it is not expected to leak. When sloped glazing leaks, it can speed degradation of the compounds that make up the sloped glazing system.

TABLE 20.1 Sloped Glazing*

Nominal thickness	Monolithic heat-strengthened or fully tempered construction (0.060 PVB)	Thickness tolerance		Maximum size, ft²	Weight, lb/ft²
		Min.	Max.		
5/16 in	1/8 in + 1/8 in	0.290	0.328	25	3.6
7/16 in	3/16 in + 3/16 in	0.420	0.458	40	5.2
9/16 in	1/4 in + 1/4 in	0.498	0.548	40	6.5

*Heat-strengthened or tempered glass is used in the indicated laminated construction.

Nominal thickness	Insulated construction (0.060 PVB)			Thickness tolerance		Maximum size, ft²	Weight, lb/ft²
	Outdoor	Air space	Indoor	Min.	Max		
1 1/16 in	1/4 in	1/2 in	1/8 in + 1/8 in	1.04	1.10	25	6.9
1 3/16 in	1/4 in	1/2 in	3/16 in + 3/16 in	1.17	1.23	40	8.5
1 5/16 in	1/4 in	1/2 in	1/4 in + 1/4 in	1.25	1.32	40	10.2

*Heat-strengthened or tempered glass is used in the indicated laminated construction.

Annealed laminated glass construction	Maximum area, ft²	Application
2 plies 1/8-in glass	12	Sloped
2 plies 3/16-in glass	18	Sloped
2 plies 1/4-in glass	25	Sloped

Water leaking through sloped glazing systems can degrade the construction perimeter, resulting in roof damage, corrosion of the steel elements within the roof, ceilings, and walls, and staining and breakdown of drywall, interior finishes, and carpeting. Water leakage through sloped glazing can follow structural elements to appear finally in remote parts of the building. When a sloped glass roof leaks onto floors and stairs, the leakage can turn into a serious safety hazard.[2]

Performance Characteristics

The architect or designer should be familiar with the nomenclature of glazing materials being considered. For flat glass and some plastic glazing, he or she should refer to the current edition of the Flat Glass Marketing Association's *Glazing Manual.*

Specific standards are covered by various federal specifications, such as DD-G-451d and DD-G-1403B. Most local, state, and model building codes specify the types of sloped glazing and skylight materials that may be used in building construction. In addition, *Spectext,*

Section 08960, Sloped Glazing is another ideal document for the architect. The architect should always be aware of the potential factors and associated costs for the various glazing material options before the specifications are released.[3]

Design Principles

Little design information is available to architects on sloped glazing. Most architects rely on manufacturers to provide the details for their systems. Manufactures develop their designs in response to the demands of the marketplace and slant them toward their particular component of the system. Aluminum manufacturers produce designs that deal with the advantages of aluminum but sometimes fall short when it comes to considerations of glass or seals. Panelized glazing and acrylic manufacturers often neglect the effect of the frame on performance.

Opinions differ on how to accomplish all the details necessary for a well-functioning system. Manufacturers take one of two approaches. The first and most prevalent is to rely on the exterior seals between the components to keep water from entering the glazing framing. These systems may provide some sort of interior seal between the frame and glazing. However, no real emphasis is placed on providing a water seal at these interior junctions of components.

The problems with such an approach include exposure of the exterior seal and sealants to ultraviolet (UV) radiation, thermal stresses, pollutants, and poor workmanship. Installers clean, prime, place, and tool sealants in various weather conditions with the hope of providing 100 percent perfect, long-lasting finished products. Millions of dollars are spent every year in recaulking the exteriors of buildings that rely on this approach. All too often, when a sloped glazing system leaks, the maintenance staff calls for another layer of a sealant on all the joints.

The second approach is based on the realization that the exterior seals will not be completely watertight. The watertight seals are designed in the interior of the system to minimize their contact with water and to drain the water in the system back to the exterior. Some manufacturers have realized the folly of the first approach and have started to change their designs. It is difficult in today's market to make a watertight system and remain competitive.

Modern vertical aluminum and glass systems used in high-rise construction are based on this second approach, as illustrated in Figure 20.1, using pressure equalization and draining of the glazing rabbet in what is referred to as a rain screen design. Air pressure differences across the exterior surface of a wetted wall can drive water through imperfections in the interior seals.

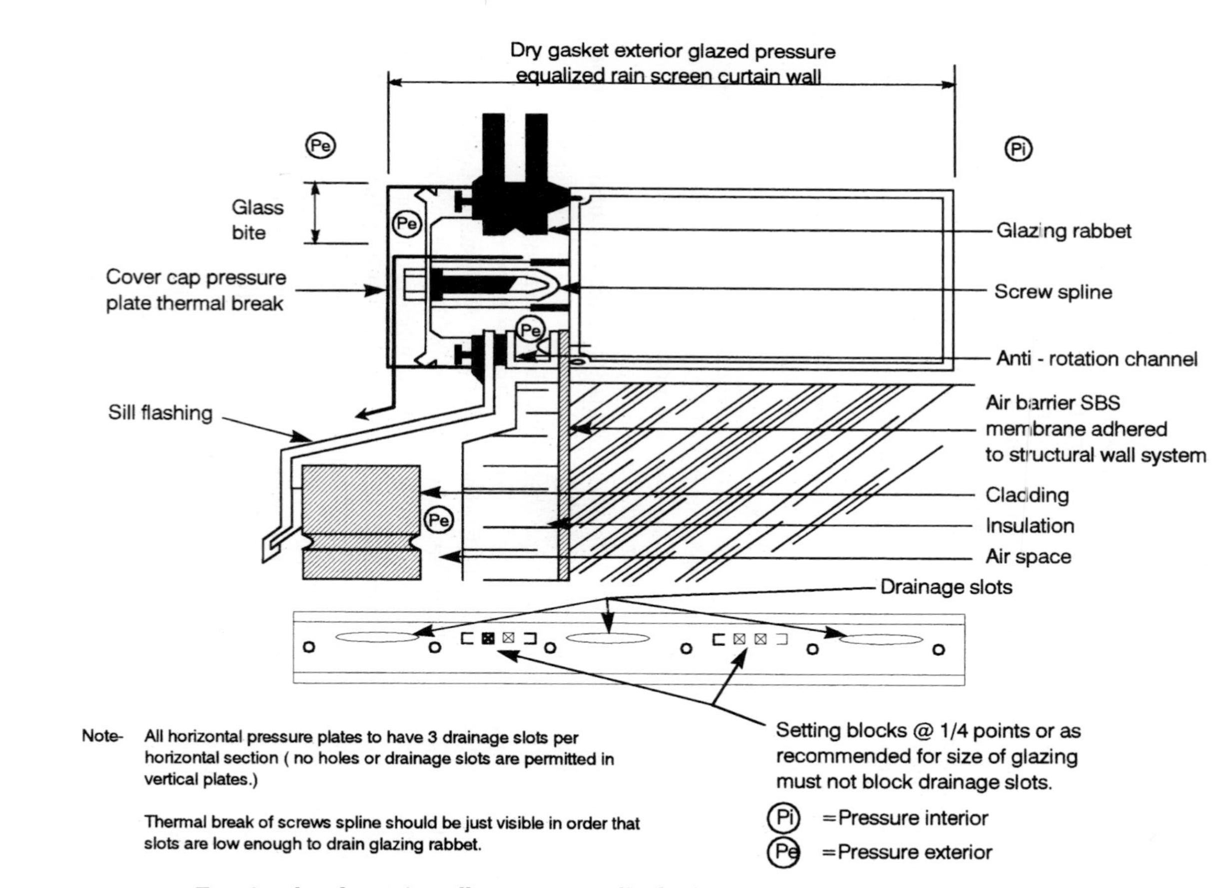

Figure 20.1 Exterior glazed curtain wall pressure-equalized rain screen.

Equalizing the pressure across the cladding components and exterior seals that act as a screen to the inner wall can eliminate these forces, hence the name pressure-equalized rain screen. Pressure equalization requires the presence of other features in the framing system: an effective air barrier, a compartmentalized glazing rabbet, and a large protected venting area through the cladding relative to the leakage area of the air barrier.

The air barrier is created by the inner glass lite, the seal between this inner lite and the frame, the frame tube face of the glazing rabbet, and the sealed joints of the inner framing sections. Compartmentalization is achieved by continuing this barrier out through the screw spline, thermal break, and pressure plate to form a pocket of air between the sealed unit edge and frame. This pocket is called the glazing rabbet.

Compartmentalization is not complete unless the small gap at the corners created by the lack of an extension of the screw spline is filled with a corner plug. Venting is provided through slots in the pressure plate, which can also act as drains should any water enter the glazing rabbet.

Interior seals of butyl tapes, sealants, plugs, and gaskets between aluminum and glass are protected from water in the interior of such a design, so that their performance is unaffected by water.

Vertical glazing walls in sloped glazing applications often fail because water is drained through the exterior face of pressure plates and covercaps but is held at a variety of interior butt joints, where it finds a hole or degrades the sealant materials until a hole is created. Leaving out the corner plugs in an effort to drain water down the verticals eliminates compartmentalization of individual units, thus reducing the effectiveness of the pressure-equalized design.

Additional water entry floods the glazing rabbet, and pumping of the system during windy conditions can drive water through the interior seals that are now exposed to water. Sealants used in joints between the frames sections or between frame and glass prematurely break down. Sealed units positioned in a nondraining glazing rabbet will lose their seals prematurely, resulting in fogging of the unit.

The components and methods used to install vertical and sloped glazing systems have a great deal in common, but differences in design to control the water that bypasses the exterior deterrent seal set them apart. How the water is drained back to the exterior will determine the performance of the systems.

The design and construction of each sloped glazing installation is different. However, following some simple concepts can make the difference between success and failure.

1. Maximize the effectiveness of the outside seal or gasket system, but do not rely on it to provide total waterproofing for the system.

2. Minimize or eliminate the ponding of water at the exterior seal locations. This will prolong the effectiveness of the seal and reduce the buildup of dirt and other contaminants that may enter the system if an imperfection exists. Water ponding at the exterior seals can be pumped into the system through movement under high-wind conditions.

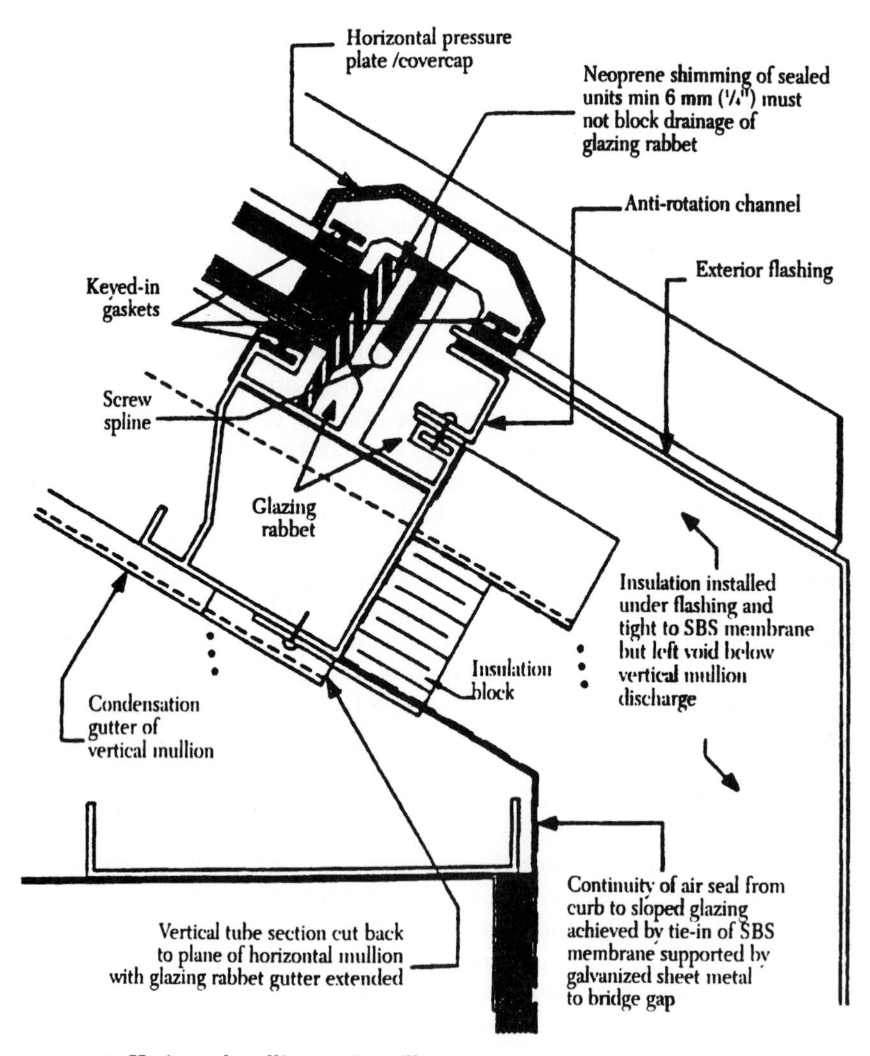

Figure 20.2 Horizontal mullion-section sill.

3. When water does get into the glazing rabbet, it must be contained, controlled, and redirected back to the exterior. The glazing rabbet provides a gutter that should adequately separate the water and the system's interior seals (see Figure 20.2). The gutter is created by the screw spline, tube face, and the glazing leg of the rabbet. The glazing leg is elevated to ensure that the glazing does not sit in water, which can degrade its components and seals.

4. Pressure equalization of the glazing rabbet will not perform to the same extent as it does in a vertical wall system, in part because the interconnected glazing rabbet of the individual glazing is not compartmentalized. It is still extremely important, however, to provide an "air barrier" plane within the system. Without this separation between the inner and outer environments, the envelope is incomplete and can provide a pathway for transporting air and water through the envelope. Infiltration of cold winter air can cool interior construction surfaces to a temperature at which they can reach the dew point of the interior air, resulting in condensation on these surfaces. Freezing of interior pipes, discomfort, and the transport of outside contaminants may occur as a result of uncontrolled air leakage into the building.

Exfiltration of moist interior air through the air barrier can result in condensation on colder surfaces within the glazing rabbet. This additional water would have to be contended with.

The air barrier location is complicated by the variety of planes providing this function. Sealants, gaskets, or other materials susceptible to water degradation should be located where they will have only limited contact or where they will be accessible for periodic replacement.

5. A means of containing, collecting, and disposing of condensation should be developed within the framing profiles to contain condensed water that might accumulate on the inner aluminum and glass surfaces. Where moderate humidity levels are maintained, adequate heat is provided, and some air movement exists over the surfaces to break up the insulating air film, condensation should not be a problem.

The use of add-on systems usually is aesthetically unappealing and leads to joints that may have to be sealed. Incorporating condensed gutters is not a safeguard against water leakage. A condensed water collection system must not drain back to the exterior through the air barrier system but from an evaporation trough at the sill or by mechanical drainage if a large amount of condensation is expected.

General profiles

All details in this chapter are based on two basic aluminum profiles—horizontal mullions as shown in Figure 20.2 and vertical mullions as in Figure 20.3. They are similar in design in that the main tube sec-

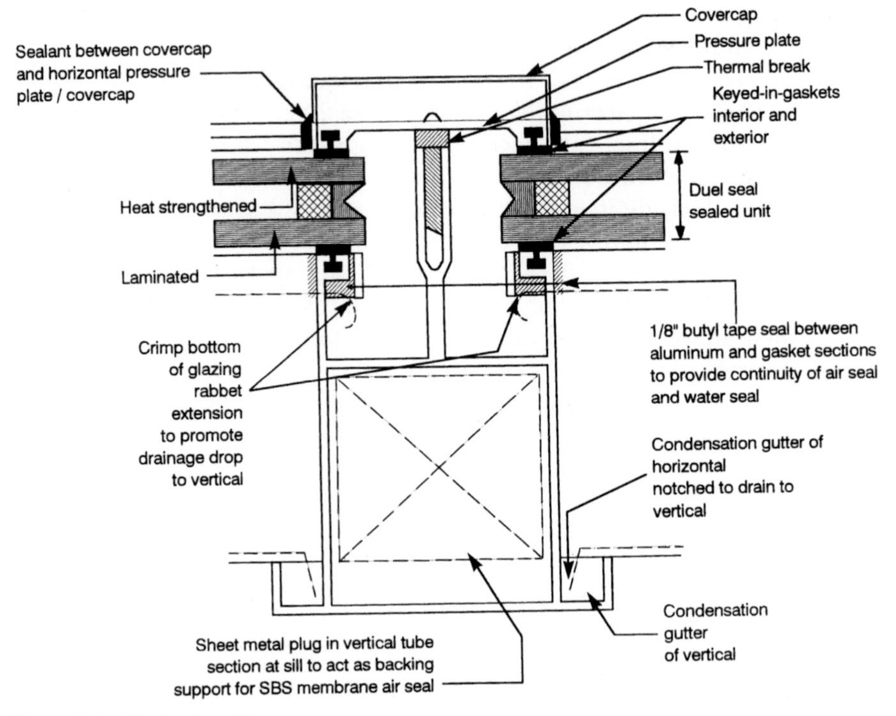

Figure 20.3 Vertical mullion.

tion provides the structure to support the live and dead loads of the systems. It may be economical to place a miscellaneous metal frame under the aluminum framing to provide support when loads increase beyond the maximum allowable loading of the aluminum profile. It still can be introduced within some aluminum profiles as well, but aesthetics and cost usually govern the approach.

The sealed units are installed from the exterior so that a minimum gap of ¼ in (6 mm) exists between the unit edge and the face of the screw plane. They are positioned by resting the sill edge of the unit on 80-durometer hardness neoprene setting blocks at quarter points of the length of the sill screw spline. The aluminum profile raises the unit's edge support from the plane of the frame's main tube face on a raised glazing leg. The raised leg of the gutter (glazing rabbet) has a keyed-in gasket to separate the glass unit from the frame.

This glazing method allows the unit to be moved during placement to properly achieve the correct bite for the edge of the unit. The height of the raised leg of the glazing rabbet should be sufficient to elevate the joints between the aluminum and gasket and the edge of the sealed unit to prevent these points from ever sitting in water. A raised leg of $^3/_8$ to $^1/_2$ in (10 to 13 mm) is sufficient as rule of thumb for a horizontal member no greater than 6 ft (1.8 m) long where the angle of the system's slope is 30° or greater.

Water at the gasket can leak through the end joints or between the gasket and glass to the interior. Water in contact with the edge of a sealed unit can cause some sealants to lose adhesion to the glass surface, resulting in seal failure. When in contact with water, the edge seal of a sealed unit can cause some sealants used to manufacture the unit to swell or lose adhesion to the glass surface, resulting in seal failure when in contact with water. The polyvinyl butyl (PVB) plastic inner layer between the plies (layers of glass) of the sealed unit's inner lite can discolor and affect the optical quality of the lite.

The units are held in place by an exterior applied pressure plate that should be thermally separated from the screw spline. This can be achieved by a polyvinyl chloride (PVC) or ethylene propylenediene monomer (EDPM) keyed-in profile that is secured to the screw spline and penetrated by the screw fasteners of the pressure plate.

The vertical pressure may or may not have an overcap similar to that of a standard curtain wall. If it does have a covercap, it should be designed to accommodate the weight of glaziers, window washers, and maintenance staff walking on it from time to time. When they do, they may damage the profile and the ability of the profile to clamp to the pressure plate. Such loss of contact can result in caps being caught by the wind and ripped from the system.

For horizontal pressure plates, a low profile with no covercap should be used to minimize the amount of water retained at the outside seal. The degree of the glazing's slope and the height of the gasketed pressure plate will affect how much water and dust is retained at the horizontals. This is visible on barreled vault systems, where the top glazing is dirty and the lower lites are cleaner. Retention of water and dirt on the upper lites may result in streaking of the lower lite, where water eventually drains.

Some manufacturers promote the use of a silicone or semicapless design for the horizontals. Although the design has some merit in minimizing the amount of water and dirt retained, it is not without cost. The performance of silicones can be affected greatly by or be a factor in degrading other elements in the glazing system. Comparability testing of the silicone with all contacting surface materials must be undertaken before and during construction and by maintenance staff if work is to be undertaken.

Preparation of component surfaces for priming, proper placement of the silicone, and tooling of the joint are essential to ensure acceptable long-term performance of this sealant. The exterior sealant weatherseal design may have to be installed in less than favorable conditions, whereas, with a dry glazing and pressure-plate system, the limitations imposed by weather are not as critical. Only when the slope of the glazing system is nearly flat does the silicone weatherseal design seem to have a small benefit.

The dry gasket and pressure-plate approach prevents the units from being uplifted from the system and provides an acceptable degree of watertightness. The drained design approach acknowledges that, whichever exterior seal approach is used, it does not have to be 100 percent watertight for the life of the installation.

The main difference between the horizontal and vertical mullion profiles is in the design height of the glazing rabbet's raised leg. Obviously, the vertical mullion must be sufficiently high to collect all the water draining from the horizontal mullions. It also must be sufficiently high to protect the seals at the junction of the horizontal to vertical mullions from coming in contact with water, which would degrade them.

Horizontal and vertical mullion intersection

Water collected and contained in the glazing rabbet must be directed back to the exterior. Otherwise, the water level will rise enough to wet the interior seals and possibly soak through them to the interior. The horizontal mullion, therefore, drains into the vertical glazing rabbet, as shown in Figure 20.4, where it can be drained at the sill of the sloped glazing system to the exterior.

If the two aluminum sections were to butt together, as in a standard curtain wall, the seal joint would constantly be in contact with water. If the horizontal glazing rabbet overlapped onto the tube face of the vertical glazing rabbet directly, the seal would still be exposed and drainage of the vertical glazing rabbet would be restricted. Jointing of the horizontal to the vertical mullion, therefore, must be overlapped and elevated.

As previously stated, the vertical glazing rabbet is deeper to ensure that it is not blocked by the overlapping extension of the horizontal glazing rabbet and to prevent water from coming into contact with seals necessary between the aluminum sections to ensure continuity of the air barrier.

Water flowing down the vertical glazing rabbet, if of sufficient quantity, will form a series of waves similar to those seen on sloped sidewalks or roads. If these waves are slowed or interrupted, the resulting water turbulence can raise the water level in the gutter. Excessive caulking of the junction joint, debris within the system, and water dripping from the horizontals into the verticals can be disruptive to this wave pattern in the vertical glazing rabbet.

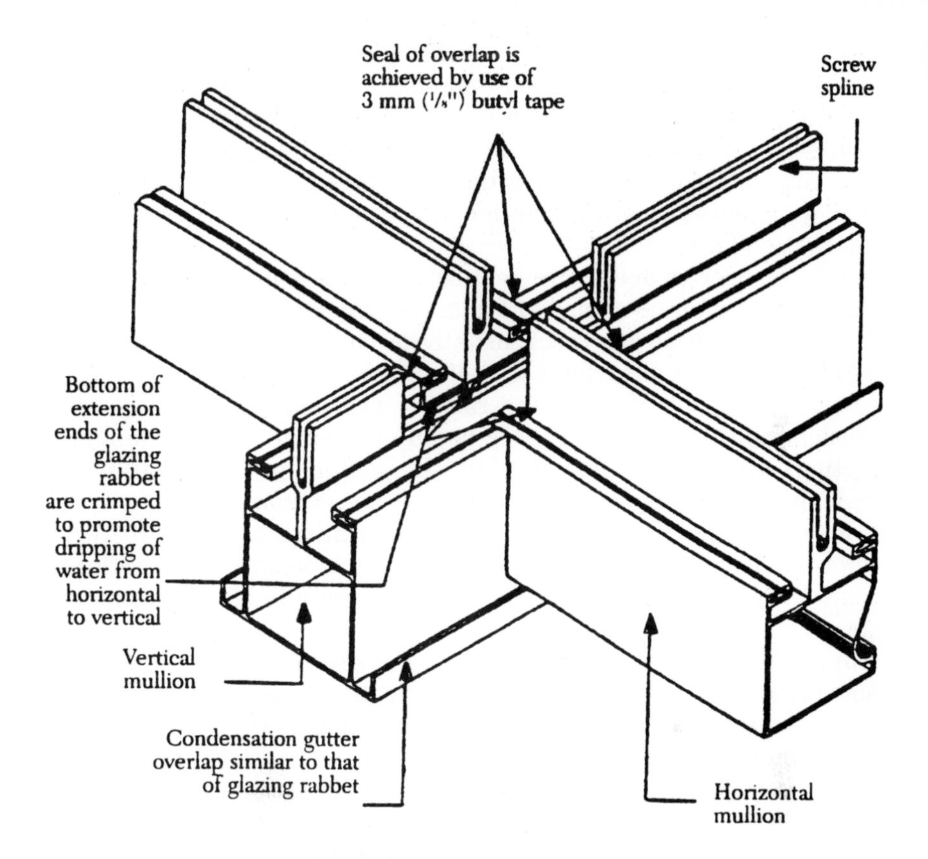

Figure 20.4 Horizontal-to-vertical junction.

To accommodate such situations, the height of the overlap should be at least ⅜ in (10 mm). This depth may seem excessive to some manufacturers. In fact, most overlaps are not as deep. When investigating leakage at these joints, one often finds bits of construction debris, fine dust, bits of vegetation, and insect bodies. If the system can clean itself, then the problems may never occur.

The extension overlap of the horizontal glazing rabbet should promote the dripping of water draining from the horizontal into the vertical. This can be accomplished by cutting or crimping the lip edge of the overlap. When there is only a small amount of water draining from the horizontal glazing rabbet, the surface tension of the water will cause the water to cling to the surfaces of the aluminum on either underside of the horizontal glazing rabbet extension or on the vertical surfaces of the downslope screw spline and the raised leg of the profile.

This causes water to flow on these surfaces and over the seal junction between the horizontal and vertical sections. If the surface tension can be broken by a drip at the termination of the horizontal, contact with the critical air barrier and seal joint can prevent water entry. See Figure 20.4.

The air barrier and water seal between the aluminum sections at this overlap often is a bead of sealant as recommended by the manufacturers. Such an approach usually is not acceptable. The glazier rarely prepares the surfaces and often uses too much sealant, which can block the vertical glazing rabbet. The joint is anything but a "design joint," and should there be any movement of the aluminum sections after the sealant has set, the sealant probably will shear, leaving an open joint for water and air to pass through.

If the sealing joint requires several applications of sealant between the frame, gaskets, and glazing, other joint problems are introduced. Instead, use ⅛-in (3-mm)-thick × ½-in (12-mm)-wide butyl tape. In sloped glazing applications opened up after 15 years, the butyl tape is still pliable. This tape may have some self-sealing ability when it gets hot in the summer. One continuous strip of tape is used to seal from the top of the gasket on one side of the section to the top edge of the gasket on the underside of the section. It is preferable to provide some design joint for the joints, but, for now, the tape approach is more forgiving. From investigations of sloped glazing projects with problems, the most common failure at this junction is the reliance of the design on a sealant.

Sill junction

Designers, manufacturers, and installers of sloped glazing systems often disagree about the sill junction. Designers want a minimal visual element. Manufacturers are reluctant to extend their systems beyond the perimeter plane of their exterior horizontal mullion sections into the grey area of trade responsibility between roofer, general contractor, and themselves. The result often is lack of room and a variety of materials detailed, which forces the glazier to make an attempt at sealing this joint with caulking.

The sill detail and the horizontal-to-vertical junctions should be no different in the way water is drained from one plane to another. The design should minimize the possibility of any water contact at the joint between the aluminum sections and the roof.

To do this, only the vertical gutter of the glazing rabbet is extended beyond the plane of the horizontal tube mullion's outer face (see

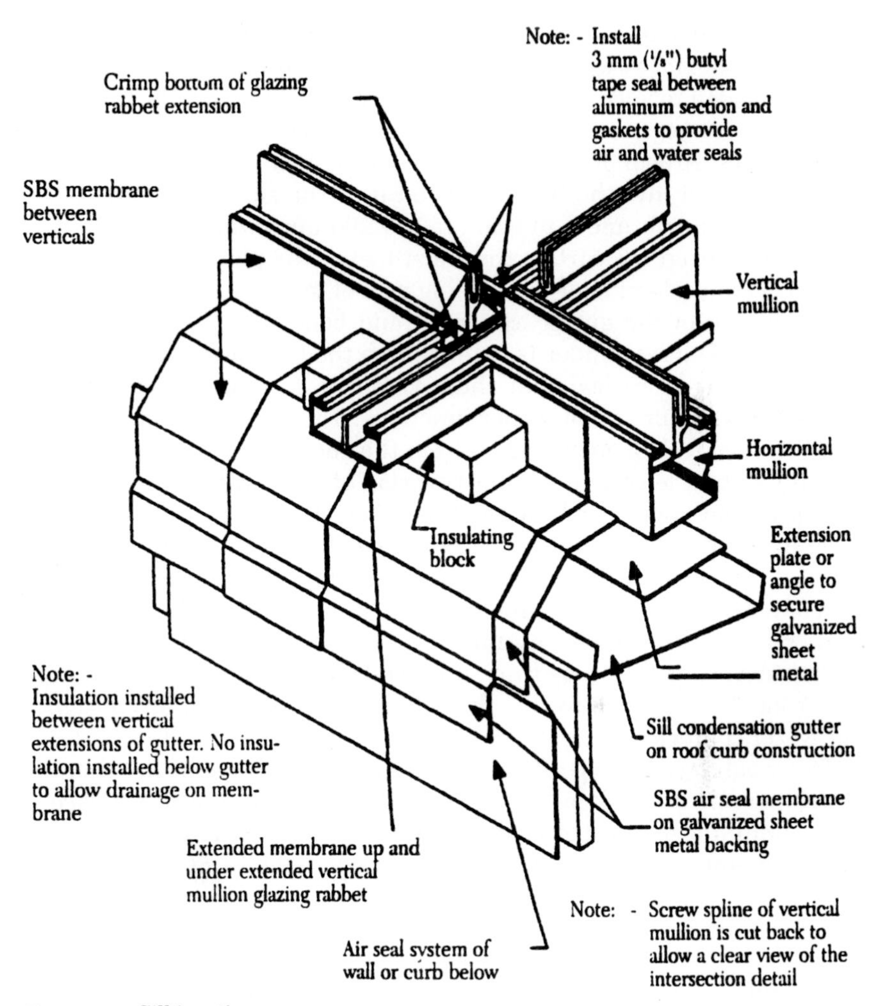

Figure 20.5 Sill junction.

Figures 20.2, 20.3, and 20.5). The remainder of the vertical tube is cut back to the exterior plane of the horizontal tube so that the tube face can be used as the plane of an air seal and waterproofing.

The air barrier and waterproofing seal of the protected membrane roof are extended up the exterior vertical face of the roof curbing. From the curb to an extension angle or plate (depending on the manufacturer's horizontal profile), a 20-gage, galvanized sheet metal backing is used to provide structural support for a torch-applied reinforced SBS membrane. The void created at the end of the vertical tube must also be plugged (see Figure 20.3). The sheet metal is not overlapped at joints to minimize the buildup of the elements. Then, the metal surfaces are primed with the membrane manufacturers' recommended primer and allowed to flash off.

Care must be taken when using a small detail torch to install the membranes. A pool of liquid SBS should be created before the membrane is rolled into the liquid to bond it to the surfaces. First, at the verticals, a membrane 6 to 8 in (150 to 200 mm) wide is extended up the surface from the curb to the sheet metal, the plane of the horizontal, and the underside of the glazing rabbet gutter extension. It is cut to the width of the gutter extension and bonded to the underside of the gutter. The remainder is extended up the side of the raised legs of the gutter. It is critical that these cuts are accurate; otherwise, air leakage will occur in the corner junctions. The remainder of the space between verticals is membraned from the curb to the plane of the horizontal with at least 2 in (50 mm) overlap of membrane joints.

A peel-and-stick membrane can be used. However, it is less workable in these tight confines, and in hot conditions the membrane may have a tendency to flow. The membrane's top termination should be mechanically fastened to prevent such slippage. This detail protects the critical joints of the aluminum sections from having water on their surface. At the same time, it extends a flexible system of materials from the roof curb to the aluminum sections, which are designed for water contact.

Rigid polystyrene type IV provides the continuation of the thermal barrier from the roof up the exterior of the membrane to the face of the aluminum sections. A void of insulation is created at each vertical at the membrane plane to allow sufficient heat to maintain water flow to the roof during colder weather.

A block of insulation is used to wedge the membrane against the underside of the glazing rabbet extension to provide some resistance to membrane slippage or sag. Flashings are installed over the insulation to protect it and provide a good appearance.

Head junction

The air barrier and waterproofing functions are achieved by galvanized sheet metal and a reinforced SBS membrane sealing from the exterior glazing gasket key of the sloped glazing system to the air seal of the rain screen wall above (see Figures 20.2 and 20.6). Here, the vertical mullion tubes are extended for anchorage to provide for the sealing of the vertical glazing rabbet.

A plug, made from aluminum or neoprene and laid into a bed of a compatible butyl sealant, is needed for the vertical glazing rabbet profile. The plug should be held back from the edge of the metal

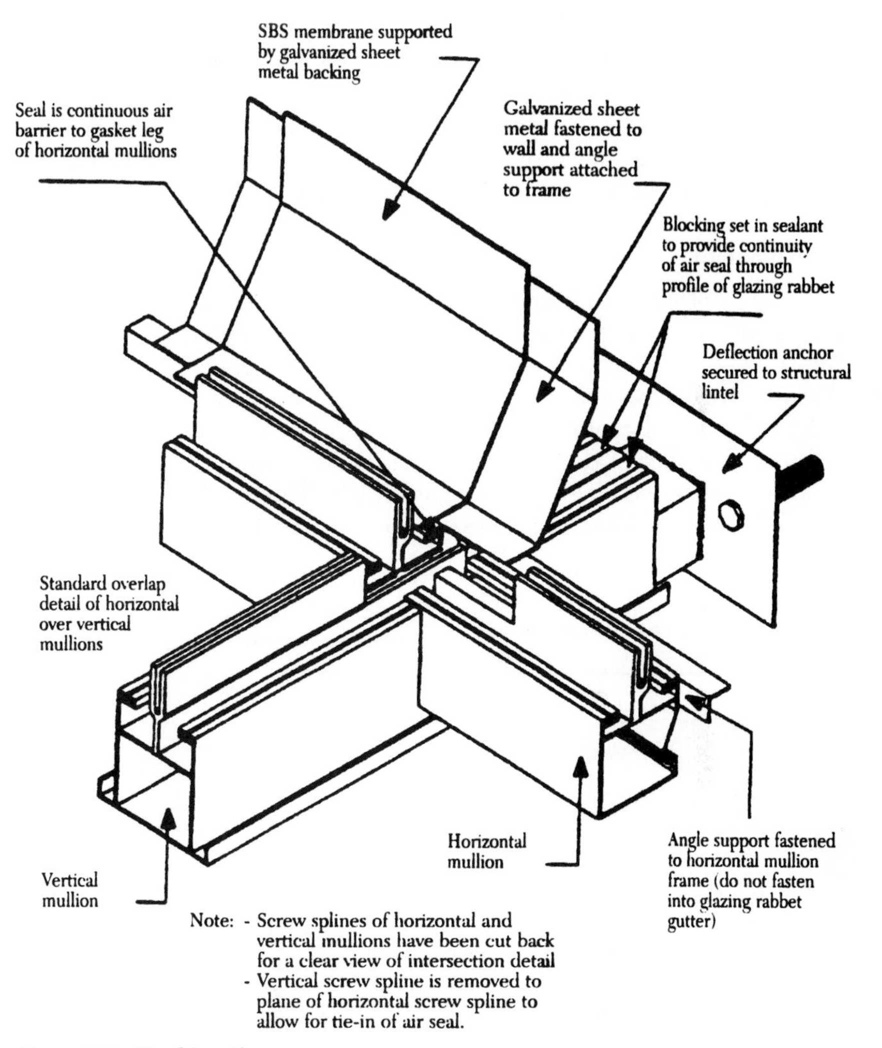

Figure 20.6 Head junction.

backer and membrane that overlaps it to create a drip. This prevents the butyl seal from getting wet should water enter the systems from construction above.

Water that enters through exterior joints or seals is contained on the membrane plane, drained to the glazing rabbet of the horizontal mullions, and, from there, into the vertical mullions. Insulation is installed on the membrane's exterior. Flashing on the exterior is again designed to shed water but is not relied on to be completely watertight. The flashing can be installed over the pressure plate and covercap system, but it must be removed should glazing replacement be necessary. Installing flashing under the pressure plates provides mechanical fastening and a finished look similar to that of the glazing itself. Only the joint created between the pressure plate gaskets at the verticals would need to be sealed. A plug or sealant can be used to minimize water entry and must be maintained.

Jamb junction

The jamb detail is similar to the other two details in that all joints that could have water on their surfaces are shingled, as shown in Figure 20.7. The vertical end wall should be designed as an exterior-glazed and pressure-equalized curtain wall framing system. Galvanized sheet metal is installed so that its exterior face is flush with the tube face of the glazing rabbet and is brought flush to the top surface of the raised leg's glazing gasket key. This can easily be accomplished with aluminum angles fastened to the typical tube profiles.

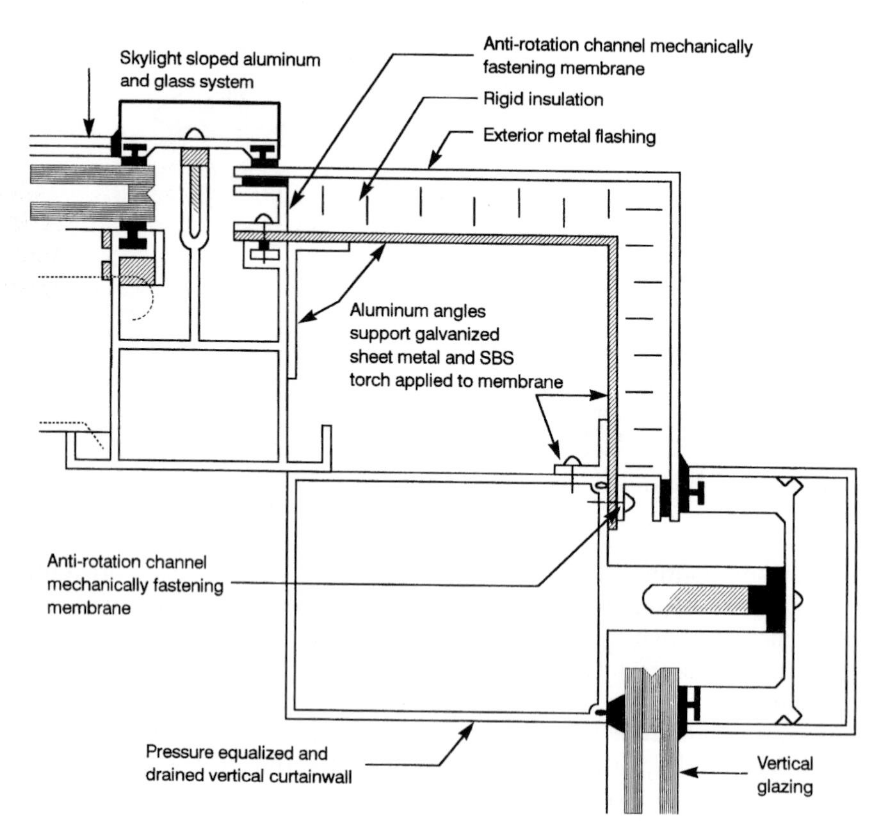

Figure 20.7 Jamb junction.

These angles must not be fastened through into the glazing rabbet gutter of the sloped glazing system. The surface of the galvanized backer and the aluminum surfaces, to which a membrane is to adhere, are primed, and a reinforced SBS membrane is installed in shingle fashion from the tube face of the curtain wall glazing rabbet to the top face of the sloped glazing framework's key profile raised leg. The membrane with the galvanized sheet metal backing now functions as an air barrier and the plane of waterproofing.

Water entering through any joint in the exterior cladding or seal is either drained into the glazing rabbet of either aluminum frame system or directed onto the slope of the membrane back to the exterior at the sill. Antirotation channels are installed to provide support for the pressure plates and to mechanically retain the membrane.

Polystyrene type IV is installed on the exterior of the membrane and then covered with finished metal. All joints from the top of the sloped glazing are shingled and drained so that the vulnerable sealing products used to join the aluminum profiles and glazing to aluminum are not subjected to wetting.

Curved Sloped Glazing

The beauty of sloped glazing in contemporary architecture is evidenced by its vast popularity. Add some curves to the slope, and it has been complimented and proves aesthetically pleasing. Closed sloped glazing has its roots in the glass enclosure family of greenhouses, skylights, walkways, and domes. Within the same family, featuring curved sloped glazing, are sidewalk cafes, patio enclosures, canopies, and passive solar collectors or sun rooms.

The most common framing systems are the wetband flange system ("I"-beam system) and the tubular system. The tubular system is gaining popularity because of its pleasing sight lines and minimally exposed fasteners. Both systems usually employ pressure glazing, often with a snap-on cap to hide exterior fasteners and for weathertightness. Pressure-glazed systems can easily be designed as thermally broken by not allowing the exterior pressure plate to directly touch any interior metal at proper compression.

The main rafters should be treated similarly to conventional sloped glazing. They must be of sufficient strength to adequately support the additional dead and live loads, such as snow and ice, rain accumulation, wind force, and exterior particles. The rafters should provide gutters to weep out any water that might have seeped through. Secondary condensation gutters are often preferred to control condensation from the glass and metal and to act as a backup for any water leakage.

The depth and wall thickness of the tube or "I" section are the main factors affecting strength and rigidity of the main rafter. Because depth and wall thickness affect bending difficulty and/or practicality, during the design stage it is important to choose the proper member for the job, one that is both strong enough and can be bent practically to the desired radii.

To maximize strength with ease of fabrication, it is preferable to have rafters bent in one piece if possible. If this is impractical because of bending difficulty or because the rafter is too long for a continuous bar, either a splice is needed or a transition member is required. The transition member should contain a gutter to weep to the exterior any water that might have accumulated internally.

Purlins (horizontal members) should have a low profile on the exterior to allow water to pass freely over them. They should contain gutter systems to transport water to the main rafter where it will eventually be weeped to the exterior.

The entire framing system must be compatible with the glazing material. It must have sufficient edge bite and minimal deflection to function properly as a weathertight and safe glazing structure. The framing must be fabricated and installed properly to maximize the beauty of the curve.[4]

Suspended Glazing

One method of providing a frameless glazing facade is to fix together a matrix of toughened glass lites, hung from the building structure.[5] A system of this type, commonly referred to as a suspended glass assembly, was designed and developed in the 1960s by Pilkington Glass Limited. It allowed designers to glaze large openings in buildings, without using metal frames or mullions, to create light and space with a minimum of visual barriers. Figure 20.8 illustrates a suspended glass assembly.

The system comprises a series of specially processed and toughened glass lites bolted together at their corners by small metal patch fittings. Pane-to-pane joints are sealed with a silicone building sealant, and toughened glass stabilizers are used at each vertical joint to provide lateral stiffness against wind loading. The assembly is suspended from the building structure by hangers bolted to its top edge and is sealed to the building in peripheral channels by neoprene strips or nonsetting mastic.

The concept of the design ensures that the facade is, at all times, "floating" in the peripheral channeling, and problems, which might arise due to the differential movement between components, are eliminated. Assemblies, therefore, can be used to advantage when the design must account for vibratory or seismic forces.

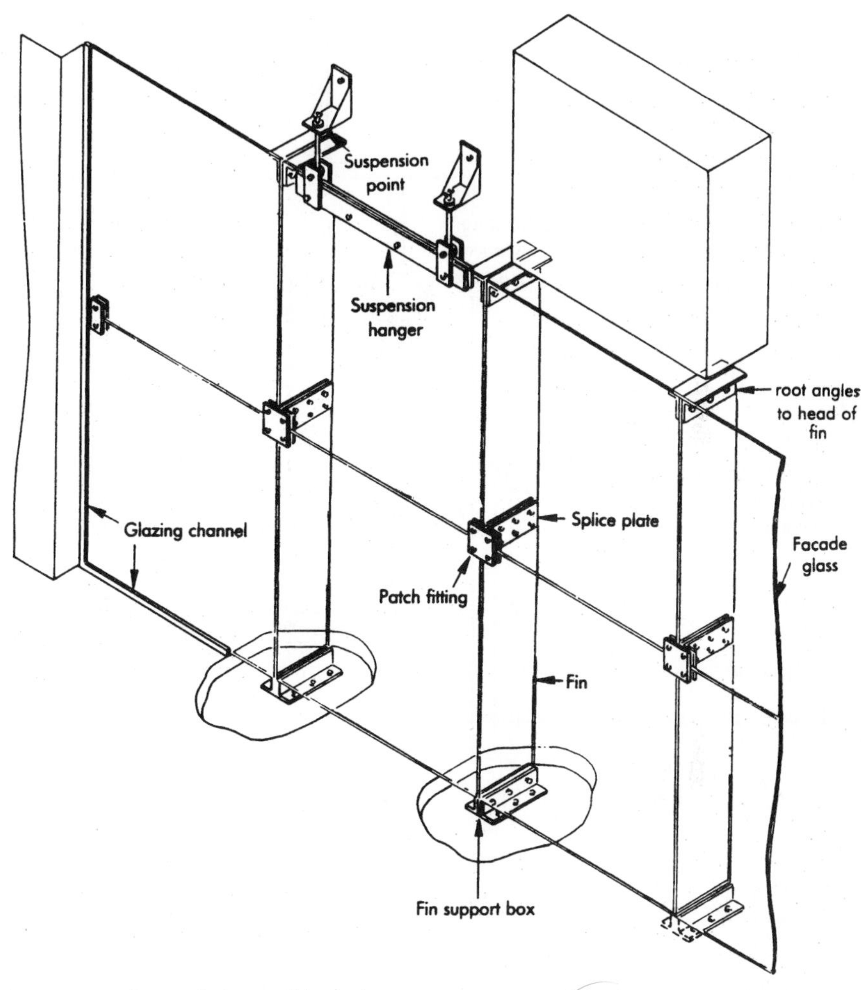

Figure 20.8 Suspended assembly design.

Weather sealing is carried out at all joints in the facade using a structural silicone building sealant. In the design calculations, the structural properties of the sealant in providing greater stiffness to the facade are not recognized. However, from extensive laboratory and on-site testing, it is known that the sealant does improve the load-bearing capabilities of the facade, and its use, therefore, is an added safety factor in the design.

The principle behind the design of the fittings for a suspended glazing assembly is that all in-place forces transferred between components are resisted by friction developed at the metal/gasket/glass

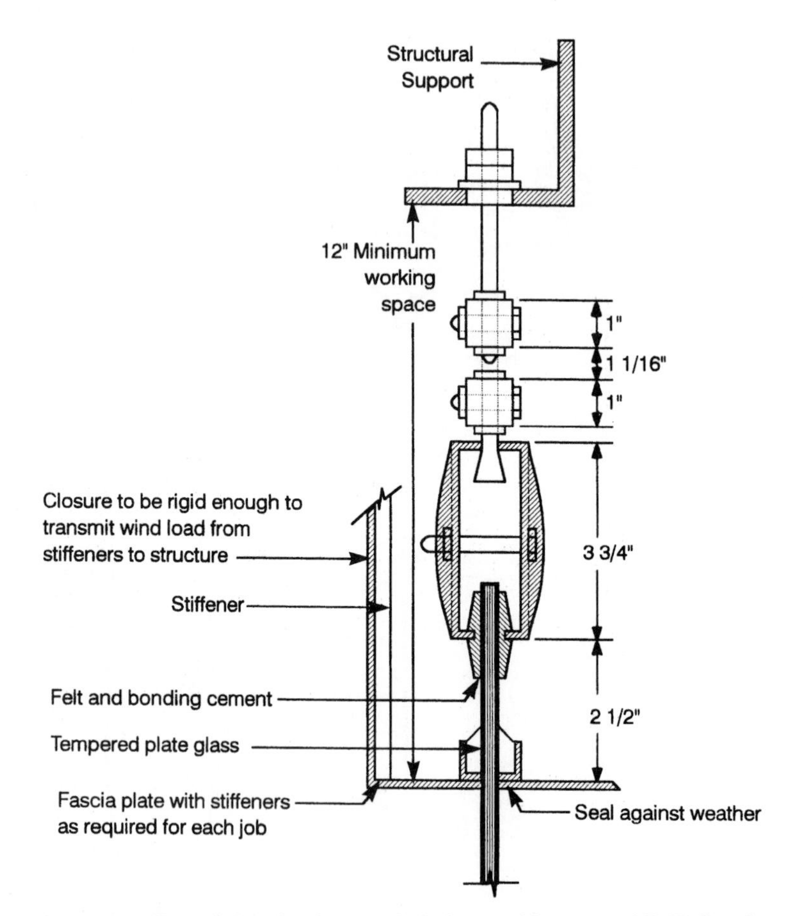

Figure 20.9 Typical details of suspended glazing. (*Courtesy of F. H. Sparks Co., Inc., New York, NY.*)

interfaces, arising from the tension developed in the fixing bolts. Figure 20.9 illustrates the details of the suspended assembly.

The use of friction to transfer forces between glass and fittings makes it essential that bolts of the correct size and quality, tightened to the specified torque, are used. Although the holes in the glass can resist considerable bearing forces from the bolt through the hard bushing, this is only taken into account in the height limitation for assembly constructions. The friction grip is of particular importance in the design splice joints and root support of the stabilizers, where the bearing strength of the holes is unlikely to be able to resist the turning moments generated in the stabilizers when an assembly is subjected to wind forces.

If required, the coefficient of friction at the metal/gasket/glass interface can be enhanced by applying a suitable adhesive.

The facade lites resist lateral wind forces through the small metal patch plates supporting the four corners of adjacent lites off the stabilizers. These metal patch plates clamp the glass at the corners of each pane, developing significant stress concentrations at the edges of the patch plate and around the bolt holes. To safely design panes supported in this way, it is essential to have a detailed knowledge of the stresses generated around and under the patch plates, for various shapes of panes subjected to different levels of lateral load. Equally essential is knowledge of the strength of the toughened glass at and around the fixing holes.

The size of suspended assembly facade lites is rarely limited by deflections. The clamping effect of the patch plates, which reduces deflections, together with the relatively high stresses generated, dictates that most assembly facade lites are stress limited rather than deflection limited in design.

Extensive research into both stress and strength angles has made the glass producer confident in the successful design of large facades built using a single system. Single assemblies can be designed up to 66 ft (20.12 m) in height on a 5-ft (1.524-m) module and up to 76 ft (23.17 m) on a 4-ft (1.22-m) module. Assuming an adequate main building structure, any height can be specified using multiple assembly design. Any length is possible, and curved facades are not unusual.

Countersunk Bolt Fixings (Planar System)

To meet designers' requirements and achieve a flush and uninterrupted glass surfaced facade, Pilkington developed the Planar system using countersunk bolt fixings, as shown in Figure 20.10.

One of the features of a suspended glazing assembly was that it could not be used in conjunction with sealed insulating glass units or nonvertical applications. The Planar system can be used for both. It is capable of fixing either single or insulating toughened glass to any structure. In some cases, glass mullions are used to form part of the substructure to which the glass is attached. The system is used for vertical or sloped glazing and can be incorporated as a complete cladding system.

The principle behind the design of the Planar fittings is almost exactly opposite to that of suspended assembly fittings. The fittings are designed to support the weight of the glass by direct bearing of the bolt through the bushing on the hole in the glass. This feature is made possible by fixing each lite separately, not supporting the

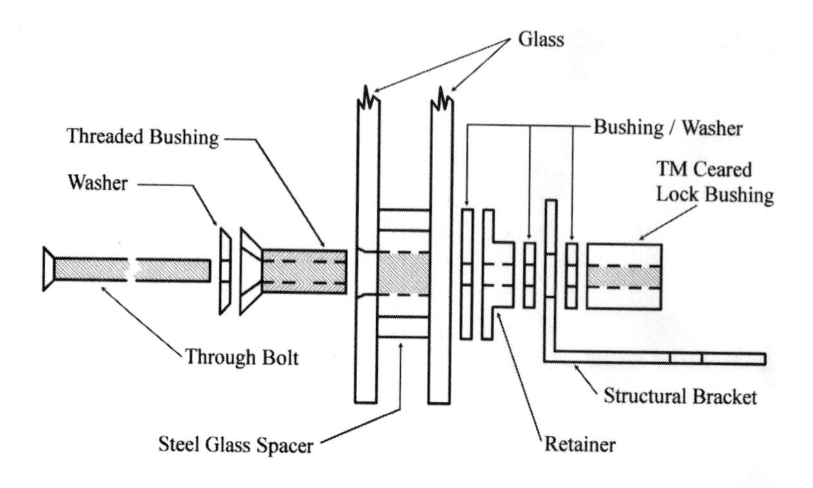

Note: All Parts Stainless Steel

Figure 20.10 Countersunk bolt fixings (Planar system).

weight of those below it. The fitting is also designed to give minimal clamping by attaching the fixing bolt to a spring plate, which is sufficiently flexible to allow rotation of the glass. The overall effect is to significantly reduce the stresses developed in the glass in the region of the Planar fitting, compared with those developed around the patch plates.

Because frictional forces and hence the clamping forces needed to generate them do not play a key role in the design of the Planar system, it has proved possible to develop an insulating glass version. In this, the outer lite provides the main load-bearing capability, whereas the inner lite is located, by ingenious design of the Planar fixing components, in a way which avoids crushing the unit edge seal.

Careful research into the stresses developed in the Planar facade lites, especially around the fixing holes in both single and double glazes of various shapes and fixing arrangements, is required to successfully design the Planar system.

The size of the Planar facade lites, particularly in single glazing, is more likely to be deflection limited than the facade lites in a suspended glazing system because of the flexibility, or lack of clamping, provided by the Planar fixing, combined with reduced stresses generated around the installation.

Because the glass lites are individually mounted to the structure, there is no restriction on the height of building which can be glazed. The specially engineered bushings, bolts, and fittings are standard for all designs, and spring plates may be designed and fabricated to suit each specific application.

References

1. Peterson, C. O., Jr., Sloped Glazing: Unconventional Curtain Walls, *Glass Digest,* 3/1980.
2. Makepeace, C. B., and J. O'Connor, The Glass Roof: Sloped Glazing Need Not Leak, *The Construction Specifier,* 11/1996.
3. Schultheis, J. A., Guidelines for Sloped Glazing. In *Architects' Guide to Glass, Metal, and Glazing,* 1984.
4. Sussman, S., Curved Sloped Glazing, Another Aspect of the Market Which Merits Your Attention, *Glass Digest,* August 1, 1982.
5. Button, D., and B. Pye, *Glass in Building,* Butterworth Architecture, 1994.

Chapter

21

Glass Washing

Joseph S. Amstock

President, Professional Adhesive and Sealant Systems

Introduction

Attractive, efficient use of glass requires a clean glass surface. Companies have developed guidelines for washing primary glass based on detailed research investigations, years of first-hand production cleaning experience for coated glass, insulating glass units, and heat-treated glass, and work in solving hundreds of customer glass-cleaning problems.

Because dirt affects glass undergoing processing, glass surfaces must be cleaned and washed effectively. Unclean glass loses many aesthetic qualities. If dirt is not removed effectively, it inhibits processes where adhesion of other materials to glass, such as sealants and coatings, is necessary, and it adversely affects processes, such as heat-strengthening, tempering, and bending.[1]

Categories of Dirt

Dirt on glass is any unwanted material on its surface. The degree of dirtiness or acceptable cleanliness depends on the end use.

There are four general categories of dirt on glass surfaces: particulates, surface residues, reaction contaminants, and surface corrosion. Glass fabricators commonly encounter only particulates and surface residues, and these will receive primary attention in this chapter.

Particulates are solid materials loosely deposited on the glass surface. They are best removed with nonabrasive mechanical methods such as a high-pressure water spray. Typical glass pack particulates include paper interleaving fibers, acrylic beads, wood flour and other powder interleaving materials, glass handling chips, wood splinters, cardboard fragments, and other packaging materials.

Surface residues are contaminants cleaned from glass with detergent washing solutions. Examples are fingerprints, paper scum, and cutting oils.

Surface corrosion technically is not dirt, although a corroded glass surface will appear semiopaque and this can be mistaken for deposited dirt. Surface corrosion is irreparable and usually results from inappropriate storage under high temperature and humidity. To prevent surface corrosion, it is recommended that glass be stored properly and a detailed inventory be maintained so that stock can be used on the first-in, first-out basis.

In its early stages, glass surface corrosion is visible only under critical light or occasionally after coating application. It cannot be observed by the naked eye. At this point and before fabricating, it may be possible to remove the corroded glass layer(s) with an abrasive (e.g., cerium oxide) to reveal underlying, undamaged glass. Once glass corrosion becomes severe, a heavy blue or white staining is visible on the glass surface, the glass matrix has undergone degradation, and glass transparency cannot be restored. The only practical remedy at this stage is to replace the glass.

Protecting Glass

To protect glass surfaces from corrosion and scratching, a chemical and physical barrier must be maintained between adjacent glass lites when stacked for shipment and storage. This dual function is performed by interleaving. Paper and powder are the two major interleaving materials used by the flat glass industry.

As interleaving, paper serves as a physical separator between lites and can be acid-treated to counteract alkaline buildup on the glass surfaces in a pack. (Soda-lime glass surfaces are alkaline.) However, it is not practical to use paper interleaving with automatic glass handling equipment. Current procedures require that paper be removed manually from a glass pack, and then stacked or collected, again by hand, for reuse or disposal.

Interleaving compatible with mechanical glass packing, which eliminates the possibility of damage from manual handling, consists of plastic beads that absorb acidic corrosion inhibitors.

During glass processing, interleaving, regardless of type, must be removed. After paper sheets are removed, glass surfaces may exhibit a haze or scum that originates from various extractable organic materials. In most cases, this scum is a surface residue that can be removed with a detergent solution. Sometimes, however, these organic materials may react with alkali in the glass surface to form a reaction contaminant, commonly called a paper stain, that is significantly more difficult to remove.

Powder interleaving can be removed by cleaning procedures for particulate contaminants. When water is applied to a powder-interleaved glass surface, the water-soluble corrosion inhibitors dissolve and the flow of water carries them away from the glass surface along with the separator beads. For this reason, a prewash system is especially suitable for powder interleaving removal.

When glass sizes preclude mechanical washing or a washer is unavailable, manual cleaning may be required. In addition, some water-insoluble contaminants, such as adhesive from tapes and labels, heavy grease, and tar, may have to be precleaned from glass surfaces prior to washer entry.

Manual glass cleaning

To hand clean incoming primary glass (particularly glass with powder interleaving), first, thoroughly wet the surface with plain water. Remove water from the glass with a sponge and dry remaining water with a clean, lint-free cloth. Incoming glass with dirty surface residues should be wet first with a water solution that contains a low-foaming, easily rinsed detergent formulated for industrial glass cleaning. Manipulate the detergent solution with a sponge or clean cloth, loosening surface residue, thoroughly rinse the glass with plain water, and use a squeegee or lint-free cloth to remove water completely.

Glass cleaners for household use are effective primarily for removing surface residue and are not the best candidates for use by glass fabricators who more frequently encounter particulates. Many commercial cleaners contain large amounts of alcohols suitable for dissolving and removing oily surface residues, but they are not always suitable for removing particulates. Particulates are more effectively removed with copious amounts of water, although nonabrasive mechanical removal is best. Alcohols in commercial cleaners may make removal more difficult, because they dissolve or partially dissolve some of these solids and subsequently smear them over glass surfaces rather than carry them away.

If a contaminant remains on glass after applying a normal detergent solution, the contaminant should be identified. Adhesives from tapes or labels, for example, may not be removed by detergent, alcohol-based cleaners or organic solvents, such as paint thinner, toluene, or mineral spirits, but a combination of materials may be needed to remove them. Care must be taken in selecting solvents for use on coated glass.

Mechanical glass washing

The key elements in mechanically cleaning glass are water, detergent, and washer system, which all must be selected carefully so that they work together effectively to clean glass.

Water is the foundation of the washer system. At high pressure, it physically removes particulates. It delivers detergent for removing surface residues, lubricates brushes contacting the glass, and carries away dirt.

Because water is important in cleaning, its quality must not be overlooked. Water used in glass cleaning should have minimum turbidity (solid content) and suitable hardness or softness to allow detergent dissolution and easy rinsing and to prevent buildup of excessive washer scale.

Appropriate filters on main water supply lines and within wash system flow lines will remove water turbidity and internally generated washer debris. Simple, large-mesh screens are not sufficient. Experience indicates that 25-μ replaceable cartridge filters, readily available from industrial plumbing suppliers, are more suitable. They reduce glass surface scratching or recontamination in the washer and lower incidence of nozzle blockage from water-borne particulates. The pH of water used in the detergent section of glass washers will be altered by cleaning additives. The pH range for incoming water cannot be specified, although a neutral pH of 6.0 to 7.0 is suggested.

Warm water dissolves detergents more readily and enhances cleaning. The wash water should be between 100 and 140°F (37.8 and 60°C).

Demineralized or deionized water should be used in the rinse water section for residue-free drying, although this water treatment may increase water corrosiveness and damage the metal washer parts. It is recommended that the washer manufacturer be consulted to determine if using demineralized or deionized water will adversely affect washer equipment.

Warm water rinses more effectively and aids drying. Rinse water does not need to be warmer than 140°F (60°C).

A detergent should be used in all washer systems. Surface residues are removed most effectively with detergent solutions, not plain water. In addition, detergents decrease the potential for glass surface scratching in washing systems.

Usually, a solution of no more than 1 percent detergent in water is sufficient to clean glass. Introducing a good detergent into a system that previously used water only, initially will liberate accumulated dirt and scale. Therefore, the detergent should be introduced during nonproduction periods. Draining and recharging prior to actual production may have to be done more than once to flush the system.

Detergents

There is no universally effective glass cleaning detergent. The best way to select a detergent is by trial and error. When choosing a detergent, consider these factors.

1. *Solubility.* Will a 1 percent (by weight for solids, by volume for liquids) mixture of the detergent and your water provide a clear, uniform solution?
2. *Foaming characteristics.* Will detergent foam excessively in your washer? To screen detergents, pour 50 cc of a 1 percent detergent solution into a 250-cc graduated cylinder, agitate vigorously, and observe foam volume. Foam volume for a low-foaming detergent should not exceed 80 cc.
3. *Rinsing characteristics.* If a 12-in-square glass sample is hand-washed with a 1 percent detergent solution, is excessive rinsing needed to remove the detergent?
4. *Environmental safety characteristics.* The chemical composition of the detergent must meet the specific environmental requirements for your water treatment area. The discharge must comply with government environmental requirements. Personnel need to wear protective equipment, such as safety glasses. This is generally required by the detergent manufacturer.
5. *Local availability.* The detergent must be readily available or a large inventory is needed.
6. *Supplier reliability.*

In general, it is always important that acidic detergents rinse from glass surfaces more readily than basic (alkaline) detergents. Acidic detergents may etch some metal components of the washer, in which case stainless construction may be required. Consult the washer manufacturer to see if acidic conditions adversely affect the washer.

After a detergent is selected, it should be tested before full-scale production is begun.

Washers

Automatic glass washers are designed to remove water-soluble surface residues but cannot remove all water-insoluble surface residues. These washers consist of a detergent application (washing) section, one or more rinse sections, and a drying section. During washing, cylindrical or oscillating cup brushes provide mechanical action that operates with the detergent to clean the glass. Figure 21.1 is a photograph of an automatic vertical glass washer.[2]

Figure 21.1 Automatic vertical washer-dryer. (*Courtesy of Castlemec, Tezze sul Brenta, Italy.*)

Figure 21.2 Horizontal flat glass washer. (*Courtesy of Sommer & Maca, Chicago, IL.*)

Proper glass loading for optimum washer performance requires glass spacing greater than the roll circumference. Conveyor speeds should provide suitable washer dwell time, and conveyors should not be stopped while glass is under air knives. Figure 21.2 illustrates a horizontal flat glass washer with "flip-top."[3]

Although a typical glass washer will clean surface residues from glass, particulates can cause system problems, especially in recirculating washers unequipped with appropriate waterflow filters. Glass-handling chips or other gritty particulates that enter the washing section will be mobilized on the glass surface by the washer brushes, causing scratches. Nonabrasive particulates accumulate on washer brushes and in solution tanks.

Prewash

Recognizing the need to optimize washer performance by preventing particulate entry, companies have conducted research in glass cleaning that led to development of a simple, compact, economical, washer add-on called the prewash.

Removing particulates from the glass surface before entering a line washer provides optimum washer performance. Incorporating a prewash into existing cleaning systems allows fabricators worry-free use of powder interleaving.

This prewash has two phases. In the first phase, a dilute detergent solution is applied to the glass surface. In the second phase, detergent residue and particulate debris are removed with a high-pressure filtered water rinse.

The dilute, filtered, detergent solution can be applied with a low-pressure (10 to 15 psi) spray or a gravity-fed drip pipe, depending on conveyor line speed. If conveyor speeds allow detergent solution to completely wet the glass surface by drip-pipe application, pumps are unnecessary.

Detergent solution application, essential to the prewash system, wets the glass so that the high-pressure water rinse will flow readily over the entire glass surface, and it decreases attractive forces the particulates may have for one another, for the glass surface, or for the system components. This facilitates particulate removal and disposal.

In the second phase of the prewash, glass enters a metal sheeting enclosure fitted with spray pipes delivering filtered water at about 150 psi. If water is recirculated in the production washer that follows the prewash, a high-pressure pipe may be needed below the conveyor to remove the particulates that transfer to the bottom surface of the glass during stacking. Otherwise, one or two high-pressure pipes spraying on the glass top surface will suffice for the prewash in a line with a nonrecirculating washer.

Water in a high-pressure spray may be from any source, as long as it is filtered through at least a 25 μ replaceable cartridge filter before traveling through delivery pipes. This room temperature water must be particulate-free. If water is acidic, stainless steel pipes, nozzles, and pumps are recommended.

Ideally, the high-pressure rinse water is not recirculated and carries away detergent residue and dirt. Water can be recirculated but must be filtered through 5- to 25-μ filters before reuse and not be reused longer than 16 to 24 production hours.

Brushes are not used in the prewash system because particulates collect on a brush and redeposit on the glass surface.

Maintenance. The prewash system requires only three basic maintenance procedures.

Filter cartridges on detergent solution and high-pressure water lines must be changed when exhausted. Blocked or corroded nozzles may require occasional cleaning or replacement.

In a prewash system using recirculating high-pressure rinse water, holding tanks must be drained and rinsed at least every 24 hours.

Using a prewash system and installing appropriate filters within the main line washer will reduce the need for particulate-induced washer maintenance. Because water-soluble dirt will accumulate within the main line washer, routine washer maintenance is required, supplementing normal maintenance for washer mechanical equipment.

Scrubbing the tank or steam cleaning prevents scale accumulation and sediment on the tank walls and bottom.

Wash and rinse brushes must be checked and repositioned as required to maintain effective glass surface contact. Worn brushes should be replaced. Brushes may need occasional steam cleaning to remove heavy residue accumulation.

Knife, filters, and lines must be clean, and knife flow velocity and alignment checked periodically.

Scheduling maintenance activities depends on line throughput and general dirtiness of glass processed. The minimum general guidelines are that every 24 hours (or after 50,000 square feet of glass is processed) the prewash rinse water should be changed if it is recirculated. Every 48 hours, washer system brush contact and condition should be checked and the brushes replaced or adjusted as necessary.

Also the system filters should be checked (and replaced, if required) and the air knife alignment and flow velocity adjusted if necessary.

Every week, steam clean the washer system brushes, the washer tanks, and clean the conveyor rolls, if necessary.

If production problems occur which seem related to glass cleanliness, an unscheduled check of some or all system parameters may be indicated.

Coated glass

Architectural glass with pyrolytic coatings, cleaned in a washer system as previously outlined, can be treated as uncoated glass because of the coating durability. Abrasive cleaners should not be used for any coated glass. Coatings applied with magnetic sputtered vacuum deposition (MSVD) processes are not as durable as pyrolytic coatings. The new low-emissivity coatings are most sensitive.

Surface residues on visibly reflective MSVD coatings can be removed with a clean, grit-free cloth and commercial solvents, such as paint thinner, mineral spirits, or isopropyl alcohol. Household dish detergents should not be used, and care must be taken to assure that the solvent and residue do not affect the coating. Sample lites should be test cleaned and carefully examined to determine if there is any damage to the coating caused by the residue and solvent.

Coated glass should never be stopped beneath washer brushes when machine washing because the coating can be damaged. Because brush stiffness among machines can vary, test lites of coated glass should be run through the washer to make sure brushes will not harm the coating. When washing MSVD, low-emissivity, coated glass, the water ideally should not be recirculated because these coatings are pH sensitive. If water is recirculated, however, its pH level should be monitored.

It is essential to use demineralized or deionized water in the final rinse for cleaning coated glass to ensure residue-free drying. Because processes used to make demineralized or deionized water may increase its corrosiveness, sample lites should be tested in the water. Once it is determined that no harmful reactions are occurring, trial runs should be made prior to full-scale production to determine if the water works in the system.

Washer solution temperature and pH are interrelated. At a lower (more acidic) pH, elevated temperatures may enhance destructive action of the acid. Trial tests of the detergent solution should be made at the same temperature used in production. As with uncoated glass, the water temperature should not be higher than 140°F (60°C).

In late 1983, PPG installed and tested a prewash system on a vacuum coating line, a fabricating line that demands the most critically clean glass surfaces. The prewash system continues to operate as an easily maintained, compact, and economical means of reducing cleaning-related product defects, thereby increasing yields.

The concept behind the prewash system may be used as a guide in designing a prewash system for individual glass fabricators. The glass washing equipment manufacturer should be consulted to develop a tailor-made system meeting specific needs.

Reaction contaminants are more tenacious than surface residues and may physically or chemically bond or interact with the glass surface. For this reason, their removal may require chemical undercutting with a cleaning solution tailored for specific dissolution or bond-breaking.

Hard-water salts are common reaction contaminants, as are inorganic compounds. If glass is washed with water containing inorganic compounds, care must be taken to rinse the glass surfaces thoroughly with demineralized water before drying or further processing, such as coating or tempering. This prevents deposition of hard-water salts that may cause visible defects in the final product. If hard water is allowed to dry on glass surfaces, off-line cleaning with acid solutions may be required to remove salts. This may not be safe, environmentally correct, or economically feasible in production.

References

1. Schlueter, J. M., *Recommended Glass Washing Techniques Improve Productivity and Product Quality.*
2. Courtesy of Castelmec, Tezze sul Brenta, Italy.
3. Courtesy of Sommer & Maca, Chicago, IL.

Chapter

22

Glazing and Installation

Joseph S. Amstock

President, Professional Adhesive and Sealant Systems

Introduction

This chapter is an aid for preparing a specification for glazing of flat glass and insulating glass units.

Buildings are constructed in specific geographical locations for specific purposes. Location on its site, the building surroundings, such as trees, hills, seas, or oceans, and other buildings and structures, the need to perform certain tasks within the building, and many other factors will enter into an architect's thinking as he or she decides on the building configuration and the enclosure for the living and working spaces. Selection of windows begins at this point.

Protection of building occupants from rain, heat, cold, insects, vandals, etc., is the basic function of the total window system. The increased use of glass and sealed insulating glass units due to new regulations and the general market requirements for newer homes require improved thermal performance.[1] This chapter is necessary because of the increased use of insulating glass in buildings, as a result of the increased thermal insulation requirements of various worldwide groups since 1988 and the general market requirements for new houses and commercial structures. Further proposed changes are a certainty for the future.

The development of worldwide standards for insulating glass units should ensure that units perform better and last longer. Improvements in service life will be achieved only if the units are glazed properly into well-designed frames, using appropriate durable sealants, with good workmanship on site and in factory, and with adequate maintenance.[2]

This chapter is intended to provide information to architects, specifiers, contractors, window manufacturers, glazing contractors, and insulating glass manufacturers and should lead to more knowledgeable selection of window frames and finishes and improvement in glazing specifications. It has become increasingly obvious during recent years that the full potential service life of insulating glass units is not being achieved because the relevant factors are not understood. The main aim is to achieve a longer service life by providing additional information on the design requirements of the window frame and the insulating glass units, together with more specific details on workmanship, glazing procedures, and maintenance.

A glazing contractor or supplier of insulating glass units may not have control over all the factors affecting service life. By providing information about the more important factors, it is hoped that the people who have control will understand their responsibilities and take appropriate action.

Scope

This chapter gives the principles underlying correct glazing of glass and insulating glass units into window frames and doors in a variety of exposure conditions and includes descriptions of the glazing methods recommended by major specifying agencies in the U.S., Canada, U.K., and other countries around the world. This chapter is applicable to factory-glazed and on-site units. It also applies to conversion of existing single-glazed glass, but it should be noted that some types of window frames may need modification to meet current code requirements.

Principles of Glazing

The fundamental principles underlying the correct glazing of sheet glass and well-made insulating glass units are the following:

1. *Prevention of prolonged contact of moisture with the edge seal of the unit.* The major enemy of insulating glass units is water. If liquid water is trapped against the edge sealant of a unit for a long time, the adhesive bond of the sealant to the glass will fail. This will allow liquid water and/or moisture vapor to pass between the edge sealant and the glass, leading to excessive moisture vapor in the unit cavity and, ultimately, to condensation on the internal glass surface. In other words, under conditions of prolonged contact with liquid water, the unit will fail.

Water in the form of moisture vapor can permeate through the edge sealant into the unit cavity. The rate of permeation of moisture vapor depends on the properties of the edge sealant and on the concentration of moisture vapor. However low the rate of moisture vapor penetration, it is inevitable that, after a period of time, excess moisture vapor in the sealed unit cavity will occur, and condensation on the internal glass surfaces will result.

Moisture can penetrate through the surface to the rabbet area, either around or through the glazing system or through the window frame joints into the glazing system, from a variety of sources, such as

a. rainwater,
b. window cleaning operations,
c. condensation on the room side glass surfaces, and
d. condensation within window frame sections.

All glazing systems must protect the edge sealants of the unit by preventing access of water which penetrates to the seal or by ensuring that any water which penetrates as far as the edge seal is soon removed by drainage of the rabbet area.

Glazing methods for insulating glass under fall into two groups, known as

a. drained methods, and
b. solid bedding methods.

Drained methods are based on the principle that some water may penetrate the glazing space, and although this is kept to a minimum, provision is made in the design of the window frame to ensure that any water which does penetrate is removed by drainage and ventilation. Solid bedding methods protect the seal of the unit by embedding the edge in compound or sealant to prevent the ingress of moisture.

2. *Compatibility between edge sealants of the unit, the glazing materials, and, if applicable, coatings such as Low-E on the glass surface.* Edge seals have different compatibilities with different glazing materials. An edge seal and a glazing material are compatible if, when they are in direct contact with each other, the performance of either is not reduced in any way from that direct contact. When using glazing methods where there is likely to be contact between the edge seal and the glazing material, it is essential to check with the insulating glass manufacturer that the edge seal and the glazing material are compatible.

Some units are provided with an edge tape. In such cases, it is also essential to check with the manufacturer that the tape is compatible with the proposed glazing material.

3. *Protection of the edge seal of the insulating glass unit against weather and ultraviolet light.* The edge sealant on most sealed units can degrade if exposed to ultraviolet light. It is important, therefore,

to ensure that the rabbet fully covers the seal on a flush-edged unit. The exception to this rule is the special silicone sealant used to manufacture insulating glass units specifically designed for structural glazing applications.

4. *Quality of glazing workmanship.* Site failures and conditions and the skills of operators vary considerably. Supervision is difficult, and, as a result, the quality of workmanship may vary widely and with it the durability and reliability of the system. On-site glazing must conform to the architect's specifications and to later portions of this chapter. Compared with on-site glazing, factory glazing has the advantages of being easier to control for quality of workmanship and of being carried out in a cleaner, dry atmosphere, without problems of access to the window frames.

Unit design

Architects or specifiers should refer to ASTM-C-773/774 or another national standard for insulating glass units. There are many component parts used in the manufacture of insulating glass units. Other chapters in this handbook cover the different components in detail and should be used by the spec writers to develop more detailed specifications. Procedures should be in place to ensure that the unit design is met constantly. Many companies have third party certification for their units.

Window frame design

To be suitable for the glazing methods described in this chapter, window frames must meet the design requirements detailed as follows. This section details the important features of a window frame which have a significant effect on the performance of the glazing system and, ultimately, on the service life of the insulating glass unit glazed into the window frame. Designers and specifiers should check with the window frame manufacturer that the frame meets these design requirements before specifying a window. When selecting a window, the specifier should ensure that

1. The materials used are suitable for the glazing proposed. For example, the wood for a wood window should comply with the recommendation of NWWDA, and for a factory-painted steel window frame, the glazing materials should adhere well to the paint.

2. The rabbet height is sufficient for adequate edge cover on the unit and is sufficient for edge clearance around the unit (both of which depend on the size and type of unit), taking into account possible combinations of frame tight sizes and unit sizes.

3. The beads of sealant have the required dimensions to position and fix them correctly, and in the case of wood beads, that the quality of the material conforms to the national specifications.

4. The rabbet platform is of sufficient size for the required thickness of the glazing material on each side of the unit and to allow for the correct positioning and fixing of the beads.

5. The design of any drainage system is adequate for the rapid removal of any water that may enter the rabbet area.

Window frame materials and construction

If a window frame does not conform to the following design features, then it is likely that the service life of the glazed unit will be reduced.

1. Steel frames should be made from hot rolled steel. The coating on the steel surface can affect the choice of the glazing sealant material.

2. PVC frames are made from extruded profiles which should conform to the appropriate standards. The window frames should be fabricated from these profiles in accordance with the recommendations of the profile supplier and conform to the applicable standards.

The thermal movement of PVC window frames is greater than the movement of steel and wood sashes. Only glazing sealants that can accommodate this degree of movement should be considered.

3. Aluminum window frames are made of hollow extruded shapes. These window profiles should be fabricated from the shapes in accordance with the recommendations of the aluminum supplier and conform to a specific standard the industry has accepted.

The thermal movement of aluminum window frames is greater than the movement of steel and wood window frames. Only glazing materials that can accommodate this degree of movement should be used.

4. Wood frames: The basic design, in terms of the requirements for materials (including preservative), frame design, workmanship, construction, security and safety, weathertightness, operation, and strength performance should be in accordance with current applicable specifications.

To achieve good contact with glazing sealants, rabbet and beads should be smooth and free from knots and holes. The up stands of the rabbet should be in the same plane at the corners; i.e., there should be no steps in the corners.

Window frames with weep holes. All glazing for weep hole systems should have internal and external seals designed and applied to prevent the passage of water around or through the seal. The weep hole system of the window frame is intended as a backup for these primary sealants and must ensure the rapid removal of the water that penetrates to the rabbet platform.

The rapid removal of water from the glazing area to the outside is best achieved by incorporating a sloping edge. A slope with a 1-to-10 ratio (6°) is recommended for the bottom edge of the wood frame. Alternately, gutters with weep holes may be used for aluminum or PVC window frames. See Figure 22.1 for examples of placement of weep holes, setting blocks, etc.

A typical weep hole system may include $^{3}/_{8}$-in (10-mm)-diameter holes or slots at least $^{3}/_{4}$ in × $^{13}/_{14}$ in (20 ×5 mm), to ensure the passage of water. It is difficult to ensure the removal of water from the glazing area by weep or drainage holes alone. Water may be trapped or may be held by surface tension, as droplets or between opposing surfaces, even though the design should prevent this from happening. Some additional ventilation is necessary to dry out these areas. This is particularly true of window frames with horizontal glazing platforms.

There may be problems where the top edge of an insulating glass unit forms a collection area for moisture. Where this is likely, the moisture cannot be removed by weep holes, and ventilation systems are essential to remove it. The design of the weep hole and ventilation system is a complex matter and must be considered in conjunction with the total sealant system used in the glazing application. The degree of ventilation achieved depends on the size of the ventilation holes or slots and also on their position and accessibility to the prevailing winds. Slots $^{51}/_{64}$ × $^{1}/_{8}$ in (20 × 3 mm) can be adequate for ventilation in exposed parts of the window, but further ventilation is properly necessary in more protected parts.

To prevent wind-driven water from being carried into the window frame section, all holes and slots should be protected by fitting holes with hoods or by forming the slots in the window frame where they are not subject to direct wind and weather.

In general, there should be no weep holes near each end of the sill member, between the corner and the setting block, and there should be a central weep or drainage hole with additional holes, as necessary. Ventilation slots may be necessary at the head of the window, at each end of the head member, to ensure ventilation at the top of the unit.

It is advisable to mark the top of some factory-glazed, fixed lite units because the difference between the head and the sill regarding drainage and ventilation may not be immediately obvious. Similarly, for on-site glazed windows, top and bottom beads should be identified. Typical weep, drainage, and ventilation holes are shown in Figure 22.2.

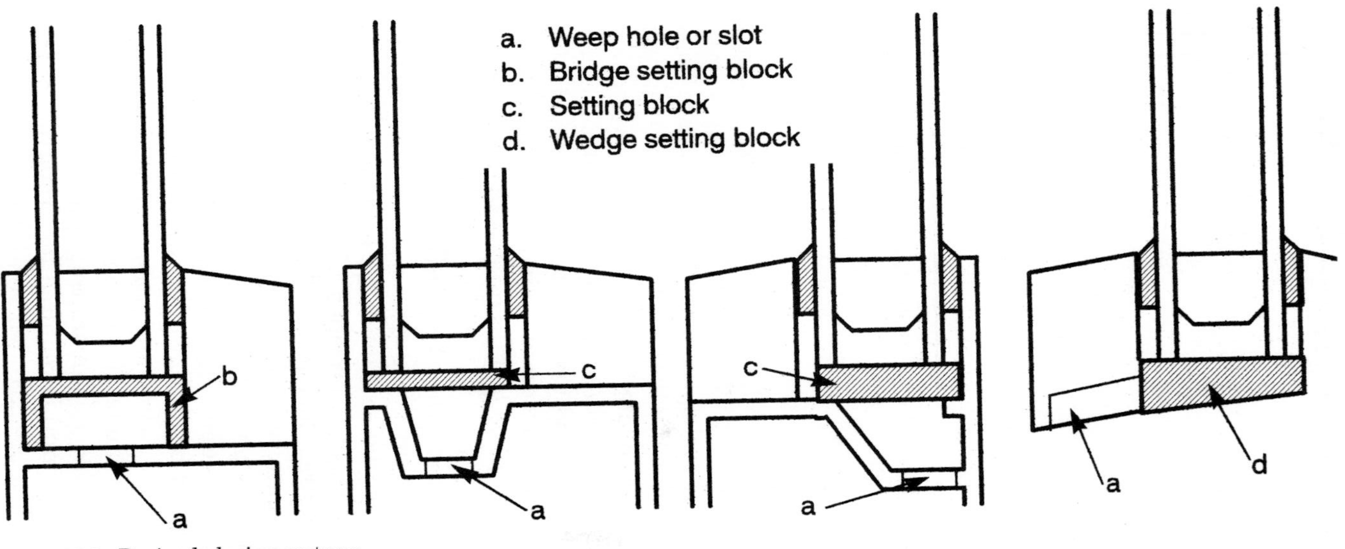

Figure 22.1 Drained glazing systems.

Figure 22.2 Weep-drainage and ventilation holes.* = vents. – = setting block.

Edge clearance. This is necessary to prevent window frame-to-glass contact and to prevent water from bridging between the rabbet platform and the edge seal of the insulating glass unit. The edge clearance should be sufficient to allow for thermal movement and should be a minimum of $^{15}/_{64}$ in (6 mm) at the sill to prevent bridging by water. For well-drained and ventilated window frames, the clearance can be reduced for the side and top rabbets to a minimum of $^{1}/_{8}$ in (3 mm) in the case of glass lengths up to $^{5}/_{64}$ in (2 mm), and $^{13}/_{64}$ in (5 mm) minimum in the case of glass lengths over $^{5}/_{64}$ in (2 mm).

Edge cover. An edge cover is required around the edge of an insulating glass unit to provide adequate mechanical support for the unit and to protect the edge sealant of the unit from ultraviolet rays.

Mechanical support is provided by the rabbet up stand and bead. The amount required depends on the type and thickness of glass and the degree of exposure of the building. Advice on the cover required can be sought from glass suppliers.

Protection of the edge sealant from ultraviolet light is also provided by the sealant and bead. However, in some cases, for example, with steel window frames, some protection can be obtained from the glazing material which can be sloped to finish several fractions of an inch above the sight line. For insulating glass units, a minimum edge cover of $^{15}/_{32}$ in (12 mm) is required for protection of the sealant. For units made with a preextruded reinforced tape spacer, a minimum edge cover of $^{23}/_{64}$ in (9 mm) is required to protect the sealant.

TABLE 22.1 Maximum Unit and Frame Deviations

Insulating glass units	Length and height	Area up to 3 m^2	± 2 mm
		Area over 3 m^2	Consult manufacturer
	Thickness	Glass < 6 mm thick	± 1 mm
		Glass ≥ 6 mm thick	± 1.5 mm
Frames (tight size)		Wood or metal	± 2 mm
		PVC-u	± 4 at 23°C mm
		Concrete	Consult manufacturer

Note: A deviation, for example, of ± 2 mm represents a tolerance of 4 mm.

Rabbet height. The rabbet height should provide adequate edge cover to protect the edge sealant from ultraviolet light, allow sufficient edge clearance for good drainage, and take into account the size deviations for both the unit and the frame given in Table 22.1.

The following example for a wood window frame illustrates how the actual clearance and hence the required rabbet height are affected by the deviation of the frame tight edge size and unit size.

For a unit less than $\frac{5}{64}$ in (2 mm) long, which requires a minimum $\frac{1}{8}$-in (3-mm) edge clearance at the sides and top, $\frac{15}{64}$ in (6 mm) in the bottom rabbet, and $\frac{15}{32}$ in (12 mm) of edge cover, a minimum rabbet height of $\frac{23}{32}$ in (18 mm) at the top and sides and $\frac{51}{64}$ in (20 mm) at the bottom is required. This gives adequate drainage and protection of the edge sealants around all four edges for most, but not all, combinations of window frame and unit sizes based on the deviations given in Table 22.1.

If the same rabbet height is required around all four sides of the window frame, then this should be $\frac{3}{4}$ in (19 mm). The required rabbet height is the sum of the required edge cover plus the actual edge clearance which results from the deviation in Table 22.1. The normal size of the unit is calculated from the normal tight size of the window frame using the following equation:

$$\text{Normal unit length or height} = F - C - du - df$$

where F = nominal window frame tight size,
C = sum of the two required edge clearances, i.e., top and bottom,
du = maximum unit deviation, and
df = maximum window frame deviation.

Du and *df* are taken from Table 22.1.

Maximum protection for the edge sealant on all four sides of a unit is provided by the frame when a unit is in the window frame. The edge clearance will be the same at each side, and the clearance at the top will be the same as at the bottom.

For a unit less than ⅛ in (3.175 m^2) in area and up to $\frac{5}{64}$ in (2 mm) in length to be glazed into a weep hole wood frame, the minimum required edge clearance at the sides of the unit is ⅛ in (3 mm), and deviations on both the unit size are ± $\frac{5}{64}$ in (2 mm). See Table 22.1.

When the unit is centralized in the window frame, the edge clearance will be the same at each side of the unit, and so the *actual* edge clearance will be half the total edge clearance.

If the normal tight size of the window frame side to side is 40 in (1000 mm), the nominal unit width from the equation above is given by

$$\text{Nominal unit width} = 1000 - 6 - 2 - 2 = 990 \text{ mm}$$

The *actual* Edge Clearance for a centralized unit is given by

$$\text{Actual Edge Clearance} = \tfrac{1}{2} \times (\text{actual tight size} - \text{actual unit size})$$

The minimum actual edge clearance occurs when the minimum unit size is combined with the minimum tight size:

$$\text{Minimum Actual Edge Clearance At Sides} = \tfrac{1}{2} \times (998 - 992) = 3 \text{ mm}$$

The maximum *actual* edge clearance occurs when the minimum unit size is combined with the maximum tight size:

$$\text{Maximum Actual Edge Clearance at Sides} \\ = \tfrac{1}{2} \times (1002 - 988) = 7 \text{ mm}$$

The rabbet height required for total protection of the edge sealant from ultraviolet for all possible combinations of window frame tight size and unit size is given by

$$\text{Required rabbet height} \\ = \text{Required edge cover} + \text{maximum Actual edge clearance}$$

$$\text{Required rabbet height for sides} \\ = \tfrac{15}{32} \text{ in} + \tfrac{19}{64} \text{ in} = \tfrac{3}{4} \text{ in } (12 \text{ mm} + 7 \text{ mm} = 19 \text{ mm})$$

For the same unit in a window frame with top and bottom nominal tight size of 1000 mm, the minimum required edge clearance at the bottom is $\frac{15}{64}$ in (6 mm) and at the top ⅛ in (3 mm). The deviations of the unit size and frame tight size from Table 22.1 are ± $\frac{5}{64}$ in (2 mm).

From the above equation, nominal unit height = 1000 − 9 − 2 − 2 = 987 mm.

The *Minimum Actual Edge Clearances* will occur when the maximum unit size is combined with the maximum tight size.

Sum of Actual top and bottom edge clearances = 998 − 989 = 9 mm

The minimum edge clearance at the bottom for good drainage should be 15⁄64 in (6 mm), and, therefore, the *Minimum Actual Edge Clearance* at the Top = 9 − 6 = 3 mm

The *Maximum Actual Edge Clearance* will occur when the minimum unit size is centralized in an opening with the maximum tight size.

$$\text{Maximum Actual Edge Clearance at Top and Bottom} = \tfrac{1}{2} \times (1000 - 985) = \tfrac{1}{2} \times 15 = 7.5 \text{ mm}$$

The rabbets height of 23⁄32 in (18 mm) for the top and sides and 25⁄32 in (20 mm) for the bottom will ensure that adequate edge cover is achieved for most combinations of actual tight size and actual unit size, but they will not give adequate edge cover when the tight size approaches its maximum deviation and the unit size approaches the minimum deviation.

The tight size deviation for a PVC window frame is larger than the deviation for wood and steel frames, and so a minimum rabbet height greater than the quoted values will be required.

Rabbet platform. The width of the rabbet platform should be sufficient to provide the required front and back clearances and to provide sufficient contact with the glazing materials on the platform. The width of the rabbet platform should be equal to the sum of the front and back clearances, the nominal thickness of the unit, the width of the bead, and an allowance for the deviations of the unit and sealant bead thicknesses. For example, to achieve xx in (xx mm) of glazing sealant on each side of the unit, for a 9⁄16-in (14-mm) unit, the rabbet platform should be 9⁄16 in (14 mm + mm + the width of the bead, plus an allowance for the deviations on the insulating glass unit thickness, say, 3⁄64 in (1 mm), making a total of 53⁄64 in (21 mm) plus the width of the bead.

Beads. Wood beads should have a thickness equal to the rabbet height and a width in contact with the rabbet platform equal to or equal than the height to achieve firm fixing of the bead.

Preferably, top and side beads should not project beyond the face of the window frame and should be flush with the face. If the top beads project, this will encourage water penetration which could lead to reduced window life.

Window frames suitable for tape systems

Edge clearance is required for any type of glazing system, and careful consideration must be given to design these systems properly. Because the data is essentially the same as in the previous discussion, it will not be repeated. There are minor deviations for PVC, wood, and aluminum, and you can consult with the window manufacturer fabricating or supplying them for a specific building.

Setting blocks. As illustrated in Figure 22.3, setting blocks are used between the bottom edge of the unit and the window frame or surround to support and centralize the insulating glass unit in the opening. Setting blocks should be made of resilient, nonabsorbent, rotproof, compatible materials.

Setting blocks are required in one or two positions, depending on the window frame. Positions for setting blocks for all types of window frames are shown in Figure 22.4. Table 22.2 describes the preferred material.

Where it is necessary to avoid undue deflection of the window frame, the window manufacturer should specify the position of setting blocks as either

1. not less than $1\frac{11}{64}$ in (30 mm) from the corner, or
2. in position to coincide with the window fixing points if these are between $1\frac{11}{64}$ in (30 mm) from the corner and the quarter points.

For designs where the actual positions are critical or where blocks are required in additional positions, the positions should be given by the architect or window manufacturer. To ensure that both lites of glass are supported, the width of the setting block, in general, should

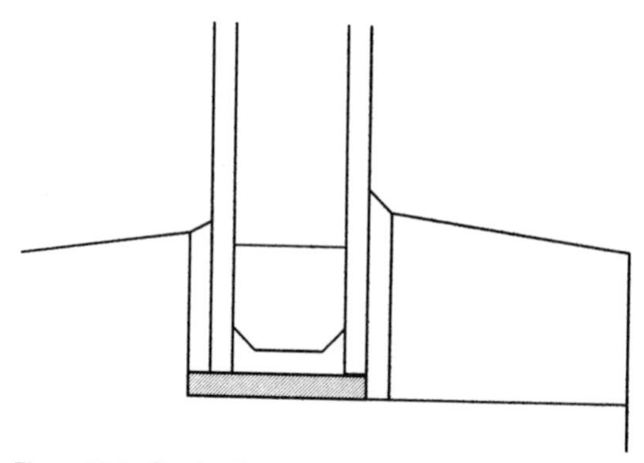

Figure 22.3 Setting blocks.

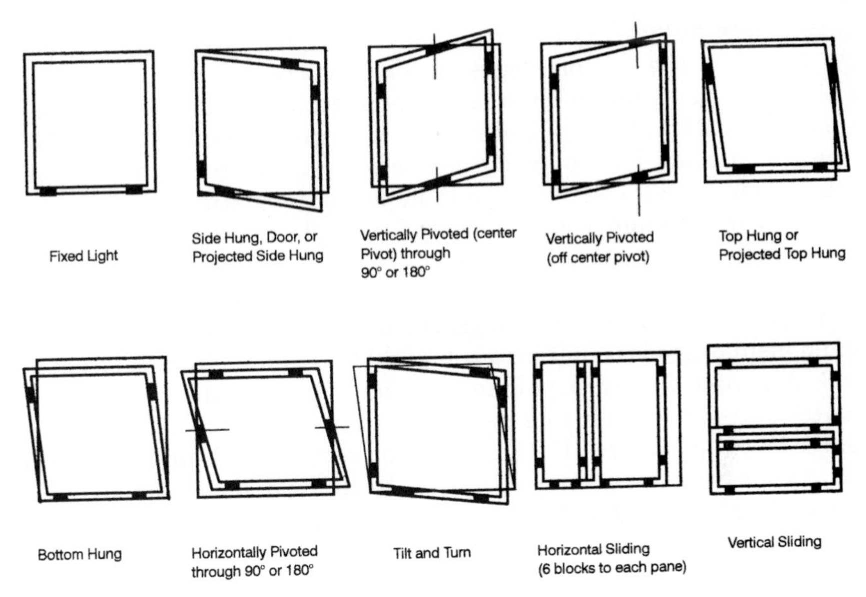

Figure 22.4 Positions of setting and location blocks.

TABLE 22.2 Setting Block Material

Block material	Max. area of insulating glass unit
Sealed hardwood, i.e., teak, mahogany	2.5 m^2
Rigid plastics, i.e., PVC, nylon, polystyrene	1.0 m^2
Plasticized PVC or neoprene (Shore A hardness 80–90)	0.8 m^2

be equal to the sum of the insulating glass unit thickness plus the back face clearance. In cases where the gasket or tape strip material is of such a dimension to prevent positioning of the setting block against the up stand, the width of the setting block should be the same as the width of the sealed unit. The thickness of the setting block is determined by the minimum edge clearance required at the bottom of the unit or by the actual edge clearance, when the unit is centralized.

The length of the setting block is determined by the area of the insulating glass unit and the block material. For handling purposes, the minimum length of each setting block should be $^{63}/_{64}$ in (25 mm). In cases where two setting blocks are required at two positions, the maximum area of the insulating glass unit for blocks of $^{63}/_{64}$-in (25-mm) length is given below:

1. Where the area of the insulating glass unit exceeds the values, the length of the setting blocks should be increased proportionately, either by a single larger block of the required length, or by using multiples of smaller blocks placed side by side. For example, $1^{31}/_{32}$-in (50-mm) plasticized PVC or neoprene setting blocks (or 2-in [50-mm] blocks side by side) would be suitable for an insulating glass area of 1.6 m^2; 4-in (100-mm) setting blocks of rigid plastics would be suitable for an insulating glass area of 4.0 m^2.

2. For vertically pivoted, side-hung and projected side-hung windows and for doors, where a single setting block is required, the maximum insulating glass unit area is half that calculated from Table 22.2.

3. Attention should be paid to the compatibility of the setting block materials with the glazing sealants of the sealed insulating glass units. When plasticized PVC or neoprene setting blocks are used with rigid window frames, compatibility with the window frame material should be also considered.

Setting block location. Location blocks are used between the edges of the insulating unit, other than the bottom edge, and the window frame or surround to prevent movement of the unit within the frame when the window or door is opened or closed, and to prevent the weight of the galls from causing the frame to become out-of-square. Minimum positions of location blocks and setting blocks for all types of frames are shown in Figure 22.4.

For designs where actual positions are critical, this should be stated by the architect or window manufacturer. To prevent movement of the unit in the window frame prior to and during installation of the factory-glazed windows, additional location blocks may be necessary.

Location blocks should be sufficiently resilient to accommodate thermal expansion and contraction of the glass and window frame without imposing stress on the glass. Blocks should be made of nonabsorbing material with a Shore A durometer range of 75 to 80.

Attention should be paid to the compatibility of the location blocks with the glazing materials and edge seal of the unit. In the case of PVC frames, compatibility with the window frame material should also be considered.

Spacers. Spacers are necessary to prevent displacement of the glazing sealants by wind pressure on the glass. Spacers, referred to as distance pieces, should be nonabsorbent material with a Shore A hardness of 74 to 80. They should be $^{63}/_{64}$ in (25 mm) long, of a height to suit the depth of the rabbet and the method of glazing, and of a thickness equal to the actual face clearances, retaining the unit firmly in the window frame, so that it cannot be displaced in service conditions.

1. Spacers are available in various sizes and shapes.
2. Spacers should be used in all cases, except where a load-bearing strip compound is used.
3. Spacers on both sides of the unit should be placed opposite each other.
4. Where beads are fixed by screws or over studs, spacer pieces should coincide with the bead fixing point.
5. Where beads fit into continuous grooves, the first spacers should be at approximately $^{31}/_{32}$ in (50 mm) from each corner and the remainder should be located at approximately $11^{13}/_{16}$-in (300-mm) centers.
6. A spacer should never coincide with a setting block or location block.

Glazing method selection

The following factors should be considered in selecting the glazing method:

1. exposure conditions,
2. window frame materials and finish,
3. window frame design,
4. glazing sealant compatibility,
5. type of glass used in the window,
6. conditions of use for the window.
7. on-site or factory glazed,
8. skills required by the glazing contractor,
9. inside or outside glazing,
10. accessibility,
11. maintenance,
12. color, and
13. economics.

Workmanship. The objective of securing the maximum service life for the glazing system can be achieved only by following the details of the specified glazing procedure. Not only is it important that the window frame design is correct and that an appropriate glazing method is selected, but also that the work is carried out to a good standard by following the glazing procedure exactly as specified. No steps should be omitted.

Storage and handling

All units must be handled with care. Insulating glass units delivered in cases must be unpacked on arrival and the glazing contractor must check that they conform to the specification. If units are found wet, dry them immediately. All units must be stored out of the sunlight to avoid thermal degradation. All units must be stored on their edges to prevent distortion and bowing. Units that absorb a considerable amount of heat, i.e., solar controlled units, are particularly vulnerable in service if the edge is damaged.

Some insulating glass units are manufactured with an aluminum barrier channel or tape around the edge to facilitate handling of the units during glazing. The contractor must ensure that such tapes, etc., are not damaged or displaced during transit or when being installed. It should be noted that damaged edge tapes may trap moisture against the seal of the unit. Units which have been factory-glazed into window frames will need care in handling due to the weight of the units. It may be advisable to label the windows or issue instructions on proper handling to avoid distortion.

Glazing methods

The methods described below are suitable for glazing flat glass and insulating glass units in normal situations into new frames or replacing failed or broken units in existing frames. In other situations, for example, when there is a higher than normal level of condensation, when the rabbet is contaminated by the previous glazing sealant, or when the rabbet had been damaged by the removal of a unit, then some minor variations of the basic methods illustrated may be acceptable. Information on suitable variations should be obtained from the sealant manufacturer.

The illustrations of the glazing methods which follow are intended to indicate the design principles and the use of glazing sealants. For clarity, setting and location blocks, spacers, bead fixings, and covers for holes and slots have been omitted from the illustrations. Where details for inside or outside glazing are substantially different, two examples are given. The glazing methods are illustrated in window frames of the material or materials most likely to be used with them, but providing the design of the rabbet and beads is suitable, window frames of other materials may also be appropriate.

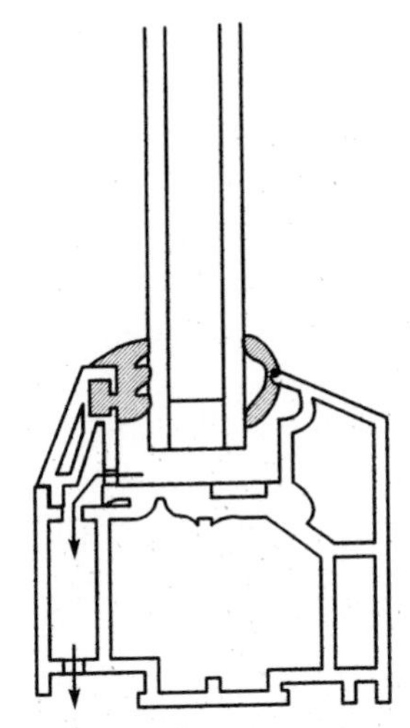

Figure 22.5 Method IG 1.

Method IG 1 insert gasket section. This is a weep hole method as seen in Figure 22.5.

The exposure rating is low, medium, or high depending on the window frame and gasket. A window frame of PVC with races, aluminum or wood, is suitable to accommodate the complementary gasket profile with appropriate weep holes.

Make sure that the window frame rabbet surfaces and the beads are clean. If the frame is mechanically jointed, apply the sealant to the window frame joints, if specified by the architect or window manufacturer.

If the gasket sections have not been supplied precut to size, cut them to the correct length. Where possible, gaskets should go around the window frame in one piece, nicked in the corners to ease the bending and with the joint at the top of the frame. Cut gaskets slightly over size to ensure that they are not under tension when assembled. Tension would cause creep-back in service. Fit the gaskets to the beads and the frame up stand where appropriate to the window frame design.

Place setting blocks in the window frame ensuring that they do not obstruct the weep holes. Clean the perimeter of the unit with a dry cloth, and insert the insulating glass unit into the frame. Center the unit in the window frame, and fit location blocks as required. Apply pressure by inserting the beads or by applying the beads and inserting a wedge-shaped gasket. The wedge-shaped gasket should be inserted at the corners first, and then worked outward from the center. Check that there are no gaps at the corners of the frame.

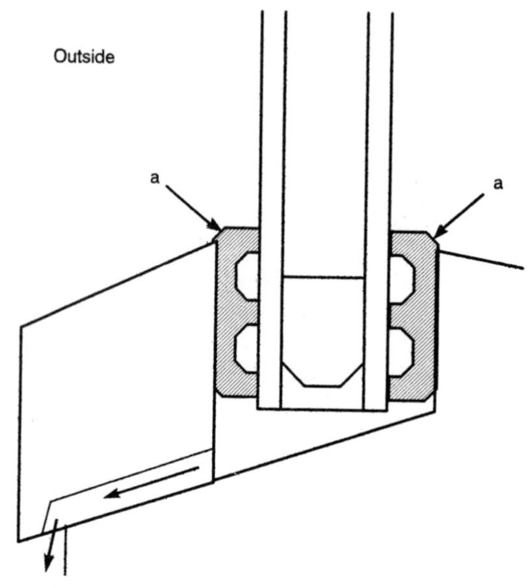

Figure 22.6 Method IG 2.

Method IG 2. This is a weep hole method that can utilize cellular adhesive sections or strips, load-bearing mastic tapes, or preshimmed mastic tapes, as illustrated in Figure 22.6.

This method also has a low, medium, or high exposure rating depending on the materials and design and is suitable with wood, metal, or PVC window frames with appropriate weep hole designs.

Front and back bedding with profiled, closed-cell synthetic rubber sections with self-adhesive backing, closed-cell synthetic rubber strip with self-adhesive backing, load-bearing mastic tapes, or preshimmed mastic tapes can be utilized in this type of application.

Ensure that the window frame rabbet surfaces and the beads and the areas of the glass unit, which will contact the glazing materials, are clean. Apply a self-adhesive cellular section, strip, or mastic tape to the rabbet up stand. Cellular sections or strips should be applied in line with the sight line. Mastic tapes should be applied slightly above the sight line.

Place setting blocks in the window frame ensuring that they do not obstruct the weep holes or other drainage provided. Insert the insulating glass into the window frame. Center and insert location blocks if required. Apply pressure to the bed of the unit against the section of tape or strip.

Apply the glazing material to the beads. Cellular sections or strips should be applied in line with the sight line. Mastic tapes should be applied slightly above the sight line. Bed the beads to the glass by applying pressure to compress both the inside and outside glazing materials, and fix the beads in position. If wood window frames are fixed with screws, these must be no more than $2^{61}/_{64}$ in (75 mm) from each corner and no more than $7^{7}/_{8}$ in (200 mm) between centers. If pins are used, they must be at least twice as long as the bead thickness, and fixed at no more than $5^{57}/_{64}$ in (150 mm) between centers. In the case of load-bearing or preshimmed mastic tapes, trim the excess material, which has been compressed above the sight line, with a sharp knife. Where possible, trim to form a proper drain-off.

Method IG 3. This is a weep hole method for load-bearing or preshimmed mastic tapes or cellular adhesive sections and sealant capping with low, medium, or high exposure depending on materials and design. It is suitable for wood and metal window frames. See Figure 22.7.

The same basic procedure can be followed as for the previous exam-

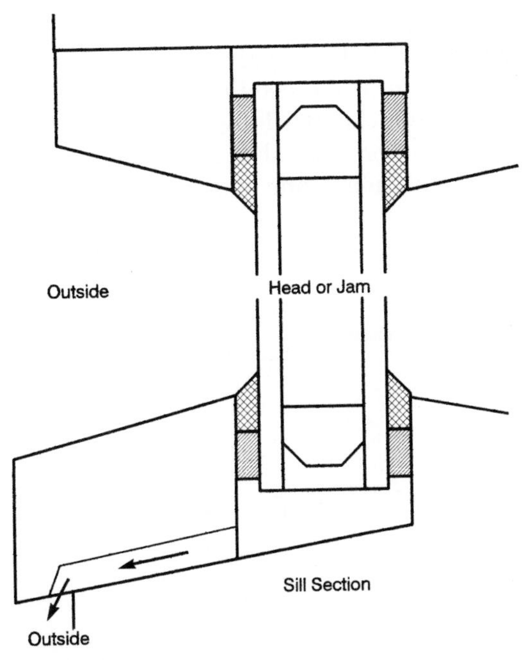

Figure 22.7 Method IG 3.

ple, Method IG 2, except that the final step is to apply a sealant capping on both sides of the glass or insulating glass and tool it to a smooth chamfer to shed the water.

Method IG 4. This is solid bedding system with no drainage or weep holes, which incorporates a moisture vapor barrier permeable sealant bedding with a low, medium, and high exposure rating depending on the materials and design. It is suitable for wood and metal window frames. In this example, shown in Figure 22.8, the outer bedding material is moisture vapor permeable silicone sealant, and the inner bedding is low-permeability glazing material, i.e., load-bearing or preshimmed mastic tape, polysulfide caulking, or nonsetting butyl mastic.

For outside glazing, ensure that the window frame rabbet surfaces, beads, and the area of the units in contact with the glazing material are clean. If necessary, treat the substrates of the frame and glass with the relevant sealer and/or sealant primer. Apply the back bedding to the rabbet up stand. Apply mastic tapes slightly above the sight line.

Insert setting blocks and spacer shims, as required. Center the glass in the frame, insert location blocks if required, and apply pressure to the unit against the back bedding.

Apply the silicone sealant around the edge of the insulating glass unit to fill the perimeter void, and then apply a substantial fillet around the perimeter of glass to form the bedding between the bead and the glass. Tool to form a smooth chamfer for water runoff.

As in previous examples, the spacers should be installed as directed by the manufacturer's recommendation, along with other details.

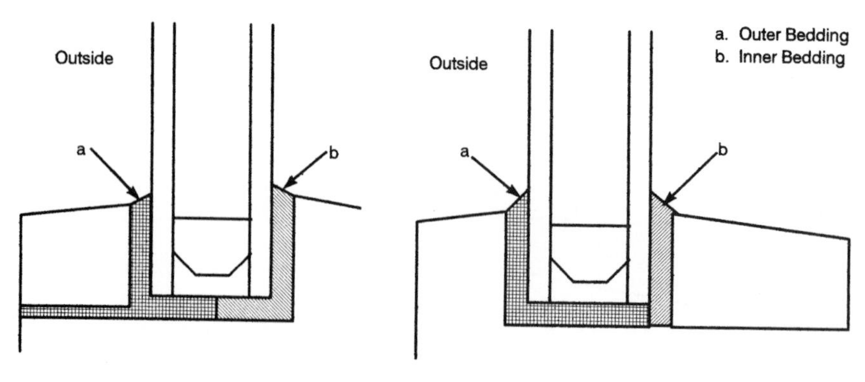

Figure 22.8 MVP sealant bedding.

On inside glazing, the system is reversed, bedding the unit in the silicone sealant and bedding the bead to the glass with the less permeable material.

Method IG 5. This is a solid bedding method incorporating a single sealant as shown in Figure 22.9.

It has an exposure rating in the low, medium, or high range, depending on materials and design, and is capable of being used with wood or metal window frames. Caulking sealants that can be used in this application are organic polysulfide, polyurethane, and polymercaptan or inorganic silicone.

Ensure that the window frame rabbet surfaces and the beads and the areas of the insulating glass unit which will come in contact with the sealants are clean. If necessary, prime the appropriate parts with primer, as suggested by the sealant manufacturer. Apply a generous amount of the sealant around the perimeter of the window frame and to the angle between the rabbet and the platform. Insert setting blocks and spacer blocks as required, center the glass, and insert location blocks, if required. Push the glass back until firmly bedded on the spacers.

Apply the sealant around the edge of the unit to fill the perimeter void completely and apply a substantial fillet around the perimeter of the unit to form the bedding between the beads and the glass lite. Insert spacers as required. Bed the beads to the glass squeezing the bulk of the sealant between the beads and the glass, and squeezing a small amount of sealant between the beads and the rabbet platform. Press the beads back until the spacers are held firmly between the beads and the glass. Fix the beads in position. If wood beads are fixed with screws, these must be no more than $2\frac{61}{64}$ in (75 mm) from each corner and no more than $7\frac{7}{8}$ in (200 mm) on centers.

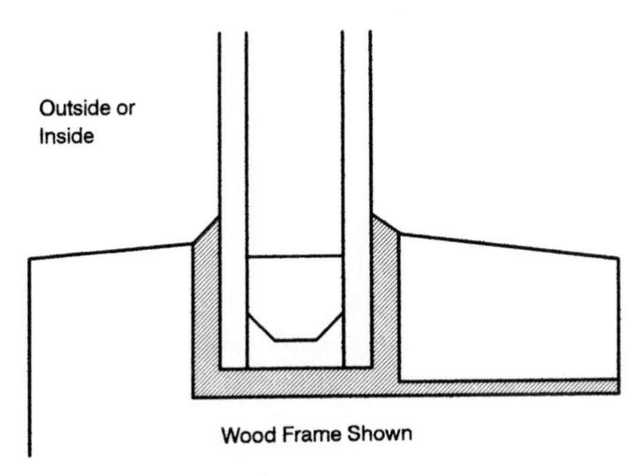

Figure 22.9 Method IG 5.

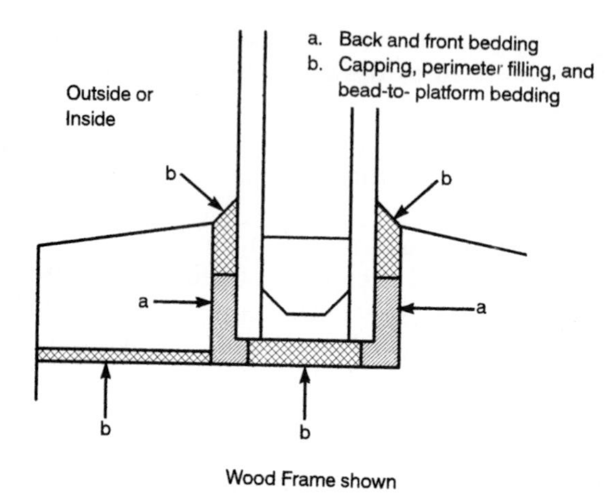

Figure 22.10 Method IG 6.

Apply additional sealant on both sides of the glass lite to completely fill any residual voids between the glass and the rabbet. Tool to a chamfer to shed water.

Method IG 6. This a solid bedding method with load-bearing or preshimmed mastic tapes or cellular adhesive sections and sealant capping with a low, medium, or high exposure rating depending on materials and design, suitable for wood and metal window frames, as illustrated in Figure 22.10.

Load-bearing mastic tapes, preshimmed mastic tapes, or closed-cell synthetic rubber strips with self-adhesive backing are used for front and back bedding. For capping, perimeter filling, and bead-to-platform bedding a suitable one- or two-part curing sealant can be utilized.

Apply all sealing materials, whether they are tapes or sealants, by following the procedure outlined in the section covering method 1G.

Structural Glazed Systems

Structural glazing is a method of bonding glass and insulating glass units to an aluminum window frame or curtain wall system utilizing a high-strength, high-performance silicone sealant.

These silicone sealant systems are specifically designed and tested for structural glazing. In structural glazing applications, dynamic wind loads are transferred from the edge of the glass to the perimeter structural support by the structural silicone sealant.

"Structural" in reference to the glazing is sometimes misinterpreted. Although major design and engineering work has been done in the structural glazing area, the basic system is quite simple.[3] There are two basic structural glazing systems in use today. These systems consist of the two-sided joint system and the four-sided system.

Advantages

Structural glazing is a fast-developing construction system offering several advantages:

1. It allows for broader architectural design flexibility.
2. It increases the thermal efficiency of buildings because the exterior exposure of the metal framing is reduced or eliminated.
3. It reduces or eliminates water and air infiltration.
4. It reduces the potential for thermal breakage of the glass.

Manufacturer review and testing

On many structural glazed buildings, the manufacturer will offer a valuation program designed to reduce the risks for all project participants. This program usually includes

1. shop drawing review to confirm the required contact widths of the sealant,
2. project name and location,
3. name and address of the architect/designer,
4. dimensions of the lites,
5. sealant contact width and joint width dimensions,
6. design wind load (psf) specified by the architect/engineer,
7. glass type and manufacturer,
8. metal framing type, finish, and manufacturer, and
9. spacer and setting block type and manufacturer.

Laboratory testing

The manufacturer of the glass or insulating glass will generally perform the lab testing of all substrates which come in contact with or are in close proximity to the structural silicone sealant. Samples of substrate should be submitted in the following quantities:

1. *Metal:* nine pieces of minimum 12-in-long production samples with specified finish identification.
2. *Glass:* nine pieces minimum 4 × 6-in minimum production lots of glass with manufacturer and type identified.

3. *Gaskets, spacers, and setting blocks:* one piece minimum, 12-in-long minimum production samples of gaskets, spacers, and setting blocks with manufacturer and type identified.

The architect will require that the drawing review and the substrate testing must be submitted in writing for review prior to the start of the glazing job.

Components of structural glazing

Structural glazing systems are basically comprised of structural framing, the glass, silicone sealant and spacers, setting blocks, and gaskets. Each of these components must be compatible with each other and be able to perform its primary functions. The following further explains each component:

Structural framing or support structure: The structural framing members are specified by the architect/engineer. The framing members may be anodized aluminum or a high-performance painted finish. If a high-performance painted finish is applied, it must be by a licensed applicator in strict conformance to the paint manufacturer's specification and quality control procedures. Anodized finishes on aluminum must be strictly monitored because they tend to vary. In addition, anodized finishes may cause sealant adhesion failure. All finishes must provide a substrate to which structural silicone will adhere on a long-term basis.

Glass: The glass manufacturer will specify the glass type and thickness in accordance with the wind load and size limitations determined by the architect. Many types of glass can be used in a structurally glazed system. The glass can be monolithic or insulating glass, tinted or treated with a high-performance finish. If the glass is treated with a reflective coating, Low-E coating, or an opacifier with some spandrel panels, the coatings' compatibility with the structural sealant must be verified.

Annealed, tempered, and laminated glass are other glass types considered by the architect. Annealed glass is used when minimal strength and safety requirements must be met and when minimal reflective distortions are desired. For greater strength or when joint design or configuration considerations are important, heat-strengthened or tempered glass may also be used.

Only high-quality dual-sealed, insulating glass units with a silicone secondary seal should be used in structural glazing systems. When ordering insulating glass units from the glass manufacturer, it must be made clear that the units will be used in structural glazed applications. If a polysulfide or butyl sealant is used as a secondary seal, the unit will fail because these sealants are incompatible with the construction silicone sealant. The compatibility of the structural sealant with the insulating glass edge seal must be verified.

Structural glazing sealants

Sealants play a major role in structural glazing. They weatherseal the exterior of the building, and they also transfer structural loads (winds and seismic) from the exterior glass to the structural framing located inside the building. In addition, the sealant is responsible for anchoring the glass to the structural framing.

Testing on a full-scale mock-up has demonstrated that silicone construction sealants perform under positive and negative wind loads because of their high tensile and shear strengths. Silicone sealants also have excellent adhesive and cohesive qualities and excellent resistance to ultraviolet radiation.

Although polyurethanes match silicones in every respect except ultraviolet resistance, it is that one property that dictates the use of silicone in structural glazing. Not all silicones, however, are acceptable for structural applications. Only silicones specifically formulated for this application may be used.

Spacers, setting blocks, and gaskets. Setting blocks, gaskets, and spacers (which consist of foam tapes and extruded gaskets) can consist of any number of different materials including neoprene, EDPM, silicone rubber, and butyls. EDPM and neoprene are organic rubbers which have been proven incompatible with silicone sealants because of discoloration and adhesion failure.

To avoid the chance of incompatibility in structural applications, it is important that silicone-based or other compatible spacers, blocks, and tapes are used. The fact that silicone and silicone-compatible materials are impossible to tell apart without testing brings up another problem. It is important that spacers and blocks do not get mixed up on projects with both structural and conventional-type systems. If for some reason you are not sure about the composition of the block or spacer in question, there is a simple test to determine if the material has a silicone base. To determine the composition of the block or spacer, test it by burning the material and observing the color of the ash. If the ash is white, then it is probably a silicone base or a silicone. If the ash is black, then it is probably some type of organic rubber, such as neoprene or EDPM.

Materials used to fabricate spacers, setting blocks, and gaskets must be compatible with the silicone sealant and must not contribute to sealant color change or affect the sealant's adhesion to the substrate when exposed to ultraviolet light.

Only "Type I" preformed silicone gaskets should be used in structural glazing. Type I rubber parts are designed to prevent adhesion between the silicone sealant and the silicone rubber gasket. "Type II" preformed silicone rubber parts permit adhesion to the silicone sealant. Preformed silicone gaskets are available in a wide range of colors.

The glass manufacturer should be consulted regarding specific design requirements for spacers, setting blocks, and gaskets, such as size, location, and hardness.

Setting blocks. It is desirable to have a bond between the spacer gasket and the silicone sealant in the structural glazed system. The opposite is true with setting blocks used in structural glazing. Setting blocks are extruded in various dimensions. Care must be taken to assure proper width and length as specified by the architect or manufacturer. Blocks are placed under the bottom edge of the lite as it is installed in the building.

The setting blocks support the total weight of the glass lite. Setting blocks differ from spacer gaskets, which act as backup materials and support only a fraction of the lite's weight. To achieve continuity of the water seal in structural glazing, it is important that the sealant bonds to the setting blocks. Special care must be taken to ensure that the setting blocks are of appropriate width, as shown in Figure 22.11.

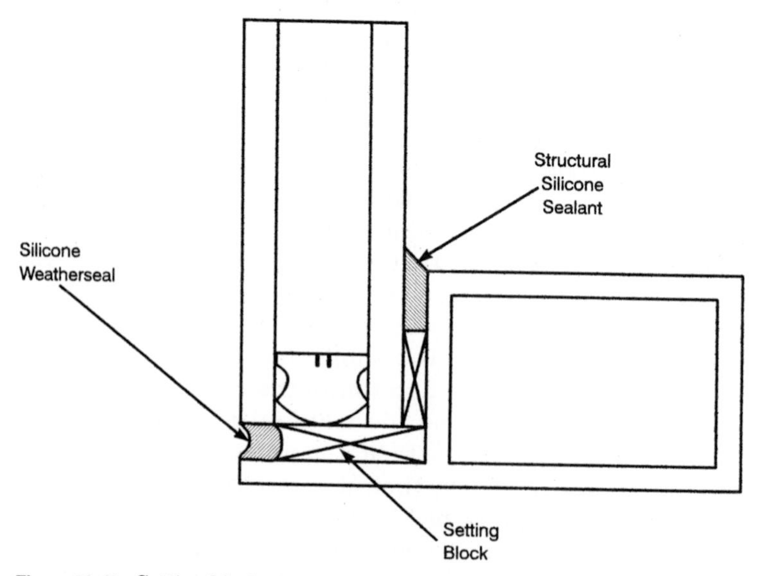

Figure 22.11 Setting blocks in structural glazing.

To distinguish between silicone rubber designated for spacer gaskets and setting blocks, the term Type II has been introduced to the industry to identify silicone rubber parts that will bond with silicone sealants.

Manufacturers' specifications. The sealant manufacturer must be consulted regarding the compatibility of preformed rubber parts and materials used to fabricate other components. Some materials may cause a color change in the sealant. This color change can indicate a chemical reaction between the gasket material and the structural silicone sealant. In the long term, this reaction may cause a complete loss of adhesion between the structural sealant and the glass and/or metal substrates when they are exposed to ultraviolet radiation.

Sealant and framing safety factors

When structural glazing was introduced to the construction industry, only high-modulus silicones could be used. These silicones were specially formulated and engineered for this application and provided an overall "safety factor" in structural applications. The formula is based on overall lite size, joint depth and width, and calculated wind loads. At that time, only high-modulus silicones could meet the safety factor criteria.

The term "safety factor" as it is applied to structural glazing is defined as the ratio of a sealant's ultimate strength (usually in tension) to the most commonly used design stress of 20 psi. Safety factors of 5:1 and 6:1 have proven adequate since the inception of structural glazing and are the most commonly specified.

Considering the number of variables which can be encountered in a structurally glazed assembly and the potential liability to all participants in the project, the higher the safety factor, the more tolerant of errors the system becomes. Today, this safety factor has been considered to be overkill. After years in the field and lab testing by different manufacturers, there is a movement toward the softer 2:1 materials. These silicones of mid-modulus or low-modulus nature can provide better adhesion to most finishes and are claimed to allow for more movement between the glass and the structural framing. This capability is thought to be desirable during seismic movement.

Generally, a high-modulus silicone sealant is used in structural glazing applications. However, if unusual movement conditions are a factor, a low-modulus silicone sealant is recommended. These sealants have a $\pm$ 50 percent continuous extension and compression rating, about double that of a high-modulus sealant. Before choosing a sealant for a structural application, consult the manufacturer about its recommendations for a structural sealant. The usual safety factor applied to the curtain wall structural members and fasteners is 1½:1. If this safety factor is applied to the sealant design stress of 20 psi, then, the ultimate stress on the sealant becomes 30 psi.

Structural Glazing Systems

Two-sided structural glazing

The two-sided structural system consists of structural metal framing (usually aluminum) at the head and sill with a backup mullion. If reflective glass is used from the outside of the building, the backup mullion is not visible, giving the effect of a continuous ribbon of glass, as shown in Figure 22.12*a* and *b*.

Although not visible, this backup mullion or stiffener acts as a structural member because high-grade silicone sealants form a structural connection between the mullion and the glass.

Should wind-load requirements exceed the strength of the primary glass, an interior mullion will be required, as illustrated in Figure 22.12*a*.

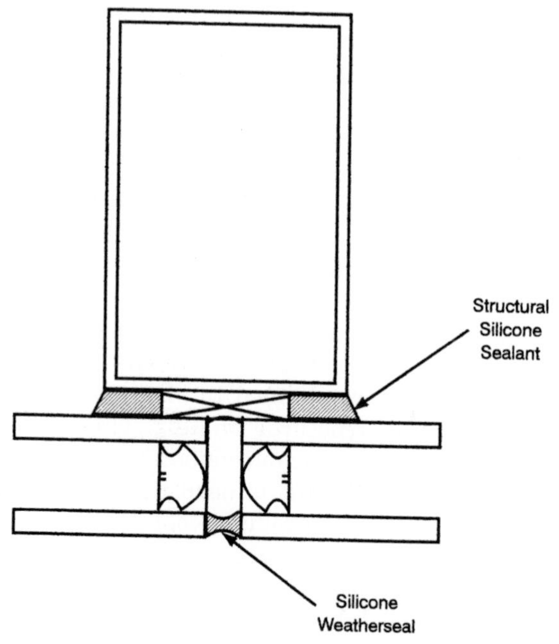

Figure 22.12(a) Mullion (structurally glazed).

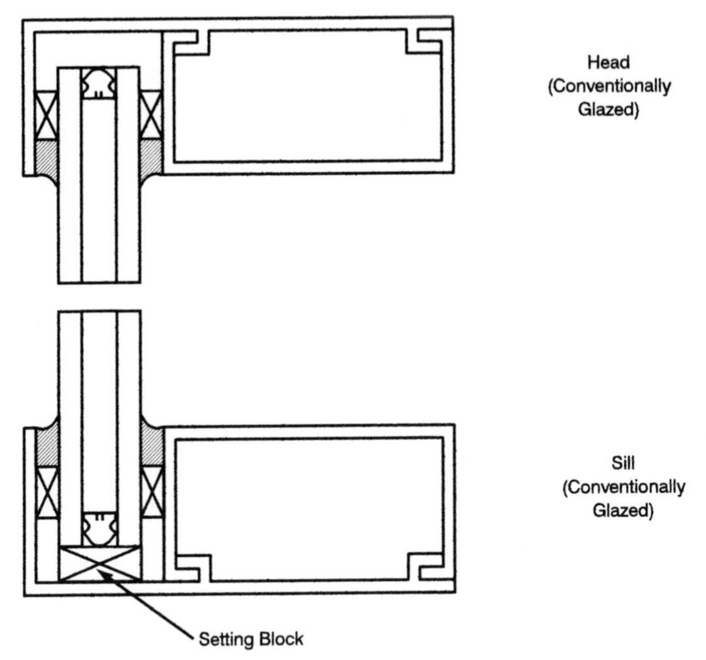

Figure 22.12(b) Two-sided silicone structurally glazed system.

Four-sided structural glazing

The four-sided structural system is used primarily in curtain walls, as shown in Figure 22.13*b*. It uses the two-sided structural design and also incorporates exterior butt joints at the horizontal line, thus eliminating any exterior framing members. The four-sided system sometimes uses a topless perimeter member eliminating the perimeter sight line. See Figure 22.13*a* and *b*.

Although one might be skeptical about the integrity of such a system, extreme testing has proven that if assembled properly, these systems perform surprisingly well. In fact, testing has shown that the sealant strength, in many cases, exceeds that of the primary glass product. In a laboratory, when tested to destruction, the perimeter glass edge remains bonded to the structural framing system.

Sealant Application

Sealants will not maintain long-term adhesion to any substrate if the surface is not properly prepared and cleaned before the sealant is applied. Using correct cleaning agents and following specified surface preparation and cleaning procedures is vital to good sealant adhesion.

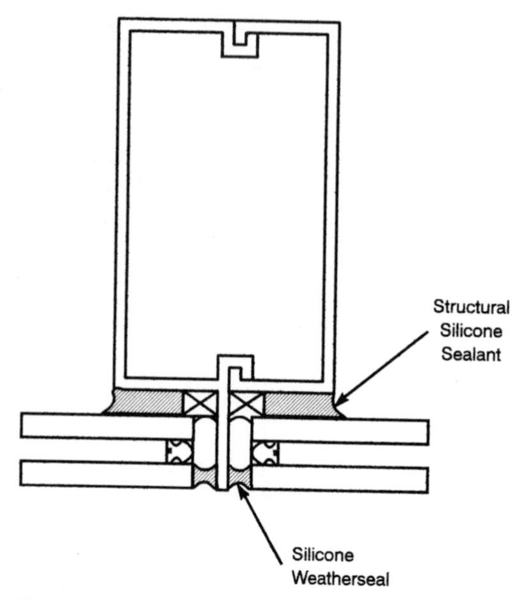

Figure 22.13(a) Mullion (structurally glazed).

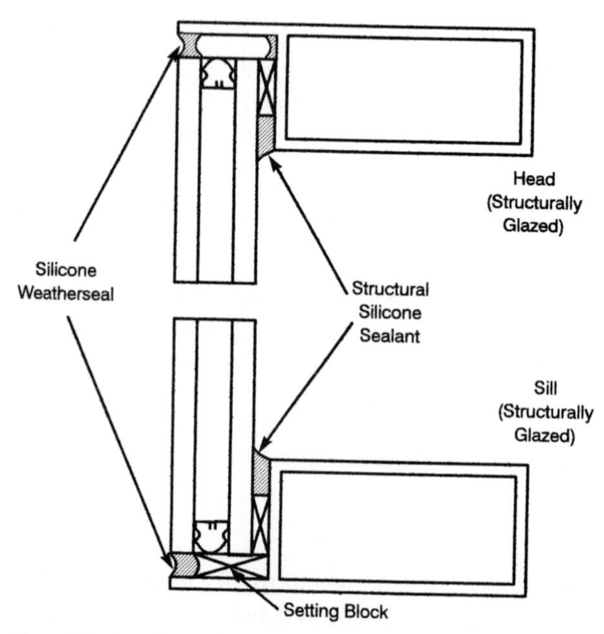

Figure 22.13(b) Four-sided silicone structurally glazed system.

When cleaning substrates, the proper materials should be used. These would include clean, fresh solvent as specified by the structural silicone manufacturer, clean, white, lint-free cloths or other approved wiping materials, and a clean narrow-blade putty knife. To begin the cleaning process, first, the glazier must remove all loose dust and dirt plus any oil, frost, or other contaminants from the substrates where structural silicone sealant adhesion is required. Next, the substrate receiving the sealant must be cleaned using the rag wiping techniques as follows:

1. Wet one clean, white rag with an approved solvent and wipe the substrate surface area.
2. Use a second rag to wipe the wet solvent from the surface BEFORE it evaporates. If the solvent is allowed to dry on the surface before it is wiped off with a second rag, the contaminants will redeposit as the solvent dries.
3. Do not spread the contaminants being removed by the solvent over the face of the area being cleaned. Any residue left may discolor or stain the face of the panels (such as metal or glass curtain walls).
4. When cleaning deep, narrow joints, wrap the cleaning cloth around a clean, narrow-blade putty knife. This permits force to be applied to the surface to be cleaned.
5. Clean only as much area as can be sealed in one hour. If cleaned areas are exposed again to rain or contaminants, the surface must be cleaned again.
6. Change rags frequently as they become soiled. It is easy to see the soiling if white rags are used.

When using solvents of any type remember the following general guidelines:

1. To prevent contaminating the entire container of solvent, do not dip soiled or used rags into the solvent container. Cleaning with contaminated solvents can result in sealant failure.
2. Always use clean containers for solvent use and for solvent storage.
3. Smoking, sparks, welding, and flames of any type must not be permitted in the areas or vicinity where solvents are being used.
4. Follow all precautions on the solvent warning label.
5. Material Safety Data Sheets (MSDS) must be obtained from your employer or solvent supplier for your use and protection. MSDS must be read before using any solvent, and all safety guidelines must be followed.

Primers

Whether a primer is required depends on the substrate and the silicone sealant being applied. Most primers are a blend of organic chemicals, resins, and solvents. Primers properly used help assure strong and consistent sealant adhesion to surfaces to which it may be difficult to bond.

A primer should never be applied to glass surfaces. A primer is not to be used in place of good surface preparation. Primers must be used in accordance with the sealant manufacturer's instruction.

When applying primer, the following materials are necessary:

1. pressure-sensitive masking tape,
2. drop cloths, and
3. clean, fresh primer as recommended by the structural sealant manufacturer.

The following general guidelines must be followed when applying a primer:

1. Mask the joint edges to prevent the primer from being applied on adjacent surfaces.

2. Apply a thin film of primer to the joint surface with a natural bristle brush, a clean, lint-free cloth, or other approved wiping material. Do not puddle the primer in the glazing pockets.

3. Allow the primer to dry before applying the sealant. Drying time depends on the ambient conditions but most primers contain fast-evaporating solvents so that the waiting time should be short, approximately 30 minutes.

4. Primers contain solvents. Smoking, sparks, welding, and flames of any type must not be permitted in the area.

5. Material Safety Data Sheets are available from the employer or the primer manufacturer. To protect you and other workers, MSDS must be read before any primer product is used.

Masking

Before applying the structural sealant, use pressure-sensitive tape to mask the exterior face of the joint. Apply the tape starting from the top and work down overlapping the runs. Use a drop cloth to cover any horizontal surfaces that may come in contact with excess sealants during the tooling operation.

Application of silicone structural sealant

Sealants should always be applied with the highest professional glazing standards. This is especially true of structurally glazed systems because these systems do not utilize internal weeping. Unlike other curtain wall systems, the exterior curtain wall seal is the only means of defense against water penetration. That is why proper field glazing must be done with the utmost care and skill and in accordance with the written instructions on the sealant container or architect's specification.

The following guidelines should be followed:

1. Material Safety Data Sheets are available upon request from the sealant manufacturer or the sealant applicator. Obtain them and read them for your own protection.
2. Keep in mind that, when field applying sealants, you must be aware of the high likelihood of contamination. Even when the two-rag wipe is performed correctly, airborne contaminants from other trades or moisture can recontaminate or otherwise adversely effect the successful outcome of the job. Extreme care and skill are required in applying structural silicone sealants.
3. When handling structurally glazed lites, extreme care must be used. The edges of the glass must be protected against clams and chips. Careful handling of reflective, Low-E, or high-performance glass is imperative to prevent edge breakage and claims.
4. In addition to the sealant manufacturer, the window or metal erecting system manufacturer should be consulted for the proper metal erecting procedures, glazing sequence, and techniques.
5. Most structurally glazed systems use a black silicone sealant. A black silicone sealant does not have the same aesthetic problems that other sealants have. For example, black silicone does not easily show bubbles or where tooling began or stopped.
6. Follow the sealant manufacturer's instruction for using a caulking gun or bulk dispensing equipment. If an air-powered caulking gun is being used, the pressure is not to exceed 45 psi.
7. Apply the sealant to the properly prepared joint as a full bead starting from the bottom of the vertical joint and working up. Apply the sealant carefully, pushing the bead ahead of the nozzle and making sure that the entire cavity is filled. Air pockets or voids along the edges are not acceptable.
8. Tool the joint immediately after the sealant is applied and before a skin begins to form on the surface.
9. Tooling should be done neatly, forcing the sealant into contact with the sides of the joint. Tooling also helps to eliminate any internal voids and assures good substrate contact. The sealant skins over quickly so repeated tooling is not recommended. Do not tool with soap or detergent solutions.

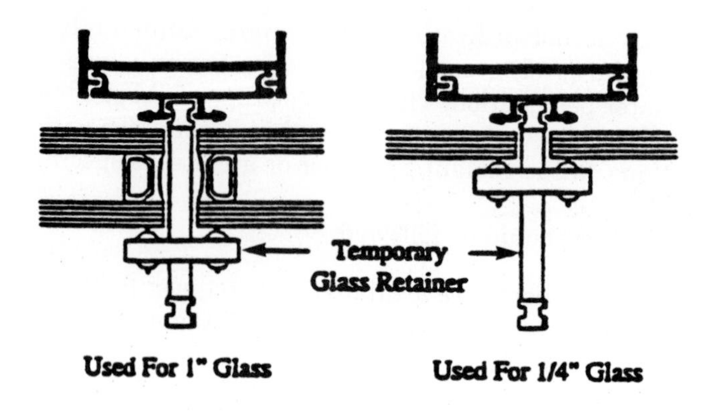

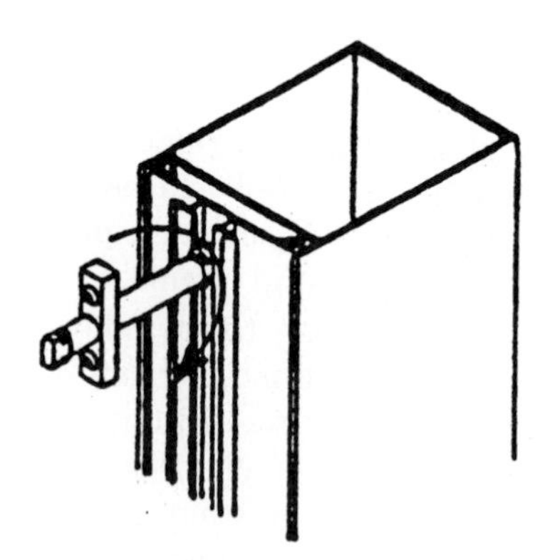

Figure 22.14 Temporary glass retainers.

Cure time

The sealant's cure time depends on the depth of the silicone joint and the relative humidity. In structurally glazed systems, temporary glass retainers must be used to hold the lites of glass immobile until the sealant has cured, generally 14 to 21 days, at room temperature above 50°F (10°C). Figure 22.14 shows temporary retainers in place until the sealant has fully cured.

Figure 22.15 is an exploded view showing a temporary retainer.

Use the "Tensile stress vs. cure time" chart (Figure 22.16) to help predict the cure time required for structural silicone sealants.

Cleaning and maintenance

Cleaning glass surfaces must be performed on a regular basis in accordance with the recommendations of the glass manufacturer.

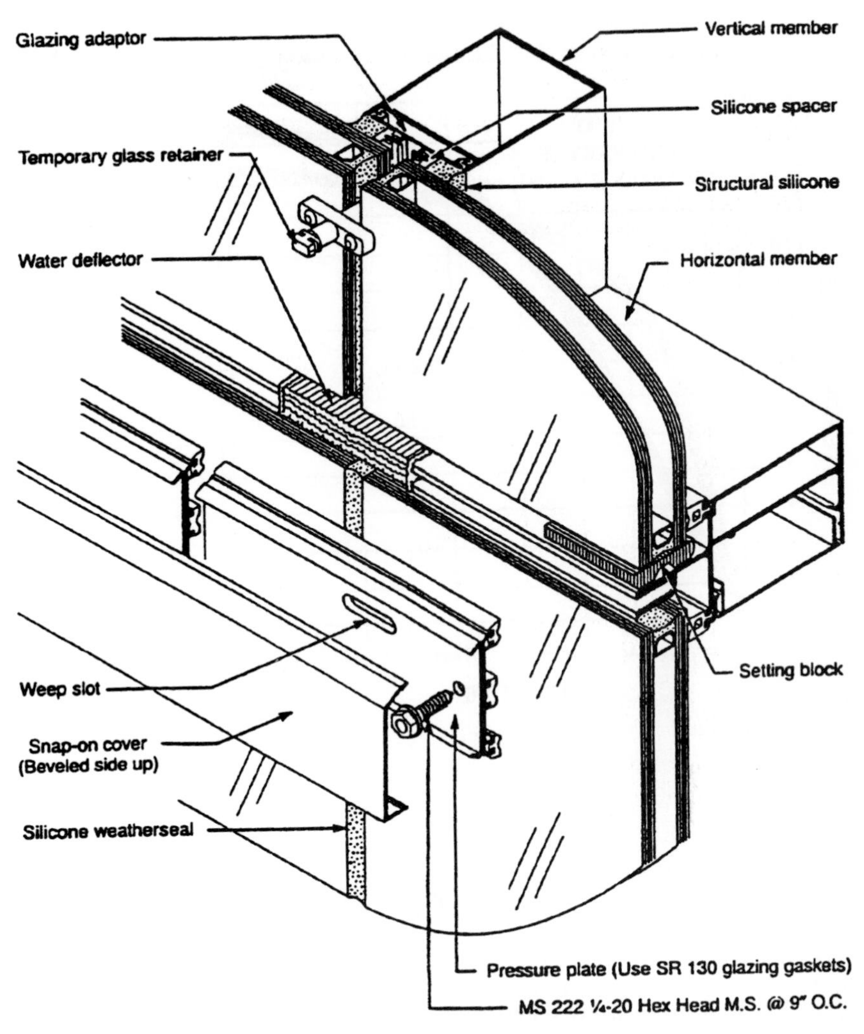

Figure 22.15 Temporary glass retainer.

All structural joinery must be inspected annually by a reliable agency approved by the architect and/or owner. During these inspections, special attention must be given to hose installation involving structurally glazed insulating glass.

Any glass or insulating glass units that exhibit evidence of condensation within the confined air space between the glass lites must be replaced as soon as possible. Failure to replace defective units may eventually affect the structural integrity of the window system and possibly cause danger to the building occupants and/or pedestrians.

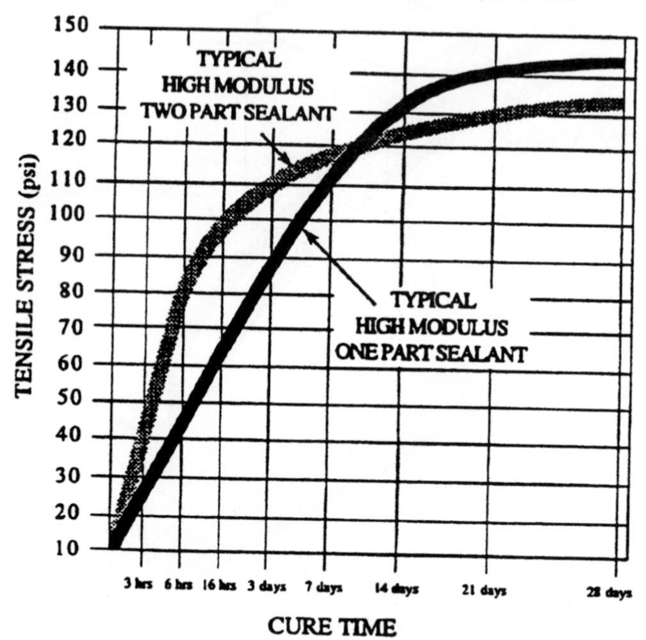

Figure 22.16 Tensile stress vs. cure time.

References

1. *Window Selection Guide,* 3rd ed., American Architectural Manufacturers Association, 1991.
2. *Products, Glazing and Techniques and Maintenance,* Section 4, Glass and Glazing Federation, London, England, 1995.
3. Structural Glazing, In *Glaziers Training Textbook,* The National Glaziers Architectural Metal and Glass Workers Industry, Washington, DC, 1995.

Chapter

23

Specifications and Quality Assurance

Joseph S. Amstock

President, Professional Adhesive and Sealant Systems

Specifications

Whether you are an architect, engineer, contractor, owner, subcontractor, material supplier, attorney, or building department official, whatever your role is in the building process, you need specifications.[1] All of the parties in the construction process need specifications for different reasons, and each can profit by his or her understanding of the specifications.

The data following illustrates reasons why the parties need specifications, and then we will look at what this means to the specification writers who prepare them. But first, we must define specifications or "specs." The term specification is the act or process of specifying a detailed, precise presentation of something or of a statement of legal particulars. It is shorthand for the project manual, which includes all the written material, bidding documents, contract documents, and general and supplementary conditions. For simplicity, we call this manual specs or specifications.

Every architectural and engineering firm faces the important task of determining how to integrate specification writing into their practice. Decisions must be made about where to obtain the necessary professional talent, what resources and training to provide, and how to organize project responsibilities. Among the variables, however, are three primary alternatives: assigning specification writing to an in-house specialist, relying on project team members to prepare specifications along with other construction documents, or retaining the services of a specification consultant.[1]

The following profiles illustrate each of these three approaches to the specification practice.

The specification writer

Although some firms view specs as overly needed or as a backroom function, others approach the discipline as a profit center. The first step in developing a good set of specifications is to hire a dedicated specification writer. To get a good product on a consistent basis, you need a person who lives and breathes in a specification environment. This procedure generates income to support the specifications studio.

Develop the strengths of its CAD capabilities. Along with the ability to process tremendous volumes of work, office automation also creates an environment in which cooperation and teamwork among the members is essential.

In an office with manual drafting, a specifier can look over the draftsman's shoulder to see what is on the boards. However, changing drawings in computer files may not be easy. Designers redline and update drawings, but they do not plot progress sets, as they often used to. Drawings could be called up on a CAD terminal.

Checklists facilitate effective communication

The alternative for the specification writer facing such a dilemma is to rely on effective communication with other project team members. Checklists are used for this purpose. When design changes occur or decisions are made, the specifier is notified in writing. This effort, combined with increased verbal communication, helps to keep the workload and schedule on track in a busy, growing organization.

During the initial service phase of a new job, the architect must interview clients intensively, questioning them about what works and what does not. Process manuals are designed specifically for the client. The manual documents the client's decision-making process to acquire and develop sites, reach agreements, develop the design, produce drawings, and put a facility in place. A set of documents called prototype paper tools is then created. These are used to record and communicate all decisions for each site adaptation. Each document is custom designed for every client, because no two clients do business exactly the same way. Too many loopholes can leave bidders wondering what the architect needed. Now, they feel that the price they quoted is the one at which they are going to build and make a profit, because they have a very buildable set of construction documents. A teamwork approach has contributed to the success of many building programs. You have set specifications, but each project must be tailored to provide a building for a specific climate, site and program, and certain goals and code divisions. The computer sorts this to pick out appropriate items and generate a shopping list of specs by zone. It is that simple. The difficult thing is including the latest trends while making sure everyone knows about modifications.

Many problems can be experienced. Keeping the specifications coordinated with the drawings is a case in point. Computer-aided drafting has to be coordinated directly with the specifications, when certain items are pulled in the basic building that are also reflected in the specs. The process eliminates the need for double procedures.

Contract drawings and specifications should be revised and should approach specifications and drawings as a prototypical situation. Work out a master specification. Every spec must be custom designed to fit the specific needs of each client. Some owners do not understand specs and do not use them as well as they should. As a result, the hard thing is to make sure that the decisions are made up front, before a production work order is issued on the specification. The specification team should meet with the client to make every decision before they edit the spec.

Computers are the future of architecture and specifications. Everything that is accomplished, from marketing to word processing, right on through to specifications and CAD drawings, is computer-related. One day, it will be possible to talk with our clients via computers, spending less time documenting and processing drawings and more time improving the quality of the design.

Training architects as specifiers

Project architects and engineers at many design firms are responsible for writing their own specifications. One reason for this approach is that the project architect and engineer probably know the job better than anyone else. This procedure allows developing the specification simultaneously with design decisions and drawings, reducing the time required for a specifier to become familiar with a project.

The drawback to this approach is that project architects and engineers are not always trained and experienced in specification writing, and without a full-time specifier to establish office specification standards, quality control may be affected.

For that reason, teaching project architects and engineers to write their own specifications can be most important. In developing the firm's master specifications coordinating and production and providing the technical expertise that only an experienced specifier can offer, techniques and options become more diverse. The same is becoming true for the task of writing construction specifications.

In small architectural and engineering firms, it is not often feasible to staff a dedicated specification writer. Yet the principal and the more experienced senior members of the firm frequently find themselves managing the office and developing new business, in addition to completing actual design work. In an effort to keep a small firm's jobs on schedule, it can be advantageous to hire a consulting specification writer on an as-needed basis.

Similarly, large firms, even those with a dedicated spec writer on staff, may find the workload so pressing that retaining an outside consultant is the best means of meeting short-term specification needs.

Owners need specifications

Without definitions of what constitutes well enough, the owner is pretty much forced to take whatever he gets. It is important that the owner do a little homework so he can benefit in any discussions with all the members of the design team he hires.

Architects need specifications

The architect uses the specs to define those things that cannot be illustrated in the drawings. The architect also needs a place for the information he develops to assist the owner in obtaining prices for the work, executing the contract, and administering the project as it is built. The location of the bidding and contract documents is the front of the specifications or, if the project is very large, a separate volume of the project manual.

Contractors need specifications

Specifications can be extremely useful to the contractor. The specs can help the contractor decide the scope of the subcontracts. The specs are not written to mandate subcontract scopes, and specifiers do not and should not assume the responsibility of doing so, but specifications divide the project into logical units of work. The specs should include everything that is indicated on the drawings, which helps the contractor in determining which of the subs is obligated to perform certain parts of the work.

The contractor uses specs when managing the project. Requests for substitutions, methods of submittal, procedures for preparation, and submissions of payment requests are a few of the administrative processes described in detail in the specifications. By understanding and following the procedures as described, the contractor greatly improves his chances of obtaining the desired results, obtaining permission to effect a substitution, making acceptable submittals, and receiving prompt payment.

Estimators need specifications

Specs are essential for the contractor's estimator. Estimators use the 16 Division system of Master format extensively in their work. Most estimating systems are based on Master format, allowing the estimators to easily compare a list of spec sections for their work.

Manufacturers need specifications

For the manufacturers and material suppliers of building products, a project's specifications can have many uses, some of which determine whether or not a particular product will be used on a specific project. With many building materials, manufacturers' representatives assist the specifier to prepare the specs. Their knowledge of their product and their skills in preparing product specs are helpful to the specifier. When assisting the specifier however, product representatives go beyond the task of spec writing. They demonstrate to the specifier and the rest of the design team the suitability of their product for the proposed application. In this connection, the manufacturer must be careful to point out options, custom features, and accessories that should be incorporated into the work. Then the product is selected and the representatives prepare the draft specification incorporating the precise features selected by the design team.

Product reps, especially those who have taken the CDT and Certified Construction Product Representative (CCPR) examinations offered by Construction Specifiers Institute (CSI), are aware of their need to study the Division requirements for the project. They will make proposed substitution requests and prepare their submittal in accordance with the requirements specified. They know that the division in which their product is specified is but a part of the whole. This increases the likelihood that their substitution proposal or their submittal will be accepted.

Attorneys need specifications

Attorneys become involved with specs primarily when preparing the contracts and resolving disputes. Many attorneys are not very proficient at reading drawings. They rely heavily on the specs and discussions with their clients to understand the project. The architects' preparation of contracts should be limited to filling in the blanks in American Institute of Architects (AIA) or similar prewritten contract forms. Modification to suit the project or the preparation of contracts in custom form should be done only by attorneys.

If disputes arise, the terminology, the punctuation, and the sentence structure all become subjects of intense scrutiny in determining what the various portions of the contract document said, what they meant, and how they were interpreted. At these times, the accuracy, completeness, and consistency of the specifications become critical.

Preparing quality specifications

The needs and uses of the specifications by the various parties described are sometimes divergent, putting several obligations on persons who prepare specifications. The Manual of Practice of CSI describes the process in detail. Here, we will touch on a few important items.[2]

1. The specifications should be accurate and complete and should not duplicate information shown on the drawings, but they should amplify, explain, and qualify that information.
2. The specifications should be consistent. Terminology used in one place should match wherever it occurs. Use of terms describing parties to the contract should be limited to the parties to the contract. Specs should be checked carefully during preparation to assure that cross references are correct.
3. Division I requirements, such as substitutions, submittals, and record drawings, should not be repeated, or worse still, specified differently in subsequent divisions.
4. The specifications should be clear. As an example, if we mean a polyurethane sealant as an edge sealant in an insulating glass unit, it must be spelled out very specifically. A casual reference to the ASTM-E-773/774 specification dealing with insulating glass is not sufficient. In fact, the precise details, method of manufacture, the issue of the latest spec, and the materials to be used should all be spelled out clearly so there is no doubt as to what is intended.
5. Clauses which are not clear, concise, and consistent should be avoided. The contractor does not bid on intent. He bids on what he sees on drawings and what he reads in the specs.

The joint publication of CSI and AIA, *Uniform Location of Subject Matter,* AIA Document A521, contains detailed descriptions of what goes in specs and what goes on drawings. By understanding the needs of each party and by placing information where it belongs in a clear, concise, consistent manner, the specifier will enhance the smooth progress of the project as a whole and the interests of each party involved.

Construction Documents Technology Program

Contractors, project architects, contract administrators, material suppliers, and manufacturers' representatives are realizing the advantages of being Construction Document Technologists (CDTs). By understanding and interpreting written construction documents, CDTs can perform their jobs more effectively and demonstrate their commitment to improving communication among all construction industry professionals.

This two-hour exam is based on CSI's Construction Documents Fundamentals and Formats module of the Manual of Practice and its appendices and general conditions in common use (AIA A-201[1987 edition] or EJCDC 1910-8 [1990 edition]).

Quality Assurance Manual*

Scope and purpose

There have been many recent advances in the technology of insulating glass manufacture. The choices of solar design, sealants, desiccants, corner design, etc., have resulted in a large number of possible combinations of materials for insulating glass construction.

This manual addresses the minimum criteria for a quality control or assurance program. The manufacturer is urged to exceed these requirements (increase the frequency of evaluation) in cases where past quality problems have occurred or where the materials or processes are new to the manufacturer.

The manual does not consider the interactions between the multitude of components that may be used. The manual considers each possible component separately and provides guidance to assure a reasonable level of quality. It also considers the proper use of each component and its limits.

Some adverse effects of changes in the manufacture and formulation of various components may be detectable only after long-term usage. These cannot be considered in a quality assurance program. As with any product fabricated from a number of components, workmanship and attention to directions furnished by the material suppliers are paramount. When the properties of a component do not comply with the standards set out in this manual, the supplier should be notified and requested to remedy and/or explain the condition. Noncomplying materials will not be used until the problem (or misunderstanding) has been resolved.

*The data listed does not represent any of the organizations involved with glass, such as IGCC, Sigma Igmac, etc.

Responsibility of the insulating glass manufacturer

The manufacturer will

1. comply with the certification guidelines and interpretations published by the Insulating Glass Certification Council,
2. retain quality control records for a period of five years,
3. in cases where a component is out of compliance, notify
   a. the component supplier, and
   b. the administrator of the IGCC Certification Program, and
4. take immediate steps to correct the causes for the noncompliance.

Tests and evaluations

The following routine tests and evaluations will be conducted and the results tabulated on a record sheet.

Sealants

Sealants include

1. extruded sealants such as polyurethane, polysulfide, silicone, and other gunnable hybrids,
2. sealants impregnated with desiccant,
3. sealants combined with desiccants and spacers, and
4. or any other variation where the edge seal of the unit is dependent upon the bond of a material to the glass and spacer surfaces.

The sealant supplier will provide documentation listing which specific glass coatings and spacer surface treatments are compatible with the sealant furnished. Compatibility includes proper adhesion and the absence of any adverse chemical reaction. No sealants will be used without this assurance and evidence.

The specific cohesive properties and resistance to shear forces of the sealants cannot be readily determined in a quality assurance program. This information must be furnished by the sealant supplier under certified conditions.

Adhesion test (test no.1.) Figure 23.1 illustrates the test sample for determining adhesion with the known or specified sealants and spacer bars.

1. This test will be conducted on specimens of glass and metal spacers (where used) that are cleaned and otherwise treated as they

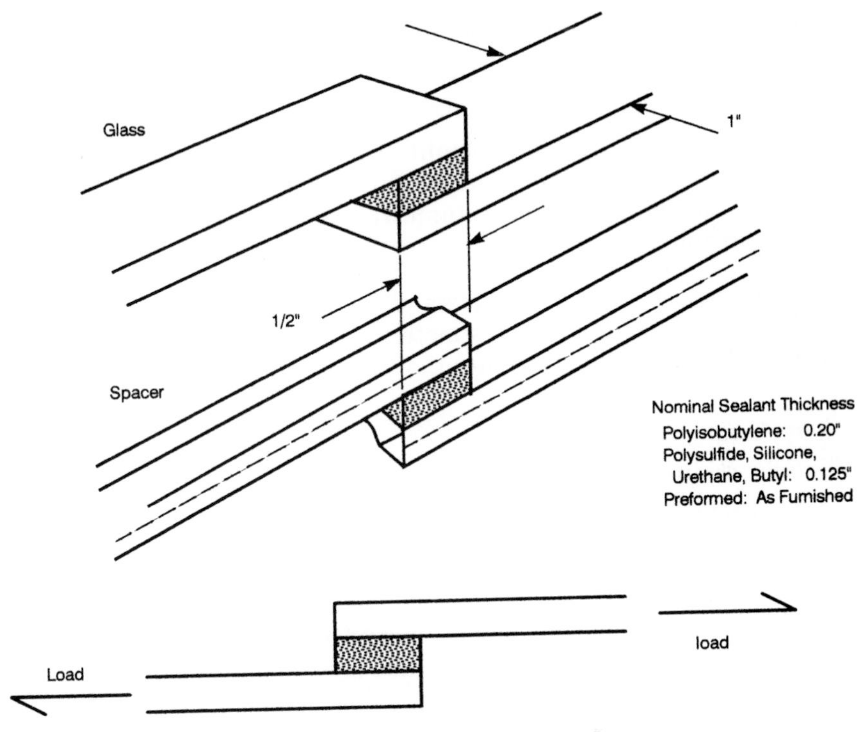

Figure 23.1 Adhesion test.

would be in production. No special surface cleaning, priming, or other treatment will be used. The sealants will be applied at temperatures and other conditions consistent with their application during the manufacture of the unit.

2. The test specimen will be taken at the start of each shift (or eight-hour period) and whenever a change in substrate (glass or spacer) or sealant is made.
3. The test specimens will be constructed as illustrated in Figure 23.1. The sealant will be cured for a minimum period as recommended by the sealant supplier.
4. The specimen will be loaded in shear as indicated in Figure 23.1, and the load will be applied until an adhesive or cohesive failure occurs. The rate or magnitude of loading is not critical to these tests. An adhesive failure (however limited in area) constitutes a failure of the sealant or of the substrate surface treatment.

Frequency of tests: at the start of each shift (or eight-hour period) and whenever a change in a substrate (glass or spacer) or a sealant occurs.

Mix of two-component sealants (test no. 2)

1. The batch number of the base material and the accelerator will be checked to be certain they correspond to each other.
2. During production, the amount of each material used from each drum will be measured. The measurement will be with a metering device or by noting the volumetric reduction of the material in each drum.

Frequency of a test: at the start of each shift (or eight-hour period) and at four-hour intervals; if a metering device is not used, the initial test will be taken one hour after the start of the period.

Spacer

Two overriding considerations are important in evaluating thermal break metal or extruded metal spacers:

1. The surface and surface treatment is compatible with the insulating glass sealants.
2. The widths of glass-to-glass tolerances are tightly controlled.

Number 1 is addressed under Sealants.

Width tolerance of spacer (test no. 3)

1. The width of the spacer will be measured at time intervals not to exceed one hour. Visual checks will be made routinely. Any apparent or seeming variation will also be checked.
2. The variation in width will not exceed $\pm$ 0.003 in based on the width specified in the order. The specimen for evaluation will consist of one short leg and one long leg of the insulating glass unit being produced.

Note: The variation in width is critical, not the specified width. The specified width varies among insulating glass manufacturers.

Note: To expedite measurements, a "go-no-go" gage is suggested. This is also appropriate for evaluating variations at corners and edge joinery. A sketch of the gage is shown in Figure 23.2.

Corner and edge joinery

Assembled corners and edge joints (for units with bent corners) are critical elements that may be the most prevalent source of failure.

For assembled corners and bent corners, there may be an increase in the width of glass-to-glass dimensions caused by

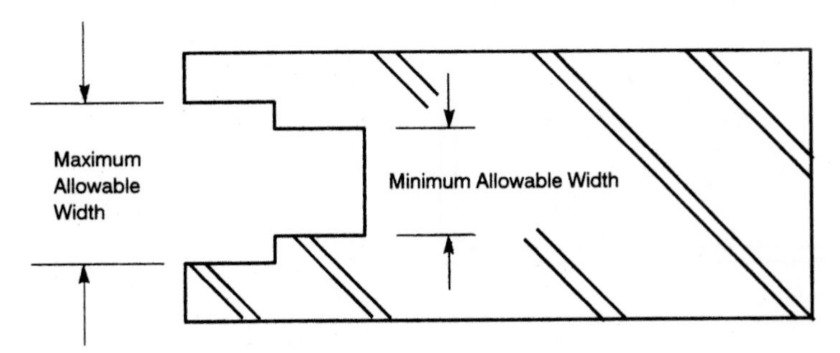

Figure 23.2 Go-no-go gauge.

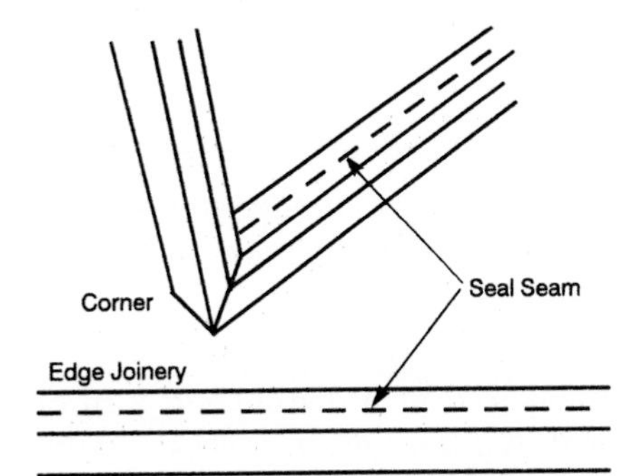

Figure 23.3 Edge joinery.

1. tolerances for corner keys inconsistent with spacer tolerances,
2. distortion of the spacer during cutting, and
3. for bent corners, deformation from the bending operation.

Frequency of test: continuous visual check; measurements at one-hour intervals.

Airtightness (test no. 4)

1. At the start of each shift (or eight-hour period) and for intervals not to exceed four hours, an assembled corner and edge specimen (if appropriate) will be extracted from an assembled unit spacer frame. The specimen shown in Figure 23.3 will be selected at random.
2. For bent, soldered, welded, and butyl-injected corners, the specimens will be tested without modifications. For force fit and interlocking joinery, sealants will be applied to the specimens to simulate, to the degree practical, the application of sealants for the completed unit.
3. The joint is immersed in water and air pressure applied as shown in Figure 23.4.

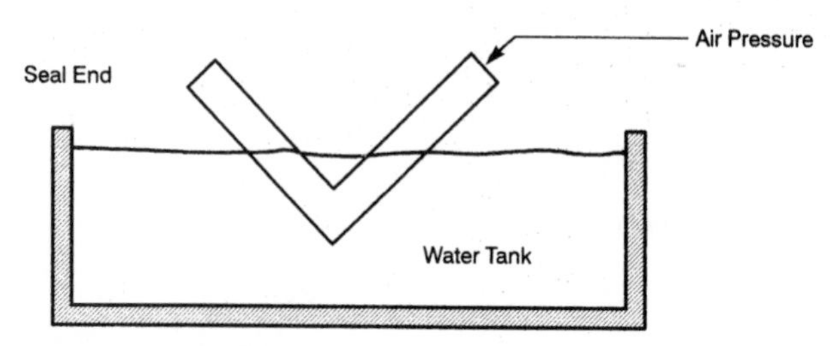

Figure 23.4 Water immersion.

4. If any bubbles appear at the joint, the joint is unsatisfactory.

Frequency of the test: the start of each shift (or eight-hour period) and at four-hour intervals.

Width tolerance (test no. 5)

1. The width at the corner (assembled or bent) and edge joinery will be measured at the start of each shift (or eight-hour period) at intervals not to exceed one hour. The measurement will be made at all four corners.
2. The width of the spacer will be measured at a distance of 2 in on either side of the corner or edge joint. A difference in excess of 0.003 in between the spacer and the corner edge joint will not be allowed.

Desiccant control (test no. 6)

The manufacturers of poured-in-place desiccants and desiccant-impregnated butyl sealants (with and without integral spacers) have each developed a proprietary method for evaluating the moisture content of the desiccant. There is a unique procedure for each variation. The desiccant manufacturer will provide a kit for evaluating its product and acceptable limits of moisture content.

The following minimum production controls will be exercised:

1. Desiccants and desiccant-impregnated butyl will be stored in hermetically sealed containers. These will be approved by the desiccant/desiccant assembly supplier. Poured-in-place desiccant will not be exposed to ambient air for more than one hour prior to encapsulation in a unit. The desiccant-impregnated butyl will not be exposed for more than four hours.
2. The supplier of the desiccant will be contacted to provide guidelines regarding the quantity of desiccant to be used. This may vary depending on the type of edge seal assembly.
3. The heat adsorption method will be used for testing poured-in-place desiccants. For other variations, the method recommended by the manufacturer will be used.

4. If the desiccant has a moisture content higher than allowed, all production since the last successful evaluation will be deemed unacceptable.

Frequency of the test: whenever a new desiccant or desiccant-impregnated butyl container is opened, at the start of each shift or eight-hour period, and at intervals of four hours.

Dimensional tolerance for units (test no. 7)

Dimensional tolerances are based on the cutting tolerances for glass specified in ASTM Standard C-1036 and the mismatch between panes. The overall tolerances, including mismatches but not including excessive sealants, are as follows. These shall apply for glass thicknesses of $\frac{1}{4}$ in (6.35 mm) or less for each pane. Proportionally greater tolerances will be allowed for thicker panes but not to exceed 1.5 times the values listed:

Edge up to 30 in	+ $\frac{1}{8}$ in, − $\frac{1}{16}$ in
Edge more than 30 in up to 60 in	+ $\frac{5}{32}$ in − $\frac{1}{16}$ in
Edge more than 60 in up to 96 in	+ $\frac{3}{16}$ in − $\frac{1}{16}$ in

For rectangular units, the difference in the diagonal measurements will not exceed $\frac{3}{4}$ in.

Frequency of tests: whenever a size change is made and at one-hour intervals.

Thickness tolerance for units (test no. 8)

The spacer width should be selected to equal the nominal air-space thickness less 0.015 in for the primary sealant at each spacer/glass interface. For single-seal units and certain other variations the 0.015 in clearance does not apply. In these cases, the nominal spacer width is the same as the air-space width. Based on these practices and the allowable ranges of glass thickness in ASTM Standard C 1036, acceptable thickness ranges will be as follows:

Glass, in	Air space, in	Thickness range, in
$\frac{1}{8}$	$\frac{1}{4}$	0.494–0.564
	$\frac{1}{2}$	0.744–0.814
$\frac{3}{16}$	$\frac{1}{4}$	0.624–0.694
	$\frac{1}{2}$	0.874–0.994
$\frac{1}{4}$	$\frac{1}{4}$	0.702–0.784
	$\frac{1}{2}$	0.952–1.034

Frequency of tests: whenever a change of glass or spacer thickness is made and at intervals of one hour.

The following minimum production controls will be exercised in addition to the listed quality checks:

1. Desiccants and desiccant-impregnated butyl will be stored in hermetically sealed containers. These will be approved by the desiccant/desiccant assembly supplier. Poured-in-place desiccant will not be exposed to ambient air for more than a one-hour period prior to encapsulation in an insulating glass unit. The desiccant-impregnated butyl will not be exposed for more than four hours.
2. The primary sealant (polyisobutylene) in a dual-sealed insulating glass unit will have a minimum moisture vapor path of $\frac{1}{8}$ in at all locations. The thickness of the primary sealant shall be controlled to 0.015 ± 0.005 in.
3. The moisture vapor path for the secondary sealant shall be a minimum of $\frac{1}{8}$ in for polysulfide and urethane sealants and $\frac{3}{16}$ in for silicone sealants.
4. The cover of the secondary sealant over the spacer shall be a minimum of $\frac{3}{32}$ in.

Note: The designs of some types of edge seal assemblies have an exposed spacer. Item 4 will not apply in these cases. However, the minimum lengths of moisture vapor paths will be met.

5. For a single-seal system, the moisture vapor path will be a minimum of $\frac{1}{4}$ in.

Note: The sealant application will be inspected on a 100 percent basis during fabrication.

Other Specifications

There are a number of specifications available to the architect, designer, manufacturer, and building owner that will enhance the programs they may be working on. Appendix G lists a variety of specifications applicable to the glass industry from the U.S. and other countries.

References

1. Drummond, J., *The Importance of Specifications—Who Needs Them?,* FCSI, CCS.
2. Chusid, M. T., Current Approach to Specification Practice, *Building, Design & Construction,* 4/1986.

Chapter

24

Glossary of Terms

Joseph S. Amstock

President, Professional Adhesive and Sealant Systems

a The symbol for a repeating unit in a polymer chain.

ABA copolymers Block copolymers with three sequences but only two domains.

Abhesive A material that resists adhesion; a film of coating applied to surfaces to prevent sticking, heat sealing, and so on, such as a parting agent or mold release agent.

Ablation A self-regulating heat and mass transfer process in which incident thermal energy is expended by sacrificial loss of material.

Abrasion The wearing away of a material surface by friction. Particles become detached by a combined cutting, shearing, and tearing action; an important factor in tire, treads, soles, and conveyor belts.

Abrasion index A value expressing abrasion resistance.

Abrasion tester A machine for determining abrasion loss quantitatively.

ABS Acrylonitrile-butadiene-styrene.

Absolute humidity The weight of water vapor present in a unit of air, such as grams per cubic foot or grams per cubic meter.

Absorption The dissipation of sound energy by heat and other forms of energy loss.

Accelerate To hasten or quicken the natural progress of a reaction or event. For example, the drying rate of an adhesive or sealer is hastened or accelerated by increasing the temperature. An accelerated test is usually a severe test that determines comparative durability in a shorter length of time than required under service conditions.

Accelerated ageing Any set of test conditions designed to determine, in a short time, the result obtained under normal conditions of ageing. In accelerated ageing tests, the usual factors considered are heat, light, and oxygen, either separately or combined. Sometimes called accelerated life. Most often accomplished by heating samples in an atmosphere of oxygen at 300 psi and 70°F (22°C) (Biermer–Davis), or by heating in an oven provided with circulating air maintained at 158 to 212°F (70 to 100°C) (Gear).

Accelerated weathering Machine-made means of duplicating or reproducing weather conditions. Such tests are particularly useful in comparing a series of products at the same time. No real correlation between test data and actual service is known for resins and rubbers used in many products.

Accelerator (of vulcanization) Any substance which hastens the vulcanization of rubber causing it to take place in a shorter time or at a lower temperature. In earlier days, basic oxides, such as lime, litharge, and magnesia, were recognized as vulcanizers. Nowadays, the important accelerators are organic substances containing either nitrogen or sulfur or both. According to potency or speed of action, accelerators are sometimes classified as slow, medium rapid, semiultra, and ultraaccelerators. Most accelerators enhance tensile properties, and many improve age resistance.

Acetone—dimethyl ketone (CH_3—CO—CH_3). Used to extract most of the nonrubber constituents in natural rubber and also the free sulfur in vulcanizates.

Acid polishing The polishing of a glass surface by acid treatment.

Acidity The property of being acid or containing ionizable hydrogen. In more common language, the property of reddening litmus or of combining with bases to form salts. The term is also used with reference to the quantity of acid substances present in any material, for example, naturally occurring organic acids in crude rubber.

Acoustical double glazing Two monolithic glass panels, set in a frame, with an air space between them usually larger than 1 in and generally not hermetically sealed.

Acrylate resins Polymerization products of certain esters of acrylic and methacrylic acid, such as methyl or ethyl acrylate. Possess great optical clarity and high degree of light transmission. Nearest approach to an organic glass.

Acrylic A group of thermoplastic resins or polymers formed by polymerizing the esters of acrylic acid.

Acrylic latex Latex-modified acrylic caulk.

Acrylonitrile A monomer with the structure CH_2:CHCN.

Activator A substance which promotes a specific chemical action by chemical interaction with a second substance. Most accelerators require activators to bring out their full effect in vulcanization (e.g., zinc oxide or other metallic oxides); some accelerators require a fatty acid, especially with zinc oxide.

Active solar heat gain Solar heat that passes through a material and is captured by mechanical means.

Adduct A chemical addition product.

Adhere That property of a sealant/compound which measures its ability to bond to the surface to which it is applied.

Adhesion The clinging or sticking of two material surfaces to each other. In rubber parlance, the strength of the bond or union between two rubber surfaces or plies, cured or uncured. The bond between a cured rubber surface and nonrubber surface, e.g., glass, metal, wood, or fabric.

Adhesion failure (1) The separation of the two surfaces with a force less than specified. (2) The separation of the two adjoining surfaces due to service conditions.

Adhesion, mechanical Adhesion due to the physical interlocking of an adhesive with the base irregularities.

Adhesion of rubber to metal The strength of a bond formed between a metal surface and natural or synthetic rubber must be known regardless of the method used in obtaining the bond. Adhesion is expressed quantitatively as the tension per unit area required to cause a rupture of the rubber-to-metal bond (ASTM-D-429).

Adhesion, peel Force required to peel a sealant, caulk, adhesive, or coating from a surface; usually expressed in pounds per inch.

Adhesion, specific Adhesion due to valence forces at the adhesive-base surface interface. Such valence or attraction forces are of the same type that give rise to cohesion.

Adhesive Substance capable of holding materials together by surface attachment. This is a general term that includes cement, glue, mucilage, paste, etc. Various descriptive adjectives are used with the term "adhesive" to indicate different types, such as

physical form: liquid adhesive, tape adhesive, etc.,
composition: resin adhesive, rubber adhesive, etc.,
end use: paper adhesive, label adhesive, etc., and
application: sprayable adhesive, hot melt adhesive, etc.

Also see Dry seal adhesive, Pressure-sensitive adhesive.

Adhesive setting Classifies the conditions to convert the adhesive from its packaged state to a more useful form.

Chemically setting: Requires addition of accelerator or catalyst to cure.
Cold setting: Sets at temperatures below 68° (120°)
Hot setting: Sets at temperatures above 212° (100°C)
Intermediate temperature setting: Sets in the temperature range of 87–212°F (31–100°C)
Room temperature setting: Sets in the temperature range of 68–86°F (20–30°C)

Adsorbent A highly porous solid which has the ability to concentrate and hold gases and vapors in contact with itself. This includes moisture as well as many other organic and inorganic molecules.

Adsorption The action of a body in condensing and holding gases, dyes, or other substances. The action is usually considered to take place only at or near the surface. The power of adsorption is one of the characteristic properties of matter in the colloidal state and is associated with surface energy phenomena of colloidally dispersed particles.

Ageing A progressive change in the chemical and physical properties of rubber, especially vulcanized rubber, usually marked by deterioration. With natural rubber, it is due primarily to oxidation and/or reversion; with GR-S it is primarily caused by continued copolymerization. May be retarded by the use of antioxidants. The verb is also used transitively to denote the setting aside of rubber goods under specified conditions for the purpose of observing their rate of deterioration (see Shelf ageing and also Accelerated ageing).

Ageing resistance Resistance to ageing by oxygen and ozone in the air, by heat, and by light. Antioxidants help, although there is no nontoxic substance to protect natural rubber, GR-S neoprene, and nitrile rubbers against ozone.

Ageing tests Accelerated tests of rubber specimens to find out their endurance by heating them in air under pressure or similarly in oxygen.

Air curing Vulcanization at ordinary room temperature or without the aid of heat.

Air drying A material is said to be air drying when it can be dried at ordinary room temperature without artificial heat.

Air infiltration The amount of air that passes between a window sash and frame or a door panel and frame. For windows it is measured in terms of cubic feet of air per minute per square foot, and for doors it is measured in cubic feet of air per minute per foot of crack.

Air side In the float process, the upper side of glass is called the air side.

Aliphatics Organic compounds (hydrocarbons) in which carbon atoms are arranged in an open or straight chain. More commonly called naphthas, they are prepared by straight-run, overhead distillation of petroleum. Familiar examples include gasoline, kerosene, natural GR-S, and butyl rubber formulas. One of the common solvents, they are about the lowest in price and the least toxic.

Alkali Substance that neutralizes acid to form salt and water. Yields hydroxyl (OH^-) ions in water solution. Proton acceptor. Turns litmus paper blue.

Alligatoring Term describing the appearance of a film that is cracked into large segments resembling the hide of an alligator. When alligatoring is fine and incomplete, it is usually called cracking. Alligatoring may be caused by one coat applied over another before the bottom coat is thoroughly dry and hard and/or having the material skin so that the lower portion of the film is still soft and elastic, or by less elastic material being applied over a more elastic undercoating. When these conditions are present and the finished article is exposed to actinic rays or changes in temperature and moisture content, expansion and contraction of the film cracks the hard crust while the softer core gives without breaking. With excessively heavy coats of dilute materials, this cracking of the outer crust can take place without temperature change by the shrinkage action of the bottom portion—much like clay mud is cracked under the summer sun. Other causes of alligatoring include too rapid evaporation of solvents or thinners and excessive air being forced into the film during spraying.

Ambient noise The all-encompassing noise associated with a given environment, usually a composite of sounds from sources near and far.

Ambient temperature The environment temperature surrounding the object.

Annealing The process which prevents glass from shattering after it has been formed. The outer surfaces of the glass shrink faster than the glass between the surfaces, causing strain which can lead to shattering. By reheating the glass and allowing it to cool slowly, this can be avoided.

Anticracking agents Softening agents are usually anticracking agents. Some antioxidants work well also. Chiefly, a good vulcanizate resists cracking; high tensile, high tear resistance, high resilience, and good resistance to heat ageing.

Antioxidant Usually organic. A substance which inhibits or retards oxidation and certain other kinds of ageing. Some antioxidants cause staining or discoloration of rubber compounds on exposure to light and are used in black or dark colored compounds. Others (phenolic), described as nonstaining, are used in white or light-colored compounds.

Antiwalk blocks Rubber blocks that prevent glass from moving sideways in the glazing rabbet because of thermal effects or vibration.

Application Term describing the act of going over the entire area to be coated or sealed only once. For example, when the operator spray-coats the entire surface without repetition, he has made one application. If he immediately goes over the same work again, he has made another application. For adhesives and coating, the principal methods of application are brushing, dipping, stenciling, flowing, stamp-padding, roll-coating, knife-coating, squeegeeing, by spatula, and by notched trowel. For sealers, the principal methods of application are spatula, caulking gun, flow gun, pressure extrusion units, and spray gun.

Art glass Art glass goes by many names. It is called opalescent glass, cathedral glass, or stained glass and is usually produced in small batch operations.

Asphalt Naturally occurring solid or semisolid mineral pitch or bitumen, more or less soluble in carbon disulfide, naphtha, and turpentine, and fusible at varying temperatures. Gilsonite, glance-pitch, manjak, Grahamite, and Trinidad pitch are examples. Also, bituminous residues left from the distillation of petroleum or coal tar.

Assembly time, closed The interval between completing the assembly and the application of pressure and/or heat.

Assembly time, open The optimum interval between spreading the adhesive on the adherent and completing the assembly for bonding.

Atmospheric or ozone cracking A fissured surface condition which develops on stretched rubber exposed to the atmosphere or to ozonized air. When the fissures are minute, the condition is called checking.

Atmospheric pressure The pressure of air at sea level exerted equally in all directions. The standard pressure is that at which mercury stands at 14.7 psi (760 mm).

Attenuation The reduction of sound pressure level, usually expressed in decibels.

Autoclave A vessel that employs heat and pressure. In the glass industry, used to produce a bond between glass and PVB or urethane sheet, thus creating a laminated sheet product.

Back-fill Placing material into the opening between glass and glazing.

Bait A webbed metal frame used to draw molten glass.

Bandage joint Sealant joint composed of bond-breaker tape over the joint movement area with an overlay of sealant lapping either side of the tape sufficient to bond well to the surfaces; often used where extreme movement occurs and conventional joint design is not possible.

Base Any substance, organic or inorganic, having an alkaline character or the ability to combine with or neutralize an acid.

Batch The mixed raw materials which are used to make glass.

Bead A sealant/compound after application in a joint. Also a molding or stop used to hold the glass product in position.

Bend test A means of testing the flexibility of a compound at a specified temperature. The compound is applied to metal, cured or dried, and after conditioning at the specific temperature, bent over a mandrel to determine the product's resistance to rupture.

Bent glass Flat glass shaped while hot into cylindrical or other curved shapes.

Benzene ring The basic structure of benzene, the most important aromatic chemical. It is an unsaturated, resonant, six-carbon ring having three double bonds. One or more of the six hydrogen atoms of the benzene may be replaced by other atoms or groups.

Bevel or compound bead In glazing, a bead of compound applied to provide a slanted top surface so that water will drain away from the glass or panel.

Beveling The process of edge-finishing flat glass to a beveled angle.

Bite The dimension by which the edge of a glass product is engaged into the glazing channel.

Bitumen Originally, mineral pitch or asphalt. Now, any of a number of flammable mineral substances, consisting mainly of hydrocarbons, including the hard, brittle varieties of asphalt, the semisolid maltha and mineral tars, the oily petroleum, or even the volatile napthtas.

Black A term referring to the several types of carbon black utilized in the rubber industry and also to the absence of color. Zero brilliance and 100 percent saturation of light waves.

Bleeding Migration of plasticizers, waxes, or similar materials to the surface forming a film or bead. Also see Bloom.

Blend A physical mixture of silica gel and molecular sieve.

Blending A step in reclaiming in which the devulcanized stock is mixed with reinforcing and processing agents prior to refining.

Blocks Rectangular, cured sections of neoprene or other approved materials, used to position a glass product in a glazing channel.

Bloom A discoloration or change in appearance of the surface of a rubber product caused by the migration of a liquid or solid to the surface. Examples are sulfur bloom and wax bloom. Not to be confused with dust on the surface from external sources.

Blowing Porosity or sponginess occurring during cure. In latex goods, a permanent deformation caused when the deposit leaves the form during curing or drying.

Blush See Bloom. Also, whitish surface appearance where moisture has condensed before solvent is completely evaporated.

Boiling point The temperature at which the vapor pressure of a liquid is equal to the pressure of the atmosphere.

Bond (noun) (1) The attachment at the interface between an adhesive and an adherent. (2) A coat of finishing material used to improve the adhesion of succeeding coats.

Bond (verb) To join materials together with adhesives. To adhere.

Bond breaker A material to prevent adhesion at a designated interface.

Bond strength The force per unit area or length necessary to rupture a bond.

Bonding agents Substances or mixtures of substances used in attaching rubber to metal. Generally, the rubber compound is vulcanized by heat in the process. Cyclized rubber or rubber isomers, halogenated rubber, rubber hydrochloride, the reaction products of natural rubber and acrylonitrile, and polymers containing diisocyanates are also used.

Boric acid A compound used in making borosilicate glass.

Bow A continuous curve of a glass sheet, either vertical or horizontal.

Breather tube A small-diameter tube placed into the space of an insulating glass unit through the perimeter wall to equalize the air pressure within the unit when shipping into areas over 5000 feet in elevation or vice versa. These tubes are to be sealed on the job site prior to installation.

Bronze glass A glare- or heat-reducing glass intended for applications where glare control and reduction of solar heat are desired or where color can contribute to design.

Brookfield synchro-lectric viscosimetre A portable form of rotating cylinder apparatus for measuring the viscosity of latex.

Brookfield test A means test for determining the viscosity of liquids or semipastes by measuring the drag produced upon a cylinder or disc rotated at a definite constant speed and at a specified temperature while immersed in the material being tested.

Btu British thermal unit. The amount of heat required to raise the temperature of 1 lb of water 1°F.

Bullet-resisting glass A multiple lamination of glass with a sheet of a tough, clear plastic usually at least $1\frac{3}{16}$ in thick overall, designed to stop bullets from ordinary firearms other than high-powered rifles.

Bull's eye The round, whorl shape in the center of old panes of glass.

Butt glazing The installation of glass products where the vertical glass edges are without structural supporting mullions.

Butt joint A joint having opposing faces which may move toward or away from one another; a joint in which the receiving surfaces stresses the sealant in tension or compression.

Buttering A method of mastic application similar to butter spreading on a piece of toast.

Butyl rubber A copolymer of about 98% isobutylene and 2% isoprene. It has the poorest resistance to petroleum oils and gasolines of any rubber. Excellent resistance to vegetable and mineral oils, to solvents, such as acetone, alcohol, phenol, and ethylene glycol, and to water and gas absorption. Heat resistance is above average. Sunlight resistance is excellent. Its abrasion resistance is not as good a natural rubber. Usually low permeability to gases.

Calcium oxide A compound used in glass making, derived from lime.

Capillary tube unit A small metal tube of specific length and inside diameter factory-placed into the spacer of an insulating glass unit to accommodate both the pressure differences at the point of installation and also the pressure differences encountered daily after installation. Capillary tubes are sealed after installation.

Catalyst Substance which markedly speeds up the cure of an adhesive when added in a minor quantity compared to the amounts of the primary reactants.

Caulk (noun) (See Sealant.) The term caulk has traditionally referred to nonelastomeric sealant compounds used where little or no movement is required (usually less than 10%); caulking compounds are a type of sealant.

Caulk (verb) To install or apply a sealant across or into a point, crack, or crevice.

Caulking The process of sealing a joint or material used. Most often refers to linseed oil and lead compounds and to cotton or oakum strands used in back of seams rather than to the more recently developed sealing materials.

Cement The dispersion or "solution" of unvulcanized rubber compound in a suitable solvent, such as petroleum naphtha or aromatic hydrocarbons. Cements are also made from latex or water dispersions with or without organic solvents.

Centipoise $\frac{1}{1000}$ of a poise, a value for viscosity. The viscosity of water at 64°F (20°C) is approximately 1 centipose.

$$\text{Millipoise} = 1/1000 \text{ poise}$$

$$\frac{\text{Millipoise} = \text{centistoke}}{10\text{dt}}$$

where dt = density of a substance at the same temperature t and centistoke = $\frac{1}{100}$ of a stoke, which is the unit of kinematic viscosity.

Certified IG unit An insulating glass unit constructed like a test model, which has successfully passed ASTM-E-773 and ASTM-E-774 testing of insulating seal durability performance at specific levels (pursuant to the administrative guidelines of a certification program).

Chain stopper A material which, when added during the polymerization process, will terminate or stop molecular growth.

Chalking Formation of a powdery surface condition caused by disintegration of surface binder or elastomer from weathering or other destructive agents.

Channel A three-sided U-shaped opening in sash or frame to receive lite or panel, with or without removable stop or stops. Contrasted to a rabbet, which is a two-sided, L-shaped section, as with face-glazed window sash.

Channel black A form of carbon black made from a natural gas by the channel combustion process. The gas is burned with insufficient air in the jets, the flames from which are allowed to impinge on a cool metallic surface (channel) from which the deposited carbon is scraped. For many years, it was the best reinforcing agent for natural rubber and made it highly abrasion resistant.

Channel depth The measurement from the bottom of the channel to the top of the stop, or the measurement from the sight line to the base of the channel.

Channel glazing The sealing of the joints around lites or panels set in a U-shaped channel employing removable stops.

Channel width The measurement between stationary stops (or stationary stop and removable stop) in a U-shaped channel.

Checking In a coating, the development of slight breaks which do not penetrate to the underlying surface. Checking may be described as visible (as seen by the naked eye) or as microscopic (as seen under a magnification of 10 diameters).

Checking sunlight The development of minute surface fissures as a result of exposing rubber articles to sunlight, generally accelerated by bending or stretching.

Chemical resistance The resistance offered by elastomer products to physical or chemical reactions from contact with or immersion in various solvents, acids, alkalies, salts, etc.

Chlorinated rubber When chlorine is passed into a solution of crude rubber, substitutions as well as the addition of chlorine occur and a white fibrous product of the appropriate formula $(C_{10}H_{13}CL_7)X$ is obtained. Soluble in benzene, chloroform, and acetone.

Clear glass As its name states, transparent or clear.

Clips Wire spring devices to hold glass in a rabbet sash, without stops or face glazing.

Coat, double Generally, two successive coats applied to one surface. In spraying, it means to spray a single coat first with vertical strokes and then across with horizontal strokes, or vice versa. In the woodworking industry, the term means one coat applied to each of two mating surfaces. In adhesives and coatings, the terms mean a single coat applied to each surface.

Coat, single One layer of material applied on a surface, by brush, spray gun, flow gun, etc. In spray jargon, it means two strokes of the gun, the first down and second up or the first to the right and the second to the left. The second stroke should cover the first completely. The third and fourth strokes should cover one-half of the first two, while the fifth and sixth should cover one-half of the third and fourth, etc. See Coat, double.

Coating A material, usually liquid, used to form a covering film over a surface. Its function is to decorate and/or protect the surface from destructive agents or environments (abrasion, chemical action, solvents, corrosion, and weathering).

Coefficient of expansion The coefficient of linear expansion is the ratio of the change in volume per degree to the length at 32°F (0°C). The coefficient of volume expansion (for solids) is three times the linear coefficient. The coefficient of volume expansion for liquids is the ratio of the change in volume per degree to the volume at 32°F (0°C).

Cohesion That form of attraction by which the particles of a body are held together, usually primary or secondary valence forces. The internal strength of film.

Cohesive failure The splitting and opening of a sealant/compound within its body, resulting in water penetration.

Cold flexibility Flexibility during exposure to a predetermined low temperature for a predetermined time.

Cold flow Permanent deformation under constant stress. Also defined as continuing dimensional change under static load following initial instantaneous deformation. If subjected to pressure long enough, no organic material will return exactly to its original shape. Compression set is the amount by which a small cylinder fails to return. See also Creep.

Cold resistant Withstands the effect of cold or low temperatures without loss of serviceability.

Color A property or quality of visible phenomena distinct from form and from light and shade; depends on the effect of different wavelengths on the retina of the eye. The term color is sometimes used inaccurately to denote hue, tint, shade, pigment, dyestuff, etc.

Color cast glass Includes many kinds of cast and rolled glass. There are more than 100 colors of *dalle* glass (*dalle* is French for "tile").

Compatibility Ability of two or more substances to mix or blend without separation or reaction.

Compound A formulation of vehicle, fillers, and polymer(s) producing an elastomeric sealant.

Compression modulus The ratio of compressive stress to the resulting compressive strain (the latter expressed as a fraction of the original height or thickness in the direction of the force). Compression modulus may be either static or dynamic.

Compression set The residual decrease in thickness of a test specimen measured 30 minutes after removal from a suitable loading device in which the specimen had been subjected for a definite time to compressive deformation under specified conditions of load application and temperature.

Method A measures compression set of vulcanized rubber under constant load.
Method B employs constant deflection (see ASTM Method D-395).

Concave bead Bead of compound with a concave exposed surface.

Condensation Moisture that forms on surfaces colder than the dew point.

Conduction The transfer of heat through matter, whether solid, liquid, or gas.

Consistency The viscosity or solidness of a semisolid or syrupy substance. It may be called the resistance to deformation. That property of a body by which it resists deformation or permanent change of shape. See Viscosity.

Container stability Period of time a compound will remain in satisfactory condition when stored in unopened containers. Synonymous with Shelf life.

Continuous film A film that is free of breaks or pinholes.

Control A product of known characteristics which is included in a series of similar service or bench tests to provide a basis for evaluating one or more unknown products.

Convection A transfer of heat through a liquid or gas, when that medium hits against a solid.

Convex bead, surface Bead of compound with a convex exposed surface.

Coolant A liquid for cooling down engineering work or equipment.

Corrosion The deterioration of a metal by chemical or electrochemical reaction resulting from exposure to weathering, moisture, chemicals, or other agents or media.

Cracking A fissured surface condition which develops on rubber articles exposed to the atmosphere, light, heat, or repeated bending or stretching. When the fissures are minute, the condition is called crazing.

Creep The dimensional change with time of a material under load, following the initial instantaneous elastic deformation. Creep at room temperature is sometimes called Cold flow.

Critical temperature A definite temperature above which a gas cannot be liquefied by pressure. The term is loosely used, sometimes to denote an approximate temperature below which a reaction such as vulcanization does not take place or proceeds only very slowly. Also used to denote a definite temperature below which an accelerator does not function properly.

Crown glass The old-fashioned way of making glass.

Crown wool A process used in making glass fiber.

Crystalline Resembling small crystals, or crystal shaped.

Cullet Recycled or waste glass.

Cure To change the properties of a material by chemical reaction, which may be condensation, polymerization, or vulcanization. Usually accomplished by the action of heat and catalysts, alone or in combination, with or without pressure. Also see Overcure and Semicure.

Cure time The time required to produce vulcanization at a given temperature. The cure time varies widely, being dependent on the type of compounding used, the thickness of the product, etc.

Curing, glass Similar to annealing.

Curing agent Generally the second of a two-part system which, when added to the base materials cures or solidifies the base material by chemical reaction.

Curing range In vulcanization, an approximate range of curing times at a given temperature, over which the physical properties of a vulcanizate do not change materially.

Curing temperature The temperature at which a rubber product is vulcanized.

Damping The dissipation of sound energy in a medium over time or distance.

Dangerous chemical Any substance or mechanical mixture or chemical compound of substances which is volatile or unstable or which tends to oxidize or decompose spontaneously, thus creating fire or explosion, or which may generate flammable or explosive gas, and which is capable of creating hazard to life or property, when designated as such by rule or regulation.

Dead load Load due to glass weight.

Decibel (dB) A unit adopted for convenience in representing vastly different sound pressures. It is 20 times the logarithm to the base 10 of the ratio of the sound to the pressure of 0.0002 dyne/cm^2. This reference pressure is considered the lowest value that the ear can detect.

Deflection The degree of inward or outward movement in lites of glass exposed to unequal pressures from either side of the glass. This condition may be permanent or temporary. Deflection can be caused by the wind or temperature, elevation, or barometric pressure changes. Air-absorbing desiccants can contribute to deflection.

Deformation Any change of form or shape produced in a body by stress or force.

Degradation Deterioration, usually in the sense of a physical or chemical process, rather than a mechanical one.

Degree of Fineness of subdivision of dispersed particles.

Dehydration Removal of water as such from a substance, or after formation from a hydrogen and hydroxyl group in a compound, by heat or dehydrating substance.

Delaminate To split a laminated material parallel to the plane of its layers. Sometimes used to describe cohesive failure of an adherent in bond strength testing.

Delamination Separation or splitting, usually lack of adhesion in plied goods.

Density The ratio of the mass of a body to its volume or the mass per unit volume of the substance. When CGS units are used, the density of a substance is numerically equal to the specific gravity of the substance referred to water at 39.5°F (4°C), the maximum density (1.000 g/cc) of water. For practical purposes, density and specific gravity may be regarded as equivalent.

Desiccants Porous crystalline substances used to absorb moisture and solvent vapors from the air space of insulating glass units. More properly called absorbents.

Dew point The temperature at which air is saturated with respect to a condensible component, such as water vapor or solvent.

Dew point (IG units) The temperature above 32°F (0°C) at which visible vapor or other liquid vapor begins to deposit on the air-space glass surface of a sealed insulating glass unit in contact with the measuring surface of the dew point apparatus.

Diaphragm A thin body that separates two areas; in sound, the skin of a partition or ceiling that separates the room from the structural space in the center of the partition or ceiling assembly.

Diluent A diluting agent. Any liquid or solid which, when added to another liquid or solid, reduces the quantity per unit volume of the base material in the total volume.

Dip coat. A thin coat on a surface obtained by dipping the material to be coated in the coating material.

Discoloration Staining. Changing or darkening in color from the standard or original.

Dispersion The act of causing colloidal particles of matter to separate and become uniformly scattered throughout a medium. Any system of matter in which finely divided colloidal particles of one or more phases (components) are uniformly scattered throughout another phase or medium; the components or phases may be solid, liquid, or gaseous (a wholly gaseous system is not considered a dispersion, but a simple mixture). Colloidal solutions are examples of liquid dispersions; the dissolved or dispersed particles are not subdivided to the molecular state as in true solutions and dispersion.

Double glazing Windows in which two pieces of glass have been installed, usually ⅛ to ½ in apart.

Double laminated insulated glazing Two laminated glass panels set on a frame that provides an air space between them. Such units may or may not be hermetically sealed. Air-space thickness can vary with acoustic and thermal requirements.

Drawing tower Used in the sheet glass process for drawing molten glass.

Dry time Lack of adhesion; more specifically, poor contact of adhering surfaces. See Starved joint.

Drying time The time required for solvent dissipation after a film is spread. The drying process includes several states. The first, known as "tacky," starts almost immediately after application. The optimum point in the tacky state is the earliest period when the adhesive may be touched lightly with a finger without transferring material to the finger. The bond should be made at this point. The second state, "dust free," is the time required for the film to reach the condition where if dust settles on it, the dust will not become embedded but may be wiped off after the material has been hardened. The "tack free" state is the time required for the film to reach the condition where it can be touched with the finger without feeling the slight retention of surface stickiness. "Print-free" is closely related to this, although one stage later. It refers to the time required for the film to reach the condition where it may be touched with a finger without retaining the imprint of the finger on the surface of the film. The final state, "hard-dry," is the time required for the film to become thoroughly hard so that it may be handled and polished if necessary.

Dry seal adhesive One which is nonblocking except to itself. Two adherents may be precoated, dried, and then bonded at any time using only normal pressure.

Dual-sealed units Sealed insulating glass units fabricated with an inner seal and an outer secondary seal. Generally, each of the two seals has been selected for its special performance characteristics, i.e., adhesion and moisture vapor transmission properties.

Durometer The hardness of rubber determined by measuring its resistance to the penetration (without puncturing) of a blunt indentor point impressed on the rubber surface against the action of a spring. A special scale indicates resistance to penetration or "hardness." The scale reads from zero to 100, zero being very soft and 100 being very hard.

Edge clearance Normal spacing between the edge surface of a glass product and the glazing channel base.

Edge joint A joint made by bonding the edge faces of two adherends.

Elastic limit The limit to which a body may be deformed and yet return to its original shape after removal of the deforming force. Steel has a well-defined elastic limit (yield point) below which it is perfectly elastic; vulcanized rubber, on the other hand, shows no definite elastic limit, but takes a "set," more or less, depending on the stretch given to it.

Elasticity The property of matter which causes it to return to its original shape after deformation, such as stretching, compression, or torsion. It is the opposite of plasticity. It is often loosely employed to signify the "stretchiness" of rubber. As applied to rubber, it usually refers to the phenomenal distance to which vulcanized rubber can be stretched without losing its ability to return very nearly to its original shape; in this respect, rubber is the most elastic substance known. Also see Modulus.

Elastomer A substance that can be stretched to at least twice its original length and, after having been stretched and the stress removed, returns to approximately its original length in a short time.

Elastomeric Having the property of returning to the original shape and position.

Elongation Increase in length expressed numerically as a fraction or percentage of initial length.

Emissivity The relative ability of a surface to radiate heat. Emissivity factors range from 0.0 (0 percent) to 1.0 (100 percent).

Emittance Heat energy radiated by the surface of a body, usually measured per second per unit area.

Enamel, glass A soft glass compound of flint or sand, soda potash, and red lead.

Endothermic A chemical reaction which absorbs heat energy is said to be endothermic; a compound the formulation of which absorbs heat is an endothermic compound. Such compounds are less stable than exothermic compounds, many of them being explosive.

ENR Exterior noise rating.

Equivalency The relationship between approved production insulating glass units and prototype test units. In the United States under IGCC regulations, a production unit must have at least as much desiccant per perimeter inch as the test unit. In Canada, equivalency regulations are under discussion and have not been fixed.

Equivalent of combined load Combination of the instant applied load of wind and factored long-term loading of glass weight and snow accumulation.

Erosion Destruction of metal or other material by the abrasive action of liquid or gas. Usually accelerated by the presence of solid particles of matter in suspension and sometimes by corrosion.

EWNR Exterior wall noise rating.

Exothermic A chemical reaction in which heat energy is liberated is termed exothermic; a compound the formation of which involves the evolution of heat, is called an exothermic compound. Exothermic compounds are more stable than endothermic compounds. Vulcanization of rubber with sulphur is an exothermic reaction.

Exterior glazed Glass set from the exterior of the building.

Exterior stop The removable molding or bead that holds the lite or panel in place when it is on the exterior side of the lite or panel, as contrasted with an interior stop located on the interior side of the lite.

Extruded Forced through a die or continuous mold for shaping.

Face glazing On rebated sash without stops, the triangular bead of compound applied with a glazing knife after bedding, setting, and clipping the lite in place.

Failed insulating glass unit An insulating glass unit failure exhibits permanent obstruction of vision through the unit due to accumulation of dust, moisture, or film on the internal surface of the glass. Surface numbers two or three in dual-pane units; surface numbers two, three, four, or five in triple-pane units.

Faying surfaces The surfaces of materials in contact with each other and joined or about to be joined.

Feeder The part of the furnace which ensures that the glass has a consistent temperature throughout.

Fiberglass An individual filament made by drawing molten glass.

Filler Relatively nonadhesive substance added to an adhesive to improve its working properties, permanence, strength, or other qualities.

Fillet A rounded bead or concave junction of sealing material over or at the edges of structural members.

Film Thin layer of material, not necessarily visible. "Free films" are not attached to any body. They are often called "unsupported films." "Supported films" have flexible backing, usually cloth or paper, to give the film greater support and structural strength.

Flame resistant Describes a material which does not burn too readily, when a flame is removed.

Flame retardants Substances mixed with rubber to retard its burning (e.g., highly chlorinated hydrocarbons). Neoprene is less flammable than natural rubber and GR-S.

Flammable A volatile liquid or gas which has a flash point of 30°F (−2°C) or lower. Flammable is synonymous with inflammable.

Flange The projection around the exterior perimeter of some sash.

Flanking The passage of sound between two rooms through a medium other than the partition dividing them. The sound may pass via the side walls, floor, ceiling, or other structures. This term also refers to air leakage paths through partitions and to sound bypassing the main leaves of a partition via the studs or other connecting framework.

Flash point The temperature to which a liquid must be heated before its vapors will flash or burn momentarily when a small flame is applied. This ignition will not take place unless there is also a spark or open flame. There are several standard methods for determining flash point, most of which may be classified as "open-cup" or "closed-cup."

Flat glass Pertains to all glasses produced in a flat form.

Float glass Transparent glass with flat, parallel surfaces formed on the surface of a pool of molten tin.

Flow out The ability of a material to level after application (whether brushed, sprayed, roll-coated, or applied through pressure units). "Orange peel" is the surface appearance when a sprayed material does not flow or level. Excessive flowing on vertical surfaces is termed "sagging," usually caused by too much material or solvent that is too slow drying.

Fluxes Oxides used to produce glass which is easy and inexpensive to shape.

Fogged unit An insulating glass unit with a permanent deposit that contaminates its interior surfaces.

Forehearth An extension of the furnace where glass is made at the same temperature throughout.

Forming Shaping or molding into shape.

Frequency The number of times an action occurs in a given time period. In sound, the number of complete vibration cycles per second, represented by the hertz (Hz).

Front putty The putty forming a triangular fillet between the surface of glass and the front edge of the rabbet.

Frost point The temperature below 32°F (0°C) at which visible frost begins to deposit on the air-space surface of a sealed insulating glass unit in contact with the measuring surface of the frost-point apparatus.

Fully tempered glass Transparent or patterned glass with a surface compression of not less than 10,000 psi or edge compression of not less than 9700 psi.

Gas-filled units Insulating glass units with a gas other than air in the air space to decrease the unit's thermal conductivity (U-value) and to increase the unit's sound insulating value.

Gasket Pre formed shape, such as a strip, grommet, etc., of rubber and rubber-like composition used to fill and seal a joint or opening, alone or in conjunction with the supplemental application of a sealant.

Gather A small amount of glass taken from a melting pot.

Gel A semisolid, jelly-like condition of matter. A form of colloidal dispersion in which the dispersed component and the dispersing medium are associated to form a jelly-like mass. A solution of gelatine or glue in warm water is liquid and is termed a sol; on cooling, the liquid changes to a jelly, termed a gel.

Glass A transparent, brittle substance formed by fusing sand with soda or potash or both; it often contains lime, alumina, or lead oxide.

Glaze A hard-fired glass finish on pottery.

Glazing The securing of glass in prepared openings in windows, door panels, screens, partitions, etc.

Glazing bead A strip surrounding the edge of the glass in a window or door; applied to the sash on the outside, it holds the glass in place.

Glazing channel A three-sided U-shaped sash detail into which a glass product is installed and retained by a removable stop.

Glazing channel depth The measurement from the bottom of the glazing channel to the top of its stops.

Glazing channel width The measurement between the stationary stop and the removable stop.

Glory hole An opening in a small furnace used to reheat glass articles.

Glue Originally a hard gelatin obtained from hides, tendons, cartilage, bones, and other connective tissues of animals. Also, an adhesive prepared from these substances by hydrolysis (application of water and heat), chemically known as a collagen.

Gobs Short lengths into which glass is cut for forming.

Green strength The mechanical strength, that, even though cure is not complete, allows removal of the mold without tearing or permanent distortion.

Greenhouse glass This is a translucent rolled glass with a special surface design to scatter light.

Grout The filler between courses of tile, especially ceramic tile.

Gun consistency Degree of softness of a compound suitable for application through the nozzle of a caulking gun.

Hardness. Property or extent of being hard. Measured by extent of failure of indentor point of any one of a numbr of standard testing instruments to penetrate the product.

Heat-absorbing glass Glass (usually tinted) formulated to absorb an appreciable portion of solar energy.

Heat break This occurs at 90° to the surface of the glass and resembles a smooth curve.

Heat-seal To bond or weld a material to itself or to another material by the use of heat. This may be done with or without the use of adhesive, depending on the nature of the materials.

Heat-strengthened glass Transparent or patterned glass with a surface compression of not less than 3500 psi or greater than 10,000 psi or an edge compression of not less than 5000 psi.

Heel bead Sealant applied at the base of channel, after setting lite or panel and before the removable stop is installed. One of its purposes is to prevent leakage past the stop. Sealant must be able to bridge the gap between the glass and frame.

Hermetically sealed glass unit A double section (two pieces) of glass sealed with a vacuum between; Thermopane®, for example.

Hertz (Hz) The unit frequency, representing cycles per second, named for Heinrich R. Hertz, the noted German scientist.

Heterogeneous Consisting of dissimilar constituents.

Hiding power The power of a paint or pigment to obscure or render invisible a surface over which it is applied. It is one of the most important physical properties of a white pigment. It is determined by the difference in index of refraction between the material and its surrounding medium.

Holiday In coated fabrics, a place not coated by coating compound.

Homogeneity Uniformity of composition throughout a material.

Homogeneous The opposite of heterogeneous. Consisting of the same element, ingredient, component, or phase throughout, or of uniform composition throughout. Crystalloids and crystalloid solutions (true solutions) are usually considered homogeneous systems of matter as opposed to colloids, which are heterogeneous or polyphasic.

Ignition point This is the temperature at which a liquid gives off sufficient vapors to burn continuously upon application of a flame, this temperature ordinarily being 40 to 80° higher than the flash point of the liquid. Most fire departments arbitrarily designate solvents as "extremely flammable," "flammable," and "combustible." These refer to solvents with flash points below 20°F (−2°C), and between 80 and 150°F (25 and 66°C), respectively. Those liquids with no flash points are termed "nonflammable."

Immersion Placing an article into a fluid, generally so it is completely covered.

Impact The single instantaneous stroke or contact of a moving body with another either moving or at rest, such as a large lump of material dropping on a conveyor belt.

Impact strength Measure of toughness of a material, as the energy required to break a specimen in one blow.

Infrared (IR) Infrared radiation is used for the spectroscopic examination of high polymers. Infrared absorption spectra of high polymers are usually obtained for wave numbers 1800 to 600 cm^{-1} and the percentage of absorption recorded on a chart. Comparisons for structure are made of low molecular weight substances which contain what are assumed to be similar groups.

Intaglio A light engraving on the surface of glass.

Intensity The rate of sound energy passing through a unit area. The intensity is measured by the energy in ergs transmitted per second through 1 cm^3 of surface. The energy in ergs per cm^3 in a sound wave is given by $E = 2\pi^2 dn^2a^2$.

Interface The common boundary between two substances. Sometimes described as two surfaces with no air space between them (e.g., where the air contacts this paper is the air-paper interface).

Interior glazed Glass set from the interior of the building.

Interior muntins Decorative grid installed between the glass lites that does not actually divide the glass.

Interior stop The removable molding or bead that holds the lite in place, when it is on the interior side of the lite (as contrasted to an exterior stop which is located on the exterior side of a lite or panel).

Interlayer The transparent damping material used in laminated glass.

Isostere A plot of dew point versus temperature for given values of water capacity of a desiccant.

Isotherm A plot of the water capacity of a desiccant versus water partial pressure or dew point at a given temperature. Isotherms are often shown as a family of courses at various temperatures.

Jambs The vertical members of a frame adjacent to the structural members of a building.

Joint The location at which two adherents are held together by an adhesive. Also see Lap joint and Starved joint.

Laminated glass Two or more lites of glass bonded together with a plastic interlayer.

Laminated insulating glazing A laminated glass panel and a monolithic glass panel set in a frame that provides an air space between them.

Lap A part that extends over itself or a like part.

Lap joint A joint made by overlapping adjacent edge areas of two adherents to provide facing surfaces which can be joined with an adhesive.

Lap seam A seam made by placing the edge of one piece of material extending flat over the edge of the second piece of material.

Ldn Day/night average sound level in decibels.

Lehr Similar to an oven, used for reheating glass and allowing it to cool slowly.

Leq Equivalent sound level.

Lifting Softening and penetration of a film by the solvents of another film resulting in raising and wrinkling.

Light reducing glass Glass formulated to reduce the transmission of visible light.

Light transmittance Clear glass, depending on its thickness, allows 75 to 92 percent of visible light to pass through.

Lite Another term for a pane of glass used in windows.

Liter A measure of capacity in the metric system equal to 61.022 in^3, 0.908 U.S. quarts dry, and 1.0567 U.S. quarts wet.

Live load Load due to the weight of nonpermanent attachments: people, glazing rigs, washing rigs.

Low-emissivity glass Glass with a transparent metallic or metallic oxide coating applied onto or into a glass surface which reflects longwave infrared energy and thus improves the U-value.

Low-temperature flexibility The ability of a rubber product to be flexed, bent, or bowed at low temperature.

Lumps Surface protrusions, usually of the basic material as distinguished from foreign material.

Mandrel A forming or shaping tool used to make glass tubing.

Manganese A metal powder used as a flux in glass making.

Marver To roll molten glass on an iron slab.

Mastic An adhesive of such a consistency that it must be applied by notched trowel, gob, or buttering.

Melting point The temperature at which a polymer loses its crystalline character as evidenced by X-ray diffraction studies. Also called the "S-ray melting point." For low molecular weight solids, it is the temperature at which a solid melts and becomes liquid.

Metal spacers Roll-formed metal shapes used at the edges of an insulating glass unit to provide the designated air-space thickness.

Migration Spreading or creeping of oil or vehicle from a compound out onto adjacent nonporous surfaces, as contrasted to bleeding which refers to absorption into adjacent porous surfaces.

Milliliter One-thousandth of a liter, equal to one cc.

Miscible Soluble or compatible with each other, i.e., capable of being mixed to form a homogeneous mass.

Mitred corners Usually a 45° mitred joint produced in some sash where vertical jamb members meet horizontal head and sill members.

Mixer A machine, other than a mill, for mixing rubber compounds or dough, or a covered chamber or trough in which two blades or rotors revolve in opposition to each other. The axes of the blades may be horizontal, as in the Banbury mixer, or vertical, as in some types of cement mixers. In the latter case, the mixing chamber may rotate as well as the blades.

Mixing The process of incorporating the ingredients of a rubber compound into rubber, usually done on a mixing mill or in an internal mixer. The mixing process consists of (1) breaking down the rubber, (2) gradual incorporation of compound ingredients, (3) final working of the rubber ("cutting back") after all the ingredients are in, and (4) removing the mixed compound from the mill in sheets. "Mixing," or simply "mix," also denotes the completed mixture.

Modulus In the physical testing of rubber, the ratio of stress to strain, i.e., the load in pounds per square inch or kilos per square centimeter of initial cross-sectional area necessary to produce a stated percentage elongation. It is a measure of stiffness, influenced by pigmentation, state of cure, quality of the rubber, and other factors.

Moisture For all practical purposes, moisture may be considered as very finely divided particles of water. Moisture in the form of steam or a jet of water is sometimes used in a kiln to regulate the humidity.

Moisture absorption The absorption of moisture by a rubber or textile product.

Moisture vapor transmission rate (MVTR) The steady water vapor flow in unit time through a unit area of a body, normal to specific parallel surfaces, under specific conditions of temperature and humidity at each surface.

Molecular sieve (13X variety) An adsorbent; any of a class of zeolites having small, precisely uniform pores in their crystal lattices that can absorb molecules small enough to pass through pores. The cage structure is such that the pore openings are approximately 8.5 Å.

Molecular sieve (4A variety) As above, but with a crystalline cage structure such that pore openings are approximately 4 Å.

Monolithic glass Glass having a single uniform thickness.

Monomer A substance or simple chemical that can be polymerized, yielding a much larger molecule called a polymer.

Mottling A film defect associated with spraying. It appears as a uniform series of approximately circular imperfections.

Mullion A horizontal or vertical member that holds together two adjacent lites of glass or units of sash or sections of curtain wall.

Multiple-glazed units Units of three lites (triple-glazed) or four lites (quadruple-glazed) with two and three air spaces, respectively.

Muntin In sash having horizontal and vertical bars that divide the window into smaller lites of glass, the bars are termed muntin bars. Similar to mullion but lighter in weight.

Natural rubber The elastomer obtained from the hevea tree. The basic polymer is also present in other shrubs and trees. The first truly elastomeric type of product known.

Needle-glazing Application of a small bead of compound at the sight line by means of a gun nozzle with an opening about $\frac{1}{4} \times \frac{1}{8}$ in.

Neoprene A synthetic rubber with physical properties closely resembling those of natural rubber but not requiring sulfur for vulcanization. It is made by polymerizing chloroprene, and the latter is produced from acetylene and hydrogen chloride.

Nitrile rubber A class of rubber-like copolymers of acrylonitrile with butadiene. There are many types, and a few of the trade names are: Buna-N, Butaprene, and Chemigum. It has high resistance to solvents and oils, greases, heat, and abrasion.

Noise reduction between rooms In decibels, the amount by which the mean square sound pressure level averaged throughout the source room exceeds the sound pressure level averaged throughout the receiving room.

Nondrying Descriptive of a compound that does not set up hard.

Nonoxidizing Descriptive of a compound that withstands accelerated weathering, the equivalent of 20 years of normal weathering without oxidizing. Does not become hard after exterior exposure.

Nonskinning Descriptive of a product that does not form a surface skin after application. Usually remains tacky and sticky.

Nonstaining Characteristic of a compound that does not stain a surface by bleeding or migration of its oils or vehicle content.

Nonvolatile Any substance which does not evaporate or volatilize under normal conditions of temperature and pressure.

NR Noise reduction.

Open time Time interval between spreading the adhesive and completing the bond.

Opaque glass Glass that transmits no light whatsoever.

Opaline glass This glass is closely related to opaque. It is an opaque cast with ground and polished surfaces.

Optimum cure State of vulcanization at which the maximum desired property is attained.

Orange peel A surface defect caused by vortex currents set up during evaporation of solvents from a coating.

Organics Compounds which consist of carbon and generally hydrogen, with a restricted number of other elements, such as oxygen, nitrogen, sulfur, phosphorous, chlorine, etc., but not containing atoms or molecules generally known as metals.

Orifice ring Bowl-shaped with a hole in the bottom.

Ornamental glass Rolled glass with the surface figured by shaping or embossing rolls.

Overcure A state of excessive vulcanization resulting from overstepping the optimum cure, i.e., vulcanizing longer than necessary to attain full development of physical strength. Manifested by softness or brittleness and impaired age-resisting quality of the vulcanizate.

Ozone resistant Withstands the deteriorating effects of ozone (generally cracking).

Paint A pigmented liquid composition which is converted to an opaque solid film by application in a thin layer. An oil base paint contains drying oil or oil varnish as the basic vehicle. A water base paint contains a water emulsion or dispersion as the vehicle. The term is loosely used, sometimes designating the whole coating field.

Pane See Lite.

Parison The rough shape of a glass item. It is sometimes known as a "blank."

Parlon The trade name of a chlorinated natural rubber (Hercules Powder Co.).

Pass Term used in spraying to refer to the movement of a spray gun in one direction. A double pass, back and forth, is actually a single coat.

Passive solar heat gain Solar heat that passes through a material and is captured naturally, not by mechanical means.

Patterned glass Rolled glass having a distinct pattern on one or both surfaces.

Peelback A method of separating a bond of two flexible materials or a flexible and rigid material, whereby the flexible material is pulled from the mating surface at a 90 or 180° angle to the plane in which it is adhered. The stress is concentrated only along the line of immediate separation. Strengths are expressed in pounds per inch width (piw).

Peeling The loosening of a rubber coating or layer from a base material, such as cloth or metal, or from another layer of rubber.

Permanent set The amount by which an elastomeric material fails to return to its original form after a deformation. In the case of elongation, the difference between the length after retraction and the original length, expressed as a percentage of the original length, is called the permanent set. Permanent set depends on the quality and type of rubber, the degree and type of filler loading, state of vulcanization, and amount of deformation. Also see Adhesive setting.

Permeability The degree of water vapor or gas transmission through a unit area of material of unit thickness induced by unit vapor pressure differences between two specific surfaces under specified temperature and humidity conditions.

Permeance The time rate of water vapor or gas transmission through a unit area of a body, normal to specific parallel surfaces, under specific temperature and humidity conditions.

Pinholing A film defect characterized by the presence of tiny holes. The term is rather generally applied to holes caused by solvent bubbling, moisture, other products, dry spraying, or the presence of extraneous particles in the applied film.

Pitch A black or dark heavy liquid or solid substance left as residue after distilling tar, oil, and similar materials; also found naturally in asphalt. Pitches are named according to the source from which they are obtained as "bone pitch" from bone oil, "petroleum pitch" from petroleum, etc.

Plastics Natural and artificially prepared organic polymers of low extensibility, as compared with rubber, which can be molded, extruded, cut, and worked into a great variety of objects, rigid or nonrigid, and used as substitutes for wood, metals, glass, rubber, leather, fibers, and textile materials. Many are also referred to as synthetic resins. The first commercial plastic was celluloid, introduced by Hyatt in 1869, and the first commercial thermosetting resin was introduced by Baekland in 1909. There are two general methods of formation—condensation polymerization, as in the case of phenol-aldehyde resins, and vinyl polymerization, as in the case of polyvinyl chloride resins. Certain plastics are derived from casein. Some of the more recent products are organo-inorganic, such as the silicones.

Points Thin, flat, triangular, or diamond-shaped pieces of zinc used to hold glass in wood sash by driving them into the wood.

Polybutene base Compound made from polybutene polymers.

Polyester There are many types of polyester resins and they are manufactured by reacting together two basic raw materials, a dicarboxylic acid and a dihydroxy alcohol. Polyesters are used in one- and two-part systems for coatings and molding compounds. The manufacture of Dacron is a well-known use for polyester fiber.

Polyethylene A straight chain polymer of ethylene (gaseous hydrocarbon) used for containers, packing, etc.

Polyisobutylene Polymer manufactured from gaseous hydrocarbons. The polymer is a major portion of butyl rubber which also contains a small percentage of isoprene.

Polymer A very long chain of monomer units prepared by addition and/or condensation polymerization. The units may be the same or different. There are copolymers, tri- or terpolymers, quadripolymers, and high polymers.

Polymerization The reaction occurring when two or more molecules of a compound are united to form a more complex compound with a higher molecular weight.

Polysulfide elastomer A synthetic rubber-like elastomer practically insoluble in oils and solvents, prepared from ethylene chloride and sodium tetrasulfide commonly called Thiokol. It was the first commercial synthetic elastomer (U.S.A., 1930). Other dichlorides used are dichloroethyl ether and di-2-chloroethyl formal. These are not vulcanized with sulfur but by heating with zinc oxide. Also see Thiokol.

Polyurethane A family of polymers ranging from rubbery to brittle. Usually formed by the reaction of a diisocyanate with a hydroxyl.

Pontil A metal rod to which a glass article is attached.

Pot furnace A pot-shaped furnace made of clay.

Pot life The rating in hours of the time interval following the addition of accelerator before a material will become too viscous to pass predetermined viscosity (consistency) requirements. Closely related to Working life.

Potash Potassium oxide (a flux).

Pressure break Occurs at angles to the surface usually starting at a corner.

Pressure-sensitive adhesive A type of adhesive that retains its tack even after complete release of solvent.

Primary sealant A sealant applied to the inner shoulders of a spacer with its principal purpose to minimize moisture, gas, and solvent migration into the unit's air space.

Primer A special coating designed to provide adequate adhesion of a coating system to a new surface. In the case of new wood, it is used to allow for exceptional absorption by the surface. Metal priming coatings for steel work contain special anticorrosive pigments or inhibitors, such as red lead, white lead, zinc powder, zinc chromed, etc.

Priming Sealing of surfaces to produce adhesion of sealants.

Profile glass A U-shaped rolled glass for architectural use.

Purlins Structural members, generally horizontal, on sloped glazing frames.

Pyrex Trade name for borosilicate glass.

Pyrolitic deposition A process where a metallic oxide is added to glass while the glass is still hot.

Rabbet A two-side L-shaped recess in sash or frame to receive lites or panels. When no stop or molding is added, such rabbets are face-glazed. Addition of a removable stop produces a three-sided, U-shaped channel.

Racking Movement and distortion of sash or frames because of the lack of rigidity, or caused by adjustment of ventilator sections. Puts excessive strain on the sealant and may result in joint failure.

Rafters Structural members; vertical in sloped glazing frames.

Radiation Energy released in the form of waves or particles because of a change in temperature within a gas or vacuum.

Raw glass Rolled glass with the surface rolled smooth or slightly figured.

Reaction A mutual action of chemical agents upon each other resulting in a chemical change.

Reactor A substance undergoing a reaction or chemical change. Also refers to the equipment used in the polymerization process.

Reflective coated glass Glass with metallic or metallic oxide coatings applied onto or into the glass surface to provide reduction of solar radiant energy, conductive heat energy, and visible light transmission.

Reglet Any slot cut into masonry or formed into poured or precast stone. May also be an open mortar joint left between two courses of bricks or stones or a slot cut or cast into other types of building materials.

Relative heat gain An energy comparison factor for glass products combining the radiant and conductive heat gain under specific conditions (200 Btus times the shade coefficient + 14° times the summer U-value).

Resilient tape A preshaped, rubbery sealing material furnished in varying thicknesses and widths in roll form. May be plain or reinforced with scrim, twine, rubber, or other materials.

Resonance The sympathetic vibration of an object when subjected to a vibration of a specific frequency. The object tends to act as a sound source.

Reverberation The continuation of sound reflections within a space after the sound has ceased.

Reverberation time Normally defined as the time that elapses for a sound level to drop 60 dB after a mean-square sound level, continuously generated at a steady pressure in a room, is abruptly stopped.

Reversion (1) The change which occurs in vulcanized rubber as the result of ageing or overcuring in the presence of air or oxygen, usually resulting in a semiplastic mass. (2) It is the basis of rubber reclaiming processes and is aided by the use of swelling solvents, chemical plasticizers, and mechanical disintegration to obtain a workable mass.

Rex hardness The hardness of a "soft" vulcanized rubber or other similar elastic material measured by a Rex hardness gage.

Rheology Science of deformation and flow of matter. Deals with laws of plasticity, elasticity, viscosity, and their connection with paints, plastics rubber, oils, glass, cement, etc.

Rigidity The property of bodies by which they can resist an instantaneous change of shape. The reciprocal of Elasticity.

Room temperature vulcanizing (RTV) Vulcanization or curing at room temperature by chemical reaction, particularly of silicones and other rubbers.

R-value The resistance of conductive heat energy transfer in one hour through 1 ft^2 of a specific insulating glass unit assembly for each 1°F temperature difference between the indoor and outdoor air. It is the reciprocal of U-value; R = 1/U.

Sagging Running or flowing in the finish of a coating caused by the application of too much material and/or material that is too thin.

Salt spray test A testing method to compare the corrosion resistance of materials, usually coatings. One of the best corrosion test media known is the vapor of salt solutions. The most common procedure is to spray a fine mist of a 20 percent by weight solution of sodium chloride (iron free) in water into a large closed container in which the test panels are suspended. The starting temperature is about 95°F (35°C).

Sand core A shape made of sand, around which strands of molten glass are wound.

Sash A frame into which glass products are glazed, i.e., the operating sash of a window.

Score side The upper side of glass coming off the float line, sometimes called the air side.

Sealant A material used to fill a joint, usually for the purpose of weatherproofing or waterproofing. It forms a seal to prevent gas and liquid entry.

Sealant spacer A permanent adhesive sealant extrusion which may contain a structural metal insert and a precompounded desiccant.

Sealants (for insulating glass units) Formulated elastomeric compounds with specific application and vapor transmission properties as well as controlled adhesion, cohesion, and resiliency.

Sealed insulating glass units Units constructed of two or more lites of glass separated and hermetically sealed to spacer frames at the glass edges with the enclosed air chamber(s) dehydrated at the plant's atmospheric pressure.

Sealer (1) A continuous film to prevent the passage of liquids or gaseous media; a high-bodied adhesive generally of low cohesive strength to fill voids of various sizes to prevent passage of liquid or gaseous media. (2) A coating used to seal the sand-scratched surface of a primer to obtain a smooth uniform paint base over rough metal. Sealants are products of low pigmentation.

Seam A line formed by joining material to form a single ply or layer. Also see Lap seam and Transverse seam.

Secondary sealant A sealant applied into the exterior glass-spacer cavity to provide elastic, structural bonding of the assembly. In single-sealed units, this sealant also has low gas and moisture transmission properties to achieve effective unit performance.

Selenium A metal powder used as a flux in glass making.

Semicure Partially cured. A term frequently used to designate the first cure of an article that is cured more than once in its manufacture.

Set A term used rather loosely to describe the point at which a film has either dried sufficiently (released enough solvent) so that it is tough or hard or has cured sufficiently after the addition of the accelerator to sustain the required load pressure. Also see Permanent set.

Setting Placement of lites or panels in sash or frames. Also the action of a compound as it becomes firmer after application.

Setting blocks Small blocks of composition, lead, neoprene, wood, etc., placed under the bottom edge of the lite or panel to prevent its settling onto the bottom rabbet or channel after setting, thus distorting the sealant.

Shade coefficient The ratio of the rate of solar heat gain through a specific unit assembly to the solar heat gain through a single lite of ⅛-in clear glass in the same situation.

Shear The progressive relative displacement of adjacent layers because of strain or a lateral motion.

Shear test A method of separating two materials by forcing (either by compression or tension) the interfaces to slide over each other. The force exerted is distributed over the entire bonded area at the same time. Strengths are recorded in psi (pounds per square inch).

Shelf ageing A method of determining the resistance of rubber articles to perishing by storing them under atmospheric conditions, either in light or in darkness, and testing them after definite lapses of time. The natural deterioration of rubber articles kept in storage or "on the shelf" under atmospheric conditions.

Shelf life The period of time a packaged adhesive or sealer can be stored under specific temperature conditions and remain suitable for use.

Shore hardness A measure of the hardness of a cured elastomeric material by a durometer.

Shrinkage The percentage loss of volume of a material when put through a particular process, for example, the washing and drying of crude rubber. The percentage diminution in area or volume of a piece of processed unvulcanized rubber compound on cooling. Also, the contraction of molded vulcanized rubber on cooling.

Sight line Imaginary line along the perimeter of lites or panels corresponding to the top edge of stationary and removable stops and the line to which sealants contacting the lites or panels are sometimes finished off.

Silica A type of sand derived from minerals, such as quartz.

Silica gel An adsorbent. An amorphous form of silica dioxide having a large internal surface area and range of pore sizes.

Silicone rubber A rubber prepared by the action of moisture on dichlordimethyl-silicone. These rubbers withstand temperatures from 120 to 500°F (47 to 260°C) and are vulcanized with benzoyl peroxide.

Skylight A glass and frame assembly installed into the roof of a building.

Sloped glazing Any installation of glass that is at a slope of 15° or more from the vertical.

Snow load Load due to snow accumulation.

Soda ash (sodium oxide) A flux used in glass making.

Softening point The temperature at which a prescribed load will cause the failure of a 1-in^2 shear bond of cloth to steel. Since softening under heat is progressive, increasing with temperature, it is rarely stated that a product is hard or soft at certain temperature. The preferred method is to report what weight per inch it can support without failure. A bond under high tension, for instance, will not withstand as much heat as one under light tension, and raising or lowering the temperature will, to a point, decrease or increase the apparent strength of an adhesive.

Solar energy Thermal radiation from the sun, as measured by the short radiation wavelengths, less than 3 μ long.

Solar energy absorption The percentage of the solar spectrum energy (ultraviolet, visible, and near-infrared) from 300 to 3000 nm, that is absorbed by a glass product.

Solids (plastic and elastic) Solid: a substance which undergoes permanent deformation only when subjected to shearing stress in excess of some finite value characteristic of the substance (yield stress). Plastic solid: a substance which does not deform under a shearing stress until the stress attains the yield stress, at which the solid deforms permanently. Elastic solid: a substance in which, for all values of the shearing stress below the rupture stress (shear strength), the strain is fully determined by the stress regardless of whether the stress is increasing or decreasing.

Solubility The degree to which a substance will dissolve in a particular solvent usually expressed as grams dissolved in 100 g of solvent.

Sound absorption The property possessed by material and objects, including air, of converting sound energy to heat energy.

Sound transmission class (STC) A single-figure rating of standardized test performance according to ASTM-E-413-73, for evaluating the effectiveness of assemblies in isolating airborne sound transmission.

Spacer corners Specific methods used in joining spacer lengths into spacer frames, including interlocking keys, bending, soldering, or welding.

Spacer depth That dimension of the spacer measured parallel to the glass surface.

Spacer width That dimension of the spacer measured perpendicular to the glass surface which establishes the unit's air space.

Spacer (IG) A hollow shape used to provide a fixed air space between two pieces of glass to obtain thermal or acoustical properties. Small blocks of composition, wood, neoprene, aluminum, etc., are placed on each side of the lite or panel to center it in the channel and maintain the width of sealant beads. Prevents excess sealant distortion.

Spacers Small blocks of composition, neoprene, etc., placed on each face of lite and panel to center them in the channel and maintain uniform width of sealant beads, preventing excessive sealant distortion.

Spandrel That portion of the exterior wall of a multistory commercial building that covers the area below the sill of the vision glass installation.

Specific gravity

$$\text{Specific gravity} = \frac{\text{weight of substance}}{\text{weight of equal volume of standard}}$$

It is the ratio of the weight of any volume of substance to the weight of an equal volume of a standard substance at stated temperatures. For solids or liquids, the standard substance is usually water, and the standard for gases is air or hydrogen.

Specific volume The reciprocal of specific gravity (1 divided by the specific gravity), and represents the volume in liters of 1 kg or the volume in cubic feet of 1000 lb (998.9 exactly). It also represents the ratio between the volume of 1 lb of water (27.72 in^3) and the volume of 1 lb of the material in question.

Sputter coating Same as vacuum deposition.

Starved joint A joint that has an insuffficient amount of adhesive to produce a satisfactory bond.

Stationary stop The permanent stop or lip of a rabbet on the side away from the side on which lites or panels are set.

Stoce A loose pack of glass weighing from 4000 to 10,000 lb.

Stop Either the stationary lip at the back of a rabbet or the removable molding at the front of the rabbet serving to hold the lite or panel in sash or frame with the help of spacers.

Straight chain polymer A polymer containing groups of molecules attached to each other in a straight line, like a string of beads. In other words, there is no cross-linking of groups. Examples are Vistanex and polybutene.

Strength The maximum stress required to overcome the cohesion of a material. Quantitative: a complex property made up of tensile strength and shearing strength. The force required to break a bar of unit cross section under tension is the tensile strength. It depends not only upon the rate of application of the load. Strength involves the idea of resistance to rupture.

Striking off The operation of smoothing off excess compound at the sight line when applying sealant around lights or panels.

Structural glazing gaskets Cured elastomeric channel-shaped extrusions used in place of a conventional sash to install glass products onto structurally supporting subframes, when the pressure of sealing exerted by the insert of separate lockstrip wedging splines.

Structural silicone glazing A system in which the glass product is bonded to the framing members of a curtain wall, using a structural silicone adhesive/sealant without the presence of the outdoor retainers or stops.

Sunlight The portion of solar energy detectable by the human eye; it accounts for about 44 percent of the total radiation wavelength spectrum.

Supercooled Frozen into shape.

Surface preparation The procedure required with respect to a foundation surface or the materials to be adhered which will promote optimum performance of an adhesive, coating, or sealer. For example, if higher bond strength is required, abrading and/or acid-etching the surface can improve the adhesion of the bonding material to the mating surface. Common methods of surface preparation are solvent washing, sandblasting, and vapor degreasing.

Swelling The property of raw or vulcanized rubber of absorbing organic liquids, such as benzene, gasoline, etc., and swelling to many times its original volume. The property is also shown by other colloids in contact with other liquids. In a general sense, it may be any increase in volume of a solid substance caused by the absorption of a liquid.

Tack, dry Property of certain adhesives, particularly nonvulcanizing rubber adhesives, to adhere on contact to themselves at a stage in evaporation of volatile constituents, even though they seem dry to the touch.

Tackiness The stickiness of a film while drying. For instance, after a paint or varnish sets up, it usually retains a sticky or tacky feel for some time until it is practically dry. Stickiness: a quality possessed by a solid having a low yield value and high mobility by which contact readily results in adhesion. For example, adhesive, varnish, printer's ink, and gold size under working conditions are tacky or sticky substances. When most of these dry out, set, gel, or harden through chemical or other change, they lose tack or stickiness. Those compounds that retain tack long after drying are said to be "permanently tacky" or to possess "after tack."

Tank A glass furnace.

Temperature The degree of heat or cold as measured in terms of °C or °F.

Tempered glass Fully tempered (FT) glass is reheated to just below the softening point (about 1300°F or 704.4°C) and then rapidly cooled.

Tensile strength The capacity of a material to resist a force tending to stretch it. Ordinarily the term is used to denote the force required to stretch a material to rupture and is known variously as "breaking load," "breaking stress," or "ultimate tensile strength." In rubber testing, it is the load in pounds per square inch or kilos per square centimeter on the original cross-sectional area supported at the moment of rupture by a piece of rubber, elongated at a constant rate.

Tension pull A term of total pull in pounds at the conclusion of a tension test. This test subjects a hose assembly, for example, to an increasing tensile load in a suitable testing machine until failure occurs by separation of the specimen from the end fittings or by rupture of the hose structure.

Tension stress-strain testing Determination of stress and strain (tensile strength and elongation) with the use of dumbbell specimens in conformance with the ASTM Method G-412 at 75°F (24°C). Rings may also be used.

Thermal break A material with a low thermal conductance used to separate exterior and interior materials. The thermal break is intended to stop the flow or transfer of heat.

Thermoplastic Capable of being repeatedly softened by heat and hardened by cooling.

Thermosetting Having the property of undergoing a chemical reaction by the action of heat, catalysts, ultraviolet light, etc., leading to a relatively infusible state.

Thiokol A commonly used name for the first commercial synthetic elastomer, produced in 1930 by Thiokol Chemical Company. Thiokol A is produced by the reaction of ethylene dichloride and sodium tetrasulfide, $CL—CH_2$, CH_2 CL + Na_2 S_4 (—CH_2 CH_2—S_4—)X. The atoms of sulfur may all lie in a straight line. Thiokol is vulcanized by heating with zinc oxide and is important because it is practically insoluble in petroleum oils. Thiokols are also prepared from other dichlor compounds: di-2-chloroethyl ether, $CL—CH_2$ CH_2 OCH_2 CH_2 Cl, and di-2-chlorethyl formal, $Cl—CH_2$ CH_2 OCH_2 OCH_2 CH_2 Cl. The latter Thiokols are somewhat more soluble than Thiokol A. See Polysulfide elastomer.

Thixotropic A term used to describe certain colloidal dispersions which, when at rest, assume a gel-like condition but, when agitated, stirred, or subjected to pressure or other mechanical action at ordinary temperatures, are transformed into liquids. The action is reversible and can be repeated at will. Thixotropic colloids occur in nature. The best known example is bentonite, a colloidal American clay. Rubber dispersions are not thixotropic colloids, such as bentonite.

Tin coat One layer of a coating system used to improve the adhesion of adjacent or succeeding coats.

Tinted glass Body-colored glass with specific ingredients formulated to produce light reducing and/or heat absorbing glass products.

TL Sound transmission loss.

Tooling Operation of pressing in and sticking a compound in a joint to press the compound against the sides of a joint and secure good adhesion. Also the finishing off of the surface of a compound in a joint so that it is flush with the surface. A narrow, blunt-bladed tool is used for this purpose.

Total heat gain/summer/daytime (Btu per hour per square foot) The sum of the radiant energy and the conductive energy transmitted into the building (shade coefficient × ASHRAE solar heat gain factors + summer U-value × the indoor/outdoor temperature differences).

Total heat gain/summer/nighttime (Btu per hour per square foot) The conductive energy transmitted into the building (summer U-value × the indoor/outdoor temperature difference).

Total heat loss/winter/nighttime (Btu per hour per square foot) The conductive energy transmitted to the outdoors (winter U-value×the outdoor/indoor temperature difference).

Toxicity A term referring to the physiological effect of absorbing a poisonous substance into the system either through the skin, mucous membranes, or respiratory system. When describing their toxic effect, solvents are usually classified as having a high, medium, or low toxic effect, depending upon whether a solvent vapor concentration of less than 100, from 100 to 400, or over 400 parts per million, respectively, is the maximum amount permissible in the air for safe or healthful working conditions.

Transmission The passage of sound from one location to another through an intervening medium, such as a partition or air space.

Transmittance The fraction of radiant energy that passes through a given material.

Transverse seam A seam joining two materials across the width of the finished product.

Tube-drawing Drawing glass tubing into shape.

Two-part compound A product which is necessarily packaged in two separate containers. It is comprised of a base and the curing agent or accelerator. The two compounds are uniformly mixed just prior to its use. When mixed, it cures and its useful life is quite limited from the standpoint of application characteristics.

Ultimate elongation The elongation at the moment of rupture.

Ultraviolet Type of radiation with wavelengths shorter than those of visible light and longer than those of X-rays.

Ultraviolet light (UV) A form of luminous energy occupying a position in the spectrum of sunlight beyond the violet and having wavelengths of less than 3900 Å, which is the limit of the visible spectrum. Ultraviolet rays are very active chemically, exhibit bactericidal action, and cause many substances to fluoresce.

Uniform bead Compound applied to a joint, with uniform width and appearance.

Unit Term normally used to refer to one single lite of insulating glass.

United inches The sum of the dimensions of one length and one width of a lite of glass.

U-value The amount of conductive heat energy (Btu's) transferred through 1 ft^2 of a specific insulating glass unit for each 1°F temperature difference between the indoor and outdoor air. It is the inverse of the R-value; $U = 1/R$.

Venting Providing circulation of air or ventilation between two walls or partitions. Venting is accomplished by the use of tubes, breather vents, or openings left in the wall.

Vinyl Derived from ethylene (hydrocarbon gas), the compounds of which are polymerized to form high molecular weight plastics and resins such as vinyl acetate, vinyl styrene, etc. It is a base material for plastisols and organisols and also widely used in emulsion form as polyvinyl acetate.

Vinyl glazing Holding glass in place with extruded vinyl channel or roll-in type.

Viscosimeter or viscometer An instrument used for measuring the viscosity or fluidity of liquids and plastic materials. Various types are used, such as the Baybolt, Redwood, and Engler, based on the rate of flow through a tube; some types (Brookfield, Stormer) are based on the torsion principle, and others on the time taken for a metal ball to fall through a column of the material in a definite time. For rubbers including GR-S, the Mooney viscometer is widely used for both the raw and compounded material.

Viscosity A manifestation of internal friction as opposed to mobility. The property of fluids which resists an instantaneous change of shape (i.e., resistance to flow). It is measured by the force required to cause two parallel liquid surfaces of a unit area and a unit distance apart to slide past each other in the liquid expressed in poises, or dyne-seconds per square centimeter. Water at 20.2°C has a viscosity of 1 centipoise and is taken as the standard of comparison. A number of terms have been proposed for special applications of viscosity. These are listed below:

Syrup: Material that slumps under its own weight (will not maintain its shape) when made into a ball with a diameter of 1 in or less.

Thin: Any material tested on a Ford cup or any material testing up to 40 on a #26 MacMichael wire.

Medium: Any material testing 40 to 300 on a #22 MacMichael wire or up to 40 on a #22 MacMichael wire.

Heavy: Any material testing 40 to 100 on a #22 MacMichael wire or from 0 to 65 on a #18 MacMichael wire.

Paste or mush: Material that will flow or slump under its own weight (not hold its shape) in a ball of diameter greater than 1 in. Viscosity is recorded in the range of 400 to 150 cone penetrometer.

Dough: Material that will generally not flow under its own weight. Viscosity is recorded in the range of 150 to 0 cone penetrometer.

Visible light transmittance The percentage of light in the visible spectrum range of 390 to 780 nm that is directly transmitted through glass.

Waisting Glass narrowing in the middle.

Water absorption The process of assimilating or soaking up water.

Water resistance The ability to withstand swelling by water for a specified time and temperature, usually 48 hours at 212°F (100°C), expressed as a percentage swelling or volume increase of the specimen.

Weatherometer An apparatus for estimating the comparative resistance of a soft vulcanized rubber compound to deteriorating when exposed to light with a frequency range approximating that of sunlight. The criterion used in estimating resistance to light, again, is the percentage decrease in tensile strength and in elongation at break. A supplementary criterion is the observed extent of surface crazing and cracking. During the test, water sprays of clean water are forced on the specimens to simulate the action of rain (ASTM-D-750).

Weep hole Opening at the base of a cavity wall to collect moisture and dispense it or a breather tube put in sealant to relieve moisture.

Weeping Failure of a compound to support its own weight in a joint, but less pronounced than sagging.

White wool A process used in making glass fiber.

Wind load Load on glass because of the speed and direction of the wind.

Wired-reinforced glass Glass having a layer of meshed wire completely embedded in the glass lite. It may have polished or patterned surfaces.

Working life The period of time during which an adhesive, after mixing with the catalyst, solvent, or other compounding ingredients, remains suitable to use. Synonym for Pot life (ASTM-D-907).

Wrinkling The formulating of wrinkles in the skin of a compound during the formation of its surface skin by oxidation after application.

Yield strength The first stress in an adhesive (or adherent) at less than maximum attainable stress, at which an increase in strain occurs without an increase in stress.

Young's modulus The ratio of normal stress to corresponding strain for tensile or compressive stress less than the proportional limit of the material (ASTM-D-1053).

Zeolite A class of compounds that are crystalline, aluminosilicates of group I and group II elements of the periodic table. They may occur in nature or be synthetically produced. The framework structure and pore size of these compounds are orderly and controlled. They are effective adsorbents with a very high surface area.

Bibliography

1. Building Research Institute (Defunct).
2. The Language of Sealed Insulating Glass Units, *SIGMA,* Volume *XV,* 12/1989.
3. Peterson, A. J., *How Glass is Made,* Facts On File Publications, Threshold Books Ltd., New York, 1985.
4. A glass primer, *Glass Digest,* 4 and 5/1990.
5. Fenestration for Sound Control, *The Construction Specifier,* Construction Specifications Institute, April 1990.

Chapter

25

Polymer-Dispersed Liquid Crystals and Suspended-Particle Device

Joseph S. Amstock

President, Professional Adhesive and Sealant Systems

Introduction

In a pivotal scene in the award-winning movie *Philadelphia,* the partners of a fictitious law firm convene in a glass-enclosed conference room to discuss an attorney's dismissal from the firm. As the door closes on the conference room, one member flips a light switch, and the glass instantaneously turns from clear to gray, blocking all view of the meeting to those outside the glass walls. Magic? Not at all. Just a quick look at one of the glazing industry's newest products, privacy glass, or as some manufacturers called it, "smart glass."

It sounds futuristic, but today it is being installed in buildings and offices worldwide. This technical switchable glazing is used primarily for areas that require privacy or security, but it also reduces glare and associated eye strain and diffuses direct sunlight to eliminate discomfort from hot spots. It can also reduce the solar heat gain, thus increasing comfort and reducing air-conditioning requirements.[1]

Smart Glass Development

Smart glass has been in the development stages since the mid-1960s, when researchers at Kent State University developed the liquid crystal technology that is the basis for one type of privacy glass product.

Liquid crystals, or PDLC (polymer-dispersed liquid crystals), have been the most widely used material, but not the only basis for smart glass. The special effects of this glazing have been achieved through three distinct technologies. PDLC is the most common approach to the product. The SPD (suspended particle device) is another highly developed application, and electrochromic technology is currently the least viable alternative because of its premature stage of development.

Interestingly, the first crude SPD was a light-controlling device simply called a "light valve" which was invented by Edwin Land, founder of Polaroid Corporation, and was patented by him in 1934.

The leading developers of the "smart glass" technology have had the best results with the PDLC and SPD methods. In the U.S., the major users of smart glass technology include 3M Corp., which has partnered with Viracon, Owatonna, MN, and Marvin Windows and Doors, Warroad, MN, in its privacy glass development, and Polytronix and Research Frontiers Inc., Woodbury, NY. Polytronix allows light to be scattered through an optically inhomogeneous medium. PDLC is a medium whose light-scattering power is adjustable by applying an electric field. Figure 25.1 shows the structure of a liquid crystal film.

Although this specialty glazing has existed since the late 1980s, only now is it rising to mainstream use. Talig, a West Coast manufacturer which was dissolved several years ago, was the first company to develop and market the product. 3M bought the technology from Talig and obtained licenses from Raychem, the company owning the rights to some of the patents on liquid crystal technology. 3M has spent the last four years developing its own product based on the original technology. Polytronix has been working on its liquid crystal product since 1988 and, in 1992, began manufacturing and selling privacy glass based on its technology. Research Frontiers, which uses the SPD method in manufacturing its smart glass, is in the development stage and expects to begin worldwide marketing of its products soon. Solar Energy Research Institute was an early participant with electrochromic glass. This is illustrated in Figure 25.2.

One of the major challenges these companies face in selling these products in commercial and residential markets is the relatively high cost of the glazing. At approximately $100 per square foot installed, it is outside the budgets of the potential users. However, as the technology has developed, the specific advantage of this glazing outweighs the cost to many customers who desire its unique benefits.

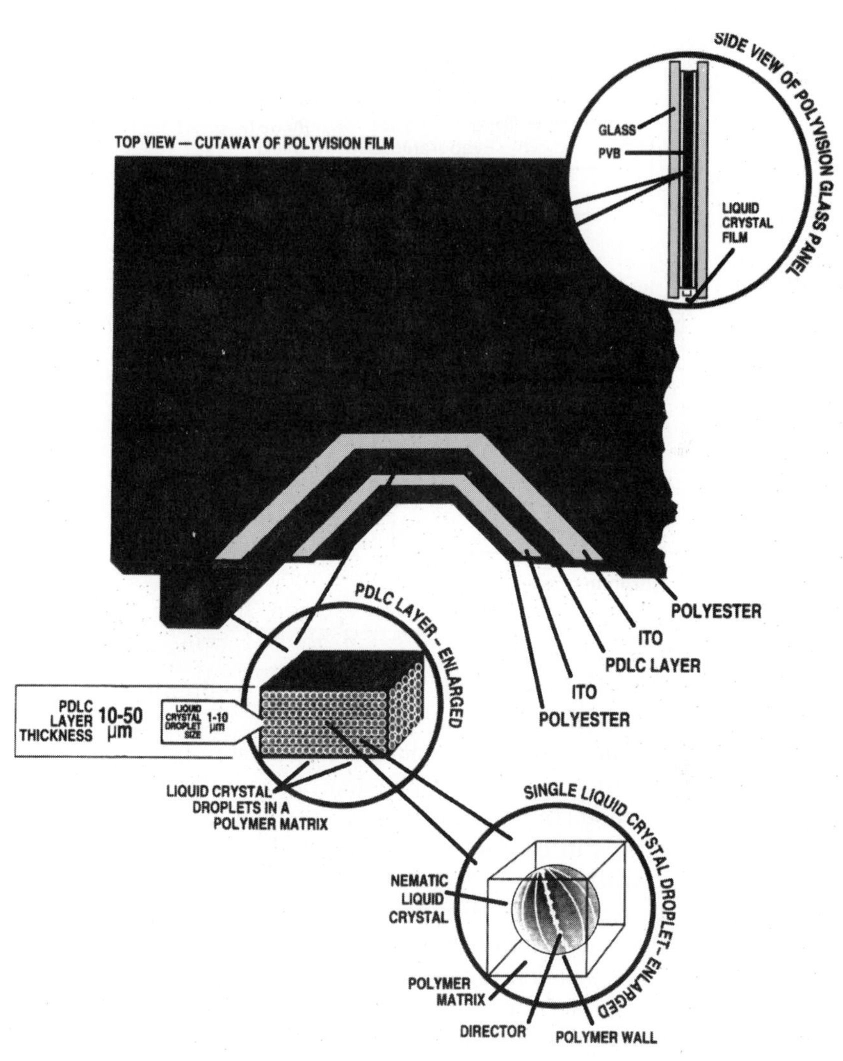

Figure 25.1 Liquid crystal film.

Functionality

A prototype of a window which can be electrically regulated to control the amount of light transmitted through it is in the testing stage. As noted earlier, it is called a "smart window" and will offer users the opportunity to control their environment in new and exciting ways. In addition to enabling one to instantly vary the amount of visible light passing through the window, Low-E coatings and other coatings within the window should enhance energy efficiency by reflecting a signifi-

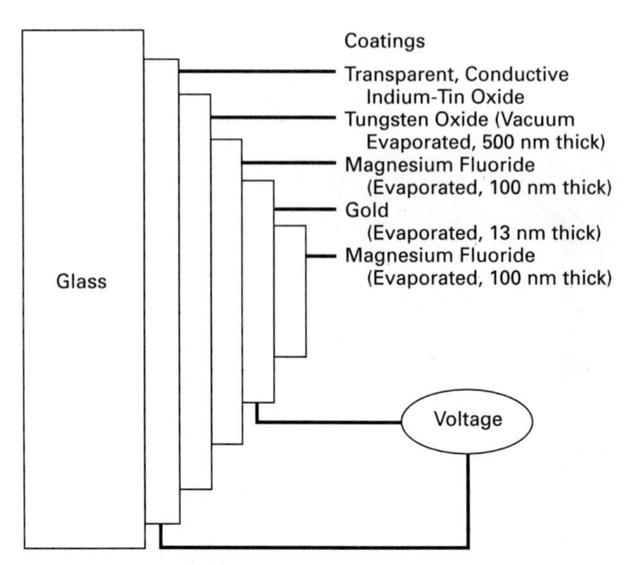

Figure 25.2 Electrochromic glass.

cant amount of radiation in the summer and trapping significant radiant heat indoors in the winter.[2]

An advanced form of such a window is constructed by sandwiching a thin film containing a liquid suspension between two lites of glass or plastic, which have transparent, electrically conductive coatings on their inside surfaces in contact with the thin film. Figure 25.3 illustrates the configuration.

The film comprises a very thin plastic layer throughout which are distributed an enormous number of tiny droplets of a dark blue liquid suspension. The suspension consists of light-absorbing, colloidal crystals (particles) dispersed randomly in a nonaqueous liquid.

To operate the "smart" window, an ac voltage is applied to the coatings, which creates an electric field that causes the anisometrically shaped suspended crystals to align (Figure 25.4). When the crystals align, the glass becomes transparent.

The schematic of figure 25.5 shows a very basic working scheme of smart films.[3] The degree of transparency depends on the amount of voltage applied. In the transparent (activated) condition, one can see through this glass very well with very little haze, looking straight through or at an angle.

To distinguish "smart glass" technology from other technologies, such as electrochromic cells or liquid crystals, the smart window is often referred to as an SPD (suspended particle device). Since this work began in 1965, new particles have been invented which will tol-

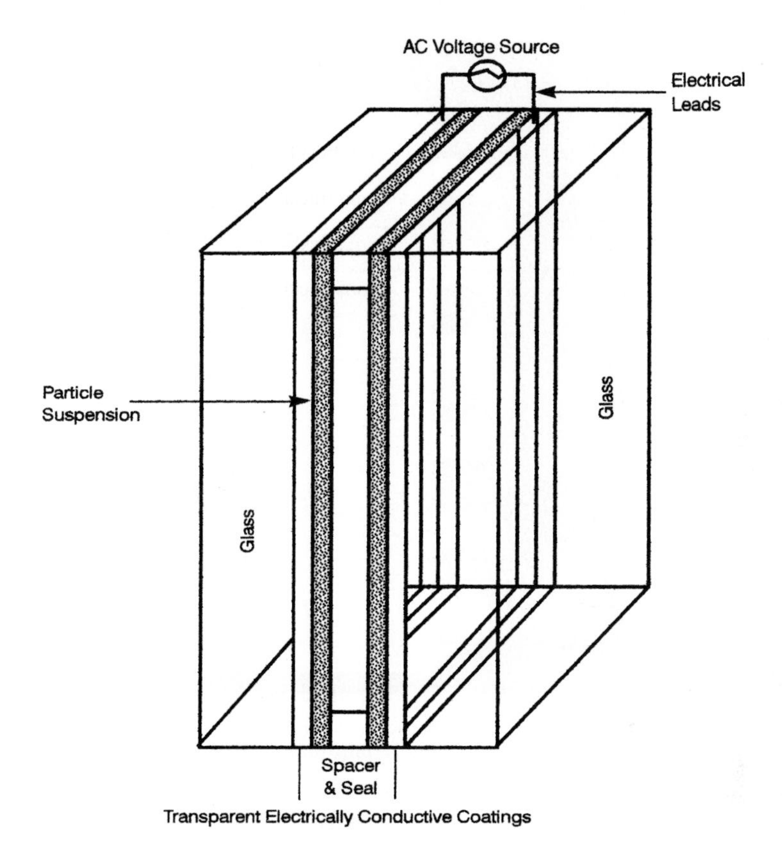

Figure 25.3 Smart windows. (*Courtesy of Research Frontiers, Inc.*)

erate temperatures as high as 250°F (121°C) for prolonged periods. Prior to 1991, only liquid suspensions were available for use as the working material in the windows. Low-frequency operation (50 to 400 hertz) at low ac voltage is now possible. The specific voltage required for a film window depends on the film thickness. Films as thin as 6 μ can be easily made and at that thickness require only 20 V less.

Operating power requirements are almost negligible. It should be possible to operate a square foot of SPD "smart glass" for only a fraction of a watt.

Various ranges of light transmission are possible. If a high particle concentration is used in the film, the off (dark) state can be very dark (e.g., 0.1 percent light transmission). Alternately, if one does not desire the off state to be so dark, a lower particle concentration can provide an off state of 10 or 15 percent light transmission, for example. Table 25.1 is a typical specification for privacy glass film.

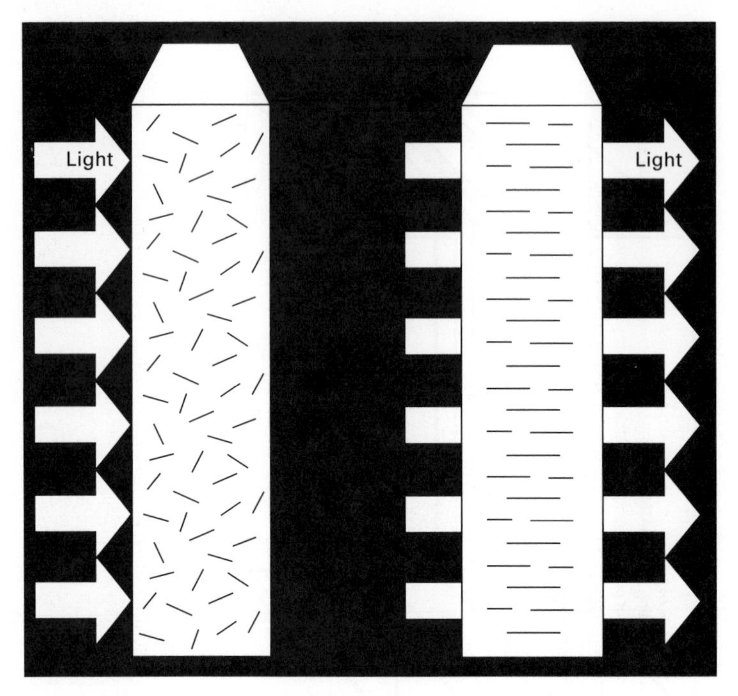

Figure 25.4 Smart glass.

The on (activated or most transparent) state, of course, is affected by the particle concentration so that various ranges of light transmission are possible. For a film with a 10 to 15 percent off state, on state light transmission of 50 to 70 percent or more can be attained depending on the amount of voltage applied. Darker films are somewhat less transmissive in the activated condition. In addition to providing visible light and the ability to reflect a significant percentage of impinging radiant heat (infrared energy), an SPD smart window should be able to block almost all impinging ultraviolet radiation.[4] Over time, ultraviolet radiation can seriously fade colors in upholstery, painting, rugs, etc.

Privacy applications

In the commercial market, the most common use of smart glass has been in conference rooms and presentation areas, especially for organizations whose office space plays a major role in their corporate image. Those who have incorporated privacy glass have reported very positive feedback.

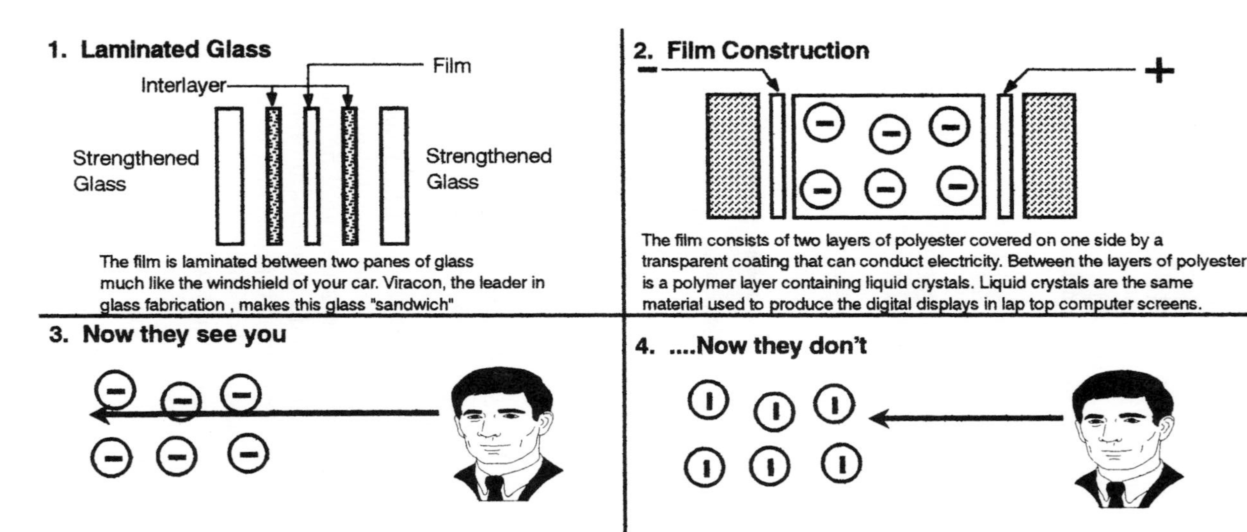

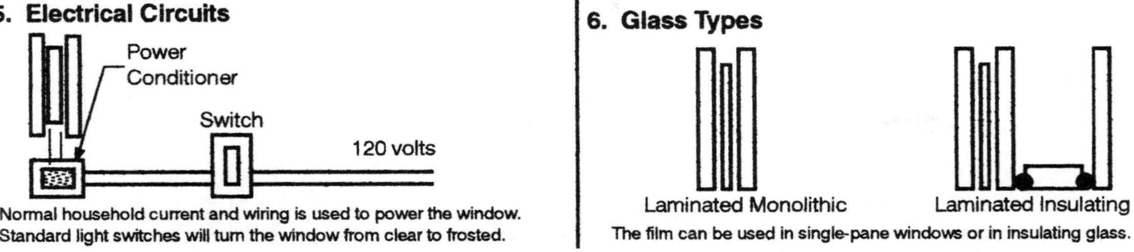

Figure 25.5 Smart glass.

TABLE 25.1 Typical Properties of a Privacy Glass Film

Glass color	Clear, bronze, gray, green
Glass type	Annealed, heat-strengthened, tempered, reflective
Thickness	Interior—$\frac{5}{16}$ in (8 mm) or $\frac{7}{16}$ in (11 mm) Door—$\frac{9}{16}$ in (14 mm) Exterior—$\frac{3}{4}$-in (19-mm) insulating glass unit $\frac{1}{8} \times \frac{1}{4}$-in air space $\times \frac{3}{8}$-in laminate 1-in (25-mm) insulating glass unit $\frac{3}{16} \times \frac{1}{2}$-in air space $\times \frac{5}{16}$-in laminate
Size	Up to 3 × 8-ft (910 × 2440 mm) For larger sizes contact your representative
Shape	Any shape, including holes anywhere
Environmental	Storage −30°F to 65°F (−22°C to 18°C) Operation −20°F to 45°F (−29°C to 7°C)
Electrical	Driving voltage 60 V ac Current less than 20 mA/ft^2 Power approximately 1 W/ft^2
Switching time	Approximately 100 milliseconds at room temperature
Optical	Transmission 80% View angle approximate 120° Scattering effectiveness approximately 1 in
Life	Greater than 5 years (indoors) (Claim is supported by manufacturer's testing data)

Areas need to be opened up with natural light so adjoining rooms can benefit. One of the most important uses for privacy glass is in a hospital where a sterile environment is critical. Glass provides a clean surface, unlike blinds and fabrics. A sterile surface like glass does not hold bacteria and is ideally suited to the hospital environment, which needs to incorporate light and views into specific operating areas and rooms, without barriers that could hold contaminants.

Skylights and sloped glazing that are often inaccessible to users will also be prime spots for privacy glass.

On the residential front, the use of smart glass is also promising. In addition to its ability to block views of certain areas, switchable glass also reduces ultraviolet light penetration, a major cause of fading in furniture and carpeting.

Precise production

Heat-strengthened glass is used to manufacture switchable glass. Because of the high sensitivity of the film encased in the glass, special care must be taken when laminating the components. The panels for smart glass or privacy glass must be laminated and built in a clean room to prevent any dust particles from contaminating the film. One of the reasons for the product's high cost is this need for an absolutely flawless surface. Even one particle of dust can render the film useless.

The raw materials are costly and a certain percentage of panels do not survive the entire production and installation process. These are figured into the final price of the product as cost of manufacturing and installation. That is why this glazing is somewhat expensive.

Some companies selling smart glass and other privacy films point out that the finished glass panel is not always crystal clear. Due to its laminated construction and liquid crystals, the privacy glasses will look different from a standard single lite of glass.

Privacy glass is incorporated into new and retrofitted interiors. Additional research is being conducted on applying the film to existing glazing, but it is not likely to be available for quite a long time, if ever. First, the film containing the crystals must be protected by glass on both sides, and existing panels must be wired to provide the electrical charge needed to activate the privacy effect of the glass.

Acceptance

Architects and particularly glazing contractors who have worked with smart glass report positive experiences. The main installation difference cited is the electrical element, which must be handled by a certified electrician. Two people need to be involved with this type of installation, one the glazier and the other the electrician. The glass is installed the same way you install any piece of glass, except that a hole must be drilled in the frame for the electrician to hook up the electricity. Promising reliability studies have been done. Because privacy glass is so new, only laboratory studies of long-term use have been conducted, but the data available demonstrates reliability after millions of transmissions.

A significant factor is the on-off frequency of several installations over a four-year period, where the results are extremely high, in the area of several million cycles. At this time, there seems to be no problem of size in typical installations.

Growth potential

Smart privacy glass is an expensive product but, according to those who have it, well worth the investment. The product volume is expected to grow dramatically. Unfortunately, the predictions are high but the manufacturers of the glass do not expect it to become more affordable. It is extremely expensive to produce right now. However, over the next 10 to 15 years, it might become more affordable to the mainstream.

Many of the principal firms have granted licenses to foreign firms in Japan, Korea, and Europe.

Conclusion

Like any new product, smart glass is sure to grow in popularity and, eventually, to drop in price, as manufacturers develop new ways to produce it more efficiently. As more architects and contractors become familiar with these products and as office tenants and homeowners request them, we may all have an opportunity to seek a little privacy beyond our looking glass.

References

1. Burns, J., Smart Glass: Switchable Glazing, *Glass Magazine,* May 1995.
2. Saxe, R. L., "Smart" Windows Reach Advanced Development State, *U.S. Glass,* March 1993.
3. Courtesy of the 3M Company.
4. Research Frontiers, Inc. has more than 19 U.S. patents as of January 31, 1996.

Chapter

26

National Fenestration Rating Council

Joseph S. Amstock

President, Professional Adhesive and Sealant Systems

Introduction

The commercial and residential building industry and government, utilities, and homeowners are getting a tighter grip on the relative energy performance of fenestration products.

To further clarify performance differences among these products, the National Fenestration Rating Council (NFRC), which has completed work on the initial procedures to determine U-values, is now addressing other energy performance characteristics of windows and doors, specifically shading coefficient, solar heat gain coefficient, visible transmittance, air infiltration, and condensation.

This is the task of NFRC's Technical Committee. It is an enormous job that this committee has undertaken to produce processes for measuring all of these window and door energy performance factors. They have been trying, for many years, to arrive at a consensus among the many industry, government, and special interest groups and disciplines involved. Now, there is positive momentum to provide the energy-related tools for rating the key performance factors that architects, contractors, specifiers, and homeowners are seeking.

In addition, several state energy commissions and Canada are currently setting mandatory energy performance standards for new commercial and residential buildings and, in some cases, are providing energy-reduction incentives to power companies.[1]

Numerous technical advances have been made by the glass industry over the past few years. Low-E glazing, inert gas fills, improved thermal-performance window frames, and advanced shading systems are just a few of these technical innovations. Despite these developments, assessing product energy performance has remained difficult. The NFRC is working to create a fair, accurate, and credible means for rating the thermal performance of fenestration products. The organization was formed to address this issue for a number of reasons.

History

In 1989, both the U.S. Senate and the House of Representatives began to circulate proposed legislation regarding global warming, known as Senate Bill 324 and often referred to as the "Global Warming Bill." This legislation reflected a far-reaching program intended to protect the global environment. What did this have to do with glass and windows? A great deal. This legislation described numerous requirements affecting the building industry, and included specific language concerning thermal-performance ratings and labels for windows and other fenestration products:

> Within 18 months after the date of enactment of this act, the Secretary (DOE), in conjunction with the National Institute of Standards and Technology, shall establish labels for thermal and optical properties and energy performance for windows and window systems.

For several years prior to this proposed federal legislation, the window industry has been struggling to resolve the issue of rating product performance. Builders tried to promote a unified system but were also unsuccessful. The scientific community had been working on more advanced computer tools and improved thermal-performance test methods to address these problems. States were about to embark on their own product rating, certification, and labeling programs. Yet, no one group was successful.

If enacted, the bills that were in Congress, however, would have required the Department of Energy and the National Institute of Standards and Technology to develop and implement a rating and labeling program for window products. It is easy to imagine the concerns of industry and others over the bureaucratic nightmare this implied.

Fortunately, the country was in the middle of two Administrations which were strong advocates of the private sector's ability to sort out problems and develop solutions, and the private sector was also able to respond.

In June 1989, at a meeting of the American Society of Heating, Refrigeration, and Air Conditioning Engineers, a small group of about a dozen people representing manufacturers, builders, state energy officials, researchers, and consumer advocates made a decision.

> The only way we are going to address these problems and keep the federal government from developing a monster is for us to all work together and beat them to the punch. If we do it right and quickly, there will be no need for federal intervention.

This group chose, as its name, the National Fenestration Rating Council. Over the next several months, individuals and groups worked feverishly on a wide variety of tasks. Some worked on test methods, others on calculation tools. Some worked on by-laws and operating guidelines and others focused on consumer-oriented information. States were surveyed to find out the level of their own activities on this topic and were encouraged to participate in this national program. Utilities were contacted to discover their activities regarding fenestration products and their weatherization program needs. The legislators proposing a uniform national rating system for fenestration product energy performance in Congress were also informed about this broad, private-sector initiative and they liked it. The challenge was made and NRFC accepted the challenge. In January 1990, the group officially convened with a 12-member board of directors representing all segments of the fenestration industry and design professionals, builders, consumer groups, state government, researchers, and utilities. Eric Elstrom of the National Wood Window Manufacturers Association was elected the first chairman of the NFRC board of directors.

The board had a unique structure. Not only was the fenestration industry involved, but also represented were the primary purchasers of these products and those developing standards and regulations. NRFC's mission was conceptually simple yet technically involved: to create a fair and accurate means for comparing the thermal performance of fenestration products. The board formed three committees to meet this objective. Each committee had its own scope and each had a different timetable.

The ratings, codes, and standards committee was assigned the task of identifying users of fenestration ratings, determining their priorities, and developing ways to assure the integrity of NFRC procedures. The technical committee was to develop the standard tools and procedures, including test methods and computer simulation, to assess fenestration energy performance. The public relations committee was to establish a format for communicating information about NFRC product ratings to consumers at all levels. NFRC has made great progress, but it still needs participation from everyone in the fenestration industry. Everyone is encouraged to become a member of NFRC and to help accomplish its objectives. Information about the organization's activities is published regularly in its newsletter. Refer to the address in Chapter 27.

Benefits

Understanding the background, it is clear that a uniform thermal-performance rating system for windows, doors, and skylights offers numerous benefits:

1. Manufacturers can establish the value of their products in the overall energy performance of a building, differentiate their products in the market, and avoid a potentially large number of tests and procedures, which their products might have to undergo to meet the requirements of individual states and local authorities.
2. Builders, contractors, and consumers are assured they are getting what they are paying for in terms of energy-saving potential and can compare products head-to-head.
3. Code officials can easily recognize which products enable buildings to meet energy performance requirements.
4. Utilities can offer incentive programs to encourage the use of energy-efficient windows and doors, as they do with other building components.

The benefits are many, but developing a single, cost-effective procedure for rating the thermal performance of fenestration products has represented a monumental task.

There are many performance parameters which impact a product's overall energy performance. A national survey of state officials was conducted to assist NFRC in prioritizing these various performance parameters. The responses clearly showed that thermal transmittance (U-value) should be the first priority. As a result, the NFRC technical committee went to work with some of the best minds in thermal testing, the computer simulation of thermal performance, and manufacturing to develop a system for determining fenestration product U-values. Simultaneously, subcommittees were also established to address the other prioritized performance parameters from the survey, including solar heat gain, air infiltration, condensation resistance, durability, and other attributes.

In August 1991, NFRC issued NFRC 100-91: Procedure for Determining Fenestration Product Properties (currently limited to U-value). This document represents the first unified method of determining the thermal performance of windows and other fenestration products, regardless of product, type, material, or size.

U-value procedure

The U-value procedure provides the first consistent method for fenestration product performance rating. The combination of computer simulation and thermal should be much more cost effective than testing every product and also provides a mechanism for rating many variations.

The NFRC procedure employs a low-cost computer simulation to determine the variety of U-values within a product line. A matrix is created for all these variations in two standard product sizes, AA and BB. Rather than test every conceivable configuration, the U-value of each variation is determined for the two standard sizes via computer simulation, as illustrated in Table 26.1.

The individual products with the highest and lowest computed U-values are known as baseline products. These two baseline products then undergo standard thermal testing under a uniform set of conditions, and the results are compared to the calculated values. If the test and compared U-values fall within a defined, acceptable range, the computed U-value for the entire line is then considered validated. The computer-generated U-values are determined by the latest versions of Windows and Frame computer programs. Windows, developed at the Lawrence Berkeley Laboratory, computes a fenestration product's overall U-value by adding up data based on center-of-glass, edge-of-glass, and frame configurations. Thermal transmittance data for many common glazing options is included in the Window 3.1 program. Frame 2.2, developed by Enermodal Engineering of Canada, analyzes the thermal transmittance of the framing portion of a fenes-

TABLE 26.1 Matrix of U-Values

Product lines—Thermally Broken Aluminum Horizontal Slider
Model Size AA: 60 × 36 in
Model Size BB: 72 × 48 in

	Representative U-Values	
Glazing Options	Size AA	Size BB
¼ inch IGU air muntins	0.76	0.72
¼ inch IGU air	0.68	0.64
½ inch IGU air	0.61	0.55
½ inch IGU gas	0.58	0.52
½ inch IGU gas, Low-E	0.50	0.46

tration product. That information then is input into Windows 3.1 to determine overall product performance. Both software programs are available through NFRC.

The test method employed to validate the U-value computed for a product line is the latest evolution of industry association and ASTM standard methods. It was developed specifically to evaluate fenestration products (unlike some previous methods) and has rigid calibration, measurement, and test condition requirements to promote accurate and dependable results.

Using product ratings

Once a uniform procedure for product ratings is developed, the next question is how to provide the rating information usefully. Existing performance rating systems for refrigerators, cars, etc., typically employ labels, directories, or other means of presenting comparative information. Fenestration manufacturers already have experience with labels on products showing certain performance attributes, such as structural strength. Now U-values can be shown similarly.

Manufacturers could affix window labels so that building code inspectors can identify products which comply with local energy codes. Homeowners, remodelers, and builders could use the labels to directly compare product performance. Rating information for various manufacturers is also likely to be presented in certified product catalogs. A builder, for example, could look in such a catalog to see which companies produce vinyl double-hung windows with a U-value of less than 0.40 needed to qualify for the rebate program sponsored by the local utility.

Cost effectiveness

The combination of computer simulation and testing, developed by NFRC, benefits manufacturers by providing a cost-effective mechanism for rating product thermal performance. In an attempt to get a better handle on the anticipated costs, several NFRC member manufacturers embarked on detailed cost analyses to illustrate the expected costs for companies ranging in size from under $10 million to over $300 million in annual sales. Each manufacturer obtained quotations for conducting the NFRC rating procedure, both simulation and testing, on their entire line.

Table 26.2 shows the cost analyses for five different-sized companies, representing producers of a variety of product types including aluminum, wood, and vinyl windows.

TABLE 26.2 NFRC Cost Analysis

(For a variety of manufacturer types and sizes)

Mfr.	Annual sales (millions)	Product lines	Glazing options	Total product options	Simulation costs (est.)	Testing costs (est.)	Total costs	Costs per product line
1	Less than $10	14	16	224	$18,200	$26,600	$44,800	$3,200
2	$20–$30	30	12	360	$30,000	$52,500	$82,500	$2,750
3	$30–$100	46	16	736	$60,000	$73,600	$133,600	$2,904
4	$100–$300	60	12	720	$85,000	$90,000	$175,000	$2,917
5	Over $300	20	4	80	$20,000	$36,000	$56,000	$2,800

This information is based on actual cost estimates from testing and simulation laboratories familiar with NFRC procedures. Where more than one laboratory provided estimates, the average value was used.

One of the most important cost benefits of NFRC to manufacturers is that it is national. As the NFRC rating system becomes recognized by almost all states and code bodies, manufacturers will be able to keep costs reasonable by eliminating expensive duplicate testing—one set of test conditions for one state or locality and a different set of conditions to meet another's requirements.

Testing and Simulation Integrity

To establish itself as a national system accepted by almost all, an important component of NFRC's scope is to develop a valid rating procedure and to ensure uniform application of its rating system. The job has been taken up by the group's ratings, codes, and standards committee, which is addressing a variety of questions. The ratings, codes, and standards committee established three working subcommittees to focus on these issues:

1. building code/utility programs,
2. certification and labeling, and
3. international liaison.

The challenge is to assist in making the rating system credible to users by providing a series of checks and balances to ensure that it is employed accurately and uniformly.

First, NFRC had to develop a mechanism to ensure that the testing and simulation laboratories are doing their jobs correctly and repeatedly. Laboratory accreditation is typically the way this is done. NFRC accreditation will include annual inspections and annual competence testing to make sure that these objectives are met.

Second, NFRC has developed the requirements by which all products may be certified and labeled for thermal performance. Any agency providing this service must follow certification and labeling rules put forward by NFRC and, similarly, will be inspected for compliance.

Finally, NFRC will retain the right to audit any product bearing the rating and certification to ensure that the system is being employed uniformly. The system must work so that it can be adopted and referred to and, more importantly, so that it has the industry's confidence.

NFRC's objectives of establishing a uniform rating system for fenestration products will not happen overnight. There is still considerable work to be done and other thermal-performance attributes to be addressed by the organization. Several states have already recognized the progress made.

Future

NFRC recognition should continue spreading. Manufacturers are promoting NFRC recognition to building code jurisdictions that do not yet use this national rating and labeling system. The growing number of utility programs promoting energy conservation can also utilize NFRC's rating system to help determine which products meet their program objectives. This will enable builders to take advantage of utility incentives for more energy-efficient products. Consumers will also be able to compare directly the thermal performance of windows and other fenestration products. The NFRC's uniform rating system provides manufacturers of windows, doors, and skylights with a standard for their product's thermal performance. This will enable them to realize greater sales and profits. Innovation will be rewarded, expanding energy-efficient technologies even further.

Although NFRC's goal of creating a thermal-performance rating system may have grown out of a need to solve a problem, the group's success will have multiple, long-term benefits for all involved and will create new opportunities for the fenestration industry.

References

1. NFRC's Energy Rating System, an article submitted on behalf of the National Fenestration Rating Council, published in *The Construction Specifier,* 3/91.

Chapter 27

Trade Organizations and Publications

Joseph S. Amstock

President, Professional Adhesive and Sealant Systems

Introduction

The trade organizations, trade publications, and governmental groups associated with the glass and window industry have contributed to making this handbook a successful tool for the industry. The author wishes to thank them for their unselfish contribution of time.

Adhesives Age
6151 Powers Ferry Rd. NW
Atlanta, GA 30339

Adhesive and Sealant Council, Inc. (ASC)
1627 K. St. NW, Suite 1000
Washington, D.C. 20006

Adhesive and Sealant Manufacturers
Association of Canada
208 Brimorton Dr.
Scarborough, Ontario M1H 2C6

Alumi–News
109-4920 de Maisonneuve Blvd. W.
Westmont, Quebec, H3Z 1N1
Canada

Adhesive Manufacturers Association (AMA)
401 N. Michigan Ave.
Chicago, IL 60611-4267

American Chemical Society (ACS)
1155 16th St., NW
Washington, D.C. 20036

American Institute of Chemical Engineers (AIChE)
345 E. 47th St.
New York, NY 10017

American Institute of Chemists
501 Wythe St.
Alexandria, VA 22314-1917

American Architectural Manufacturers Association (AAMA)
2540 E. Dundee Road
Palatine, IL 60067

Americans with Disabilities Act Information (ADA)
U.S. Dept. of Justice
Civil Rights Div.
P.O. Box 66738
Washington, D.C. 20075

American Institute of Architects (AIA)
1725 New York Ave., NW
Washington, D.C. 20006

American National Standards Institute (ANSI)
11 W. 42nd St.
New York, NY 10036

American Society of Civil Engineering (ASCE)
345 E. 47th St.
New York, NY 10017

American Society of Testing Materials (ASTM)
100 Barr Harbour Dr.
Conshohocken, PA 19428-2959

American Society of Heating, Refrigeration,
and Air Conditioning Engineers, Inc. (ASHRAE)
791 Tullie Circle
Atlanta, GA 30329

American Solar Energy Society (ASES)
2400 Central Ave. Suite B-1
Boulder, CO 80351

Architectural Testing, Inc.
130 Derry Ct.
York, PA 170402-9405

Bowser-Morner, Inc.
4518 Taylorsville Rd.
Dayton, OH 45401

British Adhesive and Sealants Association (BASA)
33 Fellowes Way
Stevenage
Hertfordshire, SG2 8BW
U.K.

British Standards Institute (BSI)
389 Cheswick High Rd.
London, W4 4AL
U.K.

Building Officials and Code Administrators (BOCA)
4051 W. Fossmoor Rd.
Country Club Hills, IL 60478

Building and Construction Information Exchange (BACIX)
3476 Valley Creek Rd.
Tallahassee, FL 32312

Building Research Establishment (BSE)
Garston Watford
Herts, WD2 7JR
U.K.

Construction Consulting Laboratory, Inc.
4751 W State St., Suite E
Ontario, CA 91762

Canadian General Standards Board (CGSB)
Standards & Specification Branch
Ottawa, Ontario
K1A 0S9
Canada

Construction Canada
46 Midland Ave.
Scarborough, Ontario
N0M 1S0

Canadian Mortgage & Housing Corp. (CHMC)
700 Montreal Rd.
Ottawa, Ontario
KIA 0PA
Canada

Chicago Committee on High Rise Buildings
111 E. Wacker Dr., Suite 3015
Chicago, IL 60601

Construction Specifications Institute (CSI)
601 Madison St.
Alexandria, VA 22314

Corning Museum of Glass
1 Museum Way
Corning, NY 14830-2253

Council of American Building Officials (CABO)
5203 Leesburg Pike, Suite 708
Falls Church, VA 22041

DIN
Burggrafen Strasse 6
D-10782 Berlin (Tiergarten)
Germany

ETL Laboratories, Inc.
3933 U.S. Route 11
Cortland, NY 14624

Factory Mutual Engineering Corp.
1151 Boston-Providence Turnpike
Norwalk, MA 02061

FEICA Association of European Adhesive Manufacturers
Ivo-Beucker Str. 43
D-40237 Dusseldorf, Germany

Flat Glass Manufacturers Association (FGMA)
3310 SW Harrison St.
Topeka, KS 66611

Glass Association of North America (GANA)
3310 SW Harrison St.
Topeka, KS 66611-2279

Glass and Glazing Federation (GGF)
44–48 Borough High St.
London, SE1 1XB
U.K.

Glass Develop AB
Box 125
S-245-22 Straffen Strop
Sweden

Glass Digest
18 East 41st St.
New York, NY 10017

Glass International
Queensway House
2 Queensway, Redhill,
Surrey RH1 1QS
U.K.

Glass Magazine
8200 Greensboro Dr., Suite 302
McLean, VA 22102

Glass Tempering Association (GTA)
3310 SW Harrison St.
Topeka, KS 66611

Glazing, California Glass Association
1154 N. Knollwood Circle
Anaheim, CA 92801

International Conference of Building Officials (ICBO)
5360 S. Workmen Mill Rd.
Whittier, CA 90601

Insulating Glass Manufacturers Association of Canada (IGMAC)
27 Goulburn Ave.
Ottawa, Ontario, H1N 8C4

Institute for Research in Construction/National Research Council (IRC/NRC)
Building M24
Ottawa, Ontario
K1A 0R6

Japan Adhesive Industry Association
1-15-10 Uchikanda
Chiyodaku, Tokyo, 101, Japan

Lawrence Berkeley Laboratory (LBL)
Window and Daylighting Group
Mail Stop 90-31-11
1 Cyclotron Rd.
Berkeley, CA 04720

Miami Testing Laboratory, Inc.
1640 W 32nd Pl.
Hialeah, FL 33012

National Association of Home Builders (NAHB)
15th and M Streets, NW
Washington, D.C. 20005

National Association of Minority Contractors (NAMC)
1333 F Street, NW, Suite 500
Washington, D.C. 20004

National Association of Women in Construction (NAWIC)
327 S. Adams St.
Fort Worth, TX 76104-1002

National Fire Protection Association
1 Batterymarch Park
Quincy, MA 02269-9101

National Glass Association (NGA)
8200 Greensboro Dr.
McLean, VA 22102

National Fenestration and Rating Council (NFRC)
952 Wayne Ave., Suite 750
Silver Spring, MD 20910

National Institute of Standards and Technology (NIST)
Bldg. 226, Room 126
Gaithersburg, MD 20899

National Institute of Building Sciences (NIBS)
1201 L Street, NW
Washington, D.C. 20005

National Industrial Research Institute (NIRI)
Nagoya, Japan

North American Association of Mirror Manufacturers (NAAMM)
9005 Congressional Ct.
Potomac, VA 20854

Norwegian Building Research Institute (NBRI)
Hogskoleringen 7
Trondheim, Norway

National Wood Window & Door Association (NWWDA)
1400 E. Toughy Ave., Suite G 54
Des Plaines, IL 60018

National Safety Council (NSC)
1121 Spring St.
Itasca, IL 60143

Occupational Safety & Health Administration (OSHA)
200 Constitution Ave.
Washington, D.C. 20210

Polyurethane Manufacturers Association (PMA)
Bldg. C, Suite 20
800 Roosevelt Rd.
Glen Ellyn, IL 60137-5833

Portland Cement Association (PCA)
5420 Old Orchard Rd.
Skokie, IL 60077-1083

Primary Glass Manufacturers Association (PGMA)
3310 SW Harrison St.
Topeka, KS 66611-2279

Professional Women In Construction 342 Madison Ave. New York, NY 10173	(PWC)
Sealant, Waterproofing & Restoration Institute 3101 Broadway, Suite 585 Kansas City, MO 64111	(SWRI)
Sealed Insulating Glass Manufacturers Association 401 N. Michigan Ave. Chicago, IL 60611-4264	(SIGMA)
Society of Plastics Engineers 400 Commonwealth Dr. Warrendale, PA 15096	(SPE)
Society of Plastics Industry 1275 K Street, NW, Suite 400 Washington, D.C. 20005	(SPI)
Southwestern Labs/Huntington 2200 Gravel Dr. Ft. Worth, TX 76118-7755	
Underwriters Laboratory, Inc. 333 Pfingsten Rd. Northbrook, IL 60062	(UL)
U.S. Glass, Metal Glazing, Auto Glass and Windows 233 Garrisonville Rd., #201 Stafford, VA 22554	
Warnock Hersey International, Inc. 530 Garcia Ave. Pittsburg, CA 94565	

Appendix

A

Temperature Conversion Chart

The column of figures in bold and which is headed "Reading in °F or °C to be converted" refers to the temperature either in degrees Fahrenheit or Centigrade which it is desired to convert into the other scale. If converting from Fahrenheit degrees to Centigrade degrees the equivalent temperature will be found in the column headed "°C," while if converting from degrees Centigrade to degrees Fahrenheit, the equivalent temperature will be found in the column headed "°F." This arrangement is very similar to that of Sauveur and Boylston, copyrighted 1920, and is published with their permission.

TABLE A.1. Temperature Conversion Table

°F.	Reading in °F. or °C. to be converted	°C.
....	−458	−272.22
....	−456	−271.11
....	−454	−270.00
....	−452	−268.89
....	−450	−267.78
....	−448	−266.67
....	−446	−265.56
....	−444	−264.44
....	−442	−263.33
....	−440	−262.22
....	−438	−261.11
....	−436	−260.00
....	−434	−258.89
....	−432	−257.78
....	−430	−256.67
....	−428	−255.56
....	−426	−254.44
....	−424	−253.33
....	−422	−252.22
....	−420	−251.11
....	−418	−250.00
....	−416	−248.89
....	−414	−247.78
....	−412	−246.67
....	−410	−245.56
....	−408	−244.44
....	−406	−243.33
....	−404	−242.22
....	−402	−241.11
....	−400	−240.00
....	−398	−238.89
....	−396	−237.78
....	−394	−236.67
....	−392	−235.56
....	−390	−234.44
....	−388	−233.33
....	−386	−232.22
....	−384	−231.11
....	−382	−230.00
....	−380	−228.89
....	−378	−227.78
....	−376	−226.67
....	−374	−225.56
....	−372	−224.44
....	−370	−223.33
....	−368	−222.22
....	−366	−221.11
....	−364	−220.00
....	−362	−218.89
....	−360	−217.78

°F.	Reading in °F. or °C. to be converted	°C.
....	−358	−216.67
....	−356	−215.56
....	−354	−214.44
....	−352	−213.33
....	−350	−212.22
....	−348	−211.11
....	−346	−210.00
....	−344	−208.89
....	−342	−207.78
....	−340	−206.67
....	−338	−205.56
....	−336	−204.44
....	−334	−203.33
....	−332	−202.22
....	−330	−201.11
....	−328	−200.00
....	−326	−198.89
......	−324	−197.78
....	−322	−196.67
....	−320	−195.56
....	−318	−194.44
....	−316	−193.33
....	−314	−192.22
....	−312	−191.11
....	−310	−190.00
....	−308	−188.89
....	−306	−187.78
....	−304	−186.67
....	−302	−185.56
....	−300	−184.44
....	−298	−183.33
....	−296	−182.22
....	−294	−181.11
....	−292	−180.00
....	−290	−178.89
....	−288	−177.78
....	−286	−176.67
....	−284	−175.56
....	−282	−174.44
....	−280	−173.33
....	−278	−172.22
....	−276	−171.11
....	−274	−170.00
−457.6	−272	−168.89
−454.0	−270	−167.78
−450.4	−268	−166.67
−446.8	−266	−165.56
−443.2	−264	−164.44
−439.6	−262	−163.33
−436.0	−260	−162.22

TABLE A.2. Temperature Conversion Table

°F.	Reading in °F. or °C. to be converted	°C.
−432.4	−258	−161.11
−428.8	−256	−160.00
−425.2	−254	−158.89
−421.6	−252	−157.78
−418.0	−250	−156.67
−414.4	−248	−155.56
−410.8	−246	−154.44
−407.2	−244	−153.33
−403.6	−242	−152.22
−400.0	−240	−151.11
−396.4	−238	−150.00
−392.8	−236	−148.89
−389.2	−234	−147.78
−385.6	−232	−146.67
−382.0	−230	−145.56
−378.4	−228	−144.44
−374.8	−226	−143.33
−371.2	−224	−142.22
−367.6	−222	−141.11
−364.0	−220	−140.00
−360.4	−218	−138.89
−356.8	−216	−137.78
−353.2	−214	−136.67
−349.6	−212	−135.56
−346.0	−210	−134.44
−342.4	−208	−133.33
−338.8	−206	−132.22
−335.2	−204	−131.11
−331.6	−202	−130.00
−328.0	−200	−128.89
−324.4	−198	−127.78
−320.8	−196	−126.67
−317.2	−194	−125.56
−313.6	−192	−124.44
−310.0	−190	−123.33
−306.4	−188	−122.22
−302.8	−186	−121.11
−299.2	−184	−120.00
−295.6	−182	−118.89
−292.0	−180	−117.78
−288.4	−178	−116.67
−284.8	−176	−115.56
−281.2	−174	−114.44
−277.6	−172	−113.33
−274.0	−170	−112.22
−270.4	−168	−111.11
−266.8	−166	−110.00
−263.2	−164	−108.89
−259.6	−162	−107.78
−256.0	−160	−106.67
−252.4	−158	−105.56
−248.8	−156	−104.44
−245.2	−154	−103.33
−241.6	−152	−102.22
−238.0	−150	−101.11
−234.4	−148	−100.00
−230.8	−146	−98.89
−227.2	−144	−97.78
−223.6	−142	−96.67
−220.0	−140	−95.56

°F.	Reading in °F. or °C. to be converted	°C.
−216.4	−138	−94.44
−212.8	−136	−93.33
−209.2	−134	−92.22
−205.6	−132	−91.11
−202.0	−130	−90.00
−198.4	−128	−88.89
−194.8	−126	−87.78
−191.2	−124	−86.67
−187.6	−122	−85.56
−184.0	−120	−84.44
−180.4	−118	−83.33
−176.8	−116	−82.22
−173.2	−114	−81.11
−169.6	−112	−80.00
−166.0	−110	−78.89
−162.4	−108	−77.78
−158.8	−106	−76.67
−155.2	−104	−75.56
−151.6	−102	−74.44
−148.0	−100	−73.33
−144.4	−98	−72.22
−140.8	−96	−71.11
−137.2	−94	−70.00
−133.6	−92	−68.89
−130.0	−90	−67.78
−126.4	−88	−66.67
−122.8	−86	−65.56
−119.2	−84	−64.44
−115.6	−82	−63.33
−112.0	−80	−62.22
−108.4	−78	−61.11
−104.8	−76	−60.00
−101.2	−74	−58.89
−97.6	−72	−57.78
−94.0	−70	−56.67
−90.4	−68	−55.56
−86.8	−66	−54.44
−83.2	−64	−53.33
−79.6	−62	−52.22
−76.0	−60	−51.11
−72.4	−58	−50.00
−68.8	−56	−48.89
−65.2	−54	−47.78
−61.6	−52	−46.67
−58.0	−50	−45.56
−54.4	−48	−44.44
−50.8	−46	−43.33
−47.2	−44	−42.22
−43.6	−42	−41.11
−40.0	−40	−40.00
−36.4	−38	−38.89
−32.8	−36	−37.78
−29.2	−34	−36.67
−25.6	−32	−35.56
−22.0	−30	−34.44
−18.4	−28	−33.33
−14.8	−26	−32.22
−11.2	−24	−31.11
−7.6	−22	−30.00
−4.0	−20	−28.89

TABLE A.3. Temperature Conversion Table

°F.	Reading in °F. or °C to be converted	°C.
−0.4	−18	−27.78
+3.2	−16	−26.67
+6.8	−14	−25.56
+10.4	−12	−24.44
+14.0	−10	−23.33
+17.6	−8	−22.22
+19.4	−7	−21.67
+21.2	−6	−21.11
+23.0	−5	−20.56
+24.8	−4	−20.00
+26.6	−3	−19.44
+28.4	−2	−18.89
+30.2	−1	−18.33
+32.0	±0	−17.78
+33.8	+1	−17.22
+35.6	+2	−16.67
+37.4	+3	−16.11
+39.2	+4	−15.56
+41.0	+5	−15.00
+42.8	+6	−14.44
+44.6	+7	−13.89
+46.4	+8	−13.33
+48.2	+9	−12.78
+50.0	+10	−12.22
+51.8	+11	−11.67
+53.6	+12	−11.11
+55.4	+13	−10.56
+57.2	+14	−10.00
+59.0	+15	−9.44
+60.8	+16	−8.89
+62.6	+17	−8.33
+64.4	+18	−7.78
+66.2	+19	−7.22
+68.0	+20	−6.67
+69.8	+21	−6.11
+71.6	+22	−5.56
+73.4	+23	−5.00
+75.2	+24	−4.44
+77.0	+25	−3.89
+78.8	+26	−3.33
+80.6	+27	−2.78
+82.4	+28	−2.22
+84.2	+29	−1.67
+86.0	+30	−1.11
+87.8	+31	−0.56
+89.6	+32	±0.00
+91.4	+33	+0.56
+93.2	+34	+1.11
+95.0	+35	+1.67
+96.8	+36	+2.22
+98.6	+37	+2.78
+100.4	+38	+3.33
+102.2	+39	+3.89
+104.0	+40	+4.44
+105.8	+41	+5.00
+107.6	+42	+5.56
+109.4	+43	+6.11
+111.2	+44	+6.67
+113.0	+45	+7.22
+114.8	+46	+7.78
+116.6	+47	+8.33
+118.4	+48	+8.89
+120.2	+49	+9.44
+122.0	+50	+10.00
+123.8	+51	+10.56
+125.6	+52	+11.11
+127.4	+53	+11.67
+129.2	+54	+12.22
+131.0	+55	+12.78
+132.8	+56	+13.33
+134.6	+57	+13.89
+136.4	+58	+14.44
+138.2	+59	+15.00
+140.0	+60	+15.56
+141.8	+61	+16.11
+143.6	+62	+16.67
+145.4	+63	+17.22
+147.2	+64	+17.78
+149.0	+65	+18.33
+150.8	+66	+18.89
+152.6	+67	+19.44
+154.4	+68	+20.00
+156.2	+69	+20.56
+158.0	+70	+21.11
+159.8	+71	+21.67
+161.6	+72	+22.22
+163.4	+73	+22.78
+165.2	+74	+23.33
+167.0	+75	+23.89
+168.8	+76	+24.44
+170.6	+77	+25.00
+172.4	+78	+25.56
+174.2	+79	+26.11
+176.0	+80	+26.67
+177.8	+81	+27.22
+179.6	+82	+27.78
+181.4	+83	+28.33
+183.2	+84	+28.89
+185.0	+85	+29.44
+186.8	+86	+30.00
+188.6	+87	+30.56
+190.4	+88	+31.11
+192.2	+89	+31.67
+194.0	+90	+32.22
+195.8	+91	+32.78
+197.6	+92	+33.33
+199.4	+93	+33.89
+201.2	+94	+34.44
+203.0	+95	+35.00
+204.8	+96	+35.56
+206.6	+97	+36.11
+208.4	+98	+36.67
+210.2	+99	+37.22
+212.0	+100	+37.78
+213.8	+101	+38.33
+215.6	+102	+38.89
+217.4	+103	+39.44
+219.2	+104	+40.00
+221.0	+105	+40.56
+222.8	+106	+41.11

TABLE A.4. Temperature Conversion Table

°F.	Reading in °F. or °C. to be converted	°C.
+224.6	+107	+41.67
+226.4	+108	+42.22
+228.2	+109	+42.78
+230.0	+110	+43.33
+231.8	+111	+43.89
+233.6	+112	+44.44
+235.4	+113	+45.00
+237.2	+114	+45.56
+239.0	+115	+46.11
+240.8	+116	+46.67
+242.6	+117	+47.22
+244.4	+118	+47.78
+246.2	+119	+48.33
+248.0	+120	+48.89
+249.8	+121	+49.44
+251.6	+122	+50.00
+253.4	+123	+50.56
+255.2	+124	+51.11
+257.0	+125	+51.67
+258.8	+126	+52.22
+260.6	+127	+52.78
+262.4	+128	+53.33
+264.2	+129	+53.89
+266.0	+130	+54.44
+267.8	+131	+55.00
+269.6	+132	+55.56
+271.4	+133	+56.11
+273.2	+134	+56.67
+275.0	+135	+57.22
+276.8	+136	+57.78
+278.6	+137	+58.33
+280.4	+138	+58.89
+282.2	+139	+59.44
+284.0	+140	+60.00
+285.8	+141	+60.56
+287.6	+142	+61.11
+289.4	+143	+61.67
+291.2	+144	+62.22
+293.0	+145	+62.78
+294.8	+146	+63.33
+296.6	+147	+63.89
+298.4	+148	+64.44
+300.2	+149	+65.00
+302.0	+150	+65.56
+303.8	+151	+66.11
+305.6	+152	+66.67
+307.4	+153	+67.22
+309.2	+154	+67.78
+311.0	+155	+68.33
+312.8	+156	+68.89
+314.6	+157	+69.44
+316.4	+158	+70.00
+318.2	+159	+70.56
+320.0	+160	+71.11
+321.8	+161	+71.67
+323.6	+162	+72.22
+325.4	+163	+72.78
+327.2	+164	+73.33
+329.0	+165	+73.89
+330.8	+166	+74.44

°F.	Reading in °F. or °C. to be converted	°C.
+332.6	+167	+75.00
+334.4	+168	+75.56
+336.2	+169	+76.11
+338.0	+170	+76.67
+339.8	+171	+77.22
+341.6	+172	+77.78
+343.4	+173	+78.33
+345.2	+174	+78.89
+347.0	+175	+79.44
+348.8	+176	+80.00
+350.6	+177	+80.56
+352.4	+178	+81.11
+354.2	+179	+81.67
+356.0	+180	+82.22
+357.8	+181	+82.78
+359.6	+182	+83.33
+361.4	+183	+83.89
+363.2	+184	+84.44
+365.0	+185	+85.00
+366.8	+186	+85.56
+368.6	+187	+86.11
+370.4	+188	+86.67
+372.2	+189	+87.22
+374.0	+190	+87.78
+375.8	+191	+88.33
+377.6	+192	+88.89
+379.4	+193	+89.44
+381.2	+194	+90.00
+383.0	+195	+90.56
+384.8	+196	+91.11
+386.6	+197	+91.67
+388.4	+198	+92.22
+390.2	+199	+92.78
+392.0	+200	+93.33
+393.8	+201	+93.89
+395.6	+202	+94.44
+397.4	+203	+95.00
+399.2	+204	+95.56
+401.0	+205	+96.11
+402.8	+206	+96.67
+404.6	+207	+97.22
+406.4	+208	+97.78
+408.2	+209	+98.33
+410.0	+210	+98.89
+411.8	+211	+99.44
+413.6	+212	+100.00
+415.4	+213	+100.56
+417.2	+214	+101.11
+419.0	+215	+101.67
+420.8	+216	+102.22
+422.6	+217	+102.78
+424.4	+218	+103.33
+426.2	+219	+103.89
+428.0	+220	+104.44
+431.6	+222	+105.56
+435.2	+224	+106.67
+438.8	+226	+107.78
+442.4	+228	+108.89
+446.0	+230	+110.00
+449.6	+232	+111.11

TABLE A.5. Temperature Conversion Table

°F.	Reading in °F. or °C. to be converted	°C.
+453.2	+234	+112.22
+456.8	+236	+113.33
+460.4	+238	+114.44
+464.0	+240	+115.56
+467.6	+242	+116.67
+471.2	+244	+117.78
+474.8	+246	+118.89
+478.4	+248	+120.00
+482.0	+250	+121.11
+485.6	+252	+122.22
+489.2	+254	+123.33
+492.8	+256	+124.44
+496.4	+258	+125.56
+500.0	+260	+126.67
+503.6	+262	+127.78
+507.2	+264	+128.89
+510.8	+266	+130.00
+514.4	+268	+131.11
+518.0	+270	+132.22
+521.6	+272	+133.33
+525.2	+274	+134.44
+528.8	+276	+135.56
+532.4	+278	+136.67
+536.0	+280	+137.78
+539.6	+282	+138.89
+543.2	+284	+140.00
+546.8	+286	+141.11
+550.4	+288	+142.22
+554.0	+290	+143.33
+557.6	+292	+144.44
+561.2	+294	+145.56
+564.8	+296	+146.67
+568.4	+298	+147.78
+572.0	+300	+148.89
+575.6	+302	+150.00
+579.2	+304	+151.11
+582.8	+306	+152.22
+586.4	+308	+153.33
+590.0	+310	+154.44
+593.6	+312	+155.56
+597.2	+314	+156.67
+600.8	+316	+157.78
+604.4	+318	+158.89
+608.0	+320	+160.00
+611.6	+322	+161.11
+615.2	+324	+162.22
+618.8	+326	+163.33
+622.4	+328	+164.44
+626.0	+330	+165.56
+629.6	+332	+166.67
+633.2	+334	+167.78
+636.8	+336	+168.89
+640.4	+338	+170.00
+644.0	+340	+171.11
+647.6	+342	+172.22
+651.2	+344	+173.33
+654.8	+346	+174.44
+658.4	+348	+175.56
+662.0	+350	+176.67
+665.6	+352	+177.78
+669.2	+354	+178.89
+672.8	+356	+180.00
+676.4	+358	+181.11
+680.0	+360	+182.22
+683.6	+362	+183.33
+687.2	+364	+184.44
+690.8	+366	+185.56
+694.4	+368	+186.67
+698.0	+370	+187.78
+701.6	+372	+188.89
+705.2	+374	+190.00
+708.8	+376	+191.11
+712.4	+378	+192.22
+716.0	+380	+193.33
+719.6	+382	+194.44
+723.2	+384	+195.56
+726.8	+386	+196.67
+730.4	+388	+197.78
+734.0	+390	+198.89
+737.6	+392	+200.00
+741.2	+394	+201.11
+744.8	+396	+202.22
+748.4	+398	+203.33
+752.0	+400	+204.44
+755.6	+402	+205.56
+759.2	+404	+206.67
+762.8	+406	+207.78
+766.4	+408	+208.89
+770.0	+410	+210.00
+773.6	+412	+211.11
+777.2	+414	+212.22
+780.8	+416	+213.33
+784.4	+418	+214.44
+788.0	+420	+215.56
+791.6	+422	+216.67
+795.2	+424	+217.78
+798.8	+426	+218.89
+802.4	+428	+220.00
+806.0	+430	+221.11
+809.6	+432	+222.22
+813.2	+434	+223.33
+816.8	+436	+224.44
+820.4	+438	+225.56
+824.0	+440	+226.67
+827.6	+442	+227.78
+831.2	+444	+228.89
+834.8	+446	+230.00
+838.4	+448	+231.11
+842.0	+450	+232.22
+845.6	+452	+233.33
+849.2	+454	+234.44
+852.8	+456	+235.56
+856.4	+458	+236.67
+860.0	+460	+237.78
+863.6	+462	+238.89
+867.2	+464	+240.00
+870.8	+466	+241.11
+874.4	+468	+242.22
+878.0	+470	+243.33
+881.6	+472	+244.44

TABLE A.6. Temperature Conversion Table

°F.	Reading in °F. or °C. to be converted	°C.
+885.2	+474	+245.56
+888.8	+476	+246.67
+892.4	+478	+247.78
+896.0	+480	+248.89
+899.6	+482	+250.00
+903.2	+484	+251.11
+906.8	+486	+252.22
+910.4	+488	+253.33
+914.0	+490	+254.44
+917.6	+492	+255.56
+921.2	+494	+256.67
+924.8	+496	+257.78
+928.4	+498	+258.89
+932.0	+500	+260.00
+935.6	+502	+261.11
+939.2	+504	+262.22
+942.8	+506	+263.33
+946.4	+508	+264.44
+950.0	+510	+265.56
+953.6	+512	+266.67
+957.2	+514	+267.78
+960.8	+516	+268.89
+964.4	+518	+270.00
+968.0	+520	+271.11
+971.6	+522	+272.22
+975.2	+524	+273.33
+978.8	+526	+274.44
+982.4	+528	+275.56
+986.0	+530	+276.67
+989.6	+532	+277.78
+993.2	+534	+278.89
+996.8	+536	+280.00
+1000.4	+538	+281.11
+1004.0	+540	+282.22
+1007.6	+542	+283.33
+1011.2	+544	+284.44
+1014.8	+546	+285.56
+1018.4	+548	+286.67
+1022.0	+550	+287.78
+1025.6	+552	+288.89
+1029.2	+554	+290.00
+1032.8	+556	+291.11
+1036.4	+558	+292.22
+1040.0	+560	+293.33
+1043.6	+562	+294.44
+1047.2	+564	+295.56
+1050.8	+566	+296.67
+1054.4	+568	+297.78
+1058.0	+570	+298.89
+1061.6	+572	+300.00
+1065.2	+574	+301.11
+1068.8	+576	+302.22
+1072.4	+578	+303.33
+1076.0	+580	+304.44
+1079.6	+582	+305.56
+1083.2	+584	+306.67
+1086.8	+586	+307.78
+1090.4	+588	+308.89
+1094.0	+590	+310.00
+1097.6	+592	+311.11

°F.	Reading in °F. or °C. to be converted	°C.
+1101.2	+594	+312.22
+1104.8	+596	+313.33
+1108.4	+598	+314.44
+1112.0	+600	+315.56
+1115.6	+602	+316.67
+1119.2	+604	+317.78
+1122.8	+606	+318.89
+1126.4	+608	+320.00
+1130.0	+610	+321.11
+1133.6	+612	+322.22
+1137.2	+614	+323.33
+1140.8	+616	+324.44
+1144.4	+618	+325.56
+1148.0	+620	+326.67
+1151.6	+622	+327.78
+1155.2	+624	+328.89
+1158.8	+626	+330.00
+1162.4	+628	+331.11
+1166.0	+630	+332.22
+1169.6	+632	+333.33
+1173.2	+634	+334.44
+1176.8	+636	+335.56
+1180.4	+638	+336.67
+1184.0	+640	+337.78
+1187.6	+642	+338.89
+1191.2	+644	+340.00
+1194.8	+646	+341.11
+1198.4	+648	+342.22
+1202.0	+650	+343.33
+1205.6	+652	+344.44
+1209.2	+654	+345.56
+1212.8	+656	+346.67
+1216.4	+658	+347.78
+1220.0	+660	+348.89
+1223.6	+662	+350.00
+1227.2	+664	+351.11
+1230.8	+666	+352.22
+1234.4	+668	+353.33
+1238.0	+670	+354.44
+1241.6	+672	+355.56
+1245.2	+674	+356.67
+1248.8	+676	+357.78
+1252.4	+678	+358.89
+1256.0	+680	+360.00
+1259.6	+682	+361.11
+1263.2	+684	+362.22
+1266.8	+686	+363.33
+1270.4	+688	+364.44
+1274.0	+690	+365.56
+1277.6	+692	+366.67
+1281.2	+694	+367.78
+1284.8	+696	+368.89
+1288.4	+698	+370.00
+1292.0	+700	+371.11
+1295.6	+702	+372.22
+1299.2	+704	+373.33
+1302.8	+706	+374.44
+1306.4	+708	+375.56
+1310.0	+710	+376.67
+1313.6	+712	+377.78

TABLE A.7. Temperature Conversion Table

°F.	Reading in °F. or °C. to be converted	°C.	°F.	Reading in °F. or °C. to be converted	°C.
+1317.2	+714	+378.89	+1533.2	+834	+445.56
+1320.8	+716	+380.00	+1536.8	+836	+446.67
+1324.4	+718	+381.11	+1540.4	+838	+447.78
+1328.0	+720	+382.22	+1544.0	+840	+448.89
+1331.6	+722	+383.33	+1547.6	+842	+450.00
+1335.2	+724	+384.44	+1551.2	+844	+451.11
+1338.8	+726	+385.56	+1554.8	+846	+452.22
+1342.4	+728	+386.67	+1558.4	+848	+453.33
+1346.0	+730	+387.78	+1562.0	+850	+454.44
+1349.6	+732	+388.89	+1565.6	+852	+455.56
+1353.2	+734	+390.00	+1569.2	+854	+456.67
+1356.8	+736	+391.11	+1572.8	+856	+457.78
+1360.4	+738	+392.22	+1576.4	+858	+458.89
+1364.0	+740	+393.33	+1580.0	+860	+460.00
+1367.6	+742	+394.44	+1583.6	+862	+461.11
+1371.2	+744	+395.56	+1587.2	+864	+462.22
+1374.8	+746	+396.67	+1590.8	+866	+463.33
+1378.4	+748	+397.78	+1594.4	+868	+464.44
+1382.0	+750	+398.89	+1598.0	+870	+465.56
+1385.6	+752	+400.00	+1601.6	+872	+466.67
+1389.2	+754	+401.11	+1605.2	+874	+467.78
+1392.8	+756	+402.22	+1608.8	+876	+468.89
+1396.4	+758	+403.33	+1612.4	+878	+470.00
+1400.0	+760	+404.44	+1616.0	+880	+471.11
+1403.6	+762	+405.56	+1619.6	+882	+472.22
+1407.2	+764	+406.67	+1623.2	+884	+473.33
+1410.8	+766	+407.78	+1626.8	+886	+474.44
+1414.4	+768	+408.89	+1630.4	+888	+475.56
+1418.0	+770	+410.00	+1634.0	+890	+476.67
+1421.6	+772	+411.11	+1637.6	+892	+477.78
+1425.2	+774	+412.22	+1641.2	+894	+478.89
+1428.8	+776	+413.33	+1644.8	+896	+480.00
+1432.4	+778	+414.44	+1648.4	+898	+481.11
+1436.0	+780	+415.56	+1652.0	+900	+482.22
+1439.6	+782	+416.67	+1655.6	+902	+483.33
+1443.2	+784	+417.78	+1659.2	+904	+484.44
+1446.8	+786	+418.89	+1662.8	+906	+485.56
+1450.4	+788	+420.00	+1666.4	+908	+486.67
+1454.0	+790	+421.11	+1670.0	+910	+487.78
+1457.6	+792	+422.22	+1673.6	+912	+488.89
+1461.2	+794	+423.33	+1677.2	+914	+490.00
+1464.8	+796	+424.44	+1680.8	+916	+491.11
+1468.4	+798	+425.56	+1684.4	+918	+492.22
+1472.0	+800	+426.67	+1688.0	+920	+493.33
+1475.6	+802	+427.78	+1691.6	+922	+494.44
+1479.2	+804	+428.89	+1695.2	+924	+495.56
+1482.8	+806	+430.00	+1698.8	+926	+496.67
+1486.4	+808	+431.11	+1702.4	+928	+497.78
+1490.0	+810	+432.22	+1706.0	+930	+498.89
+1493.6	+812	+433.33	+1709.6	+932	+500.00
+1497.2	+814	+434.44	+1713.2	+934	+501.11
+1500.8	+816	+435.56	+1716.8	+936	+502.22
+1504.4	+818	+436.67	+1720.4	+938	+503.33
+1508.0	+820	+437.78	+1724.0	+940	+504.44
+1511.6	+822	+438.89	+1727.6	+942	+505.56
+1515.2	+824	+440.00	+1731.2	+944	+506.67
+1518.8	+826	+441.11	+1734.8	+946	+507.78
+1522.4	+828	+442.22	+1738.4	+948	+508.89
+1526.0	+830	+443.33	+1742.0	+950	+510.00
+1529.6	+832	+444.44	+1745.6	+952	+511.11

TABLE A.8. Temperature Conversion Table

°F.	Reading in °F. or °C. to be converted	°C.
+1749.2	+954	+512.22
+1752.8	+956	+513.33
+1756.4	+958	+514.44
+1760.0	+960	+515.56
+1763.6	+962	+516.67
+1767.2	+964	+517.78
+1770.8	+966	+518.89
+1774.4	+968	+520.00
+1778.0	+970	+521.11
+1781.6	+972	+522.22
+1785.2	+974	+523.33
+1788.8	+976	+524.44
+1792.4	+978	+525.56
+1796.0	+980	+526.67
+1799.6	+982	+527.78
+1803.2	+984	+528.89
+1806.8	+986	+530.00
+1810.4	+988	+531.11
+1814.0	+990	+532.22
+1817.6	+992	+533.33
+1821.2	+994	+534.44
+1824.8	+996	+535.56
+1828.4	+998	+536.67
+1832.0	+1000	+537.78
+1850.0	+1010	+543.33
+1868.0	+1020	+548.89
+1886.0	+1030	+554.44
+1904.0	+1040	+560.00
+1922.0	+1050	+565.56
+1940.0	+1060	+571.11
+1958.0	+1070	+576.67
+1976.0	+1080	+582.22
+1994.0	+1090	+587.78
+2012.0	+1100	+593.33
+2030.0	+1110	+598.89
+2048.0	+1120	+604.44
+2066.0	+1130	+610.00
+2084.0	+1140	+615.56
+2102.0	+1150	+621.11
+2120.0	+1160	+626.67
+2138.0	+1170	+632.22
+2156.0	+1180	+637.78
+2174.0	+1190	+643.33
+2192.0	+1200	+648.89
+2210.0	+1210	+654.44
+2228.0	+1220	+660.00
+2246.0	+1230	+665.56
+2264.0	+1240	+671.11
+2282.0	+1250	+676.67
+2300.0	+1260	+682.22
+2318.0	+1270	+687.78
+2336.0	+1280	+693.33
+2354.0	+1290	+698.89
+2372.0	+1300	+704.44
+2390.0	+1310	+710.00
+2408.0	+1320	+715.56
+2426.0	+1330	+721.11
+2444.0	+1340	+726.67
+2462.0	+1350	+732.22
+2480.0	+1360	+737.78

°F.	Reading in °F. or C°. to be converted	°C.
+2498.0	+1370	+743.33
+2516.0	+1380	+748.89
+2534.0	+1390	+754.44
+2552.0	+1400	+760.00
+2570.0	+1410	+765.56
+2588.0	+1420	+771.11
+2606.0	+1430	+776.67
+2624.0	+1440	+782.22
+2642.0	+1450	+787.78
+2660.0	+1460	+793.33
+2678.0	+1470	+798.89
+2696.0	+1480	+804.44
+2714.0	+1490	+810.00
+2732.0	+1500	+815.56
+2750.0	+1510	+821.11
+2768.0	+1520	+826.67
+2786.0	+1530	+832.22
+2804.0	+1540	+837.78
+2822.0	+1550	+843.33
+2840.0	+1560	+848.89
+2858.0	+1570	+854.44
+2876.0	+1580	+860.00
+2894.0	+1590	+865.56
+2912.0	+1600	+871.11
+2930.0	+1610	+876.67
+2948.0	+1620	+882.22
+2966.0	+1630	+887.78
+2984.0	+1640	+893.33
+3002.0	+1650	+898.89
+3020.0	+1660	+904.44
+3038.0	+1670	+910.00
+3056.0	+1680	+915.56
+3074.0	+1690	+921.11
+3092.0	+1700	+926.67
+3110.0	+1710	+932.22
+3128.0	+1720	+937.78
+3146.0	+1730	+943.33
+3164.0	+1740	+948.89
+3182.0	+1750	+954.44
+3200.0	+1760	+960.00
+3218.0	+1770	+965.56
+3236.0	+1780	+971.11
+3254.0	+1790	+976.67
+3272.0	+1800	+982.22
+3290.0	+1810	+987.78
+3308.0	+1820	+993.33
+3326.0	+1830	+998.89
+3344.0	+1840	+1004.4
+3362.0	+1850	+1010.0
+3380.0	+1860	+1015.6
+3398.0	+1870	+1021.1
+3416.0	+1880	+1026.7
+3434.0	+1890	+1032.2
+3452.0	+1900	+1037.8
+3470.0	+1910	+1043.3
+3488.0	+1920	+1048.9
+3506.0	+1930	+1054.4
+3524.0	+1940	+1060.0
+3542.0	+1950	+1065.6
+3560.0	+1960	+1071.1

TABLE A.9. Temperature Conversion Table

°F.	Reading in °F. or °C. to be converted	°C.
+3578.0	+1970	+1076.7
+3596.0	+1980	+1082.2
+3614.0	+1990	+1087.8
+3632.0	+2000	+1093.3
+3650.0	+2010	+1098.9
+3668.0	+2020	+1104.4
+3686.0	+2030	+1110.0
+3704.0	+2040	+1115.6
+3722.0	+2050	+1121.1
+3740.0	+2060	+1126.7
+3758.0	+2070	+1132.2
+3776.0	+2080	+1137.8
+3794.0	+2090	+1143.3
+3812.0	+2100	+1148.9
+3830.0	+2110	+1154.4
+3848.0	+2120	+1160.0
+3866.0	+2130	+1165.6
+3884.0	+2140	+1171.1
+3902.0	+2150	+1176.7
+3920.0	+2160	+1182.2
+3938.0	+2170	+1187.8
+3956.0	+2180	+1193.3
+3974.0	+2190	+1198.9
+3992.0	+2200	+1204.4
+4010.0	+2210	+1210.0
+4028.0	+2220	+1215.6
+4046.0	+2230	+1221.1
+4064.0	+2240	+1226.7
+4082.0	+2250	+1232.2
+4100.0	+2260	+1237.8
+4118.0	+2270	+1243.3
+4136.0	+2280	+1248.9
+4154.0	+2290	+1254.4
+4172.0	+2300	+1260.0
+4190.0	+2310	+1265.6
+4208.0	+2320	+1271.1
+4226.0	+2330	+1276.7
+4244.0	+2340	+1282.2
+4262.0	+2350	+1287.8
+4280.0	+2360	+1293.3
+4298.0	+2370	+1298.9
+4316.0	+2380	+1304.4
+4334.0	+2390	+1310.0
+4352.0	+2400	+1315.6
+4370.0	+2410	+1321.1
+4388.0	+2420	+1326.7
+4406.0	+2430	+1332.2
+4424.0	+2440	+1337.8
+4442.0	+2450	+1343.3
+4460.0	+2460	+1348.9
+4478.0	+2470	+1354.4
+4496.0	+2480	+1360.0
+4514.0	+2490	+1365.6
+4532.0	+2500	+1371.1
+4550.0	+2510	+1376.7
+4568.0	+2520	+1382.2
+4586.0	+2530	+1387.8
+4604.0	+2540	+1393.3
+4622.0	+2550	+1398.9
+4640.0	+2560	+1404.4
+4658.0	+2570	+1410.0
+4676.0	+2580	+1415.6
+4694.0	+2590	+1421.1
+4712.0	+2600	+1426.7
+4730.0	+2610	+1432.2
+4748.0	+2620	+1437.8
+4766.0	+2630	+1443.3
+4784.0	+2640	+1448.9
+4802.0	+2650	+1454.4
+4820.0	+2660	+1460.0
+4838.0	+2670	+1465.6
+4856.0	+2680	+1471.1
+4874.0	+2690	+1476.7
+4892.0	+2700	+1482.2
+4910.0	+2710	+1487.8
+4928.0	+2720	+1493.3
+4946.0	+2730	+1498.9
+4964.0	+2740	+1504.4
+4982.0	+2750	+1510.0
+5000.0	+2760	+1515.6
+5018.0	+2770	+1521.1
+5036.0	+2780	+1526.7
+5054.0	+2790	+1532.2
+5072.0	+2800	+1537.8
+5090.0	+2810	+1543.3
+5108.0	+2820	+1548.9
+5126.0	+2830	+1554.4
+5144.0	+2840	+1560.0
+5162.0	+2850	+1565.6
+5180.0	+2860	+1571.1
+5198.0	+2870	+1576.7
+5216.0	+2880	+1582.2
+5234.0	+2890	+1587.8
+5252.0	+2900	+1593.3
+5270.0	+2910	+1598.9
+5288.0	+2920	+1604.4
+5306.0	+2930	+1610.0
+5324.0	+2940	+1615.6
+5342.0	+2950	+1621.1
+5360.0	+2960	+1626.7
+5378.0	+2970	+1632.2
+5396.0	+2980	+1637.8
+5414.0	+2990	+1643.3
+5432.0	+3000	+1648.9
+5450.0	+3010	+1654.4
+5468.0	+3020	+1660.0
+5486.0	+3030	+1665.6
+5504.0	+3040	+1671.1
+5522.0	+3050	+1676.7
+5540.0	+3060	+1682.2
+5558.0	+3070	+1687.8
+5576.0	+3080	+1693.3
+5594.0	+3090	+1698.9
+5612.0	+3100	+1704.4

Appendix B

TABLE B.1 Greek Letters

Common Uses as Symbols in Glass Technology

English spelling	Greek capital letters	Greek small letters
Alpha	Α	α—Coefficient of thermal expansion; natural or low-temperature crystal form
Beta	Β	β—Specific heat constant; secondary or high-temperature crystal form; compressibility
Gamma	Γ	γ—Surface tension; higher crystal form
Delta	Δ	δ—Increment or differential; birefringence
Epsilon	Ε	ε—Unit elongation or strain; energy potential
Zeta	Ζ	ζ—Logarithm of viscosity; deformation at breaking point
Eta	Η	η—Viscosity, poises; entropy
Theta	Θ	θ—Degrees of thermal shock; plane angle
Iota	Ι	ι
Kappa	Κ	κ—K Dielectric constant; κ conductivity
Lambda	Λ	λ—Wave length
Mu	Μ	μ—Micron (10^{-4} cm); μμ micromicron (10^{-10} cm); mμ millimicron (10^{-7} cm)
Nu	Ν	ν—Dispersion
Xi	Ξ	ξ
Omicron	Ο	ο
Pi	Π	π—Circumference ÷ diameter, 3.1416
Rho	Ρ	ρ—Electrical resistivity
Sigma	Σ	σ—Σ Summation; σ Stefan-Boltzmann constant; interfacial surface tension; Poisson's ratio
Tau	Τ	τ
Upsilon	Υ	υ
Phi	Φ	φ—Fluidity or 1/η
Chi	Χ	χ
Psi	Ψ	ψ
Omega	Ω	ω—Ohms

Appendix

C

Periodic Table of Elements

KEY

Oxidation States →	+1 +3
Atomic Number →	79
Symbol →	Au
Name →	Gold
Atomic Weight →	196.9665
Electron Configuration →	-32-18-1

Filled Shells	I	II						Transition Elements				III	IV	V	VI	VII	0
	-1 1 H Hydrogen 1.0079 1																	0 2 He Helium 4.00260 2
	+1 3 Li Lithium 6.941 2-1	+2 4 Be Beryllium 9.01218 2-2											+3 5 B Boron 10.81 2-3	+2 ±4 6 C Carbon 12.011 2-4	-1 ±2 ±3 +4 +5 7 N Nitrogen 14.0067 2-5	-2 8 O Oxygen 15.9994 2-6	-1 9 F Fluorine 18.99840 2-7	0 10 Ne Neon 20.179 2-8
	+1 11 Na Sodium 22.98977 2-8-1	+2 12 Mg Magnesium 24.305 2-8-2											+3 13 Al Aluminum 26.98154 2-8-3	+2 ±4 14 Si Silicon 28.086 2-8-4	±3 +5 15 P Phosphorus 30.97376 2-8-5	+4 +6 -2 16 S Sulfur 32.06 2-8-6	±1 +5 +7 17 Cl Chlorine 35.453 2-8-7	0 18 Ar Argon 39.948 2-8-8
2	+1 19 K Potassium 39.098 -8-8-1	+2 20 Ca Calcium 40.08 -8-8-2	+3 21 Sc Scandium 44.9559 -8-9-2	+2 +3 +4 22 Ti Titanium 47.90 -8-10-2	+2 +3 +4 +5 23 V Vanadium 50.9414 -8-11-2	+2 +3 +6 24 Cr Chromium 51.996 -8-13-1	+2 +3 +4 +7 25 Mn Manganese 54.9380 -8-13-2	+2 +3 26 Fe Iron 55.847 -8-14-2	+2 +3 27 Co Cobalt 58.9332 -8-15-2	+2 +3 28 Ni Nickel 58.70 -8-16-2	+1 +2 29 Cu Copper 63.546 -8-18-1	+2 30 Zn Zinc 65.38 -8-18-2	+3 31 Ga Gallium 69.72 -8-18-3	+2 +4 32 Ge Germanium 72.59 -8-18-4	±3 +5 33 As Arsenic 74.9216 -8-18-5	+4 +6 -2 34 Se Selenium 78.96 -8-18-6	±1 +5 35 Br Bromine 79.904 -8-18-7	0 36 Kr Krypton 83.80 -8-18-8
2-8	+1 37 Rb Rubidium 85.4678 -18-8-1	+2 38 Sr Strontium 87.62 -18-8-2	+3 39 Y Yttrium 88.9059 -18-9-2	+4 40 Zr Zirconium 91.22 -18-10-2	+3 +5 41 Nb Niobium 92.9064 -18-12-1	+6 42 Mo Molybdenum 95.94 -18-13-1	+4 +6 +7 43 Tc Technetium (97) -18-13-2	+3 44 Ru Ruthenium 101.07 -18-15-1	+3 45 Rh Rhodium 102.9055 -18-16-1	+2 +4 46 Pd Palladium 106.4 -18-18-0	+1 47 Ag Silver 107.868 -18-18-1	+2 48 Cd Cadmium 112.40 -18-18-2	+3 49 In Indium 114.82 -18-18-3	+2 +4 50 Sn Tin 118.69 -18-18-4	±3 +5 51 Sb Antimony 121.75 -18-18-5	+4 +6 -2 52 Te Tellurium 127.60 -18-18-6	±1 +5 +7 53 I Iodine 126.9045 -18-18-7	0 54 Xe Xenon 131.30 -18-18-8
2-8-18	+1 55 Cs Cesium 132.9054 -18-8-1	+2 56 Ba Barium 137.34 -18-8-2	57-71 See Lantha-nides	+4 72 Hf Hafnium 178.49 -32-10-2	+5 73 Ta Tantalum 180.9479 -32-11-2	+6 74 W Tungsten 183.85 -32-12-2	+4 +6 +7 75 Re Rhenium 186.207 -32-13-2	+3 +4 76 Os Osmium 190.2 -32-14-2	+3 +4 77 Ir Iridium 192.22 -32-15-2	+2 +4 78 Pt Platinum 195.09 -32-17-1	+1 +3 79 Au Gold 196.9665 -32-18-1	+1 +2 80 Hg Mercury 200.59 -32-18-2	+1 +3 81 Tl Thallium 204.37 -32-18-3	+2 +4 82 Pb Lead 207.2 -32-18-4	+3 +5 83 Bi Bismuth 208.9804 -32-18-5	+2 +4 84 Po Polonium (209) -32-18-6	±1 +5 +7 85 At Astatine (210) -32-18-7	0 86 Rn Radon (222) -32-18-8
2-8-18-32	+1 87 Fr Francium (223) -18-8-1	+2 88 Ra Radium 226.0254 -18-8-2	89-103 See Acti-nides	+4 104 Rf-Ku (Rutherfordium; Kurchatovium) (261) -32-10-2	105 Ha Hahnium (262) -32-11-2	106 (263) -32-12-2	107 (262) -32-13-2	108 (265) -32-14-2	109 (266) -32-15-2	(110)	(111)	(112)	(113)	(114)	(115)	(116)	(117)	(118)

Lanthanides	+3 57 La Lanthanum 138.9055 -18-9-2	+3 +4 58 Ce Cerium 140.12 -19-9-2	+3 59 Pr Praseodymium 140.9077 -21-8-2	+3 60 Nd Neodymium 144.24 -22-8-2	+3 61 Pm Promethium (145) -23-8-2	+2 +3 62 Sm Samarium 150.4 -24-8-2	+2 +3 63 Eu Europium 151.96 -25-8-2	+3 64 Gd Gadolinium 157.25 -25-9-2	+3 65 Tb Terbium 158.9254 -26-9-2	+3 66 Dy Dysprosium 162.50 -28-8-2	+3 67 Ho Holmium 164.9304 -29-8-2	+3 68 Er Erbium 167.26 -30-8-2	+3 69 Tm Thulium 168.9342 -31-8-2	+2 +3 70 Yb Ytterbium 173.04 -32-8-2	+3 71 Lu Lutetium 174.97 -32-9-2
Actinides	+3 89 Ac Actinium (227) -18-9-2	+4 90 Th Thorium 232.0381 -18-10-2	+4 +5 91 Pa Protactinium 231.0359 -20-9-2	+3 +4 +5 +6 92 U Uranium 238.029 -21-9-2	+3 +4 +5 +6 93 Np Neptunium 237.0482 -22-9-2	+3 +4 +5 +6 94 Pu Plutonium (244) -24-8-2	+3 +4 +5 +6 95 Am Americium (243) -25-8-2	+3 96 Cm Curium (247) -25-9-2	+3 +4 97 Bk Berkelium (247) -27-8-2	+3 98 Cf Californium (251) -28-8-2	+3 99 Es Einsteinium (252) -29-8-2	+3 100 Fm Fermium (257) -30-8-2	+2 +3 101 Md Mendelevium (258) -31-8-2	+2 +3 102 No Nobelium (259) -32-8-2	+3 103 Lr Lawrencium (262) -32-9-2

Note:
Atomic weights are those of the most commonly available long-lived isotopes on the 1973 IUPAC Atomic Weights of the Elements. A value given in parentheses denotes the mass number of the longest-lived isotope.

Appendix D

Conversion Chart

Table D.1.

Length:
1 ft = 12 in
1 yd = 3 ft
1 mi = 5280 ft
1 in = 25.4 mm
1 in = 2.54 cm
1 meter = 3.281 ft
1 mi = 1.609 km

Area:
1 ft^2 = 144 in^2
1 yd^2 = 9 ft^2
1 in^2 = 6.452 cm^2
1 mi^2 = 2.590 km^2
1 mi^2 = 640 acres

Volume:
1 ft^3 = 1728 in^3
1 yd^3 = 27 ft^3
1 in^3 = 16.39 cc
1 liter = 61.02 in^3
1 U.S. gal = 231 in^3
1 ft^3 = 7.481 U.S. gal

Velocity:
1 fps = 60 fpm
1 fpm = 60 fph
1 fps = 3600 fph
1 fps = 30.48 cm/s
1 mph = 1.467 fps
1 radian/s = 9.55 rpm

Volume Flow Rate:
1 cfm = 60 cfh
1 cfs = 60 cfm
1 cfm = 7.481 gal pm
1 cfs = 26,930 gal ph

Mass:
1 lb (avdp) = 16 oz (avdp)
1 short ton = 2000 lb (avdp)
1 lb (avdp) = 7000 grains
1 lb (avdp) = 453.6 g
1 kg = 2.205 lb (avdp)

Density and Specific Gravity:
Specific gravity relative to water (abbr. sgw) of 1.00 = 62.43 lb/ft^3
Specific gravity relative to air (abbr. sga) of 1.000 = 0.0763 lb/ft^3
1 g/cc = 62.43 lb/ft^3
1 lb/in^3 = 27.68 g/cc
1 lb/in^3 = 1728 lb/ft^3
1 lb/U.S. gal = 7.481 lb/ft^3
$1\ \frac{\text{short ton}}{\text{yd}^3} = 74.07\ \text{lb/ft}^3$

Pressure:
1 psi = 144 psf
1 kg/cm^2 = 14.22 psi
1 atm = 14.7 psi
1 psi = 2.309 ft wc (for water at 59°F)
1 psi = 27.71 in wc (for water at 59°F)
1 psi = 2.036 in Hg (for mercury at 32°F)
1 in Hg = 13.61 in wc (for water at 59°F and mercury at 32°F)
1 osi = 1.732 in wc

Energy, Heat, and Work:
1 Btu = 252.0 cal
1 Btu = 0.2520 kg cal
1 therm = 100,000 Btu
1 Btu = 778.2 ft · lb
1 Btu = 1055 joules

Table D.1 (*Continued*)

Energy, Heat, and Work:
1 cal = 4.187 joules
1 hp-hr = 2544 Btu
1 kwh = 3412 Btu
1 hp-hr = 1,980,000 ft · lb
1 kg-m = 7.233 ft · lb

Power and Heat Flow:
1 kw = 1.341 hp
1 hp = 550 ft · lb/sec
1 hp = 42.41 Btu/min
1 Btu/s = 1.055 kw
1 kw = 3412 Btu/h
1 hp = 2544 Btu/h

Heat Flux:
1 cal/h cm^2 = 3.687 Btu/h/ft^2
1 watt/cm^2 = 3170 Btu/h/ft^2

Thermal Conductivity:

$$1 \frac{\text{Btu ft}}{\text{h ft}^2\ {}^\circ\text{F}} = 12 \frac{\text{Btu in}}{\text{h ft}^2\ {}^\circ\text{F}}$$

$$1 \frac{\text{Btu ft}}{\text{h ft}^2\ {}^\circ\text{F}} = 14.88 \frac{\text{cal cm}}{\text{h cm}^2\,{}^\circ\text{C}}$$

$$1 \frac{\text{watt cm}}{\text{cm}^2\,{}^\circ\text{C}} = 57.79 \frac{\text{Btu ft}}{\text{h ft}^2\,{}^\circ\text{F}}$$

Heat Content:

$$1 \frac{\text{Btu}}{\text{lb}} = 0.556 \frac{\text{cal}}{\text{g}}$$

$$1 \frac{\text{Btu}}{\text{lb }{}^\circ\text{F}} = 1 \frac{\text{cal}}{\text{g }{}^\circ\text{C}}$$

Appendix

Equivalents

TABLE E.1 Units of Capacity

Fl oz	Pint	Qt	Gal	British gal	Cc	Liter	Cu in	Cu ft
1	0.625	0.0313	0.0078	0.0065	29.57	0.0296	1.805	0.00105
16	1	0.5	0.125	0.104	473	0.473	28.88	0.0167
32	2	1	0.25	0.2082	946	0.946	57.75	0.0334
128	8	4	1	0.8327	3785	3.785	231	0.1337
153.7	9.608	4.804	1.201	1	4546	4.546	277.4	0.1605
0.338	0.0021	0.00106	0.00026	0.00022	1	0.001	0.0610	0.000035
33.82	2.113	1.057	0.2642	0.2201	1000	1	61.02	0.03531
0.5541	0.0346	0.0173	0.0043	0.0036	16.39	0.0164	1	0.000579
957.4	59.84	29.92	7.480	6.232	28,320	28.32	1728	1

TABLE E.2 Conversion Factors for Concentrations

G/100 cc	G/l	Oz/gal	Grains/gal	Parts per million
1	10	1.335	584	10^4
0.1	1	0.1335	58.4	1000
0.749	7.49	1	437.5	7490
1.712×10^{-3}	1.712×10^{-2}	2.88×10^{-2}	1	17.12
10^{-4}	10^{-3}	1.35×10^{-3}	0.0584	1

TABLE E.3 Metric and Avoirdupois Weights

G	Kg	Metric ton	Grain	Ounce	Pound	Ton
1	0.001	1×10^{-6}	15.43	0.03527	0.0022	1.1×10^{-6}
1000	1	0.001	15.43×10^{3}	35.27	2.2016	0.0011
10^4	1000	1	15.43×10^{6}	35.27×10^{3}	2.2×10^{3}	1.1023
0.0648	6.48×10^{-5}	6.48×10^{-8}	1	2.285×10^{-3}	1.428×10^{-4}	7.14×10^{-8}
28.35	0.02835	2.835×10^{-5}	437.5	1	0.0625	3.125×10^{-6}
453.6	0.4536	4.536×10^{-4}	7000	16	1	5×10^{-4}
9.072×10^4	907.2	0.9072	14×10^{6}	32×10^{3}	2000	1

TABLE E.4 Fluid Measure (Small Units)

Cc	Minim	Dram	Fl oz	Gill	British fl oz	British gill
1	16.23	0.2705	0.0338	0.00845	0.352	0.00704
0.0616	1	0.0167	0.00208	0.000521	0.00217	0.000434
3.697	60	1	0.1250	0.0312	0.1302	0.0260
29.57	480	8	1	0.250	1.041	0.2082
118.3	1920	32	4	1	4.164	0.8328
28.41	461.2	7.686	0.9608	0.2402	1	0.2000
142.0	2306	37.43	4.804	1.201	5	1

Gill = 0.25 pint

TABLE E.5 Electric Heating

Volts × Amperes	= Watts
Watts × 3.41	= Btu/h
1 Amp. at 110 V	= 375 Btu/h
Amperes	= V/R (ohms)
Watt = Volts × Amp.	= V^2/R

For constant voltage, the power consumed and the heat generated in a resistance wire of given diameter vary inversely as its length.

TABLE E.6 Temperature Conversion

To change Centigrade to Fahrenheit, multiply by $^9\!/_5$ and add 32.

To change Fahrenheit to Centigrade, subtract 32 and multiply remainder by $^5\!/_9$:

$$°F = (°C \times 1.8) + 32$$

and

$$°C = \frac{°F - 32}{1.8}$$

The Centigrade degree is a longer temperature interval than the Fahrenheit degree in the ratio of 9:5.

Appendix

F

Engineering Conversion Chart

Fraction	Decimal
1/64	.016
1/32	.031
3/64	.047
1/16	.063
5/64	.078
3/32	.094
7/64	.109
1/8	.125
9/64	.141
5/32	.156
11/64	.172
3/16	.188
13/64	.203
7/32	.219
15/64	.234
1/4	.250
17/64	.266
9/32	.281
19/64	.297
5/16	.313
21/64	.328
11/32	.344
23/64	.359
3/8	.375
25/64	.391
13/32	.406
27/64	.422
7/16	.438
29/64	.453
15/32	.469
31/64	.484
1/2	.500

Fraction	Decimal
33/64	.516
17/32	.531
35/64	.547
9/16	.563
37/64	.578
19/32	.594
39/64	.609
5/8	.625
41/64	.641
21/32	.656
43/64	.672
11/16	.688
45/64	.703
23/32	.719
47/64	.734
3/4	.750
49/64	.766
25/32	.781
51/64	.797
13/16	.813
53/64	.828
27/32	.844
55/64	.859
7/8	.875
57/64	.891
29/32	.906
59/64	.922
15/16	.938
61/64	.953
31/32	.969
63/64	.984
1	1.00

Millimeters	Decimal
1.0	.039
1.5	.059
2.0	.079
2.5	.098
3.0	.118
3.5	.138
4.0	.157
4.5	.177
5.0	.197
5.5	.217
6.0	.236
6.5	.256
7.0	.276
7.5	.295
8.0	.315
8.5	.335
9.0	.354
9.5	.374
10.0	.394
10.1	.398
10.5	.413
10.8	.425
11.0	.433
11.2	.441
11.5	.453
11.7	.461
12.0	.472
12.5	.492

Millimeters	Decimal
13.0	.512
13.1	.516
13.2	.520
13.5	.531
14.0	.551
14.5	.571
15.0	.591
15.5	.610
15.9	.626
16.0	.630
16.5	650
17.0	.669
17.5	.689
18.0	.709
18.5	.728
19.0	.748
19.5	.768
20.0	.787
20.5	.807
21.0	.827
21.5	.846
22.0	.866
22.5	.886
23.0	.906
23.5	.925
24.0	.945
24.5	.965
25.0	.984

Appendix G

Specifications

TABLE G.1. Reference to ASTM Standards

	Specifications for
C552-79	Cellular Glass Block and Pipe Thermal Insulation
	Glass and Glass Products, Standard Samples (Related Material Section)
E438-80a	Glasses in Laboratory Apparatus
C599-70(1977)	Process Glass Pipe and Fittings
E211-70(1981)	Cover Glasses and Glass Slides for Use in Microscopy
D581-81	Glass Fiber Greige Braided Tubular Sleeving
D578-81	Glass Fiber Yarns
D579-81	Greige Woven Glass Fabrics
D879-62(1976)	Pin-Type Lime-Glass Insulators, Communication and Signal
C1036-90	Flat Glass
C1172-91	Laminated Architectural Flat Glass Required to Resist a Specified Load
	Test methods for
C724-81	Acid Resistance of Ceramic Decorations on Architectural Type Glass
C735-81	Acid Resistance of Ceramic Decorations on Returnable Beer and Beverage Glass Containers
C675-74(1980)	Alkali Resistance of Ceramic Decorations on Returnable Beverage Glass Containers
F218-68(1978)	Analyzing Stress in Glass
C336-71(1981)	Annealing Point and Strain Point of Glass by Fiber Elongation
C598-72(1978)	Annealing Point and Strain Point of Glass by Beam Bending
C146-80	Chemical Analysis of Glass Sand

TABLE G.1. Reference to ASTM Standards *(Continued)*

	Test methods for
C169-80	Chemical Analysis of Soda-Lime and Borosilicate Glass
C225-73(1978)	Chemical Attack, Resistance of Glass Containers to
C657-78	DC Volume Resistivity of Glass
C693-74(1980)	Density of Glass by Buoyancy
C729-75(1980)	Density of Glass by the Sink-Float Comparator
C676-74(1980)	Detergent Resistance of Ceramic Decorations on Glass Tableware
C158-80	Flexure Testing of Glass (Determination of Modulus of Rupture)
C813-75(1980)	Hydrophobic Contamination on Glass by Contact Angle Measurement
C812-75(1980)	Hydrophobic Contamination on Glass by Water Condensation
C147-76(1981)	Internal Pressure Test on Glass Containers
C730-75(1980)	Knoop Indentation Hardness of Glass
C927-80	Lead and Cadmium Extracted from the Lip and Rim Area of Glass Tumblers Externally Decorated with Ceramic Glass Enamels
C824-81	Linear Thermal Expansion of Vitreous Glass Enamels and Glass Color Frits by the Dilatometer Method
C148-77	Polariscope Examination of Glass Containers
C601-70(1977)	Pressure Test in Glass Pipe
C224-78	Sampling Glass Containers
C429-65(1977)	Sieve Analysis of Raw Materials for Glass Manufacture
C338-73(1979)	Softening Point of Glass
C777-78	Sulfide Resistance of Ceramic Decorations on Glass
C149-77	Thermal Shock Test on Glass Containers
C600-70(1977)	Thermal Shock Test on Glass Pipe
E773-88	Seal Durability of Sealed Insulating Glass Units
E774-88	Sealed Insulating Glass Units

Index

ABOUT THE EDITOR

Joseph S. Amstock is a consulting chemical engineer with over 45 years experience in the glass, glazing and construction industries. Since beginning his career he has been associated with sealants, caulks, adhesives, insulating glass and waterproofing. He has served on various committees in the ASTM and the Sealed Insulating Glass Manufacturers Association (SIGMA), and was on the board of directors of the National Wood Window and Door Association (NWWDA). His assignments included living and working in Mexico City, Mexico, Brussels, Belgium, and Galway, Ireland.

Over 17 technical papers are credited to him that have been published by trade journals and professional organizations in several countries, in addition to chapters on sealants for John Wiley & Sons, as well as McGraw-Hill. Currently, he is Editor-in-Chief of a handbook on adhesives and sealants in construction, to be published in 1998 by McGraw-Hill.